SINGLE PILES AND PILE GROUPS UNDER L

Single Piles and Pile Groups under Lateral Loading

LYMON C. REESE
Nasser I. Al-Rashid Chair Emeritus, Department of Civil Engineering, University of Texas, Austin, USA

WILLIAM F. VAN IMPE
Full Professor of Civil Engineering, Director Laboratory for Soil Mechanics, Ghent University
Professor, Catholic University Leuven, Belgium

A.A. BALKEMA PUBLISHERS   LEIDEN / LONDON / NEW YORK / PHILADELPHIA / SINGAPORE

CD-ROM with special version of Computer Programs LPILE and GROUP is enclosed in the back of the cover.

Copyright © 2001 Taylor & Francis Group plc, London, UK

All rights reserved. No part of this publication or the information contained herein may be reproduced, stored in a retrieval system, or transmitted in any form or by any means, electronic, mechanical, by photocopying, recording or otherwise, without written prior permission from the publisher.

Although all care is taken to ensure the integrity and quality of this publication and the information herein, no responsibility is assumed by the publishers nor the author for any damage to property or persons as a result of operation or use of this publication and/or the information contained herein.

Published by: Taylor & Francis / Balkema
               P.O. Box 447, 2300 AK Leiden, The Netherlands
               e-mail: Pub.NL@tandf.co.uk
               www.balkema.nl, www.tandf.co.uk, www.crcpress.com

Hardbound edition: ISBN 90 5809 340 9
Paperback edition: ISBN 90 5809 348 4

Printed in Great Britain

To our current and former graduate students, to our colleagues, and to our Universities where academic contributions are encouraged

Contents

PREFACE	XV
1 TECHNIQUES FOR DESIGN	1
1.1 Introduction	1
1.2 Occurrence of laterally loaded piles	2
1.3 Nature of the soil response	3
1.4 Response of a pile	7
1.4.1 Introduction	7
1.4.2 Static loading	7
1.4.3 Cyclic loading	7
1.4.4 Sustained loading	8
1.4.5 Dynamic loading	9
1.5 Models for use in analyses of a single pile	11
1.5.1 Elastic pile and elastic soil	11
1.5.2 Elastic pile and finite elements for soil	12
1.5.3 Rigid pile and plastic soil	12
1.5.4 Characteristic load method	13
1.5.5 Nonlinear pile and p-y model for soil	14
1.6 Models for groups of piles under lateral loading	16
1.7 Status of current state-of-the-art	18
2 DERIVATION OF EQUATIONS AND METHODS OF SOLUTION	21
2.1 Introduction	21
2.2 Derivation of the differential equation	21
2.2.1 Solution of Reduced Form of Differential Equation	25
2.2.2 Solution of the Differential Equation by Difference Equations	29
2.3 Solution for $E_{py} = k_{py}x$	35
2.3.1 Dimensional Analysis	36
2.3.2 Equations for $E_{py} = k_{py}x$	41
2.3.3 Example Solution	42
2.3.4 Discussion	46
2.4 Validity of the mechanics	47
3 MODELS FOR RESPONSE OF SOIL AND WEAK ROCK	49
3.1 Introduction	49
3.2 Mechanics concerning response of soil to lateral loading	50
3.2.1 Stress-deformation of soil	50
3.2.2 Proposed model for decay of E_s	50
3.2.3 Variation of stiffness of soil (E_s and G_s) with depth	51

	3.2.4	Initial stiffness and ultimate resistance of *p-y* curves from soil properties	53
	3.2.5	Subgrade modulus related to piles under lateral loading	59
	3.2.6	Theoretical solution by Skempton for subgrade modulus and for *p-y* curves for saturated clays	60
	3.2.7	Practical use of Skempton's equations and values of subgrade modulus in analyzing a pile under lateral loading	62
3.3	Influence of diameter on *p-y* curves		64
	3.3.1	Clay	64
	3.3.2	Sand	64
3.4	Influence of cyclic loading		65
	3.4.1	Clay	65
	3.4.2	Sand	66
3.5	Experimental methods of obtaining *p-y* curves		67
	3.5.1	Soil response from direct measurements	67
	3.5.2	Soil response from experimental moment curves	67
	3.5.3	Nondimensional methods for obtaining soil response	68
3.6	Early recommendations for computing *p-y* curves		68
	3.6.1	Terzaghi	68
	3.6.2	McClelland and Focht for clay (1958)	70
3.7	*p-y* curves for clay		70
	3.7.1	Selection of stiffness of clay	70
	3.7.2	Response of soft clay in the presence of free water	72
	3.7.3	Response of stiff clay in the presence of free water	75
	3.7.4	Response of stiff clay with no free water	82
3.8	*p-y* curves for sands above and below the water table		84
	3.8.1	Detailed procedure	84
	3.8.2	Recommended soil tests	87
	3.8.3	Example curves	87
3.9	*p-y* curves for layered soils		87
	3.9.1	Method of Georgiadis	88
	3.9.2	Example *p-y* curves	89
3.10	*p-y* curves for soil with both cohesion and internal friction		91
	3.10.1	Background	91
	3.10.2	Recommendations for computing *p-y* curves	92
	3.10.3	Discussion	96
3.11	Other recommendations for computing *p-y* curves		97
	3.11.1	Clay	97
	3.11.2	Sand	97
3.12	Modifications to *p-y* curves for sloping ground		98
	3.12.1	Introduction	98
	3.12.2	Equations for ultimate resistance in clay	98
	3.12.3	Equations for ultimate resistance in sand	99
3.13	Effect of batter		100
3.14	Shearing force at bottom of pile		101
3.15	*p-y* curves for weak rock		101
	3.15.1	Introduction	101
	3.15.2	Field tests	102
	3.15.3	Interim recommendations	102
	3.15.4	Comments on equations for predicting *p-y* curves for rock	106

3.16	Selection of p-y curves	106
	3.16.1 Introduction	106
	3.16.2 Factors to be considered	106
	3.16.3 Specific suggestions	107

4 STRUCTURAL CHARACTERISTICS OF PILES 109
4.1 Introduction 109
4.2 Computation of an equivalent diameter of a pile with a noncircular cross section 109
4.3 Mechanics for computation of m_{ult} and $e_p i_p$ as a function of bending moment and axial load 111
4.4 Stress-strain curves for normal-weight concrete and structural steel 114
4.5 Implementation of the method for a steel h-section 116
4.6 Implementation of the method for a steel pipe 118
4.7 Implementation of the method for a reinforced-concrete section 119
    4.7.1 Example computations for a square shape 119
    4.7.2 Example computations for a circular shape 121
4.8 Approximation of moment of inertia for a reinforced-concrete section 121

5 ANALYSIS OF GROUPS OF PILES SUBJECTED TO INCLINED AND ECCENTRIC LOADING 125
5.1 Introduction 125
5.2 Approach to analysis of groups of piles 126
5.3 Review of theories for the response of groups of piles to inclined and eccentric loads 126
5.4 Rational equations for the response of a group of piles under generalized loading 129
    5.4.1 Introduction 129
    5.4.2 Equations for a two-dimensional group of piles 132
5.5 Laterally loaded piles 136
    5.5.1 Movement of pile head due to applied loading 136
    5.5.2 Effect of batter 136
5.6 Axially loaded piles 137
    5.6.1 Introduction 137
    5.6.2 Relevant parameters concerning deformation of soil 137
    5.6.3 Influence of method of installation on soil characteristics 139
    5.6.4 Methods of formulating axial-stiffness curves 140
    5.6.5 Differential equation for solution of finite-difference equation for axially loaded piles. 142
    5.6.6 Finite difference equation 145
    5.6.7 Load-transfer curves 145
5.7 Closely-spaced piles under lateral loading 151
    5.7.1 Modification of load-transfer curves for closely spaced piles 151
    5.7.2 Concept of interaction under lateral loading 152
    5.7.3 Proposals for solving for influence coefficients for closely-spaced piles under lateral loading 152
    5.7.4 Description and analysis of experiments with closely-spaced piles installed in-line and side-by-side 155
    5.7.5 Prediction equations for closely-spaced piles installed in-line and side-by-side 158

X Contents

        5.7.6 Use of modified prediction equations in developing p-y curves for analyzing results of experiments with full-scale groups   160
        5.7.7 Discussion of the method of predicting the interaction of closely-spaced piles under lateral loading   173
  5.8 Proposals for solving for influence coefficients for closely-spaced piles under axial loading   173
        5.8.1 Introduction   173
        5.8.2 Concept of interaction under axial loading   174
        5.8.3 Review of relevant literature   174
        5.8.4 Interim recommendations for computing the efficiency of groups of piles under axial loading   177
  5.9 Analysis of an experiment with batter piles   178
        5.9.1 Description of the testing arrangement   178
        5.9.2 Properties of the sand   179
        5.9.3 Properties of the pipe piles   181
        5.9.4 Pile group   181
        5.9.5 Experimental curve of axial load versus settlement for single pile   182
        5.9.6 Results from experiment and from analysis   183
        5.9.7 Comments on analytical method   185

6 ANALYSIS OF SINGLE PILES AND GROUPS OF PILES SUBJECTED TO ACTIVE AND PASSIVE LOADING   187
  6.1 Nature of lateral loading   187
  6.2 Active loading   187
        6.2.1 Wind loading   187
        6.2.2 Wave loading   189
        6.2.3 Current loading   194
        6.2.4 Scour   195
        6.2.5 Ice loading   197
        6.2.6 Ship impact   198
        6.2.7 Loads from miscellaneous sources   198
  6.3 Single piles or groups of piles subjected to active loading   199
        6.3.1 Overhead sign   199
        6.3.2 Breasting dolphin   203
        6.3.3 Pile for anchoring a ship in soft soil   207
        6.3.4 Offshore platform   213
  6.4 Passive loading   223
        6.4.1 Earth pressures   223
        6.4.2 Moving soil   224
        6.4.3 Thrusts from dead loading of structures   226
  6.5 Single piles or groups of piles subjected to passive loading   226
        6.5.1 Pile-supported retaining wall   226
        6.5.2 Anchored bulkhead   231
        6.5.3 Pile-supported mat at the Pyramid Building   237
        6.5.4 Piles for stabilizing a slope   245
        6.5.5 Piles in a settling fill in a sloping valley   251

7 CASE STUDIES   259
  7.1 Introduction   259
  7.2 Piles installed into cohesive soil with no free water   260
        7.2.1 Bagnolet   260

		7.2.2	Houston	263
		7.2.3	Brent Cross	264
		7.2.4	Japan	267
	7.3	Piles installed into cohesive soil with free water above ground surface		269
		7.3.1	Lake Austin	269
		7.3.2	Sabine	272
		7.3.3	Manor	273
	7.4	Piles installed in cohesionless soil		276
		7.4.1	Mustang Island	276
		7.4.2	Garston	277
		7.4.3	Arkansas River	278
	7.5	Piles installed into layered soil		283
		7.5.1	Talisheek	283
		7.5.2	Alcácer do Sol	286
		7.5.3	Florida	288
		7.5.4	Apapa	288
	7.6	Piles installed in c-ϕ soil		290
		7.6.1	Kuwait	290
		7.6.2	Los Angeles	291
	7.7	Piles installed in weak rock		293
		7.7.1	Islamorada	293
		7.7.2	San Francisco	295
	7.8	Analysis of results of case studies		298
	7.9	Comments on case studies		299
8	TESTING OF FULL-SIZED PILES			303
	8.1	Introduction		303
		8.1.1	Scope of presentation	303
		8.1.2	Summary of method of analysis	303
		8.1.3	Classification of tests	303
		8.1.4	Features unique to testing of piles under lateral loading	304
	8.2	Designing the test program		304
		8.2.1	Planning for the testing	304
		8.2.2	Selection of test pile and test site	305
	8.3	Subsurface investigation		306
	8.4	Installation of test pile		309
	8.5	Testing techniques loading arrangements and instrumentation at the pile head		310
	8.6	Loading arrangements and instrumentation at the pile head		311
		8.6.1	Loading arrangements	311
		8.6.2	Instrumentation	313
	8.7	Testing for design of production piles		317
		8.7.1	Introduction	317
		8.7.2	Interpretation of data	317
		8.7.3	Example Computation	317
	8.8	Testing for obtaining details on response of soil		319
		8.8.1	Introduction	319
		8.8.2	Preparation of test piles	319
		8.8.3	Test setup and loading equipment	321
		8.8.4	Instrumentation	322
		8.8.5	Calibration of test piles	325
		8.8.6	Soil borings and laboratory tests	328

	8.8.7	Installation of test piles	332
	8.8.8	Test procedures and details of loading	334
	8.8.9	Penetrometer tests	335
	8.8.10	Ground settlement due to pile driving	338
	8.8.11	Ground settlement due to lateral loading	339
	8.8.12	Recalibration of test piles	339
	8.8.13	Graphical presentation of curves showing bending moment	340
	8.8.14	Interpretation of bending moment curves to obtain p-y curves	341
8.9	Summary		346

9 IMPLEMENTATION OF FACTORS OF SAFETY — 347
9.1 Introduction — 347
9.2 Limit states — 347
9.3 Consequences of a failure — 348
9.4 Philosophy concerning safety coefficient — 350
9.5 Influence of nature of structure — 351
9.6 Special problem in characterizing soil — 351
    9.6.1 Introduction — 351
    9.6.2 Characteristic value of soil parameters — 352
9.7 Level of quality control — 353
9.8 Two general approaches to selecting the factor of safety — 353
9.9 Global approach — 354
    9.9.1 Introductary comments — 354
    9.9.2 Recommendations of the American Petroleum Institute — 355
9.10 Method of partial safety factors (psf) — 356
    9.10.1 Introduction — 356
    9.10.2 Suggested values for partial factors for design of laterally loaded piles — 356
    9.10.3 Example computations — 358
9.11 Method of load and resistance factors (LRFD) — 358
    9.11.1 Introduction — 358
    9.11.2 Loads addressed by the LRFD specifications — 359
    9.11.3 Resistances addressed by the LRFD specifications — 359
    9.11.4 Design of piles by the LRFD specifications — 360
9.12 Concluding comments — 360

10 SUGGESTIONS FOR DESIGN — 363
10.1 Introduction — 363
10.2 Range of factors to be considered in design — 363
10.3 Validation of results from computations for single pile — 364
    10.3.1 Introduction — 364
    10.3.2 Solution of example problems — 364
    10.3.3 Check of echo print of input data — 364
    10.3.4 Investigation of length of word employed in internal computations — 365
    10.3.5 Selection of tolerance and length of increment — 365
    10.3.6 Check of soil resistance — 365
    10.3.7 Check of mechanics — 366
    10.3.8 Use of nondimensional curves — 366
10.4 Validation of results from computations for pile group — 366
10.5 Additional steps in design — 367
    10.5.1 Risk management — 367
    10.5.2 Peer review — 367

10.5.3 Technical contributions ... 367
10.5.4 The design team ... 368

APPENDICES

A Broms method for analysis of single piles under lateral loading ... 369
B Nondimensional coefficients for piles with finite length, no axial load, constant $E_p I_p$, and constant E_s ... 385
C Difference equations for solving the problem of step-tapered beams on foundations having variable stiffness ... 395
D Instructions for use of student versions of computer programs LPILE and GROUP ... 405
E Nondimensional curves for piles under lateral loading for case where $E_{py} = K_{py}x$ ... 409
F Tables of values of efficiency measured in tests of groups of piles under lateral loading ... 419
G Horizontal stresses in soil near shaft during installation of a pile ... 423
H Use of data from uninstrumented piles under lateral loading to obtain soil response ... 429
I Eurocode principles related to geotechnical design ... 435
J Discussion of factor of safety related to piles under axial load ... 439

REFERENCES ... 443

AUTHOR INDEX ... 457

SUBJECT INDEX ... 461

Preface

The information presented here is part of the emerging field of *soil-structure-interaction*. Not only must the engineer compute the loading at which a foundation will collapse, but the deformation at the soil-structure interaction boundary must be found for the expected loading. Further, the interaction of the foundation with the superstructure must be consistent with the details of the design and construction. The methods presented here would not have been possible except for the performance of full-scale tests on single piles and pile groups where remote-reading instruments allowed the response of the piles to be measured. Further, the computer is necessary to solve the complex models that were developed for predicting the response of the piles.

When offshore platforms were being installed in significant numbers in the 1950s, engineers quickly realized that correct solutions required that ways be found to link the soil response to the lateral deflection of a pile. A number of experiments were performed with full-sized piles, instrumented for the measurement of bending moment along their length. Static and cyclic loading was employed. Those experiments, and others performed in later years, allowed experimental *p-y* curves to be produced. Soil mechanics and structural mechanics were used to develop methods of predicting *p-y* curves for various soils that yielded excellent agreement with the response of the piles.

Solutions of the relevant differential equations in finite difference form became possible with the emergence of the main-frame digital computer for routine computations in the 1950s. Solutions on the personal computer can be obtained today with relative ease, allowing the sensitivity analysis of many significant parameters. Improvements of computer codes are occurring at a rapid pace, and experimental data on the response of single piles and pile groups to lateral loading is appearing in the technical literature with regularity.

The *p-y* method can be used to attack a wide variety of problems encountered in practice, as demonstrated by the examples given in Chapter 6. While the computer codes can be used to attack such problems and vast quantities of tabulated results and graphical results can be produced with ease, the importance of competent engineering judgement cannot be over emphasized. For example, the computer codes will lead to the required penetration of the pile supporting an overhead sign subjected to wind loading; however, the knowledgeable engineer will have to note that in many cases too few data are available to validate such a result and will take appropriate steps to ensure a safe design. One such step could be the performance of a controlled set of experiments.

Besides of the requirement of good quality and relevant soil data, important knowledge can be gained from field experiments, particularly with the use of instrumented piles, as presented in Chapter 8, or from recording the performance of completed structures. Modern methods of monitoring make entirely feasible the acquisition, under service conditions, of such information as pile-head deflection and other such data.

The writers wish to acknowledge many who contributed to various aspects of the book. Of particular importance are those authors who have made contributions of the technical literature that is referenced. The writers express thanks to Professor Heinz Brandl, Dr. William Cox, and Professor Hudson Matlock for reading the text and making important comments. Professor Michael W. O'Neill, University of Houston, used a preliminar draft of the book in teaching a course to graduate students. Thanks are extended to him and his students for helpful suggestions.

Appreciation is extended to Dr. Robert Gilbert and Dr. Alexander Avram for useful suggestions on Chapter 9. In Austin, Dr. Shin-Tower Wang, Dr. William Isenhower, and Mr. José Arréllaga were very helpful in reviewing parts of the text, in making numerous computer runs, and in preparing graphical output. Appreciation is extended to Nancy Reese, Cheryl Wawrzynowicz and Suzanne Burns for dedication and diligence in formatting and editing the text.

In Ghent, the writers appreciate the assistance of Mr. Etienne Bracke for some of the graphical output support and the help of Mrs. Linda Van Cauwenberge for the formatting and editing of the text version prepared in Ghent. Moreover the authors acknowledge Prof. J. De Rouck, Prof. R. Van Impe and Mr. Ch. Bauduin for putting available some of the information on issues related to recent European developments for structural design.

CHAPTER 1

Techniques for design

1.1 INTRODUCTION

After selecting materials for the pile foundation to make sure of durability, the designer begins with the components of loading on the single pile or the group. With the axial load, lateral load, and overturning moment, the engineer must ensure that the single pile, or the critical pile in the group, is safe against collapse and does not exceed movements set by serviceability. If the loading is purely axial, the design of a pile can frequently be accomplished by solving the equations of static equilibrium. The design of a single pile, or a group of piles, under lateral loading, on the other hand requires the solution of a nonlinear differential equation.

Linear solutions of the differential equation for single piles are available, even in some codes of practice, but are of limited value. Another simplification is to assume that raked piles in a group do not deflect laterally; the equations that result can be solved readily but such solutions are usually seriously in error. The following sections show that treating the soil, and sometimes the material in the pile, as nonlinear complicates the mathematics for the single pile and pile group but solutions can be made by numerical methods that are rational and in close to reasonable agreement with results from full-scale experiments.

The traditional technique of limit analysis, so useful in finding the ultimate capacity of many foundations, has only a marginal application to assessing the behavior of a laterally loaded pile. As will be demonstrated, acceptable solutions are only possible if explicit, nonlinear relationships are employed that give soil stiffness and resistance as a function of pile deflection, point by point, along the length of a pile. The solutions of the resulting equations, then, can be made to satisfy the required conditions of equilibrium and compatibility.

The problem involves the interaction of the soil and the pile, is one of the class of soil-structure-interaction problems, and is classified as Geotechnical Category 3 by the Eurocode 7 (1994). The resistance of the soil, in force per unit length at points along the pile, depends on the deflection of the pile and the soil resistance must be at hand in order to solve the relevant equations. Therefore, iteration is necessary to get a solution. The model, in which the pile is taken as a free body and the soil is simulated by a series of Winkler-type mechanisms, is described later in this chapter. The equations to be solved for the response of the pile come directly from ordinary mechanics.

A presentation is made initially of the methods of design for isolated or single piles. A later chapter will deal with the design of pile groups. Two classes of pile-group problems can be identified: 1. The distribution of loading to the heads of the various piles in the group; and 2. The efficiency of each of the piles in the group, a problem in pile-soil-pile interaction.

2 *Piles under lateral loading*

Both problems are addressed herein; the first is solved satisfactorily by numerical procedures; and the second is discussed fully with respect to available, empirical information.

1.2 OCCURRENCE OF LATERALLY LOADED PILES

With regard to their use in practice, laterally loaded piles may be termed *active* or *passive*. An active pile has loading applied principally at its top in supporting a superstructure, such as a bridge. A passive pile has loading applied principally along its length due to earth pressure, such as for piles in a moving slope or for a secant-pile wall. Chapter 6 will present examples of the design of both active and passive piles.

Figure 1.1 shows the installation of piles for an offshore platform. These active piles must sustain lateral loading from storm-driven waves and wind. With the advent of off-

Figure 1.1. Installation of piles for an offshore platform.

shore structures, the design of such piles was a primary concern and prompted a number of full-scale field tests. The results of some of these tests are studied in Chapter 7.

The design of the piles for an offshore platform presents interesting problems in soil-structure interaction. An example in Chapter 6 shows the elements in the design process.

Other examples of active piles are found in the design of foundations for bridges, high-rise structures, overhead signs, and piers for ships. Active piles must be designed for mooring dolphins, breasting dolphins, and pile groups that protect the foundations for bridges against ship impact. The importance of the last application is emphasized by the loss of life in the United States during failure of a portion of the Sunshine Skyway Bridge at Tampa Bay because of destruction of a bridge pier by an out-of-control ship.

The sketch in Figure 1.2 shows an example of a passive pile. While the pile will be subjected to loadings at its top, the primary concern is the influence of the sliding of the soil in the slope. Chapter 6 presents an approach for the design of such piles. In addition to use in stabilizing a slope, passive piles are used for the construction of tangent- or secant-pile walls, for soldier beams, and for supporting the base of retaining walls.

1.3 NATURE OF THE SOIL RESPONSE

The main parameter from the soil in the design of a pile under lateral loading is a *reaction modulus* (F/L), defined as the resistance (F) from the soil at a point along the pile divided by the deflection of the pile at that point (L). The reaction modulus is a function both of depth below the ground surface z and the deflection of the pile y. The reaction modulus can be defined in various ways; concepts that lead to a convenient solution of the relevant equations are presented in Figure 1.3. The sketch in Figure 1.3a shows a cylindrical pile under lateral load with a thin slice of soil shown at the depth below the ground line of z_1. (Note: the symbol z is used to show depth below the ground surface and the symbol x is used to show depth measured from the top of the pile.) The uniform distribution of unit stresses normal to the wall of the pile in Figure 1.3b is correct for the case of a pile that has been installed without bending. If the pile is caused to deflect a distance y_1 (exaggerated in the sketch for clarity), the distribution of unit stresses would be similar to that shown in Figure 1.3c. The stresses will have decreased on the back side of the pile and increased on the front side. Some of the stresses will have both a normal and a shearing component.

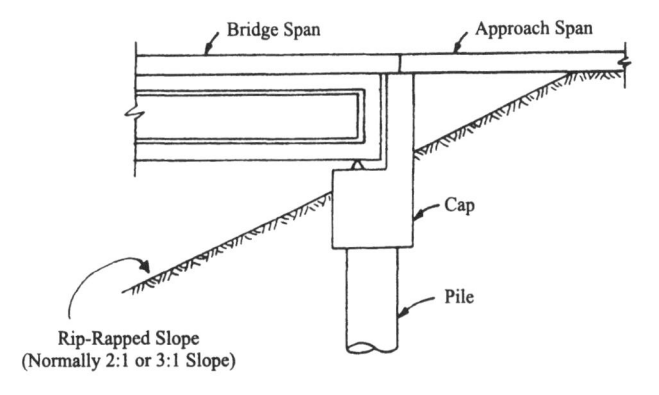

Figure 1.2. Sketch of a pile – supported bridge abutment.

4 *Piles under lateral loading*

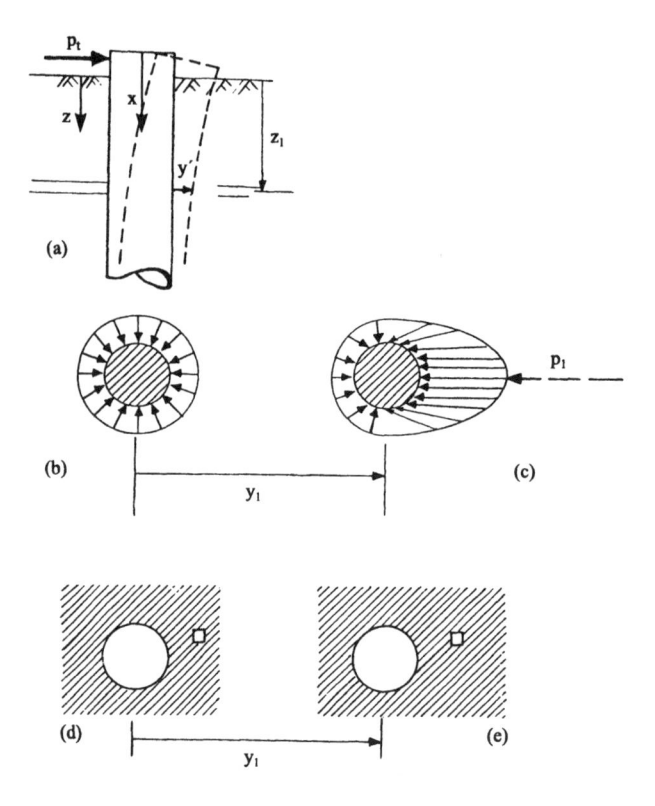

Figure 1.3. Distribution of unit stresses against a pile before and after lateral deflection.

Integration of the unit stresses will result in the quantity p_1 which acts opposite in direction to y_1. The dimensions of p are load per unit length along the pile. These units are identical to those in the solution of the ordinary equations for a beam on an elastic soil bed. The reaction modulus can be defined as the slope of the secant to a *p-y* curve.

A typical *p-y* curve is shown in Figure 1.4a, drawn in the first quadrant for convenience. The curve is one member of a family of curves that show the soil resistance as a function of depth (z). The curve in Figure 1.4b depicts the value E_{py} that is constant for small deflections for a particular depth, but decreases with increased deflection. E_{py} is properly termed *reaction modulus for a pile under lateral loading*. While E_{py} will vary with the properties of the particular soil, the term does not uniquely represent a soil property. Rather E_{py} is a simply a *parameter* for convenient use in computations. For a particular practical solution, the term is modified point-by-point along the length of the pile as iteration occurs. The iteration leads to compatibility between pile deflection and soil resistance, according to the nonlinear *p-y* curves that have been selected.

A number of authors who first wrote about piles under lateral loading used the term E_s for E_{py}, but the term E_s is used herein as describing the characteristics of the soil itself.

Chapter 3 presents recommendations for formulating equations for *p-y* curves. Experiments are cited where lateral-load tests were performed on piles that were instrumented for the measurement of bending moment in the pile as a function of depth. Differentiation and

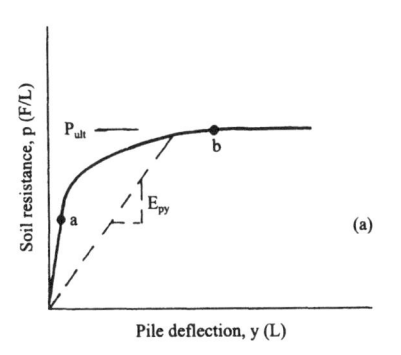

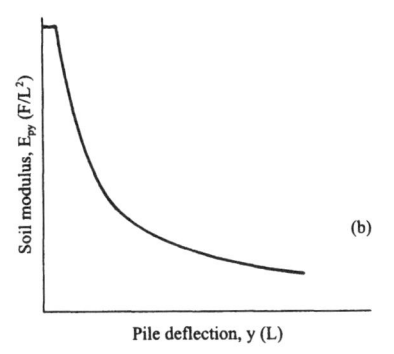

Figure 1.4. Typical *p-y* curve and resulting soil modulus.

integration of those curves yielded experimental *p-y* curves. Correlations were then developed between these experimental curves and the characteristics of the soil.

Examples of *p-y* curves obtained from a full-scale experiment with piles with a diameter of 641 mm and a penetration of 15.2 m are shown in Figures 1.5 and 1.6 (Reese et al. 1975). The piles were instrumented at close spacings for measurement of bending moment and were tested in overconsolidated clay.

The portion of the curve in Figure 1.4a from points a to b shows that the value of *p* is increasing at a decreasing rate with increasing deflection *y*. This behavior undoubtedly is reflecting the nonlinear portion of the in situ stress-strain curve. Many suggestions have been made for predicting the a-b portion of a *p-y* curve but there is no widely accepted analytical procedure. Rather, that part of the curves is empirical and based on results of full-scale tests of piles in a variety of soils with both monotonic and cyclic loading.

The straight-line, horizontal portion of the *p-y* curve in Figure 1.4a implies that the in situ soil is behaving plastically with no loss of shear strength with increasing strain. With that assumption, analytical models can be used to compute the ultimate resistance p_{ult} as a function of pile dimensions and soil properties. These models are discussed in Chapter 3.

A more direct approach to formulating *p-y* curves would be to consider the response of the soil, rather than the pile. Figure 1.3d shows an element to suggest that a solution may be obtained by the finite-element method (FEM). An appropriate solution with the FEM requires a three-dimensional model at and near the ground surface because the response of the upper soils have a dominant effect on pile behavior. The nonlinear stress-strain characteristics of the soil must be modeled, taking into account large strains. Properties must be selected for the various layers of soil around the pile. In addition, nonlinear geometry must be considered, particularly near the ground surface, where gaps in cohesive soils will occur behind a pile and upward bulging in front. For cohesionless soils, there will be settlement of the ground surface due to densification, especially under repeated loading.

A solution with the FEM must start with the constitutive modeling of the in situ soil, then the influence of the installation of the pile or piles must be modeled, and finally the modeling must address the influence of the various kinds of loading (discussed in the next section). Some solutions have been proposed with a three-dimensional FEM, for example, Shie & Brown (1991). Homogeneous soils were studied, and *p-y* curves were developed for single piles and piles in a group. The solutions were made with a super computer, computer time

6 *Piles under lateral loading*

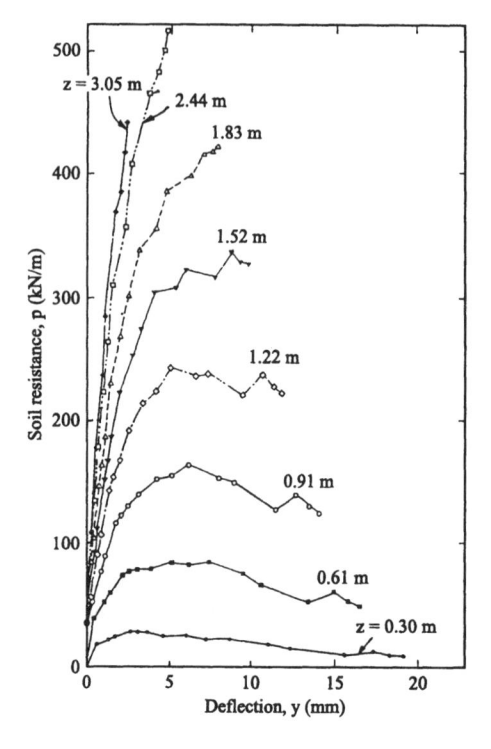

Figure 1.5. *p-y* curves developed from static load test of 641 mm-diameter pile (from Reese et al. 1975).

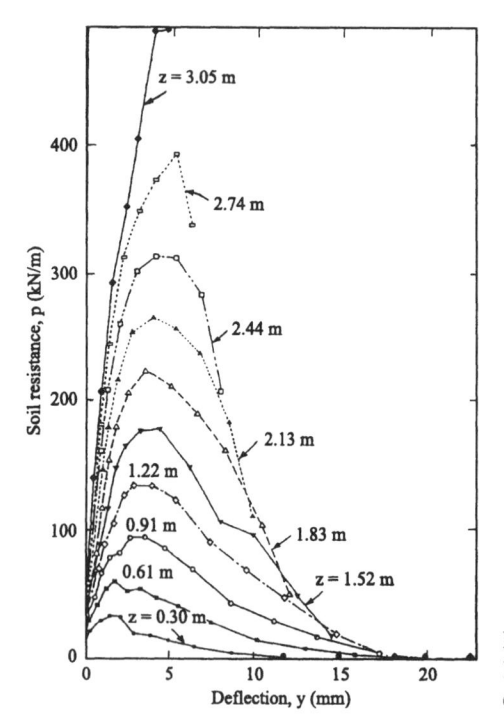

Figure 1.6. *p-y* curves developed from cyclic load test of 641 mm-diameter pile (from Reese et al. 1975).

was considerable, and some analytical difficulties were encountered. While the results were instructive, particularly with respect to pile groups, the FEM can make only a limited contribution to obtaining *p-y* curves for the generalized problem described above.

1.4 RESPONSE OF A PILE

1.4.1 *Introduction*

The nature of the loading, plus the kind of soil around the pile, are of major importance in predicting the response of a single pile or a group of piles. With respect to active loadings at the pile head, four types can be identified: short term or static, cyclic, sustained, and dynamic. In addition, passive loadings can occur along the length of a pile from moving soil, when a pile is used as an anchor. Another problem to be addressed is when existing piles are in the vicinity of pile driving or earth work. Brief, general discussions are presented below of the response of a pile to the various loadings. Analyses are presented in later chapters to illustrate the influence of some of the kinds of loading.

1.4.2 *Static loading*

The curve in Figure 1.4a represents the case for a particular value of *z* where a short-term, monotonic loading was applied to a pile. This case, called static loading for convenience, will seldom, if ever, be encountered in practice. However, static curves are useful because 1. Analytical procedures can be used to develop expressions to correlate with some portions of the curves, 2. The curves serve as a baseline for demonstrating the effects of other types of loading, and 3. The curves can be used for sustained loading for some clays and sands.

The curves in Figure 1.5 resulted from static loading of the pile. Several items are of interest: 1. The initial stiffness of the curves increases with depth; 2. The ultimate resistance increases with depth; and 3. The scatter in the curves illustrates errors inherent in the process of analyzing numerical results from measurements of bending moment with depth. Points 1 and 2 demonstrate that analyses employing soil properties can be correlated with the experimental results, emphasizing the need to do static loading when performing tests of piles.

1.4.3 *Cyclic loading*

Figure 1.7a shows a typical *p-y* curve a particular depth. Point b represents the value of p_{ult} for static loading and p_{ult} is assumed to remain constant for deflections larger that that for Point b. The shaded portion of Figure 1.7a indicates the loss of resistance due to cyclic loading. For the case shown, the static and cyclic curves are identical through the initial straight-line portion to Point a and to a small distance into the nonlinear portion at Point c. With deflections larger than those for Point c, the values of *p* decrease sharply due to cyclic loading to a value at Point d. In some experiments, the value of *p* remained constant beyond Point d. The loss of resistance shown by the shaded area is, for a given soil, plainly a function of the number of cycles of loading. As may be seen, for a constant value of deflection, the value of E_{py} is lowered significantly even at relatively low strain levels, due to cyclic loading.

8 *Piles under lateral loading*

A comparison of the curves in Figures 1.5 and 1.6 demonstrates dramatically the influence of cyclic loading, at least at a site where there is stiff clay of a given set of characteristics. As might be expected, at low magnitudes of deflection, the initial stiffnesses are only moderately affected. However, at large magnitudes of deflection, the *p*-values show spectacular decreases. The values of p_{ult} are also decreased. While the results of static loading of a pile may be correlated with soil properties, plainly the results of cyclic loading will not easily yield to analysis. Discussions in the following paragraphs will indicate the direction of some research. Of most importance are the results from carefully performed tests of full-sized piles under lateral loading in a variety of soils. Formulations for taking cyclic loading into account will be presented in a later chapter where the methods are based on the available results of testing of full-scale, fully instrumented piles.

The cyclic loading of laterally loaded piles occurs with offshore structures, bridges, overhead signs, breasting and mooring dolphins, and other structures. For stiff clays above the water table and for sands, the effect of cyclic loading is important but for saturated clays below water, which includes soft clays, the loss of resistance in comparison to that from static loading can be major. Experiments have shown that stiff clay remains pushed away near the ground surface when a pile deflects, such as shown in Figure 1.8, where two-way cyclic loading is assumed. The re-application of a load causes water to be forced from the opening at a velocity related to the frequency of loading. The usual consequence is that scour of the clay occurs with an additional loss of lateral resistance. In the full-scale experiments with stiff clay that have been performed, the scour of the soil during cyclic loading is readily observed by clouds of suspension near the front and back faces of the pile (Reese et al. 1975). The gapping around a pile not as prominent in soft clay, probably because the clay is so weak to collapse when the loading is cycled. The clouds of suspension were not observed during the testing of piles in soft to medium clays but the cycling caused a substantial loss in lateral resistance (Matlock 1970).

As may be seen in Figure 1.8, the soil resistance near the mudline would be zero up to a given deflection. No failure of the soil has occurred because the resistance is transferred to the lower portion of the soil profile. There will be an increase in the bending moment in the pile, of course, for a given value of lateral loading.

1.4.4 *Sustained loading*

Figure 1.7b shows an increasing deflection with sustained loading. The decreasing value of *p* implies the shifting of resistance to lower elements of soil. The effect of

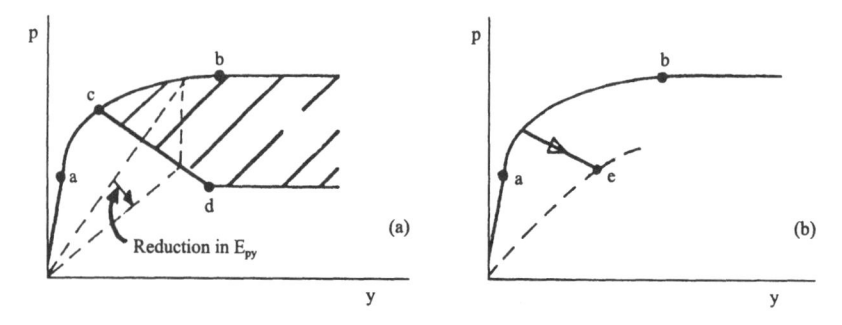

Figure 1.7. Effect of number of cycles on the *p-y* behavior at very low cyclic strain loading.

sustained loading is likely to be negligible for overconsolidated clays and for granular soils.

Sustained loading of a pile in soft clay would likely result in a significant amount of time-related deflection. Analytical solutions could be made, using the three-dimensional theory of consolidation, but the formulation of the equations depends on a large number of parameters not clearly defined physically. The generalization of such a procedure is not yet available in the literature.

The influence of sustained loading, in some cases, can be solved with reasonable accuracy by experiment. At the site of the Pyramid Building in Memphis, Tennessee, a lateral-load was applied to a CFA pile with a diameter of 430 mm in a silty clay with an average value of undrained shear strength over the top several diameters of the pile of 35 kPa. (Reuss et al. 1992). A load of 22 kN, corresponding approximately to the working load, was held for a period of 10 days, and deflection was measured. Some errors in the data occurred because the load was maintained by manual adjustment of the hydraulic pressure, rather than by a servo-mechanism. However, it was possible to analyze the data to show that soil-response curves could be *stretched* by increasing the deflection 20%, over that for static loading, to predict the behavior of the pile under sustained loading. At the Pyramid Building site, some thin strata of silt in the near-surface soils is believed to have promoted the dissipation of excess pore water pressure.

1.4.5 Dynamic loading

With respect to dynamic loading, the greatest concern is that some event will cause liquefaction to occur in the soil at the pile-supported structure. A discussion. of liquefaction

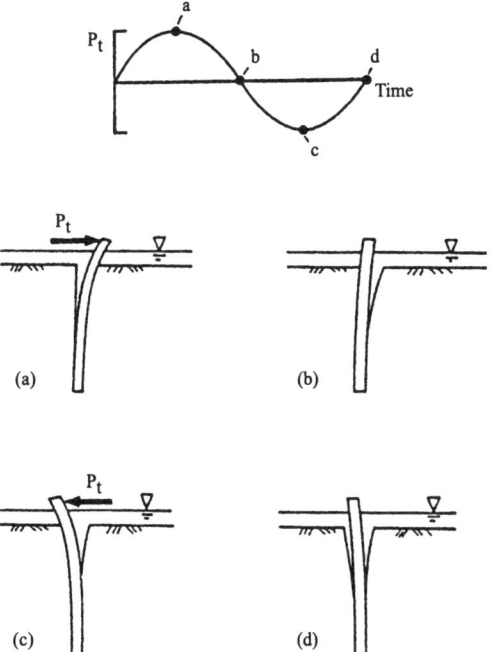

Figure 1.8. Simplified response of piles in clay due to cyclic loading (from Long 1984).

will not be presented beyond saying that liquefaction can occur in loose, granular soil below the water table.

Pile-supported structures can be subjected to dynamic loads from machines, traffic, ocean waves, and earthquakes (Hadjian et al. 1992). The frequency of loading from traffic and waves is usually low enough that p-y curves for static or cyclic loading can be used. Brief discussions are presented below about loadings from machinery and from earthquakes. In addition, some discussion is given to vibrations and perhaps permanent soil movement, as a result of the vibrations, due to installing piles in the vicinity of an existing pile-supported structure.

As noted earlier, soil resistance for static loadings can be related to the stress-strain characteristics of the soil; however, if the loading is dynamic, an inertia effect must be considered. Not only are the stress-strain characteristics necessary for formulating p-y curves for dynamic loading but the mass of the soil must be taken into account. Use of the finite element method appears promising but, if the FEM has not proven completely successful for static loading, the application to the dynamic problem appears to be doubly complex. Thus, unproven assumptions must be made if the p-y method is applied directly to solving dynamic problems.

If the loading is due to rotating machinery, the deflection is usually small, and a value of soil modulus may be used for analysis. Experimental techniques (Woods & Stokoe 1985, Woods 1978) have been developed for obtaining the soil parameters that are needed. Analytical techniques for solving for the response of a pile-supported structure and presented by a number of writers. Roesset (1988) and Kaynia & Kausel (1982) have developed techniques that are quite effective in dealing with machine-induced vibrations.

If the loading is a result of a seismic event, the analysis of a pile-supported structure will be complex (Gazetas & Mylonakis 1998). The free-field motion of the near surface soils at the site must be computed, or selected, taking micro zonation into account. A *standard* earthquake may be used with an unknown degree of approximation. The response of the piles, neglecting the superstructure, must be considered. If the soil movement is constant with depth, the piles will move with the soil without bending. Such an assumption, if valid, simplified the computations. The distributed masses of the superstructure must be employed in solving for the motion of the piles and the motion of the superstructure. Of course, p-y curves must be available with appropriate modification of the inertia effects. Not much experimental data are available on which to base a method of computation.

In the absence of comprehensive information on the response with depth of pile-supported structures that either have failed or have withstood an earthquake, and taking into account the enormous amount of computations that are needed, fully rational analyses are currently unavailable.

Various simplifying assumptions are being used: (1) pseudo horizontal load and available p-y curves are sometimes employed as a means of simulating the effects of an earthquake. If the assumption is made that the lateral soil movement during an earthquake is constant with depth, and if existing p-y curves are used, or curves perhaps modified empirically for inertia effects, the movements of elements of the superstructure can be computed by equations of mechanics.

Engineers are aware that the installation of piles near a pile-supported structure could lead to the movements of the existing structure. If the construction is near, the loss of ground from installing bored piles or the heave from installing driven piles can be detrimental. If the pile driving is some distance away, information from the technical literature can be helpful (Ramshaw et al. 1998, Drabkin & Lacy 1998). Prudent engineers can establish measurement points on existing structures and have observations made as pile installation proceeds. In cases of sensitive machinery near new construction, the installation of transducers for measurement of time-related movements can be helpful.

1.5 MODELS FOR USE IN ANALYSES OF A SINGLE PILE

A number of models have been used for the design of piles under lateral loading, and some of them can be used as supplements to the principal method proposed herein.

1.5.1 *Elastic pile and elastic soil*

The model shown in Figure 1.9a depicts a pile in an elastic soil. A similar model has been widely used. Terzaghi (1955) suggested values of the so-called subgrade modulus that can be used to solve for deflection and bending moment, but he went on to qualify his recommendations. The standard beam equation was employed in a manner that had been suggested earlier by such writers as Hetenyi (1946). Terzaghi stated that the tabulated values of subgrade modulus could not be used for lateral loads larger than the one that resulted in a soil resistance of one-half of the bearing capacity of the soil. However, no recommendation was included in regard to the computation of the bearing capacity under lateral load. Nor were any comparisons given between the results of computations and experiments.

Poulos and his colleagues have contributed extensively to developments for piles under lateral loading using the elastic model and several variations of the model (Poulos & Davis 1980, Poulos & Hull 1989). Solutions have been presented for a variety of cases of loading of single piles and for the interaction of piles with close spacings. The solutions

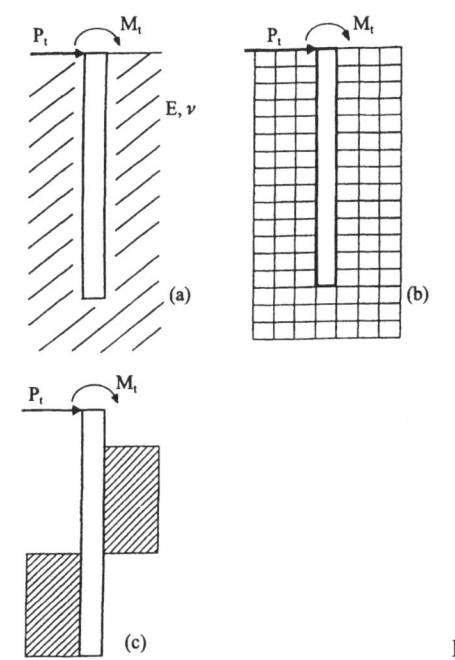

Figure 1.9. Models for a pile under lateral loading.

12 Piles under lateral loading

have gained considerable attention but cannot readily be used to compute the larger deformation or collapse of the pile in nonlinear soil.

The differential equation presented by Terzaghi required the use of values of moduli with a different format than used herein, but conversion is easily made. Values in terms of the format used herein are presented in Chapter 3 for the benefit of the reader. The recommendations of Terzaghi have proved useful and provide evidence that Terzaghi had excellent insight into the problem. However, in a private conversation with the senior writer, Terzaghi said that he had not been enthusiastic about writing the paper and only did so in response to numerous requests. The method illustrated by Figure 1.9a serves well in obtaining the response of a pile under small loads, in illustrating the various interrelationships in the response of piles, and in giving an overall insight into the nature of the problem. The method cannot be employed without modification in solving for the loading at which yielding will develop in a pile.

1.5.2 Elastic pile and finite elements for soil

The case shown in Figure 1.9b is the same as the previous case except that the soil has been modeled by finite elements. No attempt is made in the sketch to indicate an appropriate size of the map, boundary constraints, special interface elements, most favorable shape and size of elements, or other details. The finite elements may be axially symmetric with non-symmetric loading, or fully three dimensional. The elements may be selected as linear or nonlinear.

In view of computational power that is available, the model shown in Figure 1.9b appears to represent an ideal way to solve the pile problem. The elements can be fully three-dimensional and nonlinear, and nonlinear geometry can be employed. However, in addition to the problem of selecting the basic nonlinear element for the soils, some other problems are coding to disregarding tensile stresses, modeling layered soils, accounting for the separation between pile and soil during repeated loading, coding for the collapse of sand against the back of a pile, and accounting for the changes in soil characteristics associated with the various types of loading. All of these problems at present have no satisfactory solution.

Yegian & Wright (1973) and Thompson (1977) did interesting studies using two-dimensional finite elements. Thompson used a plane-stress model and obtained soil-response curves that agreed well with results at near the ground surface from full-scale experiments. Portugal & Sêco e Pinto (1993) used the finite element method based on p-y curves to obtain a good prediction of the observed lateral behavior of the foundation piles of a Portuguese bridge. Kooijman (1989) and Brown et al. (1989) used three-dimensional finite elements to develop p-y curves. Research is continuing with three-dimensional, nonlinear finite elements; for example, Brown & Shie (1991) but no proposals have been made for a practical method of design. However, a finite-element model likely will be developed that will lead to results that can be used in practice.

1.5.3 Rigid pile and plastic soil

Broms (1964a,b, 1965) employed the model shown in Figure 1.9c, or a similar one, to derive equations for predicting the loading that develops the ultimate bending moment. The pile is assumed to be rigid, and a solution is found by use of the equations of statics for the distribution of ultimate resistance of the soil that puts the pile in equilibrium. The soil resistance shown hatched in the figure is for cohesive soil, and a solution was developed for cohesionless soil as well. After the ultimate loading is computed for a pile of particular dimensions, Broms suggests that the deflection for the working load may be computed by the use of the model shown in Figure 1.9a.

The Broms method obviously makes use of several simplifying assumptions especially but can be useful for the initial selection of a pile. A summary of the Broms equations, with examples, is presented in Appendix A for the convenience of the user.

The engineer may wish to implement the Broms equations at the start of a design if the pile has constant dimensions and only if uniform characteristics can reasonably be selected for the soil.

Solution of the equations will yield the size and length of the pile for the expected loading. The pile can then be employed at the starting point for the *p-y* method of analysis. Further benefits from the Broms method are: 1. The mechanics of the problem of lateral loading is clarified, and 2. The method may be used as a check of some of the results from the *p-y* method of analysis.

It is of interest to note that the computer code for the *p-y* method of analysis, implemented in Appendix D, is so efficient that many trial solutions can be made in a short period of time. An experienced engineer can use the computer model to 'home in' rapidly on a correct solution for a particular application without the limitations imposed by the Broms equations.

1.5.4 *Characteristic load method*

Duncan et al. (1994) presented the characteristic-load method (CLM), following the earlier work of Evans & Duncan (1982). A series of solutions were made with nonlinear *p-y* curves for a range of soils and for a range of pile-head conditions. The results were analyzed with the view of obtaining simple equations that could be used for rapid prediction of the response of piles under lateral loading. Dimensionless variables were employed in the prediction equations. The authors state that the method can be used to solve for: 1. Ground-line deflections due to lateral load for free-head conditions, fixed-head conditions, and the 'flagpole' condition; 2. Ground-line deflections due to moments applied at the ground line; 3. Maximum moments for the three conditions in (1); and 4. The location of the point of maximum moment along the pile. The soil may be either a clay or a sand, both limited to uniform strength with depth.

The prediction equations have the general form of the one for clay, shown in Equation 1.1.

$$P_c = 7.34b^2 (E_p R_i) \left(\frac{c_u}{E_p R_i} \right)^{0.68} \tag{1.1}$$

where P_c = characteristic load, b = diameter of pile, E_p = modulus of elasticity of material of pile, R_i = ratio of moment of inertia of the pile to that of a solid pile of the same diameter, c_u = undrained shear strength of clay.

For a given problem of applied lateral load P_t, for a pile in clay with a constant shear strength, a value of P_c is computed by the equation above. The ratio P_t/P_c is found and becomes the argument for entering a nonlinear curve for the value of y_t/b for free-head or fixed head cases for clay.

14 *Piles under lateral loading*

Equations and nonlinear curves were developed for computing the value of the maximum bending moment and where it occurs along the pile.

Duncan and his co-workers were ingenious in developing equations and curves that give useful solutions to a number of problems where piles must sustain lateral loads. The limitations in the method with respect to applications were noted by the authors.

Endley et al. (1997) began with recommendations for the formulating *p-y* curves and developed equations similar to those of Duncan, et al., for the prediction of piles in various soils. The Endley equations were designed to deal with piles that penetrated only a short distance into the ground surface as well as with long piles.

1.5.5 *Nonlinear pile and p-y model for soil*

Interest in the model shown in Figure 1.10 developed in the late 1940s and 1950s when energy companies built offshore structures that were designed to sustain relatively large horizontal loads from waves. About the same time offshore structures were built in the United States for military defense Rutledge (1956). The relevant differential equations were stated by Timoshenko (1941) and by other writers. Hetenyi (1946) presented solutions for beams on a foundation with linear response. In 1948 Palmer and Thompson

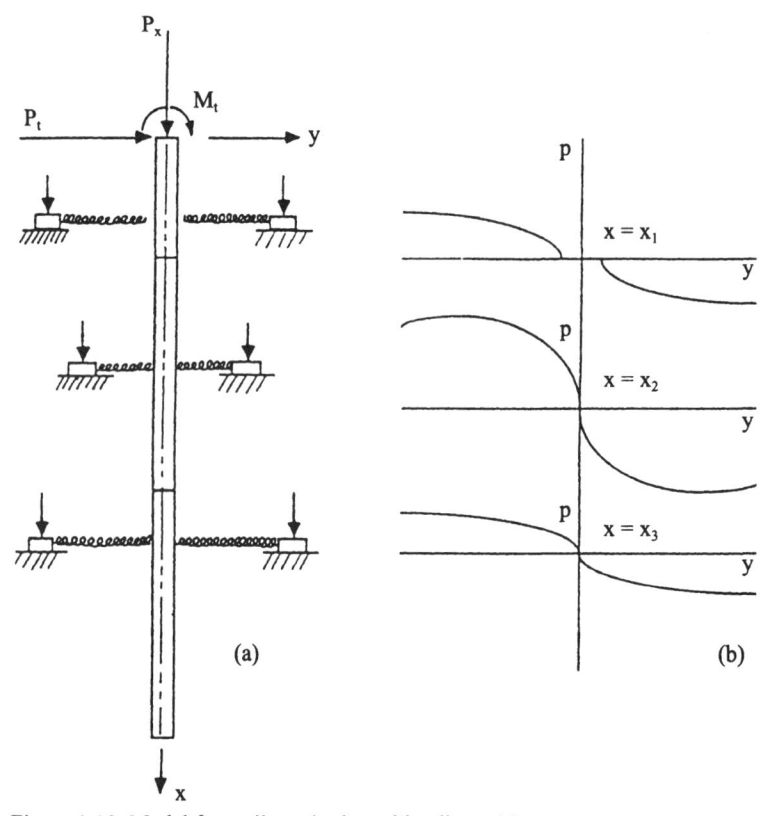

Figure 1.10. Model for a pile under lateral loading with *p-y* curves.

presented a numerical solution to the nonlinear differential equation. In 1953, the American Society for Testing and Materials sponsored a conference on the lateral loading of piles where papers by Gleser, and McCammon and Asherman emphasized full-scale testing.

The offshore industry embarked on a program of full-scale testing of fully instrumented piles in the 1950s. Fortuitously, the digital computer became widely available about the same time. Full-scale testing and the digital computer allowed the development of the method emphasized in this document; contributions continue from engineers in many countries.

As a matter of historical interest, Terzaghi (1955) wrote 'If the horizontal loading tests are made on flexible tubes or piles – values of *soil resistance* – can be estimated for any depth, if the tube or pile is equipped with fairly closely spaced strain gauges and if, in addition, provisions are made for measuring the deflections by means of an accurate deflectometer. The strain-gauge readings determine the intensity and distribution of the bending moments over the deflected portion of the tube or the pile, and on the basis of the moment diagram the intensity and distribution of the horizontal loads can be ascertained by an analytical or graphic procedure.' ... 'If the test is repeated for different horizontal loads acting on the upper end of the pile, a curve can be plotted for different depths showing the relationship between p and y.' Terzaghi goes on to write, 'However, errors in the computation of the deflections are so important that the procedure cannot be recommended.' (Meaning unchanged, but some words changed to agree with terminology used herein).

Matlock and his associates devised an extremely accurate method of measuring the bending moments and formal procedures for interpreting the data. (Matlock & Ripperger 1956, Matlock & Ripperger 1958). Two integrations of the bending-moment data yielded accurate values of deflection but special techniques were required for the two differentiations to yield adequate values of soil resistance. The result was the first set of comprehensive recommendations for predicting the response of a pile to lateral loading.

As shown in Figure 1.10, loading on the pile commonly refers to the two-dimensional case; no torsion or out-of-plane bending is assumed. The horizontal lines across the pile are meant to show that it is made up of different sections; for example, steel pipe could be used with the wall thickness varied along the length. The difference-equation method presented in detail in Chapter 2, employed for the solution of the beam-column equation, allows the different values of bending stiffness ($E_p I_p$) to be considered. Further, the method of solution allows $E_p I_p$ to be nonlinear and a function of the computed values of bending moment. For many solutions it is unnecessary to vary the bending stiffness even though the loading is carried to a point where a plastic hinge is expected to develop.

An axial load is indicated and is considered in the solution with respect to its effect on bending and not in regard to computing the required length to support a given axial load. As shown later, the computational procedure allows the determination of the rare case of the axial load at which the pile will buckle.

The soil around the pile is replaced by a set of mechanisms that merely indicate that the soil resistance p is a nonlinear function of pile deflection y. The mechanisms, and the corresponding curves that represent their behavior, are widely spaced in the sketch but are considered to be vary continuously with depth. As may be seen, the p-y curves are fully variable with respect to distance x along the pile and pile deflection y. The curve for $x = x_1$ is drawn to indicate that the pile may deflect a finite distance with no soil resistance. The curve at $x = x_2$ is drawn to show that the soil is deflection-softening. There is no reason-

16 *Piles under lateral loading*

able limit to the variations that can be employed in representing the response of the soil to the lateral deflection of a pile.

The *p-y* method is versatile and provides a practical means for design. The method was suggested over thirty years ago (McClelland & Focht 1958, Reese & Matlock 1956). Two developments during the 1950s made the method possible: the digital computer for solving the problem of the nonlinear, fourth-order differential equation for the beam-column; and the remote-reading strain gauge for use in obtaining soil-response (*p-y*) curves from experiment.

The *p-y* method evolved principally from research sponsored by the petroleum industry when faced with the design of pile-supported platforms subjected to exceptionally large horizontal forces from waves and wind. Rules and recommendations for the use of the *p-y* method for design of such piles are presented by the American Petroleum Institute (1987) and Det Norske Veritas (1977).

The use of the method has been extended to the design of onshore foundations, as for example, by publications of the Federal Highway Administration (USA) (Reese 1984). The procedure is being cited broadly, for example, Jamiolkowski (1977), Baguelin et al. (1978), George & Wood (1977), and Poulos & Davis (1980). The method has been used with success for the design of piles; however, research is continuing and improvements, particularly in the characterization on a variety of special soils, are expected.

1.6 MODELS FOR GROUPS OF PILES UNDER LATERAL LOADING

Piles are most often used in groups as illustrated in Figure 1.11, and a practical example, an offshore platform, is shown in Figure 1.1. The models that are used for the group of piles must address two problems: the efficiency of closely-spaced piles under lateral loading (and axial loading); and the distribution of the loading to each of the piles in the group, a problem in mechanics.

The efficiency of a particular pile is defined as the ratio of the load that it can sustain in close spacing to the load that could have been sustained if the pile had been isolated. Because of the variability of soil and the complex nature of constitutive models, theoretical solutions are currently unavailable for computing the efficiency of a particular pile. Methods for finding the efficiency, both under lateral and axial loading, are based on the results of experiments, most of which are from the laboratory.

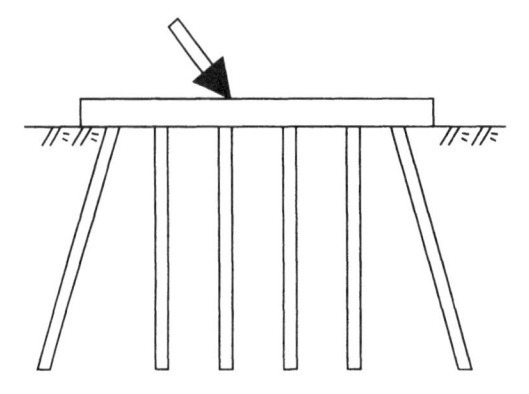

Figure 1.11. Structure supported by a group of piles.

In contrast, if one can assume that the procedures are accurate for analyzing a single pile under lateral loading (and under axial loading), the problem of the distribution of the loading to each of the piles in a group can be solved exactly. A model for the solution of the problem in mechanics is shown in Figure 1.12. As may be seen in Figure 1.12a, a global coordinate system is established for the loadings on the structure and for identifying the positions of each of the pile heads and their angle of rake. Then, a local coordinate system is utilized for each of the piles, with axial and lateral coordinates.

Figure 1.12b shows that each of the piles is replaced by nonlinear mechanisms that give the resistance to axial movement, lateral movement, and rotation as a function of pile-head movement. Also shown in Figure 1.12b is a set of movements of the origin or the global coordinate system. With these movements, the lateral and vertical movements and the rotation can be found at each pile head. The forces generated at the pile heads serve to put the structure into equilibrium. Because of nonlinearity, iteration is required to find the unique movements of the global coordinate system.

The model for the pile under lateral loading, already described, is used for finding the pile forces as a function of a lateral deflection and a pile-head rotation. Figure 1.13 shows the model that can be employed to find the axial force on the pile as a function of settlement. As may be seen, nonlinear mechanisms are used to represent the soil resistance in skin friction and in end bearing as a function of axial movement. Also, the spring representing the stiffness of the pile can be nonlinear if necessary and desirable.

A more detailed description of the method of solving for the distribution of loading to piles in a group is presented in Chapter 5.

As noted in the description presented above, all of the loading on the superstructure is assumed to be taken by the piles and none by the cap or raft supported by the piles. The sketch

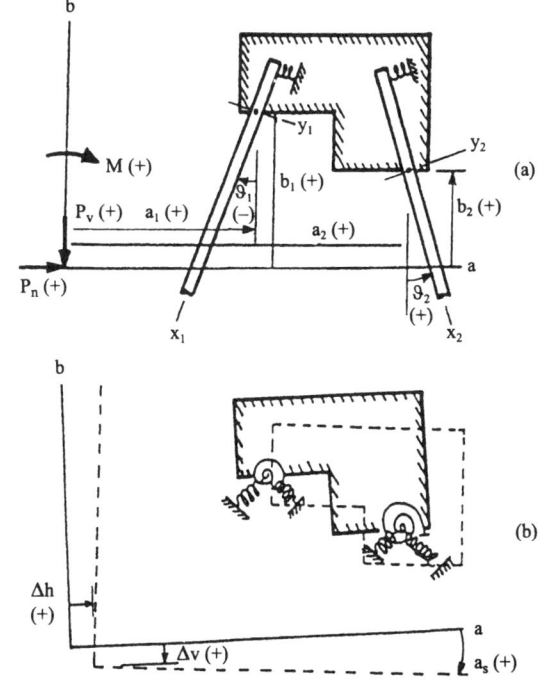

Figure 1.12. Simplified structure showing coordinate systems and sign conventions: (a) with piles shown; (b) with piles represented as springs (after Reese & Matlock 1966).

18 *Piles under lateral loading*

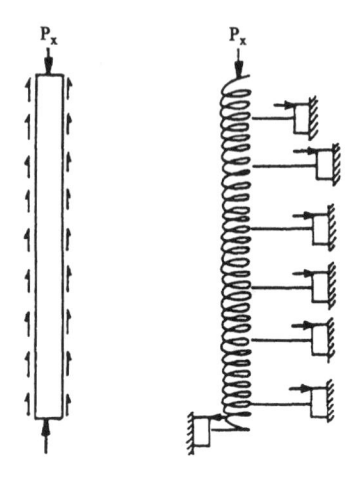

Figure 1.13. Model of a pile under axial load.

in Figure 1.11 shows that the cap is resting on the ground surface and settlement would cause some of the loading to be taken by the cap. The problem of finding the distribution of axial load to the cap (or raft) and to the piles has been addressed by a number of authors. While the proposed solutions are limited to the response of the foundation system to axial loading, a brief introduction to the technology is warranted because extensions will likely occur to the general problem, where lateral loading is an important parameter.

Van Impe & De Clercq (1994) reviewed the work of a number of authors and elected to extend the solution proposed by Randolph & Wroth (1978). Those authors employed a two-layer system for the soil. Each of the layers was characterized by a shear modulus G and a Poisson's ratio v. Using these parameters, equations were developed for the settlement of a rigid pile due to load transfer along the pile in skin friction, due to load in end bearing, and due to load on an element of the raft. The equations were solved to find the identical settlements for the elements of the system in order to obtain the distribution of load to the piles and to the cap. An important analytical difficulty was to assume a fixed radius of influence of distribution of stresses in the continuum in order to limit the magnitude of the computed settlements.

The contributions of Van Impe & De Clercq (1994) were a decay curve for the decrease in the shear modulus with increasing strain; a multi-layered soil model; and an improved method of solving for the radius of influence of stresses in the continuum. Results from the extended method were compared with experimental results from tests of an instrumented, full-sized, pile-supported bridge pier. Excellent agreement was found between computed and observed settlement and good agreement was found between computed and observed loads to the elements of the system, after an assumption was made about the distribution of the load due to the placement of concrete.

The results from the case studies suggest that benefits can be derived in extending the general method to the case of both axial and lateral loading.

1.7 STATUS OF CURRENT STATE-OF-THE-ART

As presented in Appendix D, computer programs are readily available for solving the difference equations, describing the behavior of a single pile and a group of piles, efficiently and in a user-friendly fashion. The use of computer codes will be demonstrated in Chapters 6 and 7; however, it is useful here to write briefly about the current state-of-the-art.

The computer codes allow the engineer to make solutions rapidly in order to investigate the influence of a variety of parameters. Upper-bound and lower-bound solutions can be done with relative ease. Guidance can be obtained in most cases with respect to the desirability of performing additional tests of the soil or performing a full-scale, lateral-load test at the site.

The principal advances in computational procedures in the future relate to p-y curves. Better information is needed for piles in rock of all kinds, in soils with both cohesion and a friction angle, and in silts. For piles in closely-spaced groups, relevant information is needed on pile-soil-pile interaction. In spite of these weaknesses, the technology presented herein is believed to represent a signal advance in engineering practice with respect to previously available methods.

CHAPTER 2

Derivation of equations and methods of solution

2.1 INTRODUCTION

The equation for the beam-column must be solved for implementation of the *p-y* method, and a derivation is shown in the following section. An abbreviated version of the equation can be solved by a closed-form method for some purposes, but a general solution can be made only by a numerical procedure. Both of these kinds of solution are presented.

Also presented is the use of dimensional analysis to develop nondimensional expressions for the case where $E_{py} = k_{py} x$, a solution that is most useful for sands and normally consolidated clays. An example problem is worked to show the applicability of this case to practical applications. The solutions with the linearly increasing reaction is also helpful in demonstrating the nature of the nonlinear method of analysis, and the method can be used in checking computer solutions that are presented later.

2.2 DERIVATION OF THE DIFFERENTIAL EQUATION

In most instances, the axial load on a laterally loaded pile has relatively small or little influence on bending moment. However, there are occasions when it is desirable to find the buckling load for a pile; thus, the axial load is needed in the derivation. The derivation for the differential equation for the beam-column on a foundation was given by Hetenyi (1946).

The assumption is made that a bar on an elastic foundation is subjected to horizontal loading and a pair of compressive forces P_x acting in the center of gravity of the end cross-sections of the bar.

If an infinitely small unloaded element, bounded by two horizontals a distance dx apart, is cut out of this bar (see Fig. 2.1), the equilibrium of moments (ignoring second-order terms) leads to the equation

$$(M + dM) - M + P_x dy - V_v dx = 0 \qquad (2.1)$$

or

$$\frac{dM}{dx} + P_x \frac{dy}{dx} - V_v = 0 \qquad (2.2)$$

Differentiating Equation 2.2 with respect to *x*, the following equation is obtained

22 Piles under lateral loading

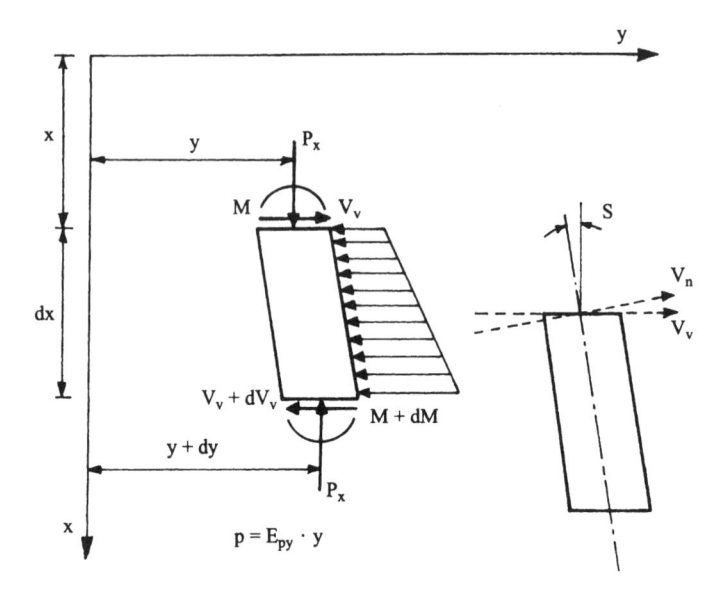

Figure 2.1. Element from beam-column (after Hetenyi 1946).

$$\frac{d^2M}{dx^2} + P_x \frac{d^2y}{dx^2} - \frac{dV_v}{dx} = 0 \tag{2.3}$$

The following identities are noted:

$$\frac{d^2M}{dx^2} = E_p I_p \frac{d^4y}{dx^4}$$

$$\frac{dV_v}{dx} = p, \text{ and}$$

$$p = E_{py} y$$

And making the indicated substitutions, Equation 2.3 becomes

$$E_p I_p \frac{d^4y}{dx^4} + P_x \frac{d^2y}{dx^2} + E_{py} y = 0 \tag{2.4}$$

The direction of the shearing force V_v is shown in Figure 2.1. The shearing force in the plane normal to the deflection line can be obtained as

$$V_n = V_v \cos S - P_x \sin S \tag{2.5}$$

Because S is usually small, $\cos S = 1$ and $\sin S = \tan S = dy/dx$. Thus, Equation 2.6 is obtained.

$$V_n = V_v - P_x \frac{dy}{dx} \tag{2.6}$$

V_n will mostly be used in computations, but V_v can be computed from Equation 2.6 where dy/dx is equal to the rotation S.

The ability to allow a distributed force W per unit of length along the upper portion of a pile is convenient in the solution of a number of practical problems. The differential equation then is given by Equation 2.7.

$$E_p I_p \frac{d^4 y}{dx^4} + P_x \frac{d^2 y}{dx^2} - p + W = 0 \tag{2.7}$$

where P_x = axial load on the pile; y = lateral deflection of the pile at a point x along the length of the pile; p = soil reaction per unit length; $E_p I_p$ = bending stiffness; and W = distributed load along the length of the pile.

Other beam formulas that are needed in analyzing piles under lateral loads are:

$$E_p I_p \frac{d^3 y}{dx^3} + P_x \frac{dy}{dx} = V \tag{2.8}$$

$$E_p I_p \frac{d^2 y}{dx^2} = M \quad \text{and} \tag{2.9}$$

$$\frac{dy}{dx} = S \tag{2.10}$$

where V = shear in the pile; M = bending moment of the pile; and S = the slope of the elastic curve defined by the axis of the pile.

Except for the axial load P_x, the sign conventions are the same as those usually employed in the mechanics for beams, with the axis for the pile rotated 90° clockwise from the axis for the beam. The axial load P_x does not normally appear in the equations for beams. The sign conventions are presented graphically in Figure 2.2. A solution of the differential equation yields a set of curves such as shown in Figure 2.3. The mathematical relationships for the various curves that give the response of the pile are shown in the figure for the case where no axial load is applied.

The assumptions that are made in deriving the differential equation are as follows.
1. The pile is straight and has a uniform cross section,
2. The pile has a longitudinal plane of symmetry; loads and reactions lie in that plane,
3. The pile material is homogeneous and isotropic,
4. The proportional limit of the pile material is not exceeded,
5. The modulus of elasticity of the pile material is the same in tension and compression,
6. Transverse deflections of the pile are small,
7. The pile is not subjected to dynamic loading, and
8. Deflections due to shearing stresses are small.

Assumption 8 can be addressed by including more terms in the differential equation, but errors associated with omission of these terms are usually small. The numerical method presented later can deal with the behavior of a pile made of materials with nonlinear stress-strain properties.

24 *Piles under lateral loading*

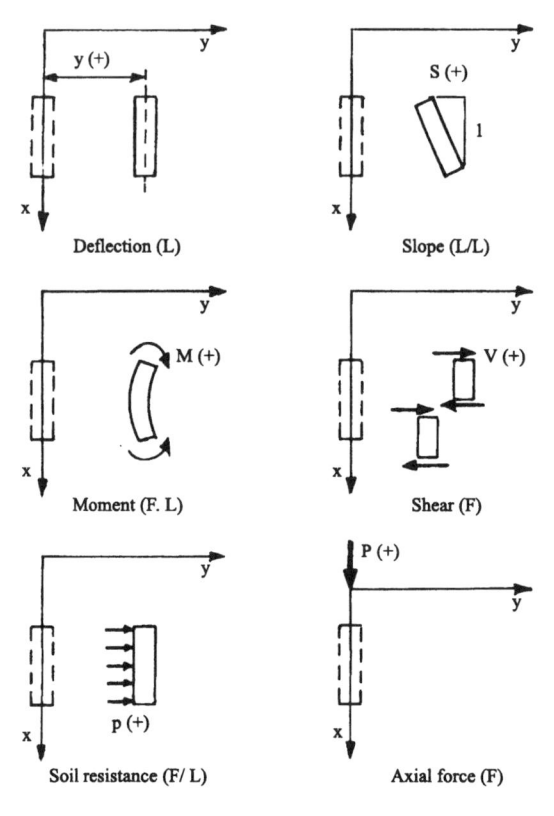

Note: All of the responses of the pile and soil are shown in the positive sense: F = force; L = length

Figure 2.2. Sign conventions.

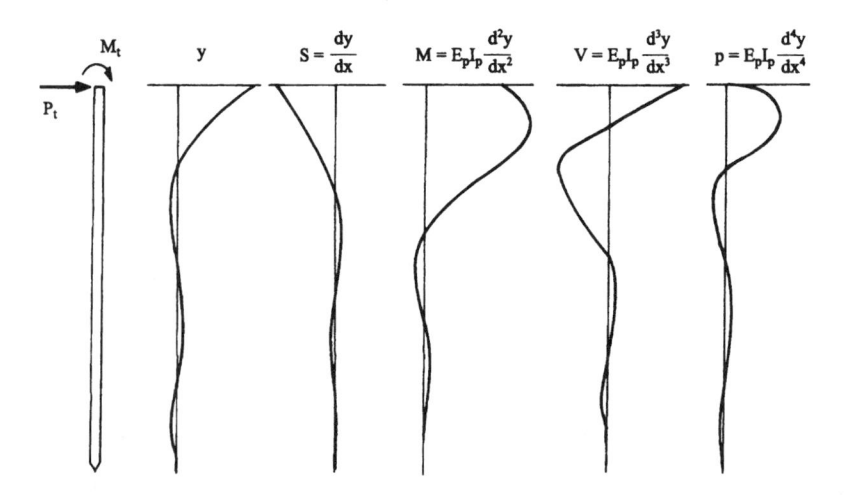

Figure 2.3. Form of the results obtained from a complete solution.

2.2.1 Solution of reduced form of differential equation

A simpler form of the differential equation results from Equation 2.4 if the assumptions are made that no axial load is applied, that the bending stiffness $E_p I_p$ is constant with depth, and that the soil reaction E_{py} is a constant and equal to α. The first two assumptions can be satisfied in many practical cases; however, the last of the three assumptions is seldom or ever satisfied in practice.

The solution shown in this section is presented for two important reasons: 1. The resulting equations demonstrate several factors that are common to any solution; thus, the nature of the problem is revealed; and 2. The closed-form solution allows for a check of the accuracy of the numerical solutions that are given later in this chapter.

If the assumptions shown above and the identity shown in Equation 2.11 are employed, a reduced form of the differential equation is shown as Equation 2.12.

$$\beta^4 = \frac{\alpha}{4 E_p I_p} \tag{2.11}$$

$$\frac{d^4 y}{dx^4} + 4\beta^4 y = 0 \tag{2.12}$$

The solution to Equation 2.12 may be directly written as:

$$y = e^{\beta x}(\chi_1 \cos\beta x + \chi_2 \sin\beta x) + e^{-\beta x}(\chi_3 \cos\beta x + \chi_4 \sin\beta x) \tag{2.13}$$

The coefficients $\chi_1, \chi_2, \chi_3,$ and χ_4 must be evaluated for the various boundary conditions that are desired. If one considers a long pile, a simple set of equations can be derived. An examination of Equation 2.13 shows that χ_1 and χ_2 must approach zero for a long pile because the term $e^{\beta x}$ will be large with large values of x.

The boundary conditions for the top of the pile that are employed for the solution of the reduced form of the differential equation are shown by the simple sketches in Figure 2.4. A more complete discussion of boundary conditions is presented in the next section. The boundary conditions at the top of the long pile that are selected for the first case are illustrated in Figure 2.4a and in equation form are:

$$\text{at } x = 0, \quad \frac{d^2 y}{dx^2} = \frac{M_t}{E_p I_p}, \text{ and} \tag{2.14}$$

$$\frac{d^3 y}{dx^3} = \frac{P_t}{E_p I_p} \tag{2.15}$$

From Equation 2.13 and substituting of Equation 2.14 one obtains for a long pile.

$$\chi_4 = \frac{-M_t}{2 E_p I_p \beta^2} \tag{2.16}$$

The substitutions indicated by Equation 2.15 yield the following:

26 *Piles under lateral loading*

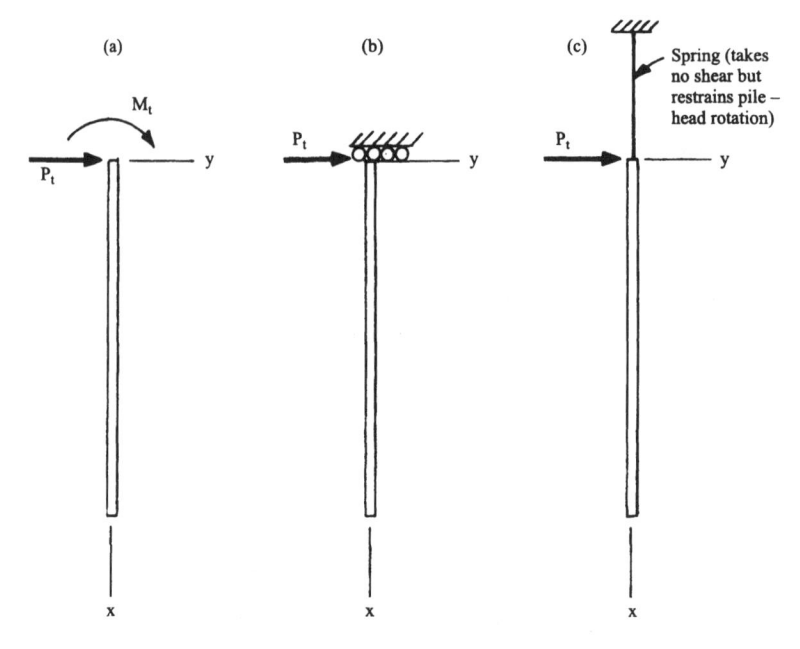

Figure 2.4. Boundary conditions at top of pile.

$$\chi_3 + \chi_4 = \frac{P_t}{2E_p I_p \beta^3} \tag{2.17}$$

Equations 2.16 and 2.17 are used and expressions for deflection *y*, slope *S*, bending moment *M*, shear *V*, and soil resistance p for the long pile can be written in Equations 2.18 through 2.22.

$$y = \frac{2\beta^2 e^{-\beta x}}{\alpha}\left[\frac{P_t}{\beta}\cos\beta x + M_t(\cos\beta x - \sin\beta x)\right] \tag{2.18}$$

$$S = e^{-\beta x}\left[\frac{2P_t\beta^2}{\alpha}(\sin\beta x + \cos\beta x) + \frac{M_t}{E_p I_p \beta}\cos\beta x\right] \tag{2.19}$$

$$M = e^{-\beta x}\left[\frac{P_t}{\beta}\sin\beta x + M_t(\sin\beta x + \cos\beta x)\right] \tag{2.20}$$

$$V = e^{-\beta x}\left[P_t(\cos\beta x - \sin\beta x) - 2M_t\beta\sin\beta x\right] \tag{2.21}$$

$$p = -2\beta^2 e^{-\beta x}\left[\frac{P_t}{\beta}\cos\beta x + M_t(\cos\beta x - \sin\beta x)\right] \tag{2.22}$$

It is convenient to define some functions for simplifying the written form of the above equations:

$$A_1 = e^{-\beta x}(\cos\beta x + \sin\beta x) \tag{2.23}$$

$$B_1 = e^{-\beta x}(\cos\beta x - \sin\beta x) \tag{2.24}$$

$$C_1 = e^{-\beta x}\cos\beta x \quad \text{and} \tag{2.25}$$

$$D_1 = e^{-\beta x}\sin\beta x \tag{2.26}$$

Using these functions, Equations 2.18 through 2.22 become:

$$y = \frac{2P_t\beta}{\alpha}C_1 + \frac{M_t}{2E_pI_p\beta^2}B_1 \tag{2.27}$$

$$S = \frac{2P_t\beta^2}{\alpha}A_1 - \frac{M_t}{E_pI_p\beta}C_1 \tag{2.28}$$

$$M = \frac{P_t}{\beta}D_1 + M_tA_1 \tag{2.29}$$

$$V = P_tB_1 - 2M_t\beta D_1 \quad \text{and} \tag{2.30}$$

$$p = -2P_t\beta C_1 - 2M_t\beta^2 B_1 \tag{2.31}$$

Values for A_1, B_1, C_1, and D_1, are shown in Table 2.1 as a function of the nondimensional distance βx along the long pile.

For a long pile whose head is fixed against rotation, as shown in Figure 2.4b, the solution may be obtained by employing the boundary conditions as given in Equations 2.32 and 2.33.

$$\text{At } x = 0, \quad \frac{dy}{dx} = 0 \tag{2.32}$$

$$\frac{d^3y}{dx^3} = \frac{P_t}{E_pI_p} \tag{2.33}$$

Using the procedures as for the first set of boundary conditions, the results are as follows:

$$\chi_3 = \chi_4 = \frac{P_t}{4E_pI_p\beta^3} \tag{2.34}$$

The solution for long piles is finally given in Equations 2.35 through 2.39.

$$y = \frac{P_t\beta}{\alpha}A_1 \tag{2.35}$$

$$S = \frac{P_t}{2E_pI_p\beta^2}D_1 \tag{2.36}$$

28 Piles under lateral loading

$$M = -\frac{P_t}{2\beta}B_1 \tag{2.37}$$

$$V = P_t C_1 \tag{2.38}$$

$$p = -P_t \beta A_1 \tag{2.39}$$

It is sometimes convenient to have a solution for a third set of boundary conditions, as shown in Figure 2.4c. These boundary conditions are given in Equations 2.40 and 2.41.

$$\text{At } x = 0, \quad \frac{E_p I_p \dfrac{d^2 y}{dx^2}}{\dfrac{dy}{dx}} = \frac{M_t}{S_t} \tag{2.40}$$

$$\frac{d^3 y}{dx^3} = \frac{P_t}{E_p I_p} \tag{2.41}$$

Employing these boundary conditions, for the long pile the coefficients χ_3 and χ_4 were evaluated and the results are shown in Equations 2.42 and 2.43. For convenience in writing, the rotational restraint M_t/L_t is given the symbol k_θ.

$$\chi_3 = \frac{P_t(2E_p I_p + k_\theta)}{EI(\alpha + 4\beta^3 k_\theta)} \tag{2.42}$$

$$\chi_4 = \frac{k_\theta P_t}{E_p I_p(\alpha + 4\beta^3 k_\theta)} \tag{2.43}$$

These expressions can be substituted into Equation 2.13, differentiation performed as appropriate, and substitution of Equations 2.23 through 2.26 will yield a set of expressions for the long pile similar to those in Equations 2.27 through 2.31 and 2.35 through 2.39.

Timoshenko (1941) stated that the solution for the *long* pile is satisfactory where $\beta L \geq 4$, however, there are occasions when the solution of the reduced differential equation is desired for piles that have a nondimensional length less than 4. The solution at whatever pile length L, can be obtained by using the following boundary conditions at the tip of the pile.

$$\text{at } x = L, \quad \frac{d^2 y}{dx^2} = 0, \, (M \text{ is zero at pile tip}) \tag{2.44}$$

$$\text{at } x = L, \quad \frac{d^3 y}{dx^3} = 0 \, (V \text{ is zero at pile tip}) \tag{2.45}$$

When the above boundary conditions are fulfilled, along with a set for the top of the pile, the four coefficients χ_1, χ_2, χ_3, and χ_4 can be evaluated. The solutions are not shown here but Appendix B includes a set of tables that were derived from the case shown in Figure 2.4a. New values of the parameters A_1, B_1, C_1, and D_1 were computed as a function of βL.

By comparing the values in Appendix B with those in Table 2.1, the influence of the length of the pile can be readily seen. If the loading at the groundline is only a lateral load P_t, the deflection along the length of the pile is given by a constant times the parameter C_1. Table 2.1 and the table for $\beta L = 10$ in Appendix B show that there will be points of zero deflection at $\beta x = 1.5+, 5.2+$, and $8.4+$, and that the deflection at the tip of the pile will be very small. The tables in Appendix B for $\beta L = 4.0, 3.5, 3.0, 2.8, 2.6, 2.4, 2.2$, and 2.0 show that the number of points of zero deflection is reduced to one and that the deflection of the bottom of the pile represents a significant portion of the deflection of the top.

The influence of the length of a pile on the groundline deflection is illustrated in Figure 2.5. Figure 2.5a shows a pile with loads applied; an axial load is shown, but the assumption is made that the load is small so the length of the pile will be controlled by the lateral load P_t and the moment M_t. Computations are made with constant loading and constant pile cross section, and with an initial length that will be in the long-pile range. The computations proceed with the length being reduced in increments; the groundline deflection is plotted as a function of the selected length, as shown in Figure 2.5b. The figure shows that the groundline deflection is unaffected until the critical length is approached. At this length, only one point of zero deflection will occur in the computations. There will be a significant increase in the groundline deflection as the length in the solution is made less than the critical. The engineer can select a length that will give an appropriate factor of safety against excessive groundline deflection. The accuracy of the solution will depend, of course, on how well the soil-response curves reflect the actual situation in the field.

The reduced form of the differential equation will not normally be used for the solution of problems encountered in design; however, the influence of pile length, pile stiffness, and other parameters, is illustrated with clarity.

2.2.2 Solution of the differential equation by difference equations

The solution of Equation 2.7 is desirable and necessary for analyses that are encountered in practice. The formulation of the differential equation in numerical terms and a solution by iteration allows improvements in the solutions shown in the previous section. The resulting equations form the basis for a computer program that is essential in practice.
- The effect of the axial load on deflection and bending moment will be considered and problems of pile buckling can be solved.
- The bending stiffness $E_p I_p$ of the pile can be varied along the length of the pile.
- And perhaps of more importance, the soil reaction E_{py} can vary with pile deflection and with distance along the pile. The concept of the soil reaction will be discussed fully in a later section and the introduction here is merely in a generic sense.

Table 2.1 Example computations for nondimensional method.

x(m)	Z	A_y	y_A(m)	B_y	Y_B(m)	y(m)	p(kN/m)	E_{py}(kN/m²)
0.00	0.00	2.40	0.0768	1.60	−0.0381	0.0387	0.00	0.00
0.76	0.190	2.07	0.0662	1.30	−0.0309	0.0353	21.3	603
1.52	0.380	1.77	0.0566	1.02	−0.0243	0.0323	40.0	1238
2.29	0.573	1.50	0.0480	0.78	−0.0186	0.0294	63.0	2143
3.81	0.953	0.99	0.0317	0.39	−0.0093	0.0224	69.5	3103
6.10	1.525	0.40	0.0128	0.06	−0.0014	0.0114	62.0	5439

30 *Piles under lateral loading*

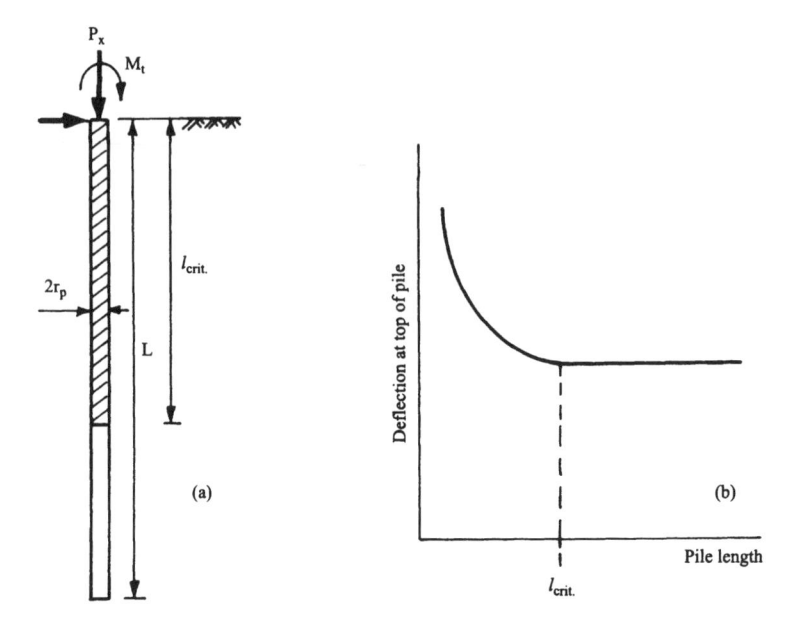

Figure 2.5. Solving for the critical length of a pile.

If the pile is subdivided in increments of length h, as shown in Figure 2.6, Equation 2.7 in difference form is as follows:

$$\begin{aligned}
& y_{m-2} R_{m-1} + y_{m-1}\left(-2R_{m-1} - 2R_m + P_x h^2\right) \\
& + y_m \left(R_{m-1} + 4R_m + R_{m-1} - 2P_x h^2 + E_{pym} h^4\right) \\
& + y_{m+1}\left(-2R_m - 2R_{m-1} + P_x h^2\right) \\
& + y_{m+2} R_{m-1} + W_m h^4 = 0
\end{aligned} \quad (2.46)$$

where $R_m = (E_p I_p)_m$ bending stiffness of pile at point m.

The assumption is implicit in Equation 2.46 that the magnitude of P_x is constant with depth. Of course, that assumption is seldom true. However, experience has shown that the maximum bending moment usually occurs a relatively short distance below the groundline at a point where the value of P_x is virtually undiminished. The value of P_x, except in cases of buckling, has little influence on the magnitudes of deflection and bending moment leads to the conclusion that the assumption of a constant P_x is generally valid.

If the pile is divided into n increments, $n + 1$ equations of the sort as Equation 2.46 can be written. There will be $n + 5$ unknowns because two imaginary points will be introduced at the top and two at the bottom of the pile. If two equations giving boundary conditions are written at the bottom and two at the top, there will be $n + 5$ equations to solve simultaneously for the $n + 5$ unknowns. The set of algebraic equations can be solved by matrix methods in any convenient way.

The two boundary conditions that are employed at the bottom of the pile are based on the moment and the shear. If the existence of an eccentric axial load that causes a moment at the bottom of the pile is discounted, the moment at the bottom of the pile is zero. The assumption of a zero moment is believed to produce no error in all cases except for short rigid piles that carry their loads in end bearing. The case where there is a moment at the pile tip is unusual and is not treated by the procedure presented herein. Thus, one of the boundary equations at the pile tip is

$$y_{-1} - 2y_0 + y_1 = 0 \tag{2.47}$$

Equation 2.47 expresses the condition that $E_p I_p (d^2y/dx^2) = 0$ at $x = L$.

The second boundary condition at the bottom of the pile involves the shear. The assumption is made that soil resistance due to shearing stress can develop at the bottom of a short pile as deflection occurs. It is further assumed that information can be developed that will allow V_0, the shear at the bottom of the pile, to be known as a function of y_0. Thus, the second equation for the boundary conditions at the bottom of the pile is

$$\frac{R_0}{2h^3}(y_{-2} - 2y_{-1} + 2y_1 - y_2) + \frac{P_x}{2h}(y_{-1} - y_1) = V_0 \tag{2.48}$$

with R_0 = flexural rigidity at the pile tip.

Equation 2.48 expresses the condition $E_p I_p (d^3y/dx^3) + P_x(dy/dx) = V_0$ at $x = L$. The value of V_0 should be set equal to zero for long piles with two or more points of zero deflection.

As presented earlier, two boundary equations are needed at the top of the pile. Equations have been derived for four sets of boundary conditions, each with two equations. The engineer can select the set that fits the physical problem.

2.2.2.1 Shear and moment at pile head

Case 1 of the boundary conditions at the top of the pile is illustrated graphically in Figure 2.7. The axial load P_x is not shown in the sketches, but P_x is assumed to be acting for each of the four cases of boundary conditions at the top of the pile. For the condition where the shear at the top of the pile is equal to P_t, and R_t the bending stiffness at the top of the pile, the following difference equation is employed.

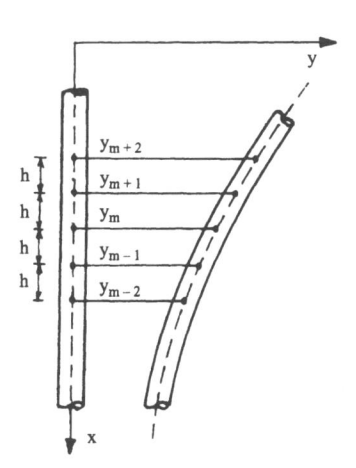

Figure 2.6. Representation of deflected pile.

32 Piles under lateral loading

$$\frac{R_t}{2h^3}(y_{t-2} - 2y_{t-1} + 2y_{t+1} - y_{t+2}) + \frac{P_x}{2h}(y_{t-1} - y_{t+1}) = P_t \tag{2.49}$$

For the condition where the moment at the top of the pile is equal to M_t, the following difference equation is employed.

$$\frac{R_t}{h^2}(y_{t-1} - 2yt + y_{t+1}) = M_t \tag{2.50}$$

2.2.2.2 Shear and rotation at pile head
Case 2 of the boundary conditions at the top of the pile is illustrated graphically in Figure 2.8. The pile is assumed to be embedded in a concrete foundation for which the rotation is known. In many cases, the rotation can be assumed to be zero, at least for the initial solutions. Equation 2.49 is the first of the two equations that are needed. The second of the two equations reflects the condition that the slope S_t at the top of the pile is known.

$$\frac{y_{t-1} - y_{t+1}}{2h} = S_t \tag{2.51}$$

2.2.2.3 Shear and rotational restraint at pile head
Case 3 of the boundary conditions at the top of the pile is illustrated in Figure 2.9. The pile is assumed to continue into the superstructure and become a member of a frame. The solution for the problem can proceed by cutting a free body at the bottom joint of the frame. A moment is applied to the frame at that joint and the rotation of the frame is computed, or estimated, for the initial solution. The moment divided by the rotation, M_t/S_t, is the rotational stiffness provided by the superstructure and becomes one of the boundary conditions. The boundary condition has proved to be very useful in some designs. An initial solution may be necessary in order to obtain an estimate of the moment at the bottom joint of the superstructure, to analyze the superstructure, and then to re-analyze the pile. One or two iterations should be sufficient in most instances.

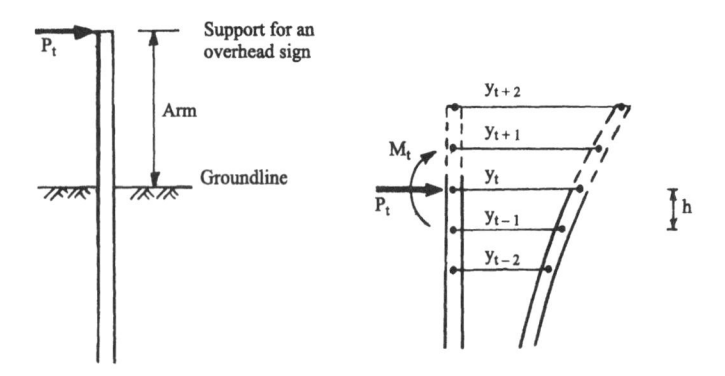

Note: P_t and M_t are known; they are shown in the positive sense in the sketches

Figure 2.7. Case 1 of boundary conditions at top of pile.

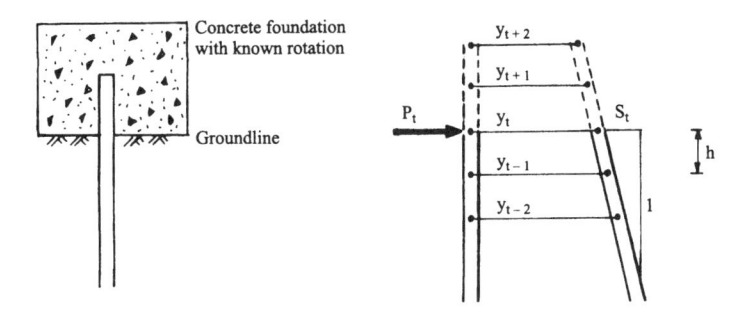

Derivation of equations and methods of solution 33

Note: P_t and S_t are known; they are shown in the positive sense

Figure 2.8. Case 2 of boundary conditions at top of pile.

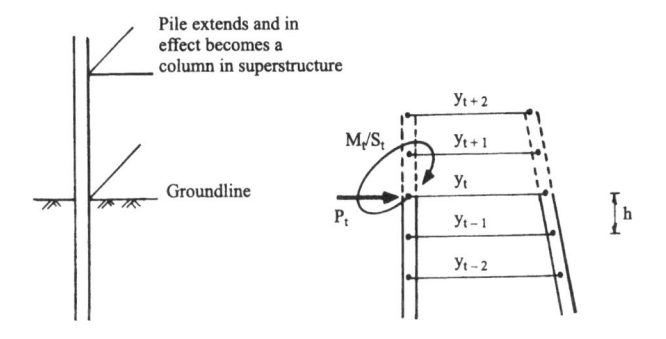

Note: P_t and M_t/S_t are shown; they are shown in the positive sense in the sketches

Figure 2.9. Case 3 of boundary conditions at top of pile.

Equation 2.49 is the first of the two equations that are needed for Case 3. Equation 2.52 expresses the condition that the rotational stiffness M_t/S_t is known.

$$\frac{\frac{R_t}{h^2}(y_{t-1} - 2y_t + y_{t+1})}{\frac{y_{t-1} - y_{t+1}}{2h}} = \frac{M_t}{S_t} \qquad (2.52)$$

2.2.2.4 Moment and deflection at pile head

Case 4 of the boundary conditions at the top of the pile is illustrated in Figure 2.10. The pile is assumed to be embedded in a bridge abutment that moves laterally a given amount; thus, the deflection y_t at the top of the pile is known. Also, the bending moment is known. If the embedment is small, the bending moment is frequently assumed to be zero. The two equations needed at the pile head for Case 4 are Equations 2.50 and 2.53

$$y_t = Y_t \qquad (2.53)$$

34 Piles under lateral loading

The four sets of boundary conditions at the top of a pile should be adequate for virtually any situation but other cases can arise. However, the boundary conditions that are available, with a small amount of effort, can produce the required solutions. For example, it can be assumed that P_t and y_t are known at the top of a pile and constitute the required boundary conditions (not one of the four cases). The Case 4 equations can be employed with a few values of M_t being selected, along with the given value of y_t. The computer output will yield values of P_t. A simple plot will yield the required value of M_t that will produce the given boundary condition P_t.

The application of the finite-difference-equation technique to the solution of the axial load at which a pile will buckle is illustrated in Figure 2.11. The pile, with a projection above the groundline, shown in Figure 2.11a, has been designed for the working or service loads that are shown. The factor of safety against buckling is found by holding P_t constant and incrementing P_x. As shown in Figure 2.11b, the increase of P_x will cause virtually no increase in y_t until the buckling load is approached. This analysis cannot be treated as an *eigenvalue* problem so the investigator must approach the buckling load with small changes in P_x. The equations become unstable at axial loads beyond the critical, with nonsensical results, and the engineer must be careful not to select an axial load that is excessive.

The computer program to solve the finite-difference equations for the response of a pile to lateral loading will be demonstrated in a subsequent chapter. The solutions of a number of example problems will be presented. Also, case studies will be shown in which the results from computer solutions are compared with experimental results. Because of the obvious approximations that are inherent in the difference-equation method, a discussion will be given of techniques for the verification of the accuracy of a solution, essential to the proper use of the numerical method. The discussion will deal with the number of significant figures to be used in the internal computations and with the selection of the increment length h. Another approximation related to the variation in the bending stiffness.

The bending stiffness $E_p I_p$, changed to R in the difference equations, is correctly represented as a constant in the second-order differential equation, Equation 2.9.

$$E_p I_p \frac{d^2 y}{dx^2} = M \qquad (2.9)$$

In finite-difference form, Equation 2.9 becomes

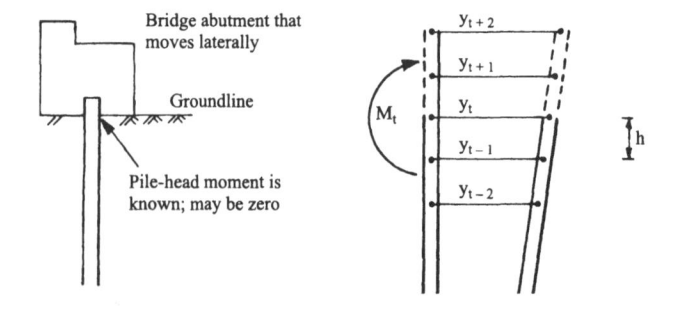

Note: P_t and y_t are shown; they are shown in the positive sense in the sketches

Figure 2.10. Case 4 of boundary conditions at top of pile.

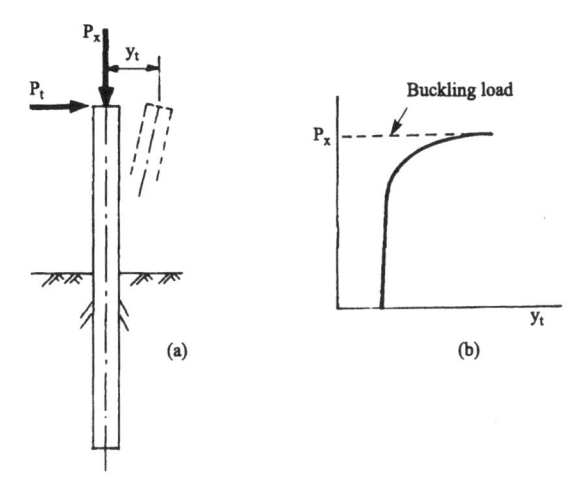

Derivation of equations and methods of solution 35

Figure 2.11. Solving for the axial load that causes a pile to buckle.

$$R_m(y_{m-1} - 2y_m + y_{m+1}) = M_m \qquad (2.54)$$

and, in building up the higher ordered terms by differentiation, the value of R is made to correspond to the central term for y in the second-order expression. The errors that are involved in using the approximation where there is a change in the bending stiffness along the length of a pile are thought to be small but must be investigated as necessary.

A derivation has been made for the case where there is an abrupt change in flexural stiffness and is shown in Appendix C. The formulation shown in Appendix C has not been incorporated into a computer program for distribution, but the method may be readily implemented if desirable.

The different equation in difference form can be solved readily by the digital computer. A compact disk with a student version of the Computer program LPILE is provided in a holder attached to the back cover of the book. The coding for the equations shown above is implemented in the computer program presented in Appendix D that solves the difference equations. The code requires iteration to achieve a solution to reflect the nonlinear response of the soil with pile deflection.

2.3 SOLUTION FOR $E_{py} = k_{py}x^1$

The previous section presents a brief exposition of the application of numerical methods to the solution of a nonlinear, fourth-order differential equation. The application of the solution will be presented in later chapters. The remainder of this chapter will show the use of the finite-difference-equation technique to produce nondimensional tables or curves that facilitate hand computations.

---

[1] The reaction modulus for the soil is referenced to the ground surface by the symbol z; however, the symbol x may be used here to reflect the ground surface and for distance along the the pile. The origin for the case discussed here must be the same for the top of the pile and the ground surface.

A solution of the reduced form of the differential equation was presented earlier for the case where E_{py} is constant (piles in heavily overconsolidated soils). Nondimensional tables were presented that have relevance in showing the nature of the pile problem and that allow for checking finite-difference-equation solutions. However, the tables are of almost no use in solving practical problems.

Dimensional analysis, along with solutions by finite-difference techniques, can lead to nondimensional curves to accommodate a wide variety of variations of soil moduli with depth, such as $E_{py} = k_1 + k_2 x$ or $E_{py} = k_{py} x$. (Matlock & Reese 1962). Solutions that assume that $k_{py} x$ do have some practical utility. Cohesionless soil and normally consolidated clay are two cases where the stiffness is zero at the groundline and increases rather linearly with depth. Furthermore, experience has shown that useful solutions can be obtained for some cases of overconsolidated clays. The following paragraphs show the development and implementation of the method where $E_{py} = k_{py} x$.

2.3.1 *Dimensional analysis*

Considering the nonlinearity of *p-y* relations at various depths, the reaction modulus for the soil E_{py} is a function of both *x* and *y*. Therefore, the form of the E_{py}-versus-depth relationship will change if the loading is changed. However, it may be assumed temporarily (subject to adjustment of E_{py} values by successive trial) that the soil reaction is some function of *x* only, or that

$$E_{py} = E_{py}(x) \tag{2.55}$$

For solution of the problem, the curve *y(x)* of the pile must be determined, together with various derivatives that are of interest. The derivatives yield values of slope, moment, shear, and soil reaction as functions of depth. The principles of dimensional analysis may be used to establish the form of nondimensional relations for the laterally loaded pile. With the use of model theory, the necessary relations will be determined between a 'prototype' having any given set of dimensions, and a similar 'model' for which solutions may be available.

For very long piles, the length *L* loses significance because the deflection may be nearly zero for much of the length of the pile. It is convenient to introduce some characteristic length as a substitute. A linear dimension *T* is therefore included in the quantities to be considered. The specific definition of *T* will vary with the form of the function for soil reaction versus depth. However, for each definition used, *T* expresses a relation between the stiffness of the soil and the flexural stiffness of the pile and is called the 'relative stiffness factor.'

For the case of a shear P_t and a moment M_t at the top of the pile, the solution for deflections of the elastic curve will include the relative stiffness factor and other terms (see Figure 2.12).

$$y = y(x, T, L, E_{py}, E_p I_p, P_t, M_t) \tag{2.56}$$

Other boundary values can be substituted for P_t and M_t.

If the assumption of linear behavior is introduced for the pile, and if deflections remain small relative to the pile dimensions, the principle of superposition may be employed. Thus, the effects of an imposed lateral load P_t and imposed moment M_t may be considered

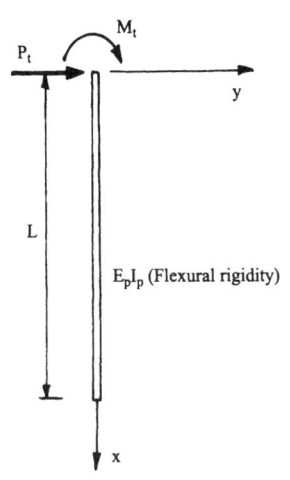

Derivation of equations and methods of solution 37

Figure 2.12. Arrangement for dimensional analysis.

separately. If y_A represents the deflection caused by the lateral load P_t and if y_B is the deflection caused by the moment M_t, the total deflection is

$$y = y_A + y_B y \tag{2.57}$$

The ratios of y_A to P_t and of y_B to M_t are sought in reaching generalized solutions for the linearly-behaving pile. The solutions may be expressed for Case A as

$$f(y_A, P_t) = f_A(x, T, L, E_{py}, E_p I_p) \tag{2.58}$$

and for Case B as

$$f(y_B, M_t) = f_B(x, T, L, E_{py}, E_p I_p) \tag{2.59}$$

The values f_A and f_B represent two different functions of the same terms. In each case there are six terms and two dimensions (force and length). There are therefore four independent nondimensional groups which can be formed. The arrangements chosen are, for Case A,

$$\frac{y_A E_p I_p}{P_t T^3}, \frac{x}{T}, \frac{L}{T}, \frac{E_{py} T^4}{E_p I_p} \tag{2.60}$$

and for Case B,

$$\frac{y_B E_p I_p}{P_t T^2}, \frac{x}{T}, \frac{L}{T}, \frac{E_{py} T^4}{E_p I_p} \tag{2.61}$$

To satisfy conditions of similarity, each of these groups must be equal for both model and prototype, as shown below.

$$\frac{x_p}{T_p} = \frac{x_m}{T_m} \tag{2.62}$$

38 Piles under lateral loading

$$\frac{L_p}{T_p} = \frac{L_m}{T_m} \tag{2.63}$$

$$\frac{E_{py_p} T_p^4}{(E_p I_p)_p} = \frac{E_{py_m} T_p^4}{(E_p I_p)_m} \tag{2.64}$$

$$\frac{y_{A_p} E_p I_{p_p}}{P_{t_p} T_p^3} = \frac{y_{A_m} E_p I_{p_m}}{P_{t_m} T_m^3} \tag{2.65}$$

$$\frac{y_{B_p} E_p I_{p_p}}{M_{t_p} T_p^2} = \frac{y_{B_m} E_p I_{p_m}}{M_{t_m} T_m^2} \tag{2.66}$$

A group of nondimensional parameters may be defined which will x/T have the same numerical value for any model and its prototype. These are shown below.

$$\text{Depth coefficient, } Z = \frac{x}{T} \tag{2.67}$$

$$\text{Maximum depth coefficient, } Z_{\max} = \frac{L}{T} \tag{2.68}$$

$$\text{Soil reaction function, } f(Z) = \frac{E_{py} T^4}{E_p I_p} \tag{2.69}$$

$$\text{Case A deflection coefficient, } A_y = \frac{y_A E_p I_p}{P_t T^3} \tag{2.70}$$

$$\text{Case B deflection coefficient, } B_y = \frac{y_B E_p I_p}{M_t T^2} \tag{2.71}$$

Thus (from definitions 2.67 through 2.71) for 1. Similar soil-pile stiffness, 2. Similar positions along the piles, and 3. Similar pile lengths (unless lengths are very great and need not be considered), the solution of the problem can be expressed from Equation 2.57 and from Equations 2.70 and 2.71, as

$$y = \left[\frac{P_t T^3}{E_p I_p}\right] A_y + \left[\frac{M_t T^2}{E_p I_p}\right] B_y \tag{2.72}$$

By the same type of reasoning, other forms of the solution can be expressed as shown below.

$$\text{Slope, } S = S_A + S_B = \left[\frac{P_t T^2}{E_p I_p}\right] A_S + \left[\frac{M_t T}{E_p I_p}\right] B_S \tag{2.73}$$

Moment, $M = M_A + M_B = [P_t T]A_m + [M_t]B_m$ (2.74)

Shear, $V = V_A + V_B = [P_t]A_v + \left[\dfrac{M_t}{T}\right]B_v$ (2.75)

Soil reaction, $p = p_A + p_B = \left[\dfrac{P_t}{T}\right]A_p + \left[\dfrac{M_t}{T^2}\right]B_p$ (2.76)

A particular set of A and B coefficients must be obtained as functions of the depth parameter, Z, by a solution of a particular model. However, the above expressions are independent of the characteristics of the model except that linear behavior and small deflections are assumed. The parameter T is still an undefined characteristic length and the variation of E_{py} with depth, or the corresponding form of f(Z), has not been specified.

From beam theory, as presented earlier, the basic equation for a pile is:

$$\dfrac{d^4 y}{dx^4} + \dfrac{E_{py}}{E_p I_p} y = 0$$ (2.77)

Where an applied lateral load P_t and an applied moment M_t are considered separately according to the principle of superposition, the equation becomes, for Case A,

$$\dfrac{d^4 y_A}{dx^4} + \dfrac{E_{py}}{E_p I_p} y_A = 0$$ (2.78)

and for Case B,

$$\dfrac{d^4 y_B}{dx^4} + \dfrac{E_{py}}{E_p I_p} y_B = 0$$ (2.79)

Substituting the definitions of nondimensional parameters contained in Equations 2.67 through 2.71, a nondimensional differential equation can be written for Case A as

$$\dfrac{d^4 A_y}{dZ^4} + f(Z)A_y = 0$$ (2.80)

and for Case B as

$$\dfrac{d^4 B_y}{dZ^4} + f(Z)B_y = 0$$ (2.81)

To produce a particular set of nondimensional A and B coefficients, 1. f(Z) must be specified, including a convenient definition of the relative stiffness factor T, and 2. The differential equations (2.80 and 2.81) must be solved. The resulting A and B coefficients may then be used, with Equations 2.71 through 2.76, to compute deflection, slope, moment, shear, and soil reaction for any pile problem which is similar to the case for which nondimensional solutions have been obtained.

40 Piles under lateral loading

To obtain the A and B coefficients that are needed to make solutions with the nondimensional method, Equations 2.80 and 2.81 can be solved by use of difference equations, presented earlier in this chapter.

In solving problems of laterally loaded piles by using nondimensional methods, the constants in the expressions describing the variation of soil reaction E_{py} with depth x are adjusted by trial until reasonable compatibility is obtained. The selected form of the soil reaction with depth should be kept as simple as possible so that a minimum number of constants needs to be adjusted.

A general form of E_{py} with depth is a power form,

$$E_{py} = k_{py}x^n \tag{2.82}$$

The form $E_{py} = k_{py}x$ is seen to be a special case of the power form. The relative stiffness factor T can be defined for any particular form of the soil reaction-depth relationship. It is convenient to select a definition that will simplify the corresponding nondimensional functions.

From the theory given above, the equation that defines the nondimensional function for soil reaction is

$$f(Z) = \frac{E_{py}T^4}{E_p I_p} \tag{2.69}$$

If the form $E_{py} = k_{py}x^n$ is substituted in Equation 2.69, the result is

$$f(Z) = \frac{k_{py}}{E_p I_p} x^n T^4 \tag{2.83}$$

It is convenient to define the relative stiffness factor T by the following expression.

$$T^{n+4} = \frac{E_p I_p}{k_{py}} \tag{2.84}$$

Substituting this definition into Equation 2.83 gives

$$f(Z) = \frac{x^n T^4}{T^{n+4}} = \left[\frac{x}{T}\right]^n \tag{2.85}$$

Because $x/T = Z$, the general nondimensional function for soil reaction is

$$f(Z) = Z^n \tag{2.86}$$

The above expression contains only one parameter, the power n. Therefore, for each value of n which may be selected, one complete set of independent, nondimensional solutions may be obtained from the solution of Equations 2.80 and 2.81. For the case where $E_{py} = k_{py}x$, n is equal to 1 and

$$f(Z) = Z$$

The nondimensional technique, along with a solution of the difference equations, allow coefficients to be computed for a variety of variations of the soil reaction with depth. However, only the case of $E_{py} = k_{py}x$ will be shown here.

2.3.2 Equations for $E_{py} = k_{py}x$

While the form of the equations has already been shown, a tabulation here is convenient. The common equations are:

$$T = \frac{E_p I_p}{k_{py}} \tag{2.87}$$

$$Z = \frac{x}{T} \quad \text{and} \tag{2.88}$$

$$Z_{max} = \frac{L}{T} \tag{2.89}$$

Where only a shear P_t is applied at the mudline, the below listed equations are for deflection y, rotation (slope) S, moment M, and shear V, respectively. (The soil resistance p is equal to E_{py} times y). The nondimensional coefficients, obtained by employing the difference-equation methods, shown in the equations, can be found in Appendix E. The coefficients are shown as a function of the nondimensional depth Z and the nondimensional length of the pile Z_{max}.

$$y_A = A_y \left[\frac{P_t T^3}{E_p I_p} \right] \tag{2.90}$$

$$S_A = A_s \left[\frac{P_t T^2}{E_p I_p} \right] \tag{2.91}$$

$$M_A = A_m [P_t T] \tag{2.92}$$

$$V = A_v [P_t] \tag{2.93}$$

Similarly, the respective equations for only a moment M_t loading at the top of the pile are:

$$y_B = B_y \left[\frac{M_t T^2}{E_p I_p} \right] \tag{2.94}$$

$$S_B = B_s \left[\frac{M_t T}{E_p I_p} \right] \tag{2.95}$$

$$M = B_m [M_t] \quad \text{and} \tag{2.96}$$

42 *Piles under lateral loading*

$$V = B_y \left[\frac{M_t}{T} \right] \tag{2.97}$$

The derivation of the fixed-head case is not shown here; however, the above principles are followed. The equation for deflection is:

$$y_F = F_y \left[\frac{P_t T^3}{E_p I_p} \right] \tag{2.98}$$

The moment may be found from the following equation:

$$M_t = F_{mt} P_t T \tag{2.99}$$

where F_{mt} = 1.06 for Z_{max} = 2; –0.97 for Z_{max} = 3; –0.93 for Z_{max} = 4; and –0.93 for all higher values of Z_{max}.

For the case where there is a rotational restraint at the top of the pile, the first step is to assign a value to the rotational stiffness. A constant conveniently can be assigned, as shown below.

$$k_\theta = \frac{M_t}{S_t} \tag{2.100}$$

The slope (rotation) at the top of the pile can be found from the following equation.

$$S_t = A_{st} \frac{P_t T^2}{E_p I_p} + B_{st} \frac{M_t T}{E_p I_p} \tag{2.101}$$

where A_{st} and B_{st} are the nondimensional coefficients at the top of the pile and can be found from the appropriate curves in Appendix E as a function of Z_{max}. With these values of the nondimensional coefficients for slope, the selection of a trial value of T allows M_t to be found; then the solution can proceed with the boundary values of P_t and M_t.

2.3.3 *Example solution*

The example selected for analysis is for a marine structure. A jacket or template, consisting of welded, tubular-steel members, is constructed onshore, transported by barge, and set in place at an offshore location. One of the legs of the jacket is shown in Figure 2.13a with the bracing for the lowest panel point at the mudline. A pile is driven after spacers are welded inside the jacket leg to ensure contacts between the pile and the jacket. Loading may be considered to arise from wave action during a storm. At this stage of the analysis, the jacket is assumed to move laterally but not rotate, which is a satisfactory assumption if the piles are relatively rigid under axial loading.

The *p-y* curves to be used are shown in Figure 2.14. The curves are typical of those for sand or normally-consolidated clay. The distance below the mudline to each of the curves is shown, and three points may be noted:

1. The *p-y* curve for zero depth shows zero soil resistance for all deflections;
2. The initial slopes to the *p-y* curves are linear and increase with depth; and

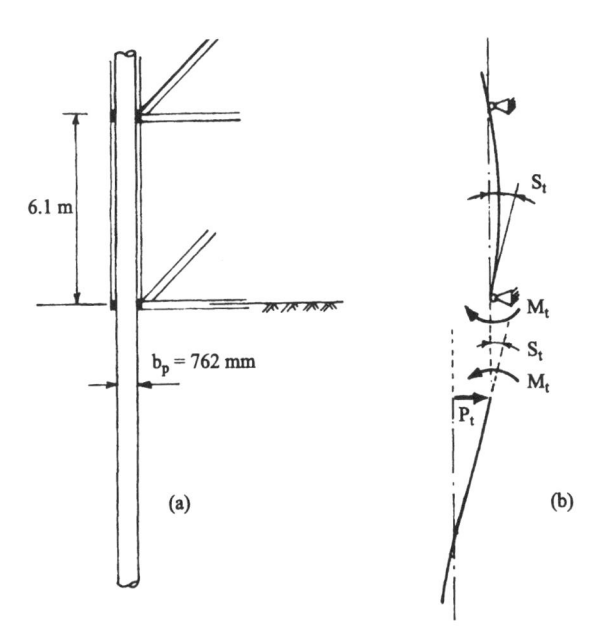

Figure 2.13. Pile at a leg of an offshore platform.

3. The ultimate resistance for each curve approaches a limiting value that increases with depth.

While the curves are for no particular soil, the character of the curves is such that the nature of the following solution is clearly indicative of a more exact solution by computer.

The portion of the pile above the mudline and within the jacket is shown in the upper sketch in Figure 2.13b. As may be seen, the pile will behave as a continuous beam with loading at its lower end. With the pile passing beyond the upper panel point in the sketch, its condition at that point is between fixed against rotation and free to rotate. Therefore, as a reasonable approximation, the relationship between M_t and S_t is given by Equation 2.102.

$$\frac{M_t}{S_t} \approx \frac{3.5 E_p I_p}{h_p} \tag{2.102}$$

where h_p = the distance between panel points (6.1 m in this case); and $E_p I_p$ = the bending stiffness of the pile within the jacket leg.

The pile is a steel pipe with an outside diameter of 762 mm and a wall thickness of 25.4 mm. The steel has a yield strength of 414,000 kN/m^2. The $E_p I_p$ is 800,000 m^{2}kN, and the ultimate bending moment was computed to be 5690 mkN, assuming no axial load. The assumption is made that no restriction exists on the deflection of the pile, and the desired solution is to find the load that will cause a plastic hinge (the ultimate bending moment). The ultimate load can then be factored to achieve a safe load. (A more comprehensive approach to achieving a given factor of safety will be discussed in Chapter 9).

Examination of the relevant equations show that there are two unknowns that must be solved by iteration. Firstly, a value of P_t must be selected and the appropriate deflected

44 Piles under lateral loading

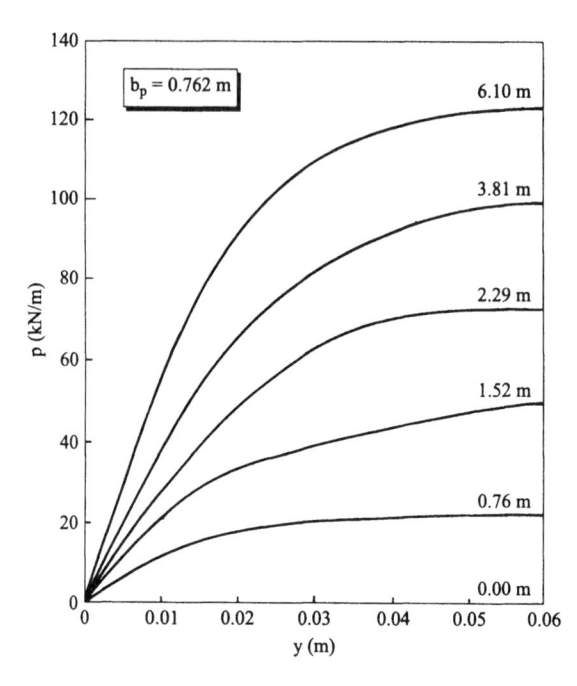

Figure 2.14. Soil-resistant curves for example solution.

shape found by varying the value of T. After selecting the value of T, trial solutions must be made by use of the p-y curves, the data for the pile, the respective equations, and the nondimensional curves in Appendix E. In using the curves in the appendix, a value of Z_{max} of 10 is selected with the view that the length of the pile is established from axial loading and is relatively long.

The value of P_t selected for the initial computations is 400 kN. The value of T is related to the loading, with T being small when the loading is relatively light, and no guidelines are possible for selecting an initial value. However, convergence is rapid. A value of T equal to about 5 pile diameters (4 m) is selected for the first trial.

A value of M_t of –1,190 mkN is found by solving Equations 2.100, 2.101, and 2.102. Equations 2.90 and 2.94 can then be used to solve for values of y. The value of Z_{max} (L/T) is needed to get values of the nondimensional coefficients from the curves in Appendix E. That value is 50/4 or 12.5, so the curves for Z_{max} of 10 or more are used.

The referenced equations, after substitutions, yield the following expression. The computations with this expression are shown in the Table 2.2 that follows.

$$y = A_y(0.032) - B_y(0.0238)$$

The values of E_{py} are plotted in Figure 2.15 and the best straight line, passing through the origin, is fitted through the points by eye. As shown for Trial 1, the solid line passes through 6,000 and 6.95, and the following value of k_{py} was obtained.

$$k_{py} = \frac{E_{py}}{x} = \frac{6000}{6.95} = 863 \frac{\text{kN}}{\text{m}^3} \tag{2.103}$$

The value of the relative stiffness factor T_{obt} can now be computed.

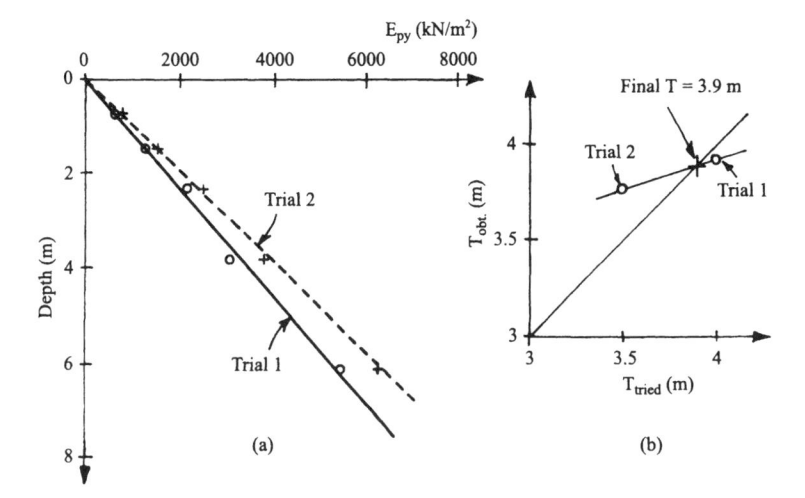

Figure 2.15. Convergence plotting for P_t = 400 kN assuming $E_s = kx$.

$$T_{obt} = \left[\frac{800000}{8.63}\right]^{0.2} = 3.92 \text{ m} \qquad (2.104)$$

The selection of the starting value of T of 4.0 turned out to be fortuitous, but it is necessary to converge to a closer result.

The second trial could have been done with a T of 3.92 m, but a better plan is to adopt a smaller value of T to obtain a point that would plot on the opposite side of the equality line (see Fig. 2.15b). Thus, the second trial was made with a T_{tried} of 3.5 m, which yielded a Z_{max} of 14.3. The computations for this case, not shown here, proceeded as before, with a computed value of M_t of –1012 mkN. Using that value, the relevant equations, and the p-y curves, the results are shown as Trial 2 in Figure 2.15a. The dashed line in the figure passes through 6000 and 5.7. The values of k_{py} and T_{obt} may be computed.

$$k_{py} = \frac{E_{py}}{x} = \frac{6000}{5.7} = 1053 \frac{\text{kN}}{\text{m}^3} \quad \text{and} \qquad (2.105)$$

$$T_{obt} = \frac{800{,}000}{1053} = 3.77 \text{ m} \qquad (2.106)$$

The point for the second trial plots across the equality line, as shown in Figure 2.15b. A straight line can be used to connect the plotted points, yielding a final value of T of 3.9 meters. It is unlikely that the line connecting the two plotted points is straight but the assumption is satisfactory for this method.

With the value of T of 3.9 m, Equations 2.100, 2.101, and 2.102 are solved and the value of M_t was found to be –1154 mkN. With values of P_t, M_t, and T, values of deflection and bending moment may be computed along the pile. Using the nondimensional curves for Z_{max} of 10, Equations 2.90 and 2.94 are used to compute deflection, and Equations 2.92 and 2.96 are used to compute bending moment. The plots are shown in Figure 2.16.

46 Piles under lateral loading

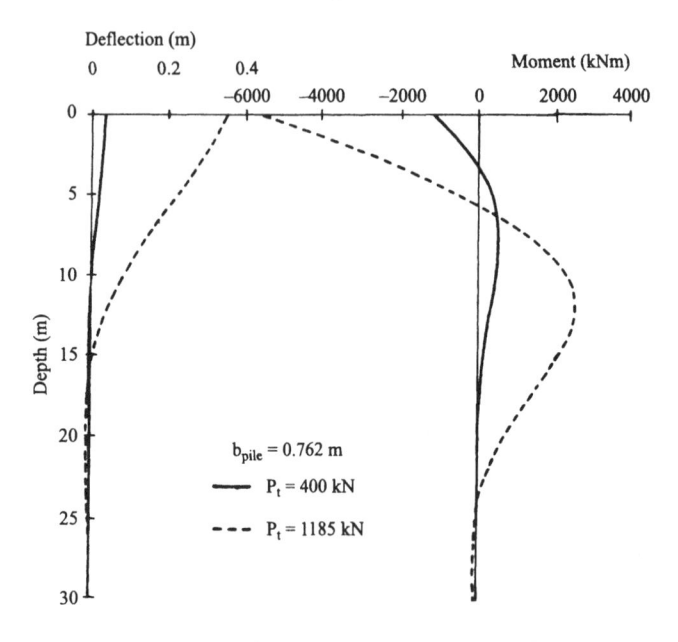

Figure 2.16. Plots of deflection and bending moment for example problem for $E_S = kx$.

In continuing with the assigned problem, other trials were made to find the P_t that would cause a plastic hinge to develop; that is, to result in a computed value of M_{max} equal to M_{ult}. These trials, following the procedures already demonstrated, yielded a P_t of 1185 kN and a value of k_{py} of 103 kN/m^3. The relative stiffness factor T was computed to be 6.0 meters. The maximum bending moment occurred at the pile head and was –5690 mkN. The Z_{max} for this value of T is 8.33, so the curves in Appendix E of Z_{max} of 10 are appropriate. The computed values of deflection and bending moment can now be found by following the procedure indicated for the P_t of 400 kN computed for the relevant equations. The plots are shown in Figure 2.16, with the curves previously plotted for the lateral load of 400 kN.

2.3.4 Discussion

A number of interesting features of the solutions can be seen. The nonlinear response of the pile to lateral load can be surmised by examining the p-y curves, but a comparison of the results of the computations reinforce this point. An increase in the lateral load from 400 to 1185 kN, a factor of less than 3, caused an increase of the groundline deflection from 0.038 to 0.348 m, a factor of over 9, and an increase in the maximum bending moment from 1170 to 5690 mkN, a factor of almost five.

The point is made, then, that a proper design requires the computation of the load that will cause a failure, either in excessive deflection or bending moment. In this particular case, applying a load factor of 3.0 yields a safe load on the pile of 395 kN, which is close to the first trial that was made.

A soil failure is not possible except for a *short* pile under lateral loading where the value of T would be 2 or less. For the *long* pile shown in the example computations, if the soil resistance near the groundline is reduced due to cyclic loading, the soil resistance merely increases at a greater depth, which then increases the deflection and bending moment.

An examination of the curves in Figure 2.16 for both of the loads shows that all values become close to zero at 25 to 30 m along the pile; therefore, the bottom 20 to 25 m of the 50-m long pile offers little or no resistance to lateral loading. Thus, if a pile carries only lateral loading, the designer may wish to give attention to the necessary penetration. However, the required penetration is found from the factored load and not the unfactored load. The nondimensional length of the pile in the example decreased from 12.8 to 8.33 as the load was increased from 400 to 1,185 kN. A convenient rule-of-thumb is that a 'long' pile is one where there are at least two points of zero deflection along its length.

Another interesting point is that the plastic hinge will develop at the top of the pile, the point where it joins the superstructure. The computed negative moment at that point is more than twice the maximum positive moment that occurs at depths of about 8 m for the P_t of 400 kN and about 12 m for the P_t of 1185 kN. Thus a larger lateral load could have been sustained had the distance between the spacers in the jacket been increased above 6.1 m, as shown in Figure 2.13. However, this adjustment may not have been possible in the structure.

The limitations in the solutions shown above are:
1. The pile has a constant wall thickness over its entire length;
2. No axial loading is applied; and
3. The hand computations are time-consuming.

These restrictions are removed with the computer code, demonstrated in Chapter 6, and the designer can have much more freedom in finding an optimum solution. For one thing, the wall thickness of the pile can be increased in the zone of maximum bending moment, and for another, the jacket leg can be extended some distance below the mudline in weak soil so that both the wall thicknesses of the pile and the jacket are effective in resisting bending moment.

2.4 VALIDITY OF THE MECHANICS

The most serious criticism directed against the p-y method is that the soil is not treated as a continuum but as a series of discrete, uncoupled resistances (the Winkler approach). Several comments can be given in response to the valid criticism of the method.

The recommendations for the prediction of *p–y* curves for use in the analysis of piles, given in a subsequent chapter, are based for the most part on the results of full-scale experiments in which the continuum effect was explicitly satisfied. Further, Matlock (1970) performed some tests of a pile in soft clay where the pattern of pile deflection was varied along the length of the pile by restraining the pile head in one test and allowing it to rotate in another test. The *p-y* curves that were derived from each of the loading conditions were essentially the same. Thus, the experimental p-y curves that were obtained from experiments with fully instrumented piles will predict within reasonable limits the response of a pile whose head is free to rotate or fixed against rotation.

The methods of predicting *p-y* curves have been used in a number of case studies, shown in Chapter 7, and the agreement between results from experiment and from computations ranges generally from good to excellent.

Finally, technology may advance so that the soil resistance for a given deflection at a particular point along a pile can be modified quantitatively to reflect the influence of the deflection of the pile above and below the point in question. In such a case, multi-valued *p-y* curves can be developed at every point along the pile. The analytical solution that is presented herein can be readily modified to deal with the multi-valued *p-y* curves.

CHAPTER 3

Models for response of soil and weak rock

3.1 INTRODUCTION

The presentation herein deals principally with the formulation of expressions for p-y curves for soils and rock under both static and cyclic loading. A number of fundamental concepts are presented that are relevant to any method of analyzing piles. Chapter 1 gave the concept of the p-y method and this chapter will present details to allow the computation of the behavior of a pile under a variety of conditions.

More than in any other deep foundation problem, the solution for a pile under lateral loading is arduous, because a successful analysis is sensitive to the stress-strain characteristics of the soil around the pile shaft. Among the concepts presented in Chapter 1 was that the soil reaction modulus is not a soil parameter but depends both on soil resistance and pile deflection. For a given soil profile, the soil reaction modulus is influenced intrinsically by the following variables:
– Pile type and flexural stiffness,
– Short term, long term, or cyclic loading,
– Pile geometry,
– Pile tip and pile cap conditions,
– Slope of the ground at the pile,
– Pile installation procedure, and
– Pile batter.

In spite of the complexities noted above, the soil reaction modulus has the advantage of analytical simplicity, and has been validated worldwide through well-documented case records. Further, and perhaps of most importance, the method has advanced to become an accepted procedure for describing the nonlinear behavior of the interaction between piles and the supporting soil.

Recommendations will be given in this chapter for the selection of a family of p-y curves for various cases of soils and loadings. The resulting curves are meant to reflect as well as possible the deflection and bending moment as a function of depth of a pile under lateral loading. Case studies, presented in Chapter 7 compare results from experiment and from computation for a range of soils and loadings.

The results of the comparisons in Chapter 7 show that bending moment with length along a pile can, in general, be computed more accurately than deflection. Thus, if the engineer must design a pile-supported structure that is sensitive to deflection, a field load test may be indicated.

The next section of the chapter presents a detailed discussion of the relevance of soil parameters and shows that any solution of a problem requires a thorough discussion of the

50 *Piles under lateral loading*

soil profile. The correspondence of, and differences between, E_s for the soil and E_{py} for the pile are discussed in detail.

3.2 MECHANICS CONCERNING RESPONSE OF SOIL TO LATERAL LOADING

3.2.1 *Stress-deformation of soil*

Some discussion was presented in Chapter 1 concerning models for the response of soil to the lateral deflection of a pile. A more detailed discussion is presented here. Plainly, any solution to a problem of lateral loading requires the determination of a range of soil properties. In the general discussion that follows and in specific recommendations at the end of this chapter, the measuring of the relevant properties of soil is an integral step in the process.

Finding the applicable properties of soil in the laboratory to lead to the solution of a particular problem requires attention to numerous details: characterization of the site; selection of representative samples; employing techniques in sampling to minimize disturbance; protecting and transporting samples to avoid loss of moisture; preparation of specimens; selection of appropriate testing procedures; and performance of tests with precise controls.

An appropriate cylindrical specimen is presumed to have been obtained and tested to represent the properties of the soil at a particular point in the continuum where an analysis is to be performed. Further, the assumption is made that any influence of the proposed construction is negligible. The specimen is initially subjected to an all-around stress σ_3 where σ_3 is selected to reflect properties at a particular point in the continuum. The load is applied to the top to cause a stress increase equal to $\Delta\sigma$, with the principal stress σ_3 on a horizontal plane becoming $\sigma_3 + \Delta\sigma$. The strain ε is defined as the shortening of the specimen Δh divided by the original height of the specimen h. Loading is assumed to continue in increments until the resulting curve of principal stress versus strain becomes asymptotic to a line parallel to the strain axis, as shown in Figure 3.1a.

A series of such tests for each of the strata that are encountered will provide data on the strength of soil for use in design. Laboratory tests are usually complemented with in situ tests in the field.

Returning to a discussion of Figure 3.1a, two lines are drawn from the origin to a point on the stress-strain curve. The slope of the lines, termed E_s, is called the soil reaction modulus and represents the stiffness of the soil. The magnitude of E_s is plainly related to the value of strain to which the line is drawn, with the largest value from a line that is tangent to the initial portion of the curve. In making some computation, investigators have used the largest value of E_s, E_{smax}, or an average value, depending on individual preference or related to some particular usage.

Careful measurements of lateral strain must be made, sometimes with special instrumentation, to obtain values of Poisson's ratio v. Of course, the value of v, similar to that of E_s will vary with the loading and vertical strain.

3.2.2 *Proposed model for decay of E_s*

The relationship between the stress-strain curves for a particular soil and *p-y* curves is undoubtedly close; there, some discussion of the change in the stiffness of a particular

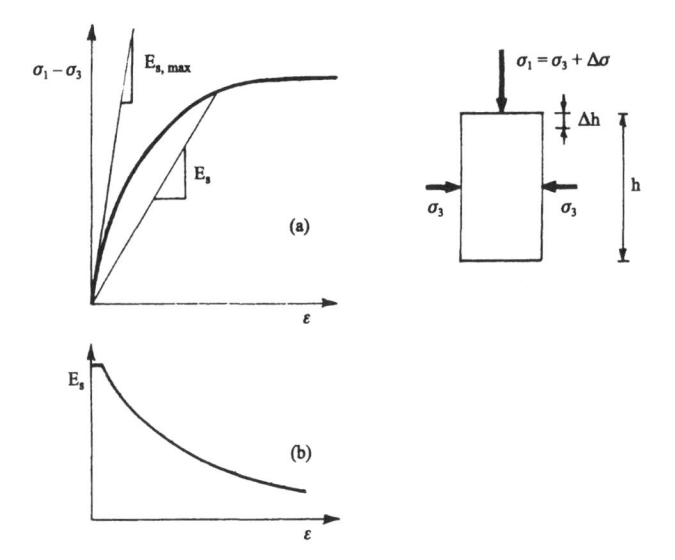

Figure 3.1. Results from testing soil specimen in the laboratory.

specimen with strain is useful. The values of E_s decrease with increasing strain, as shown in Figure 3.1b, similarly as do the values of E_{py}, shown conceptually in Figure 1.4b. Van Impe (1991) proposed a model for the decay of E_s, as shown in Figure 3.2. The figure shows G_s/G_{smax} where $G_s E_s/(2(1 + v)) =$ and also indicates approximate values of Poisson's ratio for sand and clay.

While the values of the properties as shown in Figure 3.2 are generalized and do not represent precisely the values at a specific site, the values do allow for computations that reflect the deformations of the soil in a continuum. Further, the decay of E_{py} is shown to reflect the same phenomena, but with different units, as in Figure 1.4b. Therefore, generically, the decay of E_s with increasing strain is similar to the decay of E_{py} due to pile deflection. Further, the decay in both instances is certainly due to the same phenomenon, a decrease in stiffness of an element of soil with increased strain.

3.2.3 Variation of stiffness of soil (E_s and G_s) with depth

The paragraphs above have shown close conceptual relationships between values of E_s and E_{py}; therefore, a discussion of the variation of E_s with depth is desirable. The values of E_{smax} of sand and normally consolidated clay are zero at the ground line and increase in some fashion with depth. The values of E_{smax} of some overconsolidated clays and some rocks are approximately constant with depth.

The variation of the maximum tangent shear modulus with depth z and related to the zero deflection of the pile has been proposed as shown below.

$$G_{smaxz} = G_{smax0} + a_z \quad (3.1)$$

where G_{smax0} = the maximum soil reaction modulus at ground level (zero for normally consolidated soils); and a_z = gradient of the maximum tangent soil reaction modulus with depth z.

52 *Piles under lateral loading*

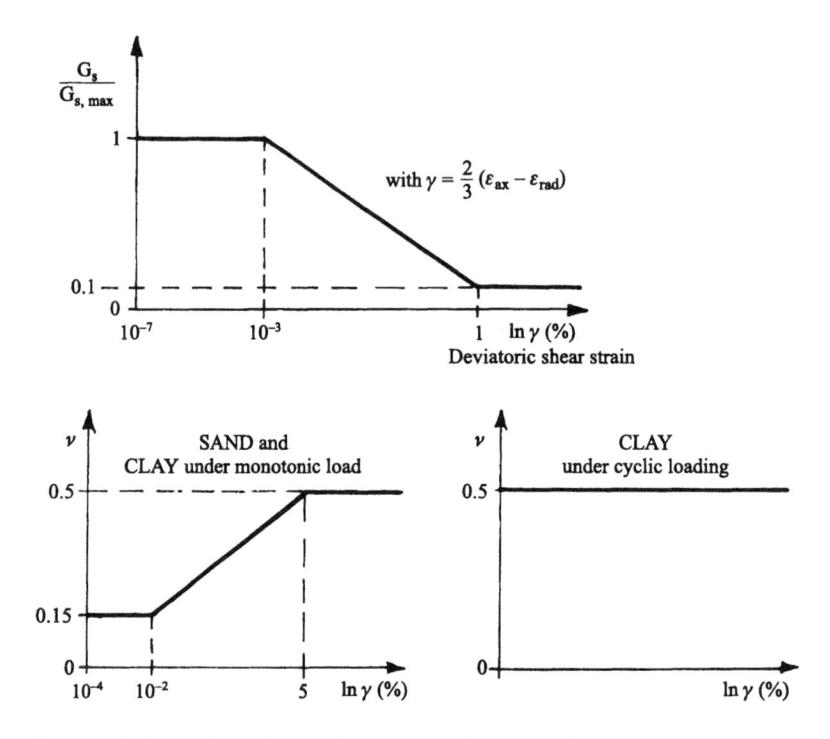

Figure 3.2. Model for Poisson ratio change and decay of E_s (Van Impe 1991).

As will be demonstrated by examples and case studies in later chapters, in most cases, the properties of the soil between the ground surface and a depth of 6 to 10 diameters will govern the behavior of a pile subjected to lateral loading. The parameters that govern the stiffness of the near-surface soils have been investigated extensively during the past two decades. Of particular significance are the works of Ladd et al. (1977) and Wroth et al. (1979), that was published in the early eighties, and the works of K.H. Stokoe et al. (1985, 1989, 1994) and M. Jamiolkowski et al. (1985, 1991, 1993), that was published from the eighties until the present.

For cohesive soils, progress has been made in describing the ratio of E_{smax}/c, where c is equal to the undrained shear strength. In these investigations, the results are dependent on the type of test, the over-consolidation ratio, and the index properties of the soil. For cohesionless soils, studies have been aimed at finding the ratio of E_{smax}/p', where p' is the mean effective stress.

Some data have been reported in the literature for cohesive soils for values of E_{smax}/c_u and much scatter in the results have been observed. Values of E_{smax}/c_u in the range of 40 to 200 were reported by Matlock et al. (1956), and Reese et al. (1968). Values probably would have been much higher had very careful attention been given to the early part of the laboratory curves. Stokoe (1989) reported that values of 2000 were routinely found for very small strains. Johnson (1982) performed tests with the self-boring pressuremeter and found values of E_{smax}/c_u that ranged from 1440 to 2840. These high values are for extremely small strains.

3.2.4 Initial stiffness and ultimate resistance of p-y curves from soil properties

The typical *p-y* curve for some depth z_1 below the ground surface, shown in Figure 1.4a, is characterized by a straight line from the origin to point *a* and a value p_{ult} for the ultimate resistance beyond point *b*. Elementary solutions, based on soil properties, are presented for these two important aspects of *p-y* curves in this section. As shown later, recommendations for the formulation of *p-y* curves are strongly based on results from full-scale testing of piles, but analytical expressions are helpful in interpreting the experiments.

3.2.4.1 Initial stiffness of p-y curves

Relevance. As may be seen in Figure 1.4b, the portion of the total *p-y* curves occupied by this initial portion of the *p-y* curve is small and may have little consequence in most analyses. Employing the concepts emphasized herein, the design of a pile under lateral loading is based primarily on limit-states, with loads limited by bending or combined stress or by deflection. The deflection of the pile where the initial portion of the *p-y* curves would be effective would occur at a considerable distance below the ground line and the resulting horizontal forces in the soil would have only a minor effect on the response. In most designs, stress will control.

The computation of the deflection under the working load would, of course, make more important the initial portion of the *p-y* curves. However, even in such cases it is unlikely that the initial portion of the *p-y* curves would play an important role.

There are some cases, on the other hand, where the early part of the *p-y* curves needs careful consideration. Two cases can be identified: the prediction of behavior under vibratory loading, and the design of piles in brittle soils. Rock frequently is a brittle material and a sudden loss of resistance can be postulated when the deflection reaches Point *a* in Figure 1.4b. Thus, the *p-y* curve would reflect a much different response than shown in the figure.

Theoretical considerations. As a means of establishing the parameters that must be evaluated in employing linear-elastic concepts in finding equations for the values of the slopes of the initial portions of *p-y* curves, some elementary concepts of mechanics can be used. The following equation is from the theory of elasticity (Skempton 1951), and provides a basis for deriving a simple expression giving the approximate slope of the initial portion of the *p-y* curve:

$$s_m = qbI_\rho \frac{1-v^2}{E_s} \tag{3.2}$$

where s_m = mean settlement of a foundation; q = foundation pressure; I_p = influence coefficient; v = Poisson's ratio of the solid; and E_s = Young's modulus of the solid.

The equation pertains to vertical loading but if the soil resistance *p* against the pile due to lateral deflection is assumed to be equal to qb and the deflection of the pile is equal to s_m, the equation can be rewritten as follows, with I_r and n taken as constants, as suggested by Skempton.

$$\frac{qb}{s_m} = \frac{p}{y} = E_{py} = \frac{E_s}{I_\rho(1-v^2)} = \xi E_s$$

54 Piles under lateral loading

where ζ is a constant. While the above equation suggests that E_{py} has a defined relationship with E_s, (discussed below with a fuller discussion of Skempton's suggestions), the relationship would appear to relate more accurately to the initial portion of the values of moduli.

$$E_{pymax} = \xi_i E_{smax} \tag{3.3}$$

where E_{pymax} and ξ_{smax} are the initial slopes of the p-y curves and the stress-strain curves, respectively, and ξ_i is given a subscript to indicate the initial value.

The values of E_{pymax} from the p-y curves in Figures 1.5 and 1.6 are plotted in Figure 3.3 and the influence of the ground surface is obvious. These data suggest that the value of z must reflect the location of the ground surface. The differences in the values of E_{pymax} for static and cyclic loading are striking.

3.2.4.2 Computation of values of p_{ult}

The prediction of the ultimate values of p as a function of the kind of soil and the depth below ground surface is of obvious importance. Analytical methods, applied here, are important and allow comparisons with values of p_{ult} determined by experiment, such as shown in Figure 1.5.

Some serious mistakes were made in early years in the analysis of piles under lateral loading when engineers failed to heed the admonition of Terzaghi that the ultimate resistance against a pile could not exceed one-half the bearing capacity of the soil. In the absence of a more rational method of computing a limiting value of the ultimate resistance p_{ult}, which may be considered as the bearing capacity, two simple models have been employed for solving the problem by limit equilibrium.

The first of the models is shown in Figure 3.4. The force F_p may be computed by integrating the horizontal components of the resistances on the sliding surfaces, taking into

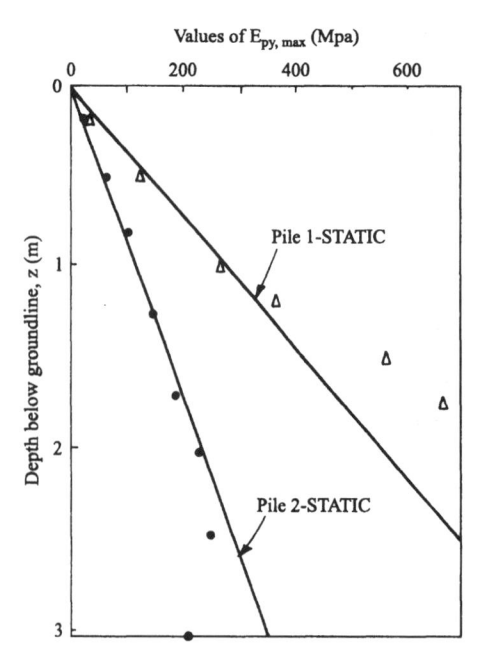

Figure 3.3. Values of E_{pymax} from experiment (see Figs. 1.5 and 1.6).

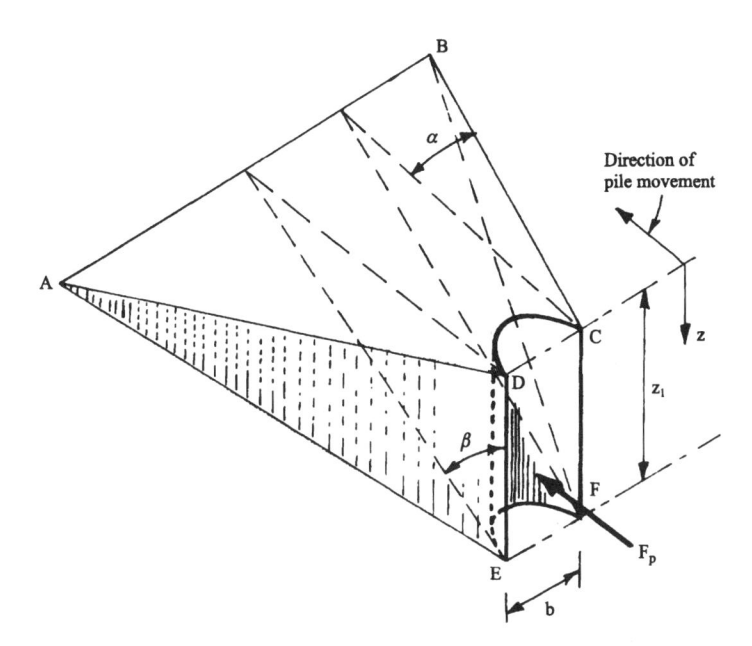

Figure 3.4. Model of soil at ground surface for computing pult.

account the weight of the wedge. Integration of F_p with respect to the depth z below the ground surface will yield an expression for the ultimate resistance along the pile, p_{ult}. The simplicity of the model is obvious; however, the resulting solutions will illustrate the form of the equation because the diameter of the pile, the depth below the ground surface, and the properties of the soil enter into the solution.

If three-dimensional constitutive relationships become available for all soils and rocks, data such as shown in Figure 3.5 will be helpful in developing a more realistic model than shown in Figure 3.4. The contours show the heave of the ground surface in front of a steel-pipe pile with a diameter of 641 mm in over-consolidated clay (Reese et al. 1968). With a lateral load of 596 kN, ground-surface movement occurred at a distance from the axis of the pile of about 4m (Fig. 3.5a). When the load was removed, the ground surface subsided somewhat, as shown in Figure 3.5b. The *p-y* curves that were derived from the loading test are shown in Figure 1.5.

One can reason that, at some depth, the resistance against a wedge will increase to the point that horizontal movement of the soil will occur. The second model, shown in Figure 3.6a, depicts a cylindrical pile and five blocks of soil. The assumption is made that the movement of the pile will cause a failure of Block 5 by shearing. Soil movement will cause Block 4 to fail by shearing, Block 3 to slide, and Blocks 2 and 1 to fail by shearing. The ultimate resistance p_{ult} can be found by observing the difference in the stresses σ_6 and σ_1. The model is plainly crude but again should indicate the form of the equation for p_{ult}. The Mohr-Coulomb diagrams shown in Figure 3.6 will be used to develop equations for flow-around failure for the soil for piles in cohesive soil and in cohesionless soil.

Cohesive soil. Two assumptions are made: 1. The soil is assumed to be saturated, 2. The undrained-strength approach will yield useful answers. Partially saturated clays can

56 Piles under lateral loading

change in water content with time, so saturation appears to be justified. The introduction of drainage of clays into the analysis, as noted in Chapter 1, introduces the necessity of formulating a three-dimensional model for consolidation and will not be addressed.

Equations for forces on the sliding surfaces in Figure 3.4, assuming the angle α to the zero, are written and solved for F_p, and F_p is differentiated with respect to z to solve for the soil resistance p_{c1} per unit length of the pile.

$$p_{c1} = c_a b[\tan\beta + (1+\kappa_c)\cot\beta] + \gamma bz + 2c_a z(\tan\beta\sin\beta + \cos\beta) \tag{3.4}$$

where κ_c = a reduction factor for the shearing resistance along the face of the pile; z = depth below the ground surface, and c_a = the average undrained shear strength over the depth of the wedge.

The value of κ_c can be set to zero with some logic for the case of cyclic loading because one can reason that the relative movement between pile and soil would be small under repeated loads. The value of β can be taken as 45° if the soil is assumed to behave in an undrained mode. With these assumptions, Equation 3.4 becomes

$$p_{c1} = 2c_a b + \gamma bz + 2.83 c_a z \tag{3.5}$$

Thompson (1977) differentiated Equation 3.5 with respect to z and evaluated the integrals numerically. His results are shown in Figure 3.7 with the assumption that the value of the term γ/c_a is negligible. Plots are shown for the case where κ_c is assumed equal to zero or equal to 1.0. Also shown in Figure 3.7, is a plot of Equation 3.5 with the same assumption with respect to γ/c_a. As may be seen, the differences in the plots are not great above a nondimensional depth of about 3.2.

The second of the two models for computing the ultimate resistance p_{ult} is investigated. The ultimate soil resistance p_{ult} can be found from the difference between σ_6 and σ_1 in Figure 3.6b. Four of the blocks are assumed to fail by shear and that resistance due to

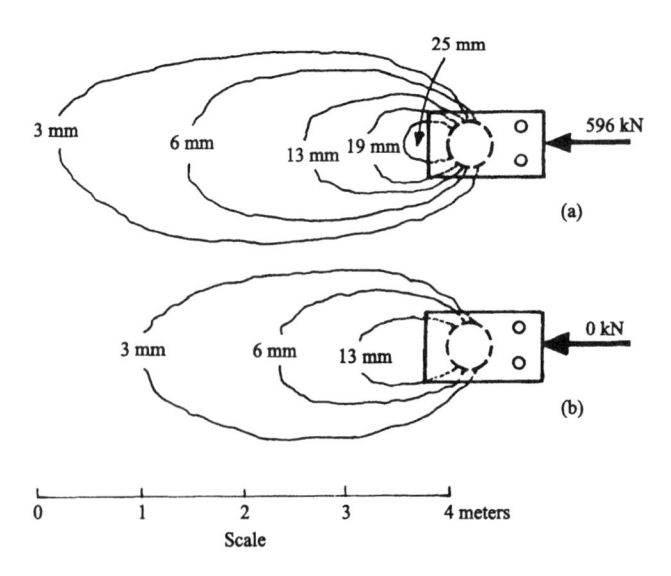

Figure 3.5. (a) Ground heave due to static loading of a pile; (b) Residual heave.

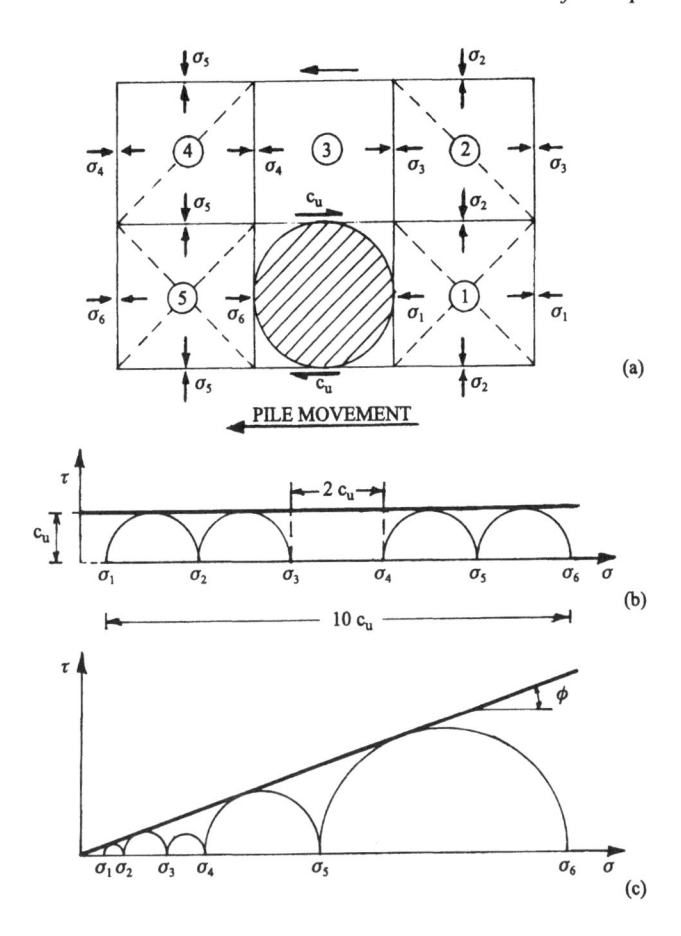

Figure 3.6. Assumed mode of failure of soil by lateral flow around a pile (a) section through pile; (b) Mohr-Coulomb diagram for pile in clay (c) Mohr-Coulomb diagram for pile in sand.

sliding is assimed to occur on both sides of Block 3. The value of $(\sigma_6 - \sigma_1)$ is found to be $10c_u$. Other work, not shown here, shows that there is justification of using the following equation for p_{ult}.

$$p_{ult} = (\sigma_6 - \sigma_1)b = 11c_u b \tag{3.6}$$

The value from Equation 3.6 is also shown plotted in Figure 3.7 below an intersection with Equation 3.5. Solving for the intersection between Equations 3.5 and 3.6, ignoring the influence of the term γ/c_a and assuming that the undrained shear strength is constant with depth, Equation 3.5 will control to where z/b is equal to about 3.2. In practice, the importance of the reduced ultimate resistance at the groundline is significant.

Thompson (1977) noted that Hansen (1961a,b) formulated equations for computing the ultimate resistance against a pile at the ground surface, at a moderate depth, and at a great depth. Hansen considered the roughness of the wall of the pile, the friction angle, and unit weight of the soil. He suggested that the influence of the unit weight be neglected and proposed the following equation for the $\phi = 0$ case for all depths.

58 *Piles under lateral loading*

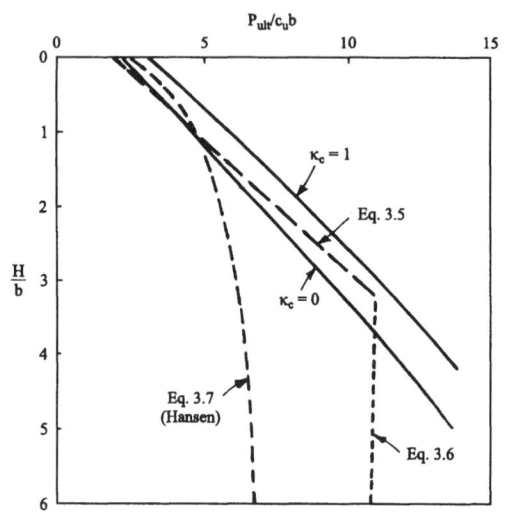

Figure 3.7. Ultimate lateral resistance for cohesive soils.

$$\frac{p_{ult}}{c_u b} = \frac{2.567 + 5.307\frac{z}{b}}{1 + 0.652\frac{z}{b}} \quad (3.7)$$

Equation 3.7 is also shown plotted in Figure 3.7. The agreement with the 'block' solutions is satisfactory near the ground surface, but the difference becomes significant with depth.

Equations 3.5 and 3.6 were used in analyzing the results from full-scale experiments and the *form* of the expressions appears to be valid. The recommended methods of computing the *p-y* curves for clays are presented later in this chapter.

Cohesionless soil. Full drainage is assumed in the analyses that follow. For most granular soils, this assumption is valid. The two models presented earlier are employed, following a similar procedure to that used for clay, except that no reduction factor was considered for shearing resistance along the face of the pile. The ultimate soil resistance near the ground surface per unit length of the pile is obtained by finding the total force against an upper portion of the pile and by differentiating the results with respect to z.

$$(p_{ult})_{sa} = \gamma z \left[\frac{K_0 z \tan\phi \sin\beta}{\tan(\beta-\phi)\cos\alpha_s} + \frac{\tan\beta}{\tan(\beta-\phi)}(b + z\tan\beta\tan\alpha_s) \right]$$
$$+ \gamma z \left[K_0 z \tan\beta(\tan\phi\sin\beta - \tan\alpha_s) - K_a b \right] \quad (3.8)$$

where K_0 = coefficient of earth pressure at rest; K_a = minimum coefficient of active earth pressure; and $\alpha\alpha_s$ = value of a for sand.

Bowman (1958) performed some laboratory experiments with careful measurements and suggested values of a_s from $\phi/2$ to $\phi/3$ for loose sand up to ϕ for dense sand. The value of β is approximated by the following equation.

$$\beta = 45 + \frac{\phi}{2} \qquad (3.9)$$

The model for computing the ultimate soil resistance at some distance below the ground surface was implemented. The stress σ_1 at the back of the pile must be equal or larger than the minimum active earth pressure; if not, the soil could fail by slumping. The assumption is based on two-dimensional behavior; thus, it is subject to some uncertainty. If the states of stress shown in Figure 3.6c are assumed, the ultimate soil resistance for horizontal movement of the soil is

$$(p_{\text{ult}})_{sb} = K_a b \gamma z (\tan^8 \beta - 1) + K_o b \gamma z \tan\phi \tan^4 \beta \qquad (3.10)$$

The equations for $(p_{\text{ult}})_{sa}$ and $(p_{\text{ult}})_{sb}$ are admittedly approximate because of the elementary nature of the models. However, the equations serve a useful purpose in indicating the form, if not the magnitude, of the ultimate soil resistance.

3.2.5 Subgrade modulus related to piles under lateral loading

The concept to the subgrade modulus has been presented in technical literature from early days and values have been tabulated in textbooks and other documents. Engineers performing analyses of piles under lateral loading, prior to developments reported herein, sometimes relied on tabulated values of the subgrade modulus in getting the soil resistance. Numerical values of the subgrade modulus are certainly related to values of E_s; and to E_{py} in some ways; therefore, a brief, elementary explanation of the term *subgrade modulus* by way of a simple experiment, is desirable.

Figure 3.8a shows a plan view of the plate with m and n indicating the lengths of the sides. If a concentrated vertical load is applied to the plate at the central point, the resulting settlement is shown by Section A-A in Figure 3.8b, along with an assumed uniform distributed load. If increasingly larger loads are applied, a unit load-settlement curve is subsequently developed, as shown by the typical curve in Figure 3.8c. The figure indicates that the magnitude of the unit load reached a point where settlement continued without any increase in load.

Several lines are drawn in Figure 3.8c from the origin of the curve to points on the curve. The slopes of these lines with units of F/L^3 are defined as the subgrade modulus, and is a measure of the stiffness of the soil under the particular loading. As shown, the maximum value is for a line drawn through the initial portion of the curve, with the other lines giving lower values.

If a plate with dimensions larger or smaller than given by m and n is employed in the same soil, one could expect a different result. Further, the stiffness of the plate itself can affect the results, because the plate would deform in a horizontal plane, depending on the method of loading. Also, soils with a friction angle will exhibit an increased stiffness with depth. As can be understood, except in some special cases, values of such type of subgrade moduli are of limited value in the solution of a problem of soil-structure interaction but are only useful in merely differentiating the stiffness of various soils (and rocks) such as soft clay, stiff clay, loose sand, dense sand, sound limestone, or weathered limestone.

60 Piles under lateral loading

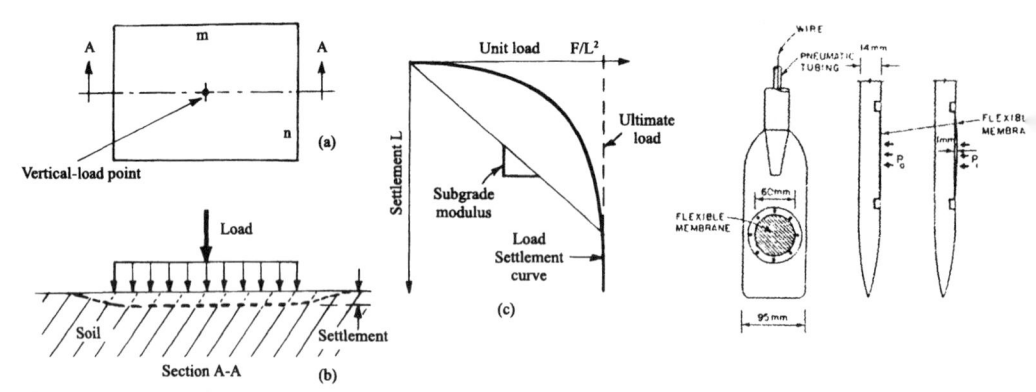

Figure 3.8. Description of experiment leading to definition of subgrade modulus.

More recent in situ testing research revealed the possibility to estimate for example the lateral subgrade modulus from Menard pressuremeter tests (Y. Ikeda et al. 1998, Imai T. 1970) and from Marchetti dilatometertests.

From the work of Baldi et al. (1986) and Robertson et al. (1989), one could in this respect at least for displacement piles, go out from flat dilatometertests (DMT) in order to estimate directly the E_{py} at a given depth from the dilatometer modulus $E_{DMT} = 34.7\ (p_1 - p_0)$; p_1 & p_0 are DMT readings (Fig. 3.8d). In our proposal, we would implement a simplified relation for the case of lateral loading of displacement piles:

$$E_{py} \text{ (at the DMT testing depth)} = F \cdot E_{DMT} \qquad (3.11)$$

with: $F = 2$ for N.C. sands; $F = 5$ for O.C. dense sands; $F = 10$ for N.C. clays.

3.2.6 Theoretical solution by Skempton for subgrade modulus and for p-y curves for saturated clays

Skempton (1951) wrote that 'simple theoretical considerations' were employed to develop a prediction for load-settlement curves. Even a limited solution, for saturated clays, is useful to reflect the practical application of theory. The theory has some relevance to *p-y* curves because the resistance to the deflection of a loaded area is common to both a horizontal plate and a pile under lateral loading.

As noted earlier, the mean settlement of a foundation, s_m, of width b on the surface of a semi-infinite solid, based on the theory of elasticity is given by Equation 3.2.

$$s_m = qbI_\rho \frac{1-v^2}{E_s} \qquad (3.2)$$

where q = foundation pressure; I_ρ = influence coefficient; v = Poisson's ratio of the solid; and E_s = Young's modulus of the solid.

In Equation 3.2, Poisson's ratio can assumed to be 1/2 for saturated clays if there is no change in water content. For a rigid circular footing on the ground surface I_ρ can be taken as $\pi/4$ and the failure stress q_f may be taken as equal 6.8 c, where c is the undrained shear strength. Making the substitutions indicated and setting $s_m = s_{m1}$ for the particular case

$$\frac{S_{m1}}{b} = \frac{4}{\frac{E_s}{c_u}} \frac{q}{q_f} \qquad (3.12)$$

Skempton noted that the influence value I_r decreases with depth below the surface but the bearing capacity factor increases; therefore, as a first approximation Equation 3.11 is valid for any depth.

In an undrained compression test, the axial strain is given by the following equation.

$$\varepsilon = \frac{\sigma_1 - \sigma_3}{E_s} = \frac{\Delta\sigma}{E_s} \qquad (3.13a)$$

where E_s = Young's modulus at the stress $(\sigma_1 - \sigma_3)$ level.

For saturated clays with no water content change, Equation 3.13 may be rewritten as follows.

$$\varepsilon = \frac{2}{\frac{E_s}{c_u}} \frac{(\sigma_1 - \sigma_3)}{(\sigma_1 - \sigma_3)_f} \qquad (3.13b)$$

where $(\sigma_1 - \sigma_3)_f$ = failure stress.

Equations 3.12 and 3.13 show that, for the same ratio of applied stress to ultimate stress, the strain in the footing test (or pile under lateral loading) is related to the strain in the laboratory compression test by the following equation.

$$\frac{S_{m1}}{b} = 2\varepsilon \qquad (3.14)$$

Skempton's arguments based on the theory of elasticity and also on the actual behavior of full-scale foundations led to the following conclusion:

> Thus, to a degree of approximation (20%) comparable with the accuracy of the assumptions, it may be taken that Equation 3.14 applies to a circular or square footing.

As may be seen in the analyses shown above, Skempton allowed the Young's modulus of the soil, E_s, to be nonlinear and to assume values from E_{smax} to much lower values when the soil was at failure. The assumption of a nonlinear value of E_s is remarkable because of varying state of stress of elements below the footing.

Skempton pointed out that the value of I_r for a footing with a length to width ratio of 10 was reported by Terzaghi (1943) and Timoshenko (1934) to be 1.26. If the bearing capacity factor is taken as 5.3 c_u, Equation 3.14 can be written as follows.

$$\frac{S_{m1}}{b} = 2.5\varepsilon \qquad (3.15)$$

Skempton stated that the failure stress for a footing reaches a maximum value of $9c_u$. A curve of resistance as a function of deflection could be obtained for a long strip footing, then, by taking points from a laboratory stress-strain curve and using Equation 3.15 to obtain deflection and $4.5\Delta\sigma$ to obtain soil resistance.

3.2.7 *Practical use of Skempton's equations and values of subgrade modulus in analyzing a pile under lateral loading*

Skempton's equations appear attractive with respect to solving the problem of the laterally loaded pile in saturated clays. However, two factors need addressing, particularly in respect to soils in general: 1. The applicability of the equations, and tabulated values of the subgrade modulus, to portions along a real pile; and 2. Accounting for the difference in units between q and p. These two factors are discussed below, and while saturated clay is used principally in the discussion, the concepts presented are general.

The first of the above problems can be dealt with by referring to Figure 3.9. The sketch in Figure 3.9 depicts a vertical cut that has been made into a clay with sufficient strength that the clay will stand without support. A side view of two loaded areas is shown in Figure 3.9a and a plan view in Figure 3.9b. The lower area is a considerable distance below the ground surface. If the areas are loaded until failure occurs, sliding surfaces will develop in the clay somewhat as shown by the dotted lines in Figure 3.9a. Soil mechanics, and simple logic, will show that the load to cause failure for the upper area is much lower that for the lower area. The simple presentation illustrates ....that p-y curves are affected by the distance from the curve to the ground surface. A more formal implementation of the concept is presented in later sections where equations are presented for the critical distance where the ground surface no longer has an influence on the magnitude of the quantity p.

Unfortunately, experimental results, and elementary theory, show that p-y curves at and near the ground surface have a dominant affect on the computation of pile deflection and bending moment. The question becomes, then, what is the effect on pile response of the soil where elements move only horizontally due to pile deflection? The question cannot be answered in general terms and requires the solution of specific problems. Therefore, the subgrade modulus and similar theories should not be implemented or otherwise used with care in solving for pile behavior, even in the regions unaffected by the presence of the ground surface.

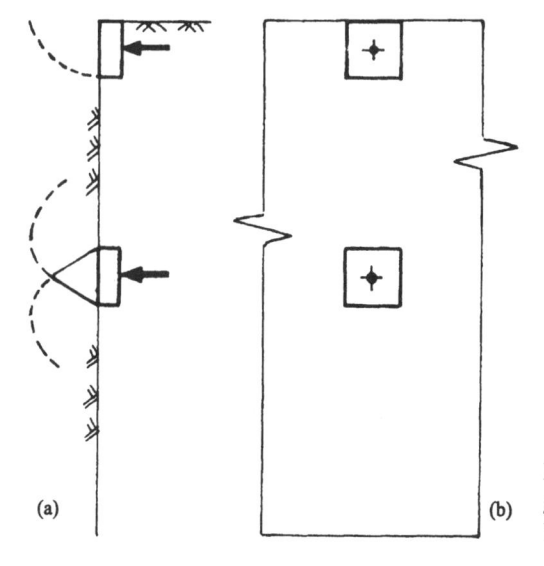

Figure 3.9. Concept showing importance of ground surface to response of soil to lateral loading.

Converting data presented in terms of q for the subgrade modulus, or for the load versus deflection for a strip or a pile, to values of p for solutions by use of the differential equation, requires the following steps.

1. Ascertaining that the values are for a strip footing, or can be converted to the strip footing. If the data are for a square plate, for example, the state of stress beneath the plate and the consequent deflection will be much different than for the portion of a pile.
2. If the data are for a strip footing, a constant value is frequently presented and information is required on whether of not the stiffness given is the maximum value or for some other value, perhaps an average. While Skempton gives such information for saturated clay, other authors rarely give such data.
3. If the tabulated data are judged to be usable, values of subgrade modulus should be computed for the width b of a strip. The values of the unit load q should then be multiplied by b to obtain values of p, along with the corresponding stiffness or deflection.

The Skempton equations can be used to make a comparison with experimental curves, shown in Figure 1.5. Those curves were derived from the experiments at Manor, described in Chapter 7, where the pile had a diameter of 641 mm. In order to estimate the influence of the ground surface, data from the Manor experiments were used and Equations 3.30 and 3.31 were solved by trial. A depth of 4.0 m was found at which the ground surface was judged to have no influence on the p-y curves; that is, where the soil was assumed to be displaced in a vertical plane by a pile under lateral loading. The greatest depth for the curves in Figure 1.5 is 3.05 m; therefore, the elementary analysis shows that all of the curves are influenced by the ground surface. However, the slopes of the initial portion of the p-y curves appear to be approaching a maximum value, so the assumption is made that the slope of the curve for the depth of 4.0 m would have been the same as shown by the curve for 3.05 m.

The initial stiffness of the final curve in Figure 1.5 can be computed by using Equation 3.11. A value of p_{ult} may be computed using Equation 3.31 and a value of undrained shear strength from the Manor test for 4.0 m, or 320 kPa.

$$p_{ult} = (11)(320) = 3520 \text{ kN/m}$$

Following the reasoning presented earlier, the assumption is made that the state of stress for the plates in Skempton's presentation is the same as for the pile, $q_f = p_{ult}/b = 3520/0.641 = 5490$ kPa. For convenience, the initial slope of the curve is taken as $q_f/20$ or where q is equal to 275 kPa and the corresponding value of p is 176 kN/m. Referring to Figure 1.5, the deflection from the curve at 3.05 m at a value of p of 176 kN is 2.5 mm. Equation 3.11 can now be employed, which is rewritten and solved as follows.

$$\frac{E_{s\,max}}{c_u} = \frac{4b}{s_{m1}} \frac{q}{q_f} = \frac{(4)(0.641)}{(2.5 \times 10^{-3})} \frac{(1)}{(20)} = 51$$

The computed value of 51 is within the range of values that were listed above and probably would have been somewhat higher if the curve had been available for the greater depth.

The result of the elementary exercise is that the Skempton equations have some utility in producing p-y curves at depth below where the soil is influenced by the ground surface. However, the exercise shows that, for the case that was examined, the response of the pile to lateral loading was almost completely controlled by the p-y curves for the near-surface soils.

64 *Piles under lateral loading*

3.2.8 *Application of the Finite Element Method (FEM) to obtaining p-y curves for static loading.*

The above discussion, illustrating the importance of the soils at and near the ground surface, leads to the suggestion to apply the finite element method to developing *p-y* curves for static loading. The topic was discussed in Chapter 1 and re-visited here because of the relevance of the FEM. The problems of characterizing the soil at a test site; selecting the appropriate nonlinear, three-dimentional constitutive model; performing the three-dimensional analyses for the full range of pile deflection for an elastic pile; and taking the nonlinear soil and nonlinear geometry into account is beyond current capabilities.

However, the predicting of *p-y* curves by the FEM is not beyond the capabilities of a comprehensive research effort. Computing power is growing rapidly and tools are available for performing the physical research. Of particular importance is that several sites are available where *p-y* curves have been determined experimentally, such as shown in Figure 1.5, giving the researchers the data needed to confirm the analytical solutions. Improved and more advanced methods of soil investigation will likely be at hand and additional soil testing at sites of previous tests of piles under lateral loading may be required.

3.3 INFLUENCE OF DIAMETER ON *p-y* CURVES

3.3.1 *Clay*

Analytical expressions for *p-y* curves indicate that the term for the pile diameter appears to the first power. Reese et al. (1975) describe tests of piles with diameters of 152 mm and 641 mm at the Manor site. The *p-y* formulations developed from the results from the larger piles were used to analyze the behavior of the smaller piles. The computation of bending moment led to good agreement between analysis and experiment but the computation of groundline deflection showed considerable disagreement, with the computed deflections being smaller than the measured ones. No explanation could be made to explain the disagreement.

O'Neill & Dunnavant (1984) and Dunnavant & O'Neill (1985) report on tests performed at a site where the clay was overconsolidated and where lateral-loading tests were performed on piles with diameters of 273 mm, 1220 mm, and 1830 mm. They found that the site-specific response of the soil could best be characterized by a nonlinear function of the diameter. These studies and subsequent studies can perhaps provide a basis for specific recommendations.

There is good reason to believe that the diameter of the pile should not appear as a linear function in *p-y* curves for cyclic loading of piles in clays below the water table. The influence of cyclic loading on *p-y* curves is discussed in the next section.

3.3.2 *Sand*

No special studies have been reported on an investigation of the influence of diameter on *p-y* curves. Case studies of piles, some of which are of large diameter, do not reveal any particular influence of the diameter. However, virtually all of the tests that have been performed in sand used only static loading.

3.4 INFLUENCE OF CYCLIC LOADING

3.4.1 *Clay*

Cyclic loading is used in the design of piles for some of the structures mentioned in Chapter 1; a notable example is an offshore platform. Therefore, a number of the field tests employing fully instrumented piles have employed cyclic loading. The first such tests were preformed by Matlock (1970). Cyclic loading has invariably resulted in increased deflection and bending moment above the respective values obtained in short-term loading. A dramatic example of the loss of soil resistance due to cyclic loading may be seen by comparing the two sets of *p-y* curves in Figures 1.5 and 1.6.

The following paragraphs deal with the phenomena of gapping and scour of clay below the water surface. As noted in Chapter 1, clouds of suspension were observed in testing of piles in stiff clay under cyclic loading, but scour was not observed by Matlock (1970) in cyclic tests in soft to medium clays. However, at the conclusion of one set of cyclic-loading tests, Matlock placed pea-sized gravel beneath the water and around the pile. Cycling was continued and a considerable quanity of the gravel worked down around the pile, indicating that the clay near the wall of the pile, with no pronounced gap, was weakened to allow the gravel to penetrate by gravity.

Wang (1982) and Long (1984) did extensive studies of the influence of cyclic loading on *p-y* curves for clays. Some of the results of those studies were reported by Reese et al. (1989). Two reasons can be suggested for the reduction is soil resistance from cyclic loading: the subjection of the clay to repeated strains of large magnitude, and scour from the enforced flow of water in the vicinity of the pile. Long (1984) studied the first of these factors by performing triaxial tests with repeated loading, using specimens from sites where piles had been tested. The second of the effects is present when water is above the ground surface and its influence can be severe.

Welch & Reese (1972) report some experiments with a bored pile under repeated lateral loading in an overconsolidated clay with no free water. During the cyclic loading, the deflection of the pile at the groundline was in the order of 25 mm. After a load was released, a gap was revealed at the face of the pile where the soil had been pushed back. Also, cracks a few millimeters in width radiated away from the front of the pile. Had water covered the ground surface, it is evident that water would have penetrated the gap and the cracks. With the application of the loads, the gap would have closed and the water carrying soil particles would have been forced to the ground surface. This process was dramatically revealed during the soil testing in overconsolidated clay at Manor (Reese et al. 1975) and at Houston (O'Neill & Dunnavant 1984).

The sketch in Figure 3.10 illustrates the phenomenon of scour. A space has opened in the overconsolidated clay in front of the pile and has filled with water as load is released. With the next excursion of the pile, the water is forced upward from the space at a velocity that is a function of the rate of load application. The water exits with turbulence and particles of clay are scoured away.

Wang (1982) constructed a laboratory device to investigate the scouring phenomenon. A specimen of undisturbed soil from the site of a pile test was brought to the laboratory, placed in a mold, and a vertical hole about 25 mm in diameter was cut in the specimen. A rod, of the same size as the hole, was placed and attached to a hinge at the base to the specimen. Water, a few millimeters deep, was kept over the surface of the specimen and

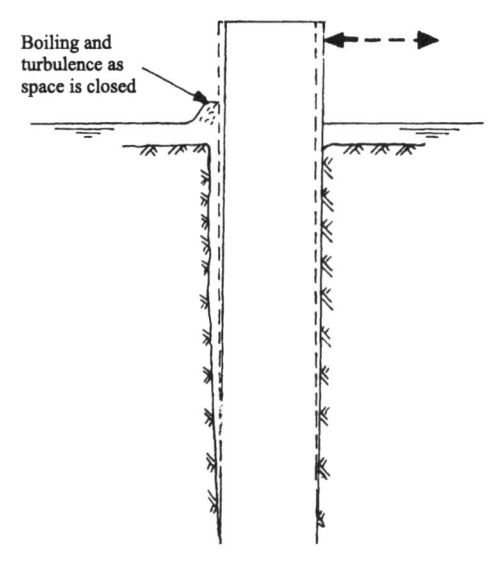

Figure 3.10. Illustration of scour around a pile in clay during cyclic loading.

the rod was pushed and pulled by a machine at a given period and at a given deflection for a measured period of time. The soil that was scoured to the surface of the specimen was carefully collected, dried, and weighed. The deflection was increased and the process was repeated. A curve was plotted showing the weight of soil that was removed as a function of the imposed deflection. The characteristics of the curve were used to define the scour potential of that particular clay.

The Wang device was found to be far more discriminating about scour potential of a clay than was the pinhole test (Sherard et al. 1976), but the results of the test could not explain fully the differences in the loss of resistance experienced at different sites where lateral-load tests were performed in clay with water above the ground surface. At one site where the loss of resistance due to cyclic loading was relatively small, it was observed that the clay included some seams of sand. The sand would not have been scoured readily and particles of sand could have partially filled the space that was developed around the pile. Another experiment showed that pea gravel placed around a pile during cyclic loading was effective in restoring most of the loss of resistance; however, O'Neill & Dunnavant (1984) report that 'placing concrete sand in the pile-soil gap formed during previous cyclic loading did not produce a significant regain in lateral pile-head stiffness'.

While the work of Long (1984) and Wang (1982) developed considerable information about the factors that influence the loss of resistance in clays under free water due to cyclic loading, their work did not produce a definitive method for predicting this loss of resistance. The analyst, thus, should use the numerical results for cyclic loading presented herein with caution. Full-scale experiments with instrumented piles at a particular site are indicated for those cases where behavior under cyclic loading is a critical feature of the design.

3.4.2 *Sand*

Very few tests of piles under cyclic lateral loading have been reported. There is evidence that the repeated loading on a pile in predominantly one direction will result in a perma-

Models for response of soil and weak rock 67

nent deflection in that direction. When a relatively large load is applied, the top of the pile will deflect a significant amount and allowing the particles of cohesionless soil to fall into at the back of the pile, preventing the pile from returning to its initial position.

Observations of the behavior of the mass of sand near the ground surface during cyclic loading support the idea that the void ratio of this sand is approaching the critical value. That is, dense sand apparently loosens during cycling and loose apparently densifies.

3.5 EXPERIMENTAL METHODS OF OBTAINING *p-y* CURVES

Methods of getting *p-y* curves from field experiments with full-sized piles will be presented prior to discussing the use of analysis in obtaining soil response. The strategy that has been employed for acquiring design criteria is to make use of the theoretical methods, to obtain *p-y* curves from full-scale field experiments, and to derive such empirical factors as necessary so that there is close agreement between results from adjusted theoretical solutions and those from experiments. Thus, an important procedure is obtaining experimental *p-y* curves.

3.5.1 *Soil response from direct measurement*

A number of attempts have been made to make direct measurement of *p* and *y* in the field. Measuring the deflection involves the conceptually simple process of sighting down a hollow pile from a fixed position at scales that have been placed at intervals along the length of the pile. The method is cumbersome in practice and has not been very successful.

The measurement of soil resistance directly involves the design of an instrument that will integrate the soil stress at a point along the pile. The design of such an instrument has been proposed but none has yet been built. Some attempts have been made to measure the soil pressure at a few points around the exterior of a pile with the view that the soil pressures at other points can be estimated. This method has met with little success.

3.5.2 *Soil response from experimental moment curves*

Almost all of the successful experiments that yielded *p-y* curves have involved the measurement of bending moment by the use of strain gauges. The deflection can be obtained with considerable accuracy by two integrations of the moment curves. The deflection and the slope at the groundline have to be measured accurately and it is helpful if the pile is long enough so that there are at least two points of zero deflection along the pile.

The computation of soil resistance involves two differentiations of a bending moment curve. Matlock (1970) made extremely accurate measurements of bending moment and was able to do the differentiations numerically (Matlock & Ripperger 1958). However, most other investigators have fitted analytical curves through the points of experimental bending moment and have performed the differentiations mathematically.

With families of curves showing the distribution of deflection and soil resistance, *p-y* curves can be plotted. A check may be made of the accuracy of the analyses by using the experimental *p-y* curves to compute bending-moment curves. The computed bending moments should agree closely with those from experiment.

Examples of *p-y* curves that were obtained from a full-scale experiment with pipe piles with a diameter of 641 mm and a penetration of 15.2 m are shown in Figures 1.5 and 1.6

68 Piles under lateral loading

(Reese et al. 1975). The piles were instrumented for measurement of bending moment at close spacing along the length and were tested in overconsolidated clay.

3.5.3 Nondimensional methods for obtaining soil response

Reese & Cox (1968) described a method for obtaining p-y curves for those instances where only pile-head measurements were made during lateral loading. They noted that nondimensional curves can be obtained for many variations of soil reaction modulus with depth. Equations for the soil reaction modulus involving two parameters were employed, such as shown in Equations 3.16 and 3.17.

$$E_{py} = k_1 + k_2 x \tag{3.16}$$

$$E_{py} = k_1 x^n \tag{3.17}$$

(Note: depth below the ground surface is denoted by the symbol z but nondimensional curves have the origin at the top of the pile; hence, the symbol x may be used for depth, as noted in Chapter 2.)

From measurement of pile-head deflection and rotation at the groundline, the two parameters may be computed for a given applied load and moment. With an expression for soil reaction modulus for a particular load, the soil resistance and deflection along the pile may be computed.

The procedure may be repeated for each of the applied loadings and soil resistance. While the method is approximate, the p-y curves computed in this fashion do reflect the measured behavior of the pile head. Soil response derived from a sizable number of such experiments will add significantly to the existing information.

As previously indicated, the major field experiments that have led to the development of the current criteria for p-y curves have involved the acquisition of experimental moment curves. However, nondimensional methods of analyses have assisted in the development of p-y curves in some instances.

3.6 EARLY RECOMMENDATIONS FOR COMPUTING p-y CURVES

The early methods were based on intuition and insight into the problem of the pile under lateral loading. Also, some experimental results may have been available to Terzaghi. McClelland and Focht did have the results from a test but the results were less than complete.

3.6.1 Terzaghi

In a notable paper, and one that is still being used, Terzaghi (1955) discussed a number of important aspects of subgrade reaction, including the resistance of the soil to lateral loading of a pile. Unfortunately, while his numerical recommendations reveal that his knowledge of the problem of the pile was extensive, he failed to give any experimental data or analytical procedure to validate his recommendations.

The units of the quantity in the basic differential equation (Eq. 2.7) are force per unit of length (F/L) and the definition of p is presented graphically in Figure 1.3. Terzaghi (1955) addressed the problem of units by introducing the quantity $1/b$ on the left-hand side of his differential equation. However, the fundamental nature of the problem of the pile under lateral loading was not changed; that is, the required solution of the differential equation required that the quantity p have the units of force per unit length, where force is defined as shown by Figure 1.3c; therefore, Terzaghi's formulation implicitly assumes that $p = qb$.

3.6.1.1 *Stiff clay*

Terzaghi's recommendations for the coefficient of subgrade reaction for piles in stiff clay were based on his notion that the deformational characteristics of stiff clay are 'more or less independent of depth.' Thus, he proposed, in effect, that the p-y curves should be constant with depth and that the ratio between p and y should be defined by the constant a_T. Therefore, his family of p-y curves (though not defined in so specific terms) consists of a series of straight lines, all with the same slope, and passing through the origin of the coordinate system.

Terzaghi recognized, of course, that the pile could not be deflected to an unlimited extent with a linear increase in soil resistance. He stated that the linear relationship between p and y was valid for values of p that were smaller than about one-half of the ultimate bearing stress.

Table 3.1 presents Terzaghi's recommendations for stiff clay. The units have been changed to reflect current practice. The values of a_T are independent of pile diameter, which is consistent with theory for small deflections.

3.6.1.2 *Sand*

Terzaghi's recommendations for sand are similar to those for clay in that his coefficients probably are meant to reflect the slope of secants to p-y curves rather than the initial soil reaction modulus. As noted above, Terzaghi recommended the use of his coefficients up to the point where the computed soil resistance was equal to about one-half of the ultimate bearing stress.

In terms of p-y curves, Terzaghi recommends a series of straight lines with slopes that increase linearly with depth, as indicated in Equation 3.18.

$$E_{py} = k_T z \tag{3.18}$$

where k_T = constant giving variation of soil reaction modulus with depth; and z = depth below ground surface.

Terzaghi's recommended values in terms of the appropriate units are given in Table 3.2.

Table 3.1. Terzaghi's recommendations for soil modulus a_T for laterally loaded piles in stiff clay.

Consistency of clay	Stiff	Very Stiff	Hard
q_u, kPa	100-200	200-400	> 400
a_T, MPa	3.2-6.4	6.4-12.8	12.8 up

Table 3.2. Terzaghi's recommendations for values of k for laterally loaded piles in sand.

Relative density of sand	Loose	Medium	Dense
Dry or moist, k_{py}, MN/m^3	0.95-2.8	3.5-10.9	13.8-27.7
Submerged sand, k_{py}, MN/m^3	0.57-1.7	2.2-7.3	8.7-17.9

3.6.2 McClelland & Focht for clay (1958)

One of the first papers, giving the concept of *p-y* curves, was presented and curves were included from the analysis of the results of a full-scale, instrumented, lateral-load test. The paper showed conclusively that E_{py} is not just a soil property but is a function of pile diameter, deflection, and soil properties.

The paper recommended the performance of consolidated-undrained triaxial test, with confining pressure equal to the overburden pressure, at various depths below the groundline. The soil curves could be converted to *p-y* curves, point by point, by use of the following equations.

$$p = 5.5b\Delta\sigma \tag{3.19}$$

$$y = 0.5b\varepsilon \tag{3.20}$$

where b = diameter of pile; $\Delta\sigma = (\sigma_1 - \sigma_3)$ or deviator stress from the stress-strain curve; and ε = strain from the stress-strain curve. The above equations are similar in form but different in magnitude from those that can be derived from the work of Skempton.

In a discussion of the paper, Reese (1958) pointed out that the ultimate value of *p* is similar to that for bearing capacity well below the ground surface. Limit-analysis was used to derive an equation for clays for the ultimate resistance of p at and near the ground surface. The influence of the ground surface will be implemented in the recommendations for *p-y* curves given later.

3.7 *p-y* CURVES FOR CLAYS

Three sets of recommendations are presented for obtaining *p-y* curves for clay. All are based on the analysis of the results of full-scale experiment with instrumented piles. A comprehensive soil investigation was performed at each site and the best estimate of the undrained shear strength of the clay was found. The dimensions and stiffness of the piles were determined accurately. Experimental *p-y* curves were obtained by one or more of the techniques given earlier. Theory was used to the fullest extent and analytical expressions were developed for *p-y* curves which, when used in a computer solution, yielded curves of deflection and bending moment versus depth that agreed well with the experimental values.

Loading in all three cases were both short-term (static) and cyclic. The *p-y* curves that result from the two tests performed with water above the ground surface have been used extensively in the design of offshore platforms.

3.7.1 Selection of stiffness of clay

A review of the recommendations for *p-y* curves for clay soils reveals that the stiffness of the curves is dependent on the value of ε_{50}, the strain corresponding to one-half the compressive strength of the specimen of clay. The sketch in Figure 3.11a depicts an undrained compressive test of a specimen of saturated clay where strain ε is defined as $\Delta h/h$. Typical stress-strain curves are plotted in Figure 3.11b for both overconsolidated (O.C.) clay and normally consolidated (N.C.) clay.

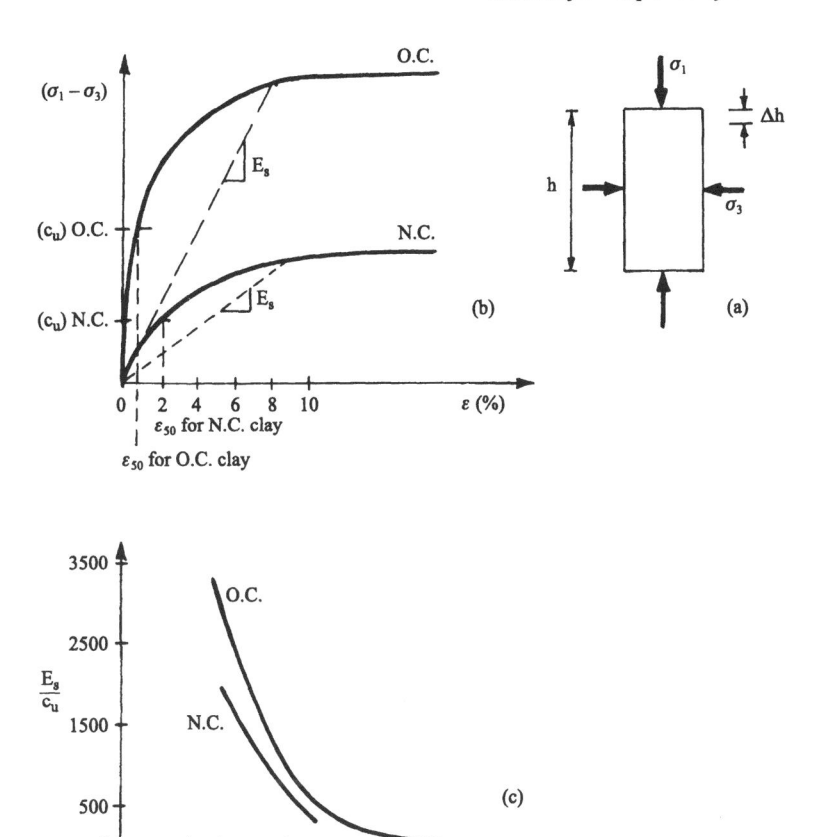

Figure 3.11. Examples of undrained stiffness to undrained shear strength for clays with low plasticity.

The undrained shear strength of the clay, c_u, is indicated for both of the tests as one-half of the maximum compressive strength, $(\sigma_1 - \sigma_3)$. The characteristic strain, ε_{50}, is the strain corresponding to c_u and is indicated in Figure 3.11b. Also shown in Figure 3.11b are secants drawn to the stress-strain curves, with the slopes of the secants defined as E_s. Values strain ε, in percent, are plotted in Figure 3.11b as a function of E_s divided by c_u. The value of c_u for a particular test is a constant; therefore, when E_s is large at the beginning of a stress-strain curve, the value of ε is small. The sharp increase in strain ε as E_s becomes smaller is evident in the figure. The decrease in the parameter E_s/c_u with increasing values of strain ε is shown in Figure 3.11c, where experimental plots are given for both normally consolidated and over-consolidated clay. Because c_u is a constant in a particular case, the curves reflect the decay in E_s.

Analytical studies presented earlier revealed the relevance of the stress-strain curve to p-y curves and ε_{50} was selected as the single parameter to characterize the stress-strain curve. The value appears in the following recommendations for formulation of p-y curves, and should be found from laboratory data when possible. As shown in the recommendations given below, and as would be expected, the *p-y* curves are closer to the *p*-axis for smaller values of ε_{50}. Thus, computed values of pile deflection will be smaller for smaller values of ε_{50}, especially at relatively lighter loads. For the relatively heavier loads, the values of the ultimate soil resistance, based on values of c_u, are controlling the results so that the computed value of ultimate bending moment is affected not at all or only slightly by the selection of ε_{50}. Therefore, the selection of the value of ε_{50} has not been considered a matter of much importance if the engineer is mainly interested in the computation of the maximum bending moment, as is the case for the design of an offshore platform. However, the reverse is true if the engineer is mainly interested in the computation of the pile deflection, especially at relatively small loads.

If triaxial tests are unavailable on the clay for a particular site, there is some guidance in the literature on the selection of values of ε_{50}. Skempton (1951) presented a study of the settlement of footings on clay and noted that E_s/c_u, corresponding to the results from testing a number of footings ranged from 50 to 200. Because $E_s = c_u/\varepsilon_{50}$ as used by Skempton, his values of ε_{50} range from 0.02 to 0.005. Skempton's plot of the settlement of footings from various experiments showed, of course, that the settlement was less for E_s/c_u in the range of 200 (ε_{50} of 0.005) than when E_s/c_u was in the range of 50 (ε_{50} of 0.02). These values of ε_{50} are consistent with the experimental results shown in Figure 3.11c.

Matlock (1970) performed experiments in soft clays and suggested that 'values of ε_{50} may be assumed between 0.005 and 0.20, the smaller value being applicable to brittle or sensitive clays and the larger to disturbed or remolded soil or unconsolidated sediments. An intermediate values of 0.01 is probably satisfactory for most purposes.' Reese et al. (1975) performed experiments on piles in stiff clays and recommended somewhat smaller values of ε_{50} than the values suggested by Matlock. Based on experimental results and the work of Skempton (1951), values of ε_{50} are recommended for normally consolidated clays, as shown in Table 3.3, and for overconsolidated clays, as shown in Table 3.5.

In summary, the stiffness of clay is an important parameter and should be determined specifically by laboratory tests or by appropriate in situ tests for each site. In the absence of specific data, values of ε_{50} may be obtained from Tables 3.3 and 3.5 from values of undrained shear strength. Proper selection of values of ε_{50} is important when computing deflection under lateral loading, especially for relatively small values of load, and less important in computing the value of the maximum bending moment.

3.7.2 *Response of soft clay in the presence of free water*

3.7.2.1 *Detailed procedure*
The procedure described below is for short-term loading, for cyclic loading, and for after-cyclic loading as illustrated by Figures 3.12a, 3.12b, and 3.12c (Matlock 1970).

Table 3.3. Representative values of ε_{50} for normally consolidated clays.

Consistency of clay	Average value of kPa[*]	ε_{50}
Soft	<48	0.020
Medium	48-96	0.010
Stiff	96-192	0.005

[*]Peck et al. 1974, p. 20.

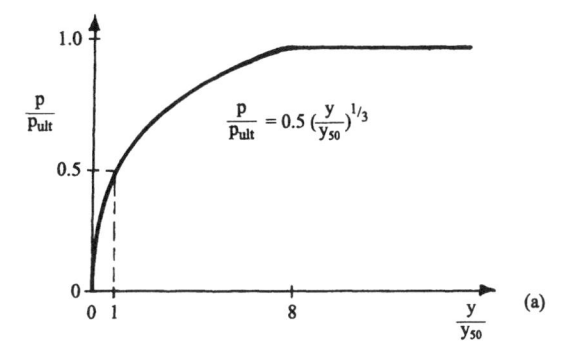

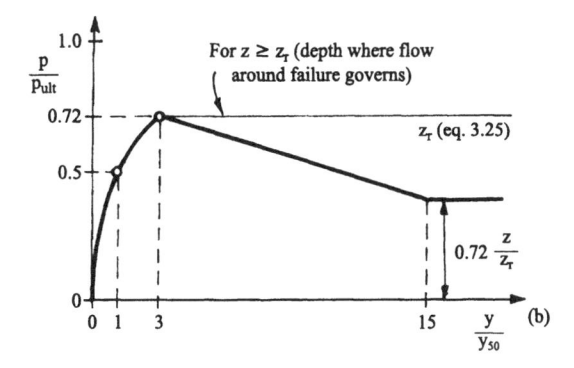

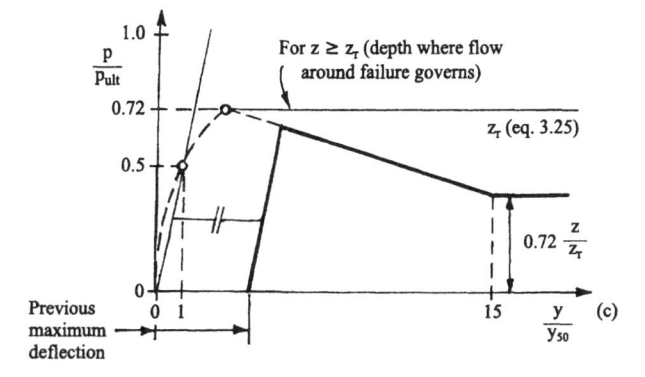

Figure 3.12. Characteristic shapes of p-y curves for soft clay in the presence of free water, (a) static loading; (b) cyclic loading; (c) after cyclic loading (after Matlock 1970).

The procedure for after-cyclic loading is shown in Figure 3.12c. The value of p is zero from the origin to the point of the previous maximum deflection and then the second branch of the curve rises to intersect with the curve for cyclic loading. As shown in the sketch, the slope of the second branch is parallel to a secant through the early part of the static curve. The ability to formulate p-y curves for after-cyclic loading is important for a structure, such as an offshore platform, that has undergone a storm. The stiffness of the

74 Piles under lateral loading

foundation would be reduced and the structure would be more likely to vibrate under loads from wind or from some machines.

The tests from which the criteria were developed were done at sites where the clay had been submerged for some time and where the clay was only slightly overconsolidated. The clay extended to several diameters below the ground surface. The data shown with the case studies for Lake Austin and Sabine in Chapter 7 provided the basis for the formulation of the *p-y* curves.

A later section will present a method of formulating *p-y* curves for layered soils, taking into account the interaction between layers with varying values of p_{ult}. However, special attention must be given to thin layers at the ground surface with water and cyclic loading. For example, if the profile consists of a layer of sand over clay, the engineer would need to use discretion in formulating the *p-y* curves for cyclic loading for the soft clay or for stiff clay. Undoubtedly, *p-y* curves for cyclic loading of clays below the water surface are affected considerably by the kind of soil at the ground surface and above the clay. Future experimental data undoubtedly will provide valuable guidance in formulating *p-y* curves for cyclic loading for clays below water.

1. Obtain the best possible estimate of the variation of undrained shear strength c and submerged unit weight with depth. Also obtain the value of e_{50}, the strain corresponding to one-half the maximum principal stress difference. If no stress-strain curves are available, typical values of ε_{50} are given in the Table 3.3.
2. Compute the ultimate soil resistance per unit length of pile, using the smaller of the values given by the equations below.

$$p_{ult} = \left[3 + \frac{\gamma'}{c_u}z + \frac{J}{b}z\right]c_u b \qquad (3.21)$$

$$p_{ult} = 9c_u b \qquad (3.22)$$

where γ' = average effective unit weight from ground surface to *p-y* curve; z = depth from the ground surface to *p-y* curve; c_u = shear strength at depth z; and b = width of pile.

Matlock (1970) stated that the value of *J* was determined experimentally to be 0.5 for a soft clay and about 0.25 for a medium clay. A value of 0.5 is frequently used for *J*. The value of p_{ult} is computed at each depth where a *p-y* curve is desired, based on shear strength at that depth.

3. Compute the deflection, y_{50}, at one-half the ultimate soil resistance from the following equation:

$$y_{50} = 2.5\varepsilon_{50}b \qquad (3.23)$$

4. Points describing the *p-y* curve are now computed from the following relationship.

$$\frac{p}{p_{ult}} = 0.5\left(\frac{y}{y_{50}}\right)^{1/3} \qquad (3.24)$$

The value of *p* remains constant beyond $y = 8y_{50}$. The following procedure is for cyclic loading and is illustrated in Figure 3.12b.

1. Construct the p-y curve in the same manner as for short-term static loading for values of p less than $0.72p_u$.
2. Solve Equations 3.21 and 3.22 simultaneously to find the depth, z_r, where the transition occurs. If the unit weight and shear strength are constant in the upper zone, then

$$z_r = \frac{6c_u b}{(\gamma' b + Jc_u)} \tag{3.25}$$

If the unit weight and shear strength vary with depth, the value of z_r should be computed with the soil properties at the depth where the p-y curve is desired.
3. If the depth to the p-y curve is greater than or equal to z_r, then p is equal to $0.72p_{ult}$ for all values of y greater than $3y_{50}$.
4. If the depth of the p-y curve is less than z_r, then the value of p decreases from the $0.72p_{ult}$ at $y = 3y_{50}$ to the value given by the following expression at $y = 15y_{50}$.

$$p = 0.72 p_{ult} \left(\frac{z}{z_r}\right) \tag{3.26}$$

The value of p remains constant beyond $y = 15y_{50}$.

3.7.2.2 Recommended soil tests
For determining the various shear strengths of the soil required in the p-y construction, Matlock (1970) recommended the following tests in order of preference.
1. In-situ vane-shear tests with parallel sampling for soil identification,
2. Unconsolidated-undrained triaxial compression tests having a confining stress equal to the overburden pressure with c being defined as one-half the total maximum principal-stress difference,
3. Miniature vane tests of samples in tubes, and
4. Unconfined compression tests.
5. Unit weight determination.

3.7.2.3 Example curves
An example set of p-y curves was computed for soft clay for a pile with a diameter of 610 mm. The soil profile that was used is shown in Figure 3.13. The submerged unit weight was assumed to be 6.3 kN/m³. In the absence of a stress-strain curve for the soil, ε_{50} was taken as 0.020 for the full depth of the soil profile. The loading was assumed to be static. The p-y curves were computed for the following depths below the mudline: 1.5, 3, 6, and 12 m. The plotted curves are shown in Figure 3.14.

3.7.3 Response of stiff clay in the presence of free water

3.7.3.1 Detailed procedure
The following procedure is for short-term static loading and is illustrated by Figure 3.15 (Reese et al. 1975). As may be seen from a study of the p-y curves that are recommended for cyclic loading, the results for the Manor site showed a very large loss of soil resistance. The data from the tests (see case study for Manor tests in Chapter 7) have been

76 Piles under lateral loading

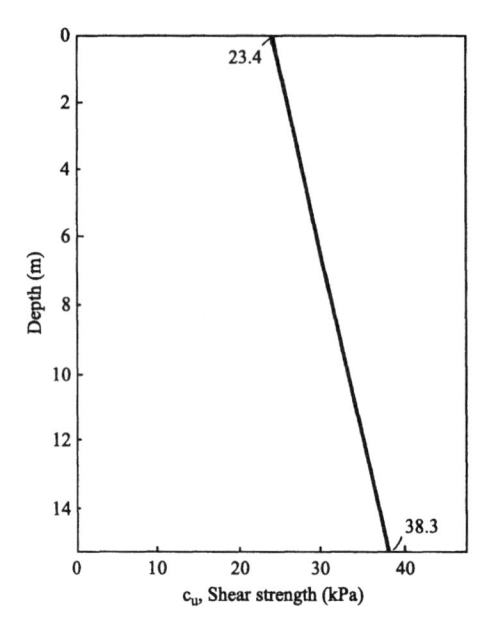

Figure 3.13. Shear strength profile for example p-y curves for soft clay.

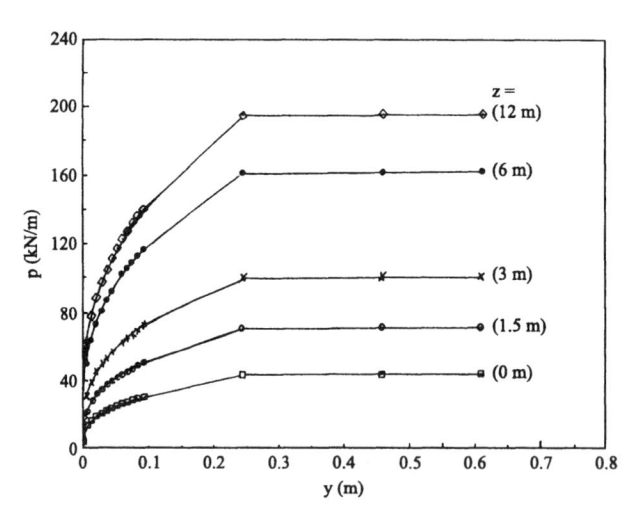

Figure 3.14. Example p-y curves for soft clay with presence of free water.

studied carefully and the recommended p-y curves for cyclic loading reflect accurately the behavior of the soil at the site. Nevertheless, the loss of resistance due to cyclic loading at Manor is much more than has been observed elsewhere, probably because the Manor soil was expansive and continued to imbibe water as cycling progressed. Therefore, the use of the recommendations in this section for cyclic loading will yield conservative results for many clays. The work of Long (1984) was unable to show precisely why the loss of resistance during cyclic loading occurred. One clue was that the clay from Manor was found to lose volume by slaking when a specimen was placed in fresh water; thus, the clay was

quite susceptible to erosion from the hydraulic action of the free water as the pile was pushed back and forth.
1. Obtain values of undrained shear strength c_u, soil submerged unit weight γ', and pile diameter b.
2. Compute the average undrained shear strength c_a over the depth z.
3. Compute the ultimate soil resistance per unit length of pile using the smaller of the values given by the equations below:

$$p_{ct} = 2c_a b + \gamma' bz + 2.83 c_a z \tag{3.27}$$

$$p_{cd} = 11 c_u b \tag{3.28}$$

4. Choose the appropriate value of A_s (static case) from Figure 3.16 for the particular non-dimensional depth.
5. Establish the initial straight-line portion of the p-y curve,

$$p = (k_s z)y \tag{3.29}$$

Use the appropriate value of k_s from Table 3.4.
6. Compute the following:

$$y_{50} = \varepsilon_{50} b \tag{3.30}$$

Use an appropriate value of ε_{50} from results of laboratory tests or, in the absence of laboratory tests, from Table 3.5.
7. Establish the first parabolic portion of the p-y curve, using the following equation and obtaining p_c from Equations 3.27 or 3.28.

$$p = 0.5 p_c \left(\frac{y}{y_{50}} \right)^{0.5} \tag{3.31}$$

Table 3.4. Representative values of k_{py} for overconsolidated clays.

	Average undrained shear strength kPa		
	50-100	100-200	300-400
k_{pys}(static) MN/m³	135	270	540
k_{pyc} (cyclic) MN/m³	55	110	540

*The average shear strength should be computed from the shear strength of the soil to a depth of 5 pile diameters. It should be defined as half the total maximum principal stress difference in an unconsolidated undrained triaxial test.

Table 3.5. Representative values of ε_{50} for overconsolidated clays.

	Average undrained shear strength kPa		
	50-100	100-200	300-400
ε_{50}	0.007	0.005	0.004

78 Piles under lateral loading

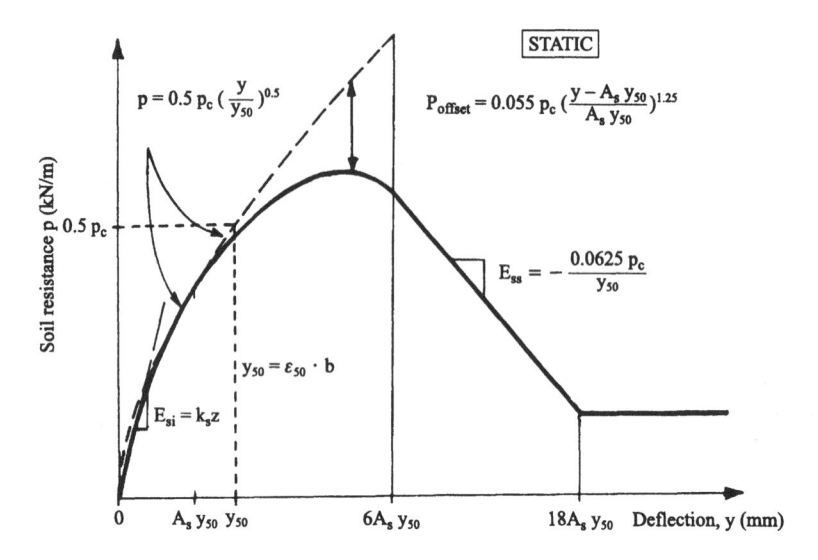

Figure 3.15. Characteristic shape of p-y curves for static loading in stiff clay in the presence of free water.

Equation 3.31 should define the portion of the *p-y* curve from the point of the intersection with Equation 3.29 to a point where *y* is equal to $A_s y_{50}$ (see note in Step 10).

8. Establish the second parabolic portion of the *p-y* curve,

$$p = 0.5 p_c \left(\frac{y}{y_{50}}\right)^{0.5} - 0.055 p_c \left(\frac{y - A_s y_{50}}{A_s y_{50}}\right)^{1.25} \tag{3.32}$$

Equation 3.32 should define the portion of the *p-y* curve from the point where *y* is equal to $A_s y_{50}$ to a point where *y* is equal to $6 A_s y_{50}$ (see note in Step 10).

9. Establish the next straight-line portion of the *p-y* curve,

$$p = 0.5 p_c (6 A_s)^{0.5} - 0.411 p_c - \frac{0.0625}{y_{50}} p_c (y - 6 A_s y_{50}) \tag{3.33}$$

Equation 3.33 should define the portion of the *p-y* curve from the point where *y* is equal to $6 A_s y_{50}$ to a point where *y* is equal to $18 A_s y_{50}$ (see note in Step 10).

10. Establish the final straight-line portion of the *p-y* curve,

$$p = 0.5 p_c (6 A_s)^{0.5} - 0.411 p_c - 0.75 p_c A_s, \tag{3.34}$$

or

$$p = p_c (1.225 \sqrt{A_s} - 0.75 A_s - 0.411) \tag{3.35}$$

Equation 3.34 should define the portion of the *p-y* curve from the point where *y* is equal to $18 A_s y_{50}$ and for all larger values of *y* (see following note). Note: The step-by-step proce-

Models for response of soil and weak rock 79

dure is outlined, and Figure 3.15 is drawn, as if there is an intersection between Equation 3.29 and 3.31. However, there may be no intersection of Equation 3.29 with any of the other equations defining the *p-y* curve. Equation 3.29 defines the *p-y* curve until it intersects with one of the other equations or, if no intersection occurs, Equation 3.29 defines the complete *p-y* curve.

The following procedure is for cyclic loading and is illustrated in Figure 3.17.
1. Steps 1,2,3,5, and 6 are the same as for the static case.
4 Choose the appropriate value of A_c from Figure 3.16 for the particular non-dimensional depth. Compute the following:

$$y_p = 4.1 A_s y_{50} \tag{3.36}$$

7. Establish the parabolic portion of the *p-y* curve,

$$p = A_c p_c \left[1 - \left(\frac{y - 0.45 y_p}{0.45 y_p} \right)^{2.5} \right] \tag{3.37}$$

Equation 3.38 should define the portion of the *p-y* curve from the point of the intersection with Equation 3.29 to where *y* is equal to $0.6y_p$ (see note in step 9).
8. Establish the next straight-line portion of the *p-y* curve,

$$p = 0.936 A_c p_c - \frac{0.085}{y_{50}} p_c (y - 0.6 y_p) \tag{3.38}$$

Equation 3.38 should define the portion of the *p-y* curve from the point where *y* is equal to $0.6y_p$ to the point where *y* is equal to $1.8y_p$ (see note on Step 9).
9. Establish the final straight-line portion of the *p-y* curve,

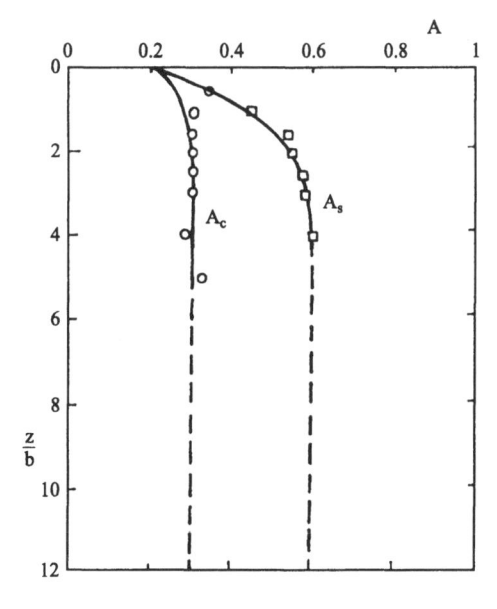

Figure 3.16. Values of constants A_s and A_c.

80 *Piles under lateral loading*

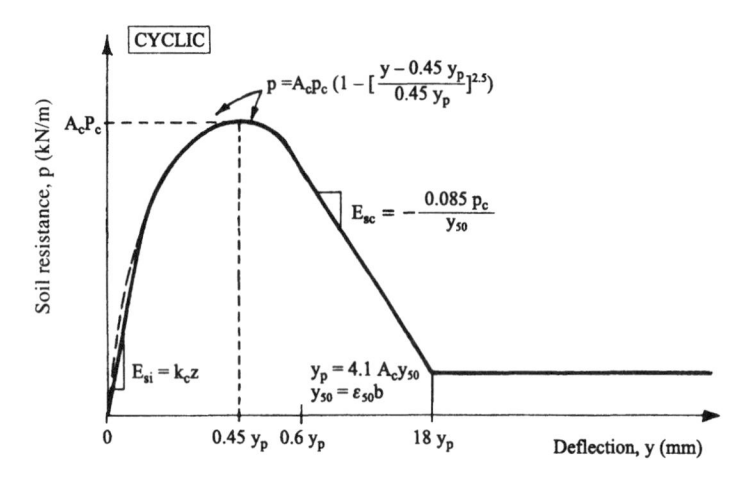

```
                              CYCLIC
                                              y - 0.45 y_p
                         p = A_c p_c (1 - [ ─────────── ]^2.5)
                                                0.45 y_p
    A_c P_c ─────

                                              0.085 p_c
                                    E_sc = - ─────────
                                                y_50

    E_si = k_c z              y_p = 4.1 A_c y_50
                              y_50 = ε_50 b

    0         0.45 y_p  0.6 y_p      18 y_p            Deflection, y (mm)
```

Figure 3.17. Characteristic shape of *p-y* curves for cyclic loading in stiff clay in the presence of free water.

$$p = 0.936 A_c p_c - \frac{0.102}{y_{50}} p_c y_p \tag{3.39}$$

Equation 3.39 should define the portion of the *p-y* curve from the point where *y* is equal to $1.8 y_p$ and for all larger values of *y* (see following note).

Note: The step-by-step procedure is outlined, and Figure 3.16 is drawn, as if there is an intersection between Equation 3.29 and Equation 3.37. There may be no intersection of Equation 3.29 with any of the other equations defining the *p-y* curve. If there is no intersection, the equation should be employed that gives the smallest value of *p* for any value of *y*.

3.7.3.2 *Recommended soil tests*
Triaxial compression tests of the unconsolidated-undrained type with confining pressures conforming to in situ pressures are recommended for determining the shear strength of the soil. The value of ε_{50} should be taken as the strain during the test corresponding to the stress equal to one-half the maximum total principal stress difference. The shear strength, *c*, should be interpreted as one-half of the maximum total-stress difference. Values obtained from triaxial tests might be somewhat conservative but would represent more realistic strength values than other tests. The unit weight of the soil must be determined.

3.7.3.3 *Example curves*
An example set of *p-y* curves was computed for stiff clay for a pile with a diameter of 610 mm. The soil profile that was used is shown in Figure 3.18. The submerged unit weight of the soil was assumed to be 7.9 kN/m^3 for the entire depth.

In the absence of a stress-strain curve, ε_{50} was taken as 0.005 for the full depth of the soil profile. The slope of the initial portion of the *p-y* was established by assuming a value of *k* of 135 MN/m^3. The loading was assumed to be static. The *p-y* curves were computed for the following depths below the mudline 0.6, 1.5, 3, and 12 m. The plotted curves are shown in Figure 3.19.

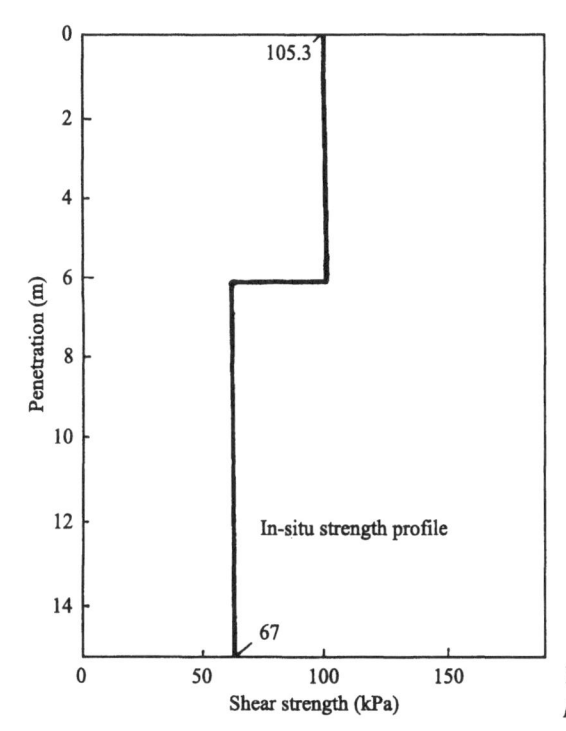

Figure 3.18. Soil profile used for example *p-y* curves for stiff clay.

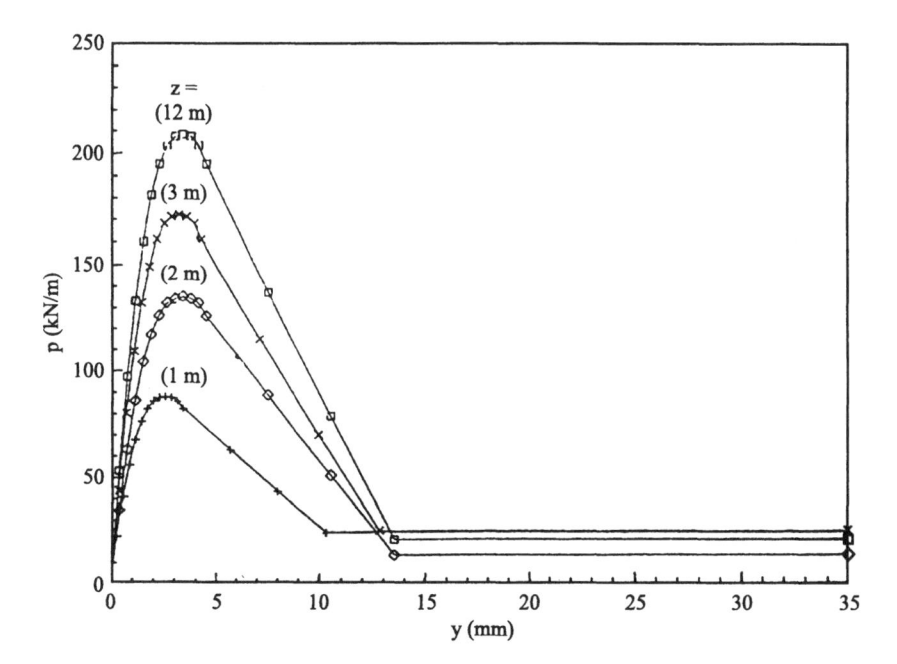

Figure 3.19. Example *p-y* curves for cyclic loading in stiff clay in the presence of free water.

82 Piles under lateral loading

3.7.4 Response of stiff clay with no free water

3.7.4.1 Detailed procedure
The following procedure is for short-term static loading and is illustrated in Figure 3.20 (Welch & Reese 1972). The data shown with the case study for the test at Houston in Chapter 7 provided the basis for the formulation of the *p-y* curves.
1. Obtain values for undrained shear strength c_u, soil unit weight γ, and pile diameter b. Also obtain the values of ε_{50} from stress-strain curves. If no stress-strain curves are available, use a value for ε_{50} of 0.010 or 0.005 as given in Table 3.5, the larger value being more conservative.
2. Compute the ultimate soil resistance per unit length of pile, p_{ult}, using the smaller of the values given by Equations 3.21 and 3.22. (In the use of Equation 3.21 the shear strength is taken as the average from the ground surface to the depth being considered and J is taken as 0.5. The unit weight of the soil should reflect the position of the water table.)
3. Compute the deflection, y_{50}, at one-half the ultimate soil resistance from Equation 3.23.
4. Points describing the *p-y* curve may be computed from the relationship below.

$$\frac{p}{p_{ult}} = 0.5 \left(\frac{y}{y_{50}} \right)^{0.25} \tag{3.40}$$

5. Beyond $y = 16 y_{50}$, p is equal to p_{ult} for all values of y.

The following procedure is for cyclic loading and is illustrated by typical curves in Figure 3.21.
1. Determine the *p-y* curve for short-term static loading by the procedure previously given.
2. Determine the number of times the design lateral load will be applied to the pile.
3. For several values of p/p_{ult} obtain the value of C, the parameter describing the effect of repeated loading on deformation, from a relationship developed by laboratory tests, (Welch & Reese 1972), or in the absence of tests, from the following equation.

$$C = 9.6 \left(\frac{p}{p_{ult}} \right)^4 \tag{3.41}$$

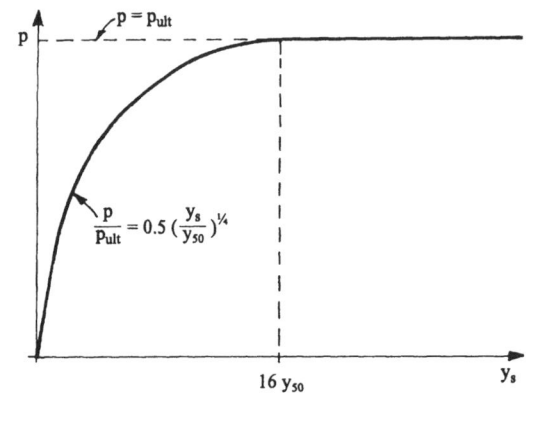

Figure 3.20. Characteristic shape of *p-y* curves for static loading in stiff clay with no free water.

Models for response of soil and weak rock 83

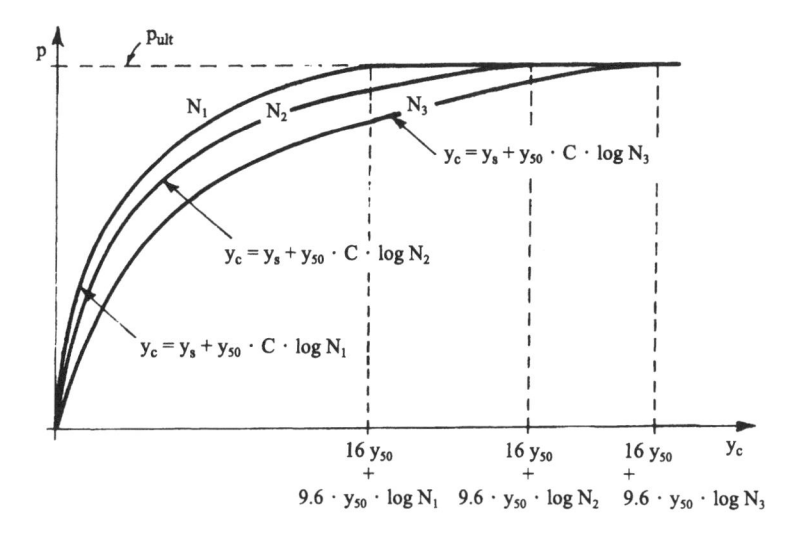

Figure 3.21. Characteristic shape of *p-y* curves for cyclic loading in stiff clay with no free water.

4. At the value of *p* corresponding to the values of p/p_{ult} selected in Step 3, compute new values of *y* for cyclic loading from the following equation.

$$y_c = y_s + y_{50}C \log N \qquad (3.42)$$

where y_c = deflection under *N*-cycles of load; y_s = deflection under short-term static load; and y_{50} = deflection under short-term static load at one-half the ultimate resistance.

5. The *p-y* curve defines the soil response after *N*-cycles of load.

3.7.4.2 Recommended soil tests

Triaxial compression tests of the unconsolidated-undrained type with confining stresses equal to the overburden pressures at the elevations from which the samples were taken are recommended to determine the shear strength. The value of ε_{50} should be taken as the strain during the test corresponding to the stress equal to half the maximum total-principal-stress difference. The undrained shear strength, c_u, should be defined as one-half the maximum total-principal-stress difference. The unit weight of the soil must also be determined.

3.7.4.3 Example curves

An example set of *p-y* curves was computed for stiff clay above the water table for a pile with a diameter of 305 mm. The soil profile that was used is shown in Figure 3.18. The unit weight of the soil was assumed to be 19.0 kN/m³ for the entire depth. In the absence of a stress-strain curve, ε_{50} was taken as 0.005. Equation 3.42 was used to compute values for the parameter *C* and it was assumed that there were to be 100 cycles of load application.

The *p-y* curves were computed for the following depths below the groundline: 0.3, 0.6, 1.2, and 5 m. The plotted curves are shown in Figure 3.22.

84 *Piles under lateral loading*

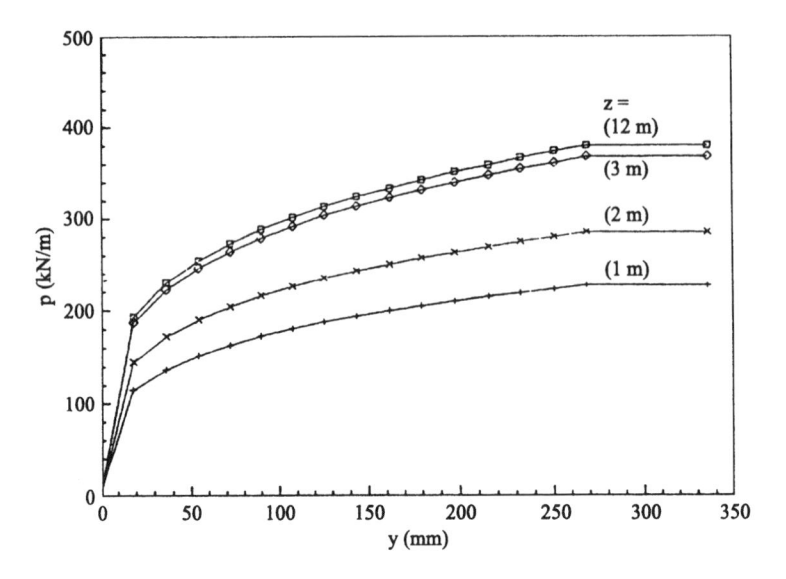

Figure 3.22. Example *p-y* curves for cyclic loading in stiff clay with no free water.

3.8 *p-y* CURVES FOR SANDS ABOVE AND BELOW THE WATER TABLE

3.8.1 *Detailed procedure*

The following procedure is for short-term static loading and for cyclic loading and is illustrated in Figure 3.23 (Reese et al. 1974). The data shown with the case study for the test at Mustang Island in Chapter 7 provided the basis for the formulation of the *p-y* curves.

1. Obtain values for the friction angle ϕ, the soil unit weight γ, and pile diameter b (Note: use buoyant unit weight for sand below the water-table and total unit weight for sand above the water table).
2. Make the following preliminary computations.

$$\alpha = \frac{\phi}{2}; \quad \beta = 45 + \frac{\phi}{2}; \quad K_0 = 0.4; \quad K_a = \tan^2\left(45 - \frac{\phi}{2}\right)$$

3. Compute the ultimate soil resistance per unit length of pile using the smaller of the values given by the following equations (for derivation from analysis of a wedge, see Equations 3.8 and 3.9).

$$p_{st} = \gamma z \left[\frac{K_0 z \tan\phi \sin\beta}{\tan(\beta-\phi)\cos\alpha} + \frac{\tan\beta}{\tan(\beta-\phi)}(b + z\tan\beta\tan\alpha) \right. \\ \left. + K_0 z \tan\beta(\tan\phi\sin\beta - \tan\alpha) - K_a b \right] \quad (3.44)$$

$$p_{sd} = K_a b\gamma z(\tan^8\beta - 1) + K_0 b\gamma z \tan\phi \tan^4\beta \quad (3.45)$$

Models for response of soil and weak rock 85

4. In making the computation in Step 3, find the depth z_t at which there is an intersection at Equations 3.44 and 3.45. Above this depth use Equation 3.44. Below this depth use Equation 3.45.
5. Select a depth at which a *p-y* curve is desired.
6. Establish y_u as $3b/80$. Compute p_{ult} by the following equation:

$$p_{ult} = \overline{A}_s p_s \quad \text{or} \quad p_{ult} = \overline{A}_c p_s \tag{3.46}$$

Use the appropriate value of $\overline{A}_s$ or $\overline{A}_c$ from Figure 3.24 for the particular non-dimensional depth, and for either the static or cyclic case. Use the appropriate equation for p_s, Equation 3.44 or Equation 3.45, by referring to the computation in Step 4.
7. Establish y_m as $b/60$. Compute p_m by the following equation:

$$p_m = B_s p_s \quad \text{or} \quad p_m = B_c p_s \tag{3.47}$$

Use the appropriate value of B_s or B_c from Figure 3.25 for the particular non-dimensional depth, and for either the static or cyclic case. Use the appropriate equation for p_s. The two straight-line portions of the *p-y* curve, beyond the point where y is equal to $b/60$, can now be established.
8. Establish the initial straight-line portion of the *p-y* curve,

$$p = (k_{py}z)y \tag{3.48}$$

Use the appropriate value of k_{py} from Table 3.6 or 3.7.
9. Establish the parabolic section of the *p-y* curve,

$$p = \overline{C} y^{1/n} \tag{3.49}$$

Fit the parabola between points k and m as follows:

a. Get the slope of the line between points m and u by,

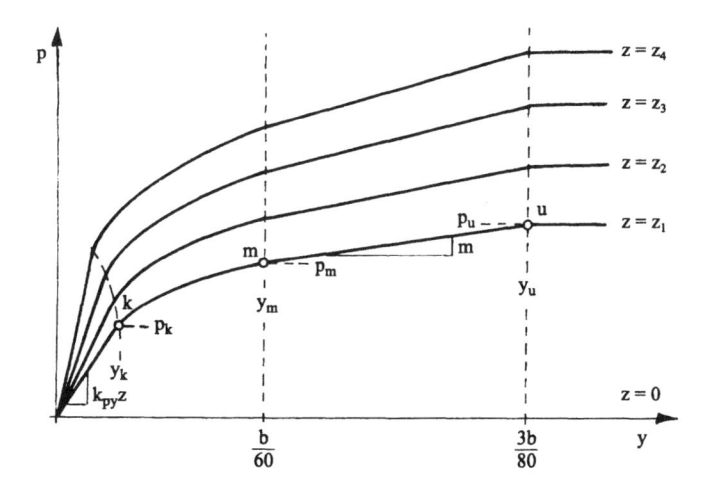

Figure 3.23. Characteristic shape of *p-y* curves for static and cyclic loading in sand.

86 *Piles under lateral loading*

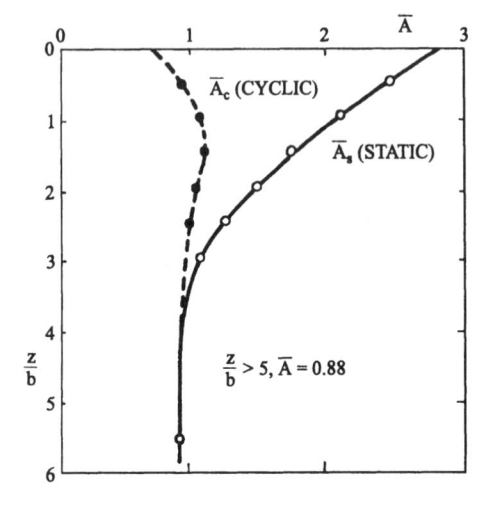

Figure 3.24. Values of coefficients $\bar{A}_c$ and $\bar{A}_s$.

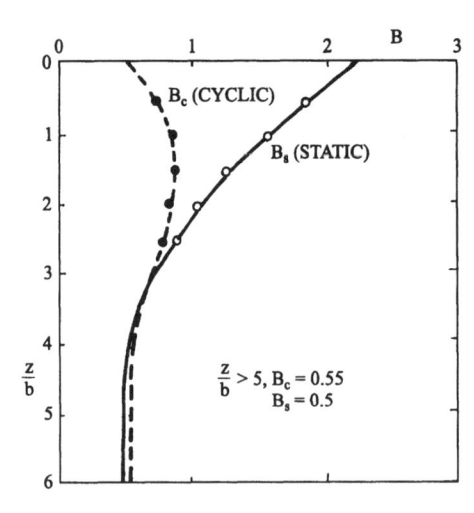

Figure 3.25. Nondimensional coefficient B for soil resistance versus depth.

$$m = \frac{p_u - p_m}{y_u - y_m} \tag{3.50}$$

b. Obtain the power of the parabolic section by,

$$n = \frac{p_m}{my_m} \tag{3.51}$$

c. Obtain the coefficient $\overline{C}$ as follows:

$$\overline{C} = \frac{p_m}{y_m^{1/n}} \tag{3.52}$$

d. Determine point k as,

Table 3.6. Representative values of k_{py} for submerged sand.

Relative density	Loose	Medium	Dense
Recommended k_{py} (MN/m³)	5.4	16.3	34

Table 3.7. Representative values of k_{py} for sand above the water table (Static and Cyclic Loading).

Relative Density	Loose	Medium	Dense
Recommended k_{py} (MN/m³)	6.8	24.4	61

$$y_k = \left(\frac{\bar{C}}{k_{py} z}\right)^{\frac{n}{n-1}} \tag{3.53}$$

e. Compute appropriate number of points on the parabola by using Equation 3.49.

Note: The step-by-step procedure is outlined, and Figure 3.23 is drawn, as if there is an intersection between the initial straight-line portion of the *p-y* curve and the parabolic portion of the curve at point *k*. However, in some instances there may be no intersection with the parabola. Equation 3.48 defines the *p-y* curve until there is an intersection with another branch of the *p-y* curve or if no intersection occurs, Equation 3.48 defines the complete *p-y* curve. This completes the development of the *p-y* curve for the desired depth. Any number of curves can be developed by repeating the above steps for each desired depth.

3.8.2 *Recommended soil tests*

Triaxial compression tests are recommended for obtaining the friction angle of the sand. Confining pressures should be used which are close or equal to those at the depths being considered in the analysis. Tests must be performed to determine the unit weight of the sand. However, it may be impossible to obtain undisturbed samples, and frequently the friction angle is estimated from results of some type of in situ test.

3.8.3 *Example curves*

An example set of *p-y* curves was computed for sand below the water table for a pile with a diameter of 610 mm. The sand is assumed to have a friction angle of 35° and a submerged unit weight of 9.80 kN/m³. The loading was assumed to be static.

The *p-y* curves were computed for the following depths below the mudline: 1.5, 3, 6, and 12 m. The plotted curves are shown in Figure 3.26.

3.9 *p-y* CURVES FOR LAYERED SOILS

There are numerous cases where the soil near the ground surface is not homogeneous but is layered. If the layers are in the zone where the soil would move up and out as a wedge,

88 Piles under lateral loading

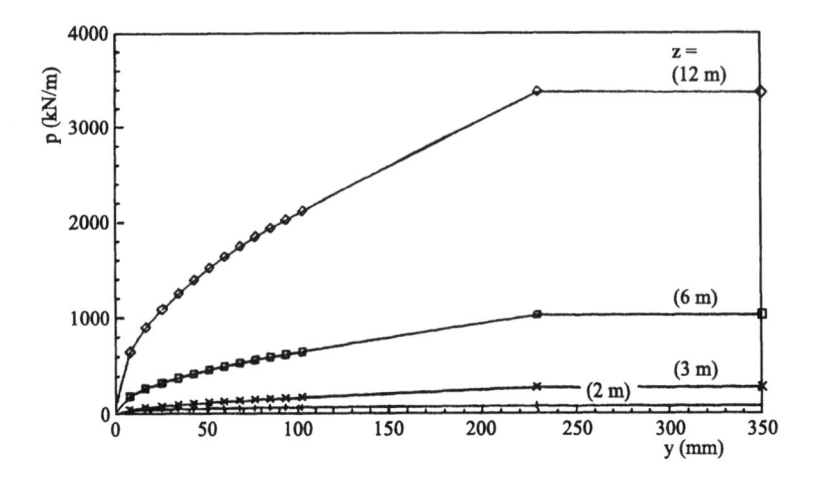

Figure 3.26. Example p-y curves for cyclic loading in sand below the water table.

some modification would plainly be needed in the way to compute the ultimate soil resistance p_u, and consequently modifications would be needed in the p-y curves.

The problem of the layered soil has been given intensive study by Allen (1985); however, Allen's formulations require the use of several computer codes. Integrating the methods of Allen with the methods shown herein must be delayed until a later date when his research can be put in a readily usable form.

3.9.1 Method of Georgiadis

The proposal of Georgiadis (1983) was selected for the purpose of providing a procedure that can be used in analyses. The method of Georgiadis is based on the determination of the 'equivalent' depth of all the layers existing below the upper layer. The p-y curves of the upper layer are determined according to the methods presented herein for homogeneous soils. To compute the p-y curves of the second layer, the equivalent depth H_2 to the top of the second layer has to be determined by summing the ultimate resistances of the upper layer and equating that value to the summation as if the upper layer had been composed of the same material as in the second layer. The values of p_{ult} are computed according to the equations given earlier. Thus, the following two equations are solved simultaneously for H_2.

$$F_1 = \int_0^{H_1} p_{ult1} dH \text{ , and} \tag{3.54}$$

$$F_2 = \int_0^{H_2} p_{ult2} dH \tag{3.55}$$

The equivalent thickness H_2 of the upper layer along with the soil properties of the second layer, are used to compute the p-y curves for the second layer.

The concepts presented above can be used to get the equivalent thickness of two or more dissimilar layers of soil overlying the layer for which the p-y curves are desired.

3.9.2 Example p-y curves

The example problem to demonstrate the manner in which layered soils are modeled is shown in Figure 3.27. As seen in the sketch, a pile with a diameter of 610 mm is embedded in soil consisting of an upper layer of soft clay, overlying a layer of loose sand, which in turn overlays a layer of stiff clay. The water table is at the ground surface and the loading is assumed to be static.

Three p-y curves are shown in Figure 3.28. The curve at a depth of 1.0 m (Curve A) falls in the upper zone of soft clay; the curve for the depth of 3.0 m (Curve B) falls in the sand below the soft clay; and the curve for depths of 5.0 m (Curve C) falls in the lower

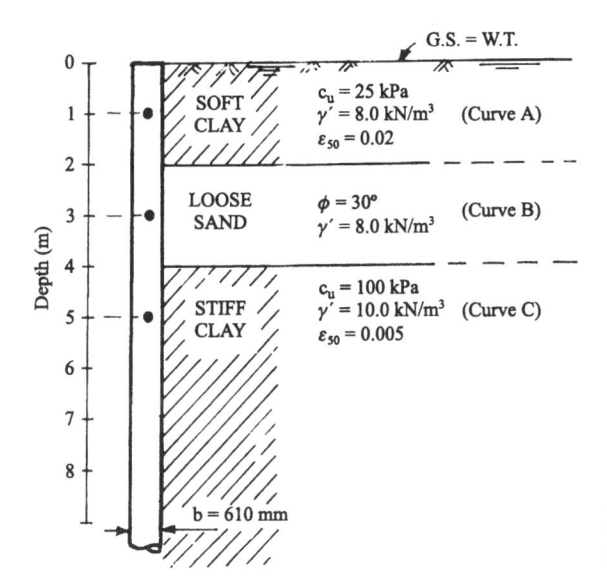

Figure 3.27. Sketch of pile and soil for solution of p-y curves for layered soil.

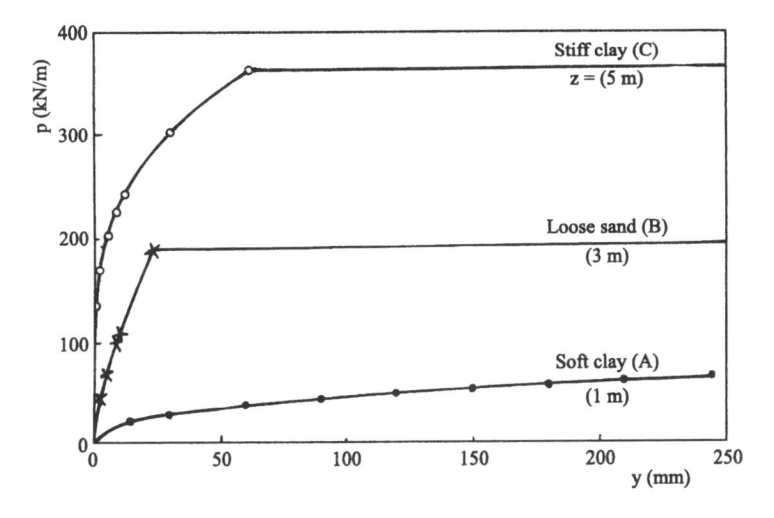

Figure 3.28. Example of p-y curve for layered soil.

zone of stiff clay. Following the procedure suggested by Georgiadis (1983), the p-y curve for soft clay can be computed as if the profile consists altogether of that soil. The value of p_{ult} was computed to be 63.1 kPa and the points on the curve were computed by equations presented earlier.

When dealing with the sand, an equivalent depth of sand is found such that the value of the sum of the ultimate soil resistance for the equivalent sand and the soft clay are equal at the interface. The first step is to employ Equations 3.21 and 3.22 for the values of p_{ult} for the clay and to compute the sum of the values of p_{ult} at the depth of 2 m. Preliminary analyses showed that the value of z_r, the intersection of the two equations, occurred at a depth of 5.26 m; therefore, Equation 3.21 would be used over the full depth of 2 m. The value of p_{ult} varies linearly with depth and values of 45.7 kPa and 80.5 kPa were computed for depths of 0 and 2 m, respectively. The sum of the values of p_{ult} was computed to be 126.2 kN at a depth of 2 m.

The next step is to compute the depth for the sand with the properties shown in Figure 3.27 such that the integral of the computed values of p_{ult} for the sand will attain a value of 126.2 kN. The equations for the values of p_{ult} for sand are highly nonlinear and the integration must be done numerically. Equations 3.44 and 3.45 are employed, along with values of A_s, to compute values as a function of depth. Integration is done numerically. Figure 3.24 was employed and values of A_s are tabulated for ease in computation in Table 3.8. Preliminary analysis showed that the intersection of Equations 3.44 and 3.45 occurred at 8.3 m below the ground surface; therefore, Equation 3.44 was used for all of the computation of values of p_{ult}. Tabulated values of p_{ult} are shown in Table 3.9 for each 0.5 m of depth. Integrating the values in Table 3.9 from a plot showed that the integral of p_{ult} was found at a depth in the sand of 2.35 m. Thus, Curve B (see Figure 3.28) was computed as if the sand existed from the ground surface to a depth of 3.35 m, where p_{ult} is equal to 189.4 kPa. The points on the curve were computed by equations presented earlier.

The next step in the analysis is to find the equivalent depth of stiff clay to equal the integral of the ultimate resistance of the sand to a depth of 4.35 m, a depth which equals the equivalent of 2 m of soft clay and 2 additional m of sand. Integrating the values of p_{ult} in Table 3.9 yields a value of 507.06 kN.

An interesting point about the stiff clay in Figure 3.27 is that if the loading had been cyclic instead of static, even though the stiff clay is below the water table, the clay will act as if the water was not present because of the intervening sand. Initially, the assumption is

Table 3.8. Tabulated values of A_s as a function of z/b.

z/b	$\overline{A}_s$
0	2.88
0.5	2.497
1.0	2.113
1.5	1.73
2.0	1.47
2.5	1.24
3.0	1.05
3.5	0.95
4.0	0.90
4.5	0.88
below 4.5	0.88

Table 3.9. Computed values of p_{ult} for the sand in Figure 3.26 as a function of depth below and assumed ground surface.

Depth	p_{ult}
m	kPa
0	0
0.5	23,26
1.0	47.33
1.5	67.95
2.0	87.20
2.5	115.31
3.0	155.48
3.5	204.95
4.0	261.14
4.5	324.07

made that Equation 3.22 can be used for the computation of p_{ult} for the stiff clay; therefore, a value of p_{ult} at the ground surface was computed to be 183.0 kPa and the linear increase in p_{ult} with depth was found to be 56.1 kPa per m. Integrating the values of p_{ult} for the stiff clay yielded a value of 2.10 m of clay to equate with the 4.35 m of loose sand (equivalent to 2 m of soft clay and 2 m of sand). Therefore, Curve C falls at an equivalent depth of 3.10 m into the stiff clay. The value of p_{ult} was computed to be 360 kPa and the points on the curve were computed by equations presented earlier.

3.10 p-y CURVES FOR SOIL WITH BOTH COHESION AND A FRICTION ANGLE

3.10.1 Background

With regard to soil conditions, previous methods are for soils that can be characterized as either cohesive or cohesionless (clay or sand, for example). There are currently no generally accepted recommendations for developing p-y curves for c-φ soils.

Among the reasons for the limitation on soil characteristics are the following. Firstly, in foundation design, where the p-y analysis has been used mostly, the characterization of the soil by either a value of c_u or φ, but not both, has been used. Secondly, the major experiments on which the p-y predictions have been based have been performed in soils that can be described either by c_u or φ. However, there are now numerous occasions when it is desirable, and perhaps necessary, to describe the characteristics of the soil more carefully.

An example of the need to have predictions for p-y curves for c-φ soils is when piles are employed to improve the stability of a slope. The analysis will generally conform to the following procedures: analyze the slope with no piles present and find a factor of safety against slope failure; select a kind of pile to be used in the slope and find the driving forces above an estimated sliding surface and the resisting forces below the surface; consider a number of piles in a slope; re-evaluate the factor of safety; and modify the position of the sliding surface and the forces on the pile to achieve compatibility. It is well known that most of the currently accepted methods of analysis of slope stability characterize the soils in terms of c_u and φ for long-term or drained analysis. Therefore, it is in-

92 Piles under lateral loading

consistent, and either unsafe or unconservative, to assume the pile to be in soil that is characterized either by c_u or ϕ.

There are other instances in the design of piles under lateral loading where it is desirable to have methods of prediction for p-y curves for c-ϕ soils. The shear strength of unsaturated, cohesive soils generally is represented by strength components of both c and ϕ. In many practical cases, however, there is the likelihood that the deposit might become saturated because of rainfall and rise of the ground water table. But there could well be times when the ability to design for dry seasons is critical.

Cemented soils are frequently found in subsurface investigations. It is apparent that cohesion from the cementation will increase soil resistance significantly, especially for soils near the ground surface.

The strength envelope for consolidated-drained clay is represented by components of both c and ϕ. Therefore, soil criteria for c-ϕ soils may be needed for drained analysis. A complication for such an analysis, that can yield to mechanics, is that there will be some lateral deflection of the pile as drainage occurs. There were no tests of instrumented piles specifically aimed at developing data for the formulation of p-y curves for c-ϕ soils. However, the tests at Kuwait and at Los Angeles described in Chapter 7 allowed development described below.

3.10.2 Recommendations for computing p-y curves

The following procedure is for short-term static loading and for cyclic loading and is illustrated in Figure 3.29. As will be noted, the suggested procedure follows closely that which was recommended earlier for sand and also includes ideas presented by Ismael (1990).

Conceptually, the ultimate soil resistance (p_{ult}) is taken as the passive soil resistance acting on the face of the pile in the direction of the horizontal movement, plus any sliding resistance on the sides of the piles, less any active earth pressure force on the rear face of the pile. The force from active earth pressure and the sliding resistance will generally be small compared to the passive resistance, and will tend to cancel each other out. Evans & Duncan (1982) recommended an approximate equation for the ultimate resistance of c-ϕ soils as:

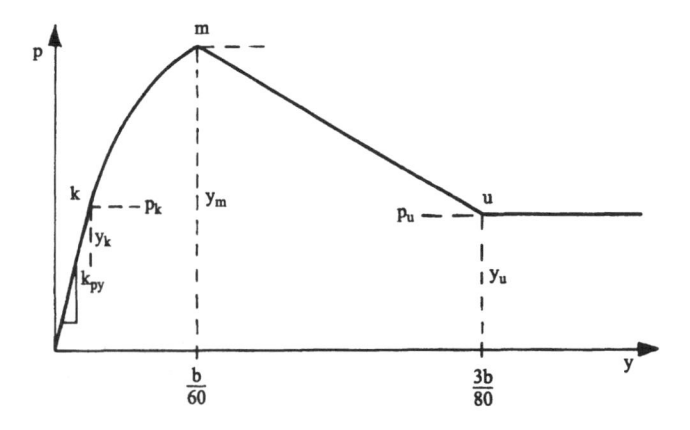

Figure 3.29. Characteristic shape of p-y curves proposed for c-ϕ soil.

$$p = \sigma_p b = C_p \sigma_h b \tag{3.56}$$

where σ_p = passive pressure including the three-dimensional effect of the passive wedge (F/L²); b = pile width (L);

$$\sigma_h = \gamma z \tan^2\left(45 + \frac{\phi}{2}\right) + 2c \tan\left(45 + \frac{\phi}{2}\right) \tag{3.57}$$

the Rankine passive pressure for a wall of infinite length (F/L²); γ = unit weight of soil (F/L³); z = depth at which the passive resistance is considered (L); ϕ = friction angle (degrees); c = cohesion (F/L²); and C_p = dimensionless modifying factor to account for the three-dimensional effect of the passive wedge.

The modifying factor C_p can be divided into two terms: $C_{p\phi}$ to modify the frictional term of Equation 3.56 and C_{pc} to modify the cohesion term of Equation 3.57. Equation 3.58 can then be written as:

$$p_{ult} = \left[C_{p\phi} \gamma z \tan^2\left(45 + \frac{\phi}{2}\right) + C_{pc} c \tan\left(45 + \frac{\phi}{2}\right)\right] b \tag{3.58}$$

The relatively straightforward derivation of equations for developing *p-y* curves for *c-φ* soil proceeds by using the concept proposed by Evans & Duncan (1982). Equation 3.58 will be rewritten as

$$p_{ult} = \overline{A} p_{ult\phi} + p_{ultc} \tag{3.59}$$

where $\overline{A}$ can be found from Figure 3.24. The friction component ($p_{ult\phi}$) will be the smaller of the values given by the equations below.

$$p_{ult\phi} = \gamma z \left[\frac{K_0 \tan\phi \sin\beta}{\tan(\beta - \phi)\cos\alpha} + \frac{\tan\beta}{\tan(\beta - \phi)}(b + z\tan\beta\tan\alpha)\right] \tag{3.60}$$
$$+ \gamma z \left[K_0 z \tan\beta(\tan\phi\sin\beta - \tan\alpha) - K_a b\right]$$

$$p_{ult\phi} = K_a b \gamma z(\tan^8\beta - 1) + K_0 b \gamma z \tan\phi \tan^4\beta \tag{3.61}$$

The cohesion component (p_{ultc}) will be the smaller of the values given by the equations below.

$$p_{ultc} = \left(3 + \frac{\gamma'}{c}z + \frac{J}{b}z\right) cb \tag{3.62}$$

$$p_{ultc} = 9cb \tag{3.63}$$

To develop the p versus y curves, the procedures described earlier for sand by Reese et al. (1974) will be used because the stress-strain behavior of *c-φ* soils is believed to be closer to that of cohesionless soil than of cohesive soil. The following procedures are used to develop the *p-y* curves.

94 *Piles under lateral loading*

1. Establish y_u as $3b/80$. Compute p_{ult} by one of the following equations:

$$p_{ult} = \bar{A}_s p_{ult\phi} + p_{ultc} \text{ or}$$

$$p_{ult} = \bar{A}_c p_{ult\phi} + p_{ultc} \tag{3.64}$$

Use the appropriate value of $\bar{A}_s$ or $\bar{A}_c$ from Figure 3.24 for the particular non-dimensional depth, and for either the static or cyclic case.

2. Establish y_m as $b/60$. Compute p_m by the following equation:

$$p_m = B_s p_s \text{ or } p_m = B_c p_s \tag{3.65}$$

Use the appropriate value of B_s or B_c from Figure 3.25 for the particular non-dimensional depth, and for either the static or cyclic case. Use the appropriate equation for p_s from Equations 3.44 or 3.45. The two straight-line portions of the *p-y* curve, beyond the point where y is equal to $b/60$, can now be established.

3. Establish the initial straight-line portion of the *p-y* curve,

$$p = (k_{py} z) y \tag{3.66}$$

The value of k_{py} for Equation 3.66 may be found from the following equation and by reference to Figures 3.30 and 3.31.

$$k_{py} = k_c + k_\phi \tag{3.67}$$

For example, if c is equal to 20 kPa and ϕ is equal to 35 degrees for a layer of c-ϕ soil above the water table under static loading, the recommended value of k_c for static

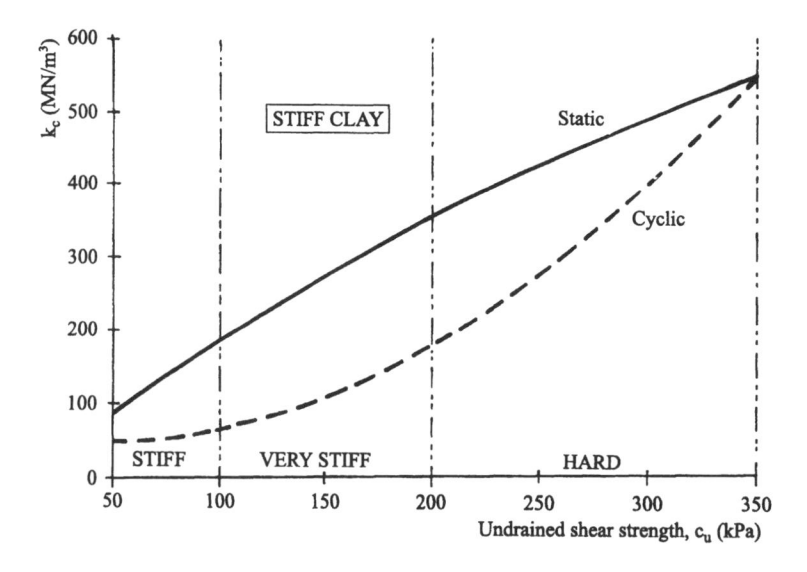

Figure 3.30. Values of k_c for stiff clays.

Models for response of soil and weak rock 95

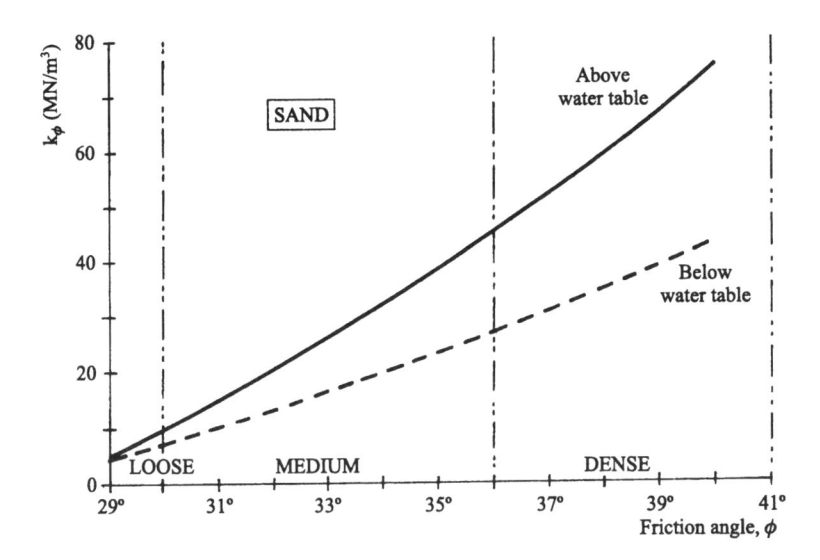

Figure 3.31. Values of k_ϕ for sand.

loading from Figure 3.30 is 90,000 kN/m³ (at beginning of curve). The recommended value of k_ϕ from Figure 3.31 is 38 MN/m³, yielding a value for k_{py} of 128,00 kN/m³.
4. Establish the parabolic section of the *p-y* curve,

$$p = \overline{C} y^{1/n} \qquad (3.68)$$

Fit the parabola between points *k* and *m* as follows:
a. Get the slope of the line between points *m* and *u* by,

$$m = \frac{p_{ult} - p_m}{y_u - y_m} \qquad (3.69)$$

b. Obtain the power of the parabolic section by,

$$n = \frac{p_m}{m y_m} \qquad (3.70)$$

c. Obtain the coefficient $\overline{C}$ as follows:

$$\overline{C} = \frac{p_m}{y_m^{1/n}} \qquad (3.71)$$

d. Determine point y_k as,

$$y_k = \left(\frac{\overline{C}}{kz} \right)^{n/n-1} \qquad (3.72)$$

e. Compute appropriate number of points on the parabola by using Equation 3.68.

96 *Piles under lateral loading*

Note: The step-by-step procedure is outlined as if there is an intersection between the initial straight-line portion of the *p-y* curve and the parabolic portion of the curve at point *k*. However, in some instances there may be no intersection with the parabola. Equation 3.66 defines the *p-y* curve until there is an intersection with another branch of the *p-y* curve or if no intersection occurs, Equation 3.66 defines the complete *p-y* curve. This completes the development of the *p-y* curve for the desired depth. Any number of curves can be developed by repeating the above steps for each desired depth.

3.10.3 *Discussion*

An example of *p-y* curves was computed for *c-*ϕ soils for a pile with a diameter of 0.3 meters. The *c* value is 20 kPa and the ϕ value is 35 degrees. The unit weight of soil is 18 kN/m^3, and the loading is assumed to be static. The *p-y* curves were computed for depths of 1 m, 2 m, and 3 meters. The *p-y* curves computed by using the simplified procedure are shown in Figure 3.32. The *p-y* curves show an initial peak strength, then drop to a residual strength at a large deflection, as is expected.

The point was made clearly at the beginning of this section that data are unavailable from a specific set of experiments that was aimed at the response of *c-*ϕ soils. Such experiments would have made use of fully instrumented piles. Further, little information is available in the literature on the response of piles under lateral loading in such soils where response is given principally by deflection of the pile at the point of loading.

Data from one such experiment, however, was available and the writers have elected to use that data in a case study presented in Chapter 7. A comparison was made there between results from experiment and results from computations.

The reader will note that the procedures presented above does not reflect a severe loss of soil resistance under cyclic loading that is a characteristic for clays below a free-water surface. Rather, the procedures are for a material that is principally granular which does

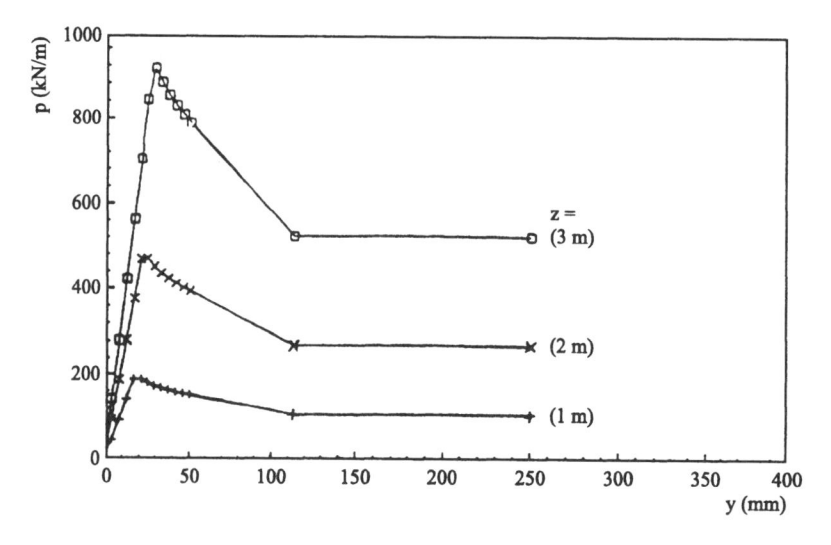

Figure 3.32. Example *p-y* curves for *c-*ϕ soil based on simplified procedure.

not reflect such loss of resistance. Therefore, if a c-ϕ soil has a very low value of ϕ and a relatively large value of c_u the user is advised to ignore the ϕ and to use the recommendations for p-y curves for clay. Further, a relatively large factor of safety is recommended in any case, and a field program of testing of prototype piles is certainly in order for jobs that involve a large number of piles.

3.11 OTHER RECOMMENDATIONS FOR COMPUTING p-y CURVES

3.11.1 *Clay*

As noted earlier in this chapter, the selection of the set of p-y curves for a particular field application is a critical feature of the method of analysis. The presentation of three particular methods for clays does not mean the other recommendations are not worthy of consideration. Some of these methods are mentioned here for consideration and their existence is an indication of the level of activity with regard to the response of soil to lateral deflection.

Sullivan et al. (1980) studied data from tests of piles in clay when water was above the ground surface and proposed a procedure that unified the results from those tests. While the proposed method was able to predict the behavior of the experimental piles with excellent accuracy, two parameters were included in the method that could not be found by any rational procedures. Further work could develop means of determining those two parameters.

Stevens & Audibert (1979) re-examined the available experimental data and suggested specific procedures for formulating p-y curves. Bhushan et al. (1979) reported field tests of bored piles under lateral load and recommended procedures for formulating p-y curves for stiff clays. Briaud et al. (1982) suggested a procedure for use of the pressuremeter in developing p-y curves. A number of other authors have also presented proposals for the use of results of pressuremeter tests is obtaining p-y curves.

O'Neill & Gazioglu (1984) reviewed all of the data that were available on p-y curves for clay and presented a summary report to the American Petroleum Institute. The research conducted by O'Neill and his co-workers (O'Neill & Dunnavant 1984, Dunnavant & O'Neill 1985) at the test site on the campus of the University of Houston developed a large volume of data on p-y curves. This work will most likely result in specific recommendations in due course.

3.11.2 *Sand*

A survey of the available information of p-y curves for sand was made by O'Neill & Murchison (1983) and some changes were suggested in the procedure given above. Their suggestions were submitted to the American Petroleum Institute and modifications were adopted by the API review committee.

Bhushan et al. (1981) reported on lateral load tests of bored piles in sand. A procedure for predicting p-y curves was suggested.

98 *Piles under lateral loading*

3.12 MODIFICATIONS TO *p-y* CURVES FOR SLOPING GROUND

3.12.1 *Introduction*

The recommendations for *p-y* curves presented to this point are developed for a horizontal ground surface. In order to allow designs to be made if a pile is installed on a slope, modifications must be made in the *p-y* curves. The modifications involve revisions in the manner in which the ultimate soil resistance is computed. In this regard, the assumption is made that the flow-around failure will not be influenced by sloping ground; therefore, only the equations for the wedge-type failure need modification.

The solutions presented herein are entirely analytical and must be considered as preliminary. Additional modifications may be indicated if it is possible to implement an extensive laboratory and field study.

3.12.2 *Equations for ultimate resistance in clay*

The ultimate soil resistance near the ground surface for saturated clay where the pile was installed in ground with a horizontal slope was derived by Reese (1958) and is shown in Equation 3.73.

$$(p_{ult})_{ca} = 2c_a b + \gamma b H + 2.83 c_a H \tag{3.73}$$

If the ground surface has a slope angle θ as shown in Figure 3.33, the soil resistance in the front of the pile, following the Reese approach is:

$$(p_{ult})_{ca} = (2c_a b + \gamma b H + 2.83 c_a H) \frac{1}{1 + \tan \theta} \tag{3.74}$$

The soil resistance in the back of the pile is:

$$(p_{ult})_{ca} = (2c_a b + \gamma b H + 2.83 c_a H) \frac{\cos \theta}{\sqrt{2} \cos(45° + \theta)} \tag{3.75}$$

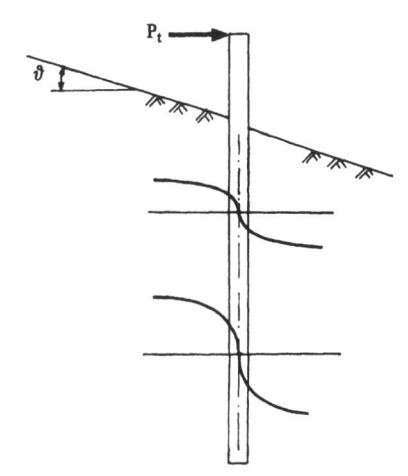

Figure 3.33. Sketch of loaded pile in sloping ground along with set of conceptual *p-y* curves.

where $(p_{ult})_{ca}$ = ultimate soil resistance near ground surface; c_a = average undrained shear strength; b = pile diameter; γ = average unit weight of soil; H = depth from ground surface to point along pile where soil resistance is computed; and θ = angle of slope as measured from the horizontal. The quantity $(p_{ult})_{ca}$ is taken as the passive soil resistance acting on the face of the pile.

A comparison of Equations 3.74 and 3.75 shows that the equations are identical except for the terms at the right side of the parenthesis. If θ is equal to zero, the equations become equal to the original equation.

3.12.3 Equations for ultimate resistance in sand

The ultimate soil resistance near the ground surface for sand where the pile was installed in ground with a horizontal slope was derived earlier and is:

$$(p_{ult})_{sa} = \gamma H \left[\begin{array}{c} \dfrac{K_0 H \tan\phi \sin\beta}{\tan(\beta-\phi)\cos\alpha} + \dfrac{\tan\beta}{\tan(\beta-\phi)}(b + H\tan\beta\tan\alpha) \\ + K_0 H \tan\beta(\tan\phi\sin\beta - \tan\alpha) - K_a b \end{array} \right] \quad (3.76)$$

If the ground surface has a slope angle θ, the ultimate soil resistance in the front of the pile is

$$(p_{ult})_{sa} = \gamma H \left[\left(\dfrac{K_0 H \tan\phi \sin\beta}{\tan(\beta-\phi)\cos\alpha}(4D_1^3 - 3D_1^2 + 1) \right) + \dfrac{\tan\beta}{\tan(\beta-\phi)}\left(bD_2 + H\tan\beta\tan\alpha D_2^2\right) \right]$$

$$+ \gamma H \left[K_0 H \tan\beta(\tan\phi\sin\beta - \tan\alpha)(4D_1^3 - 3D_1^2 + 1) - K_a b \right] \quad (3.77)$$

where

$$D_1 = \dfrac{\tan\beta\tan\theta}{\tan\beta\tan\theta + 1} \quad (3.78)$$

$$D_2 = 1 - D_1, \text{ and} \quad (3.79)$$

$$K_a = \cos\theta \dfrac{\cos\theta - \sqrt{\cos^2\theta - \cos^2\phi}}{\cos\theta + \sqrt{\cos^2\theta - \cos^2\phi}} \quad (3.80)$$

(θ is defined in Figure 3.33)

The ultimate soil resistance in the back of the pile is:

$$(p_{ult})_{sa} = \gamma H \left[\left(\dfrac{K_0 H \tan\phi \sin\beta}{\tan(\beta-\phi)\cos\alpha}(4D_3^3 - 3D_2^2 + 1) \right) + \dfrac{\tan\beta}{\tan(\beta-\varphi)} \right]$$

$$+ \gamma H \left[\left(bD_4 + H\tan\beta\tan\alpha D_4^2\right) + K_0 H \tan\beta(\tan\phi\sin\beta - \tan\alpha) \right] \quad (3.81)$$

$$+ \gamma H \left[\left(4D_3^3 - 3D_3^2 + 1\right) - K_a b \right]$$

where

100 *Piles under lateral loading*

$$D_3 = \frac{\tan\beta \tan\theta}{1-\tan\beta \tan\theta}, \text{ and} \tag{3.82}$$

$$D_4 = 1 + D_3 \tag{3.83}$$

This completes the necessary derivations for modifying the equations for clay and sand to analyze a pile under lateral load in sloping ground.

3.13 EFFECT OF BATTER

The effect of batter on the behavior of laterally loaded piles was investigated in a test tank. The lateral soil-resistance curves for a vertical pile in a horizontal ground surface were modified by a constant to account for the effect of the inclination of the pile. The values of the modifying constant as a function of the batter angle were deduced from the results in the test tank (Awoshika & Reese 1971) and also from the results from full-scale tests (Kubo 1964). The modifier to be used is shown by the solid line in Figure 3.34.

This modifier is to be used to increase or decrease the value of p_{ult}, which in turn will cause each of the p-values to be modified. While it is likely that the values of p_{ult} for the deeper soils is not affected by batter, the behavior of a pile is affected only slightly by the resistance of the deeper soils; therefore, the use of the modifier for all depth is believed to be satisfactory.

As shown in Figure 3.34, the agreement between the empirical curve and the experiments for the out-batter piles (*q* is positive) agrees somewhat better than for the in-batter piles. The data indicate that the use of the modifier will yield results that are somewhat

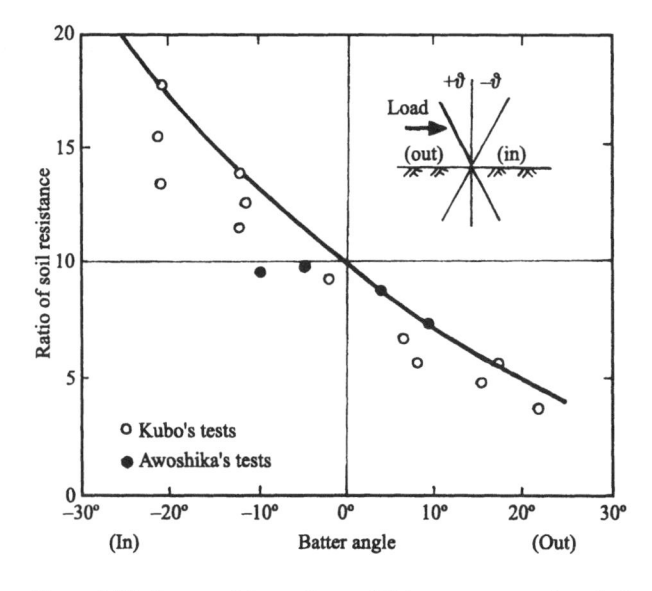

Figure 3.34. Proposed factor for modifying *p-y* curves for raked or battered piles (after Kubo 1964, and Awoshika & Reese 1971).

questionable; therefore, on an important project, the responsible engineer may wish to recommend full-scale testing.

3.14 SHEARING FORCE AT BOTTOM OF PILE

As was indicated in the development of the difference equations, a provision is made to allow a shearing force to be introduced at the bottom of the pile. The shearing force would be applicable only to those cases where the pile is short; that is, where there is only one point of zero deflection.

The equations to use to compute the shearing force, which should be computed as a function of deflection, are currently unavailable. The geotechnical engineer can make an estimate of the necessary force-deflection curve in consideration of pile geometry and soil properties.

Experimental results are available from a number of tests of short piles and the shearing force at the base of the pile can be implied for an analysis of results. Methods will be developed in due course to estimate the curves of V_o versus y_o.

3.15 *p-y* CURVES FOR WEAK ROCK

3.15.1 *Introduction*

Rock has been given little attention in the *p-y* method of analysis. In many practical designs, rock exists under a stratum or strata of overburden of sufficient thickness that the computed deflection of a pile at the rock in so small that the resistance of the rock is small and the response of the pile is little affected, regardless of the stiffness of the rock. However, the combination of rock near or at the surface and a significant magnitude of lateral loading does occur frequently. In such cases, even though the axial load is substantial, lateral loading may dictate penetration because axial capacity from end bearing is usually large.

The theory of elasticity has been used by Kulhawy and his co-workers (Carter & Kulhawy 1988, 1992) with useful results and their ideas have influenced the development shown in the remainder of this section. A serious problem with regard to applying any analytical method to the response of rock is the dominant role played by the secondary structure of rock. The Canadian Foundation Manual (1978) has addressed secondary structure by basing the behavior of rock, in addition to compressive strength, on the spacing and thickness of soil-filed cracks and joints.

The senior writer served on a panel to consider the foundations for a bridge at Nothumberland Strait Crossing, Canada. The panel member included a number of experts on rock mechanics. The water depth was moderate but rock existed at the floor of the Strait. One of the schemes under consideration was to install bored piles into pre-drilled sockets. During winter, the columns, which were to be extensions of the piles, would be subjected to large lateral loads from moving sheets of ice. One of the panel members opined that each of the piles should be proof-tested under lateral load because a soil-filled joint could exist near the surface of the rock. The weak joint would allow a mass of the rock to slide away from the pile under very low lateral resistance. The validity of the comment was ap-

102 *Piles under lateral loading*

parent if, in fact, soil-filled joints existed at the site. The question was not resolved because the planning was abandoned for other reasons.

For the design of piles under lateral loading in rock, special emphasis is necessary in the coring of the rock. Experience has shown that careful attention is required to establish procedures and specifications for the field work. The values for RQD, percent of recovery, and compressive strength can probably be more seriously in error from improper procedures than are the corresponding properties of soil. The recommended procedures that follow are based on results from field tests of piles in rock that differ significantly in character. In neither case, however, was there soil-filled joints. Methods of investigation should reveal such a condition and designers must address the potential weakness of the rock in a site-specific manner. Thus, reliance on the methods presented will be limited. No precise procedure is presented to differentiate a rock from a strong soil. In general, a rock must be sampled by coring rather than by pushing in a thin-walled tube. There are some intermediate materials, of course, and designers may wish to compare analysis performed by the stiff-clay methods and the methods given below.

3.15.2 *Field tests*

Results from two tests of full-scale, bored piles in rock are available for analysis (Reese & Nyman 1978, Roberts 1995). In both cases, data were available on geometry of piles, magnitude and point of application of loads, and characteristics of the rock. Curves showing deflection versus lateral load were reported for both of the tests. Comparisons of the results from analyses, using the procedures described herein, and results from the experiments are presented in Chapter 7.

3.15.3 *Interim recommendations for computing p-y curves for rock*

3.15.3.1 *Concepts*
An analysis of the results from the tests noted above, and a study of other information, formed the basis for the recommendations given here. The recommendations are termed *interim* for a number of reasons, and comments on their appropriate use in analysis and design are given.

The following concepts and procedures establish the framework for the recommendations.
1. The secondary structure of rock; related to joints, cracks, inclusions, fractures, and any other zones of weakness; can strongly influence the behavior of the rock.
2. The *p-y* curves for rock and the bending stiffness $E_p I_p$ for the pile must both reflect nonlinear behavior in order to predict loadings at failure.
3. The initial slope K_{ir} of the *p-y* curves must be predicted because small lateral deflections of piles in rock can result in resistances of large magnitudes. For a given value of compressive strength, K_{ir} is assumed to increase with depth below the ground surface.
4. The reaction modulus of the rock E_{ir}, for correlation with K_{ir}, may be taken from the initial slope of a pressuremeter curve. Alternatively, the results of compressive tests of intact specimens may be used to obtain values of E_{ir}. The data in Figure 3.35 (Horvath & Kenny 1979, Peck 1976, and Deere 1968) may be useful but, as may be seen, the E_{ir}

values for samples of the same type of rock may vary by several orders of magnitude. Therefore, Figure 3.35 can be expected to yield only approximate correlations between compressive strength and reaction modulus. Figure 3.36 (Bieniawski 1984) shows a correlation between E_{mass}/E_{core} and RQD. Values of E_{mass} may be estimated if tests have been performed of cored specimens (E_{mass} and E_{ir} are assumed to be equivalent). Again, scatter is significant. The reaction modulus for the mass of rock is assumed to be implemented in the expressions that follow.

5. The ultimate resistance p_{ur} for the p-y curves will rarely, if ever, be developed in practice, but the prediction of p_{ur} is necessary in order to reflect nonlinear behavior.
6. The component of the strength of rock from unit weight is considered to be small in comparison to that from compressive strength q_{ur}; therefore, unit weight is ignored. Furthermore, no influence from the differing strength of layers is considered.
7. The compressive strength of the rock q_{ur} for computing a value of p_{ur} may be obtained from tests of intact specimens.
8. The assumption is made that fracturing will occur at the surface of the rock under small deflections, therefore, the compressive strength of intact specimens is reduced by multi-

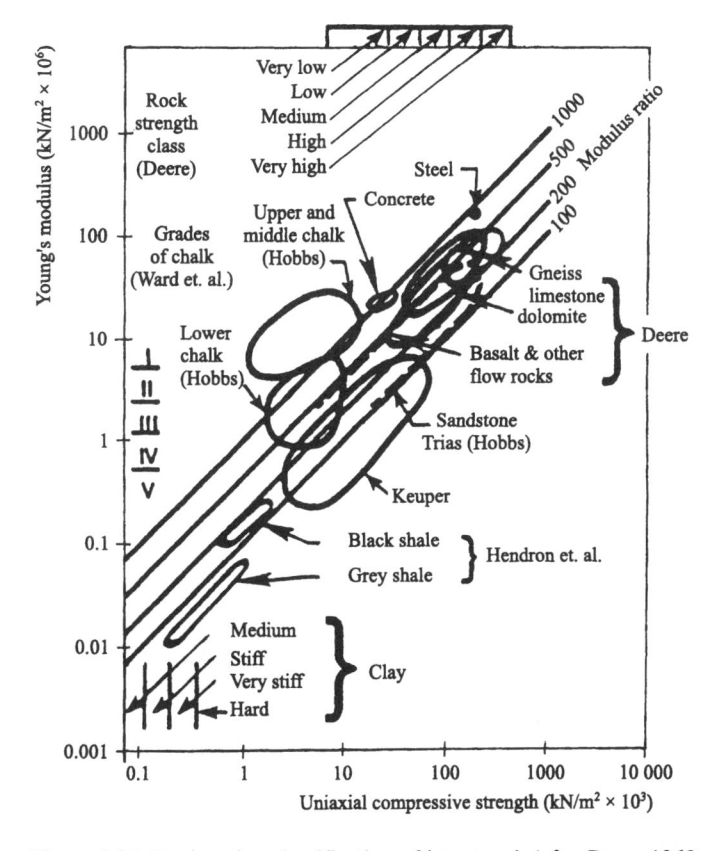

Figure 3.35. Engineering classification of intact rock (after Deere 1968 and Peck 1976 as presented by Horvath & Kenny 1979).

104 *Piles under lateral loading*

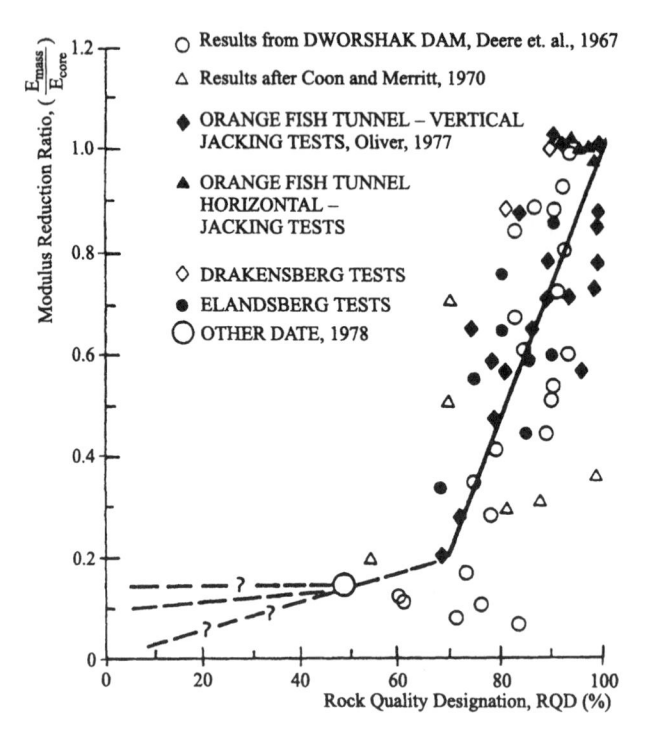

Figure 3.36. Molulus-reduction ratio for rock as a function of RQD (after Bieniawski 1984).

plication by α_r to account for the fracturing. The value of α_r is assumed to be 1/3rd for RQD of 100 and to increase linearly to unity at RQD of zero. If RQD is zero, the compressive strength may be obtained directly from a pressuremeter curve, or approximately from Figure 3.35, by entering with the value of the pressuremeter modulus.

3.15.3.2 *Ultimate resistance of the rock*
The following expression for the ultimate resistance p_{ur} for rock is based on limit equilibrium and reflects the influence of the surface of the rock.

$$p_{ur} = \alpha_r q_{ur} b \left(1 + 1.4 \frac{x_r}{b}\right) \qquad 0 \le z_r \le 3b \qquad (3.84)$$

$$p_{ur} = 5.2 \alpha_r q_{ur} b \qquad z_r > 3b \qquad (3.85)$$

where q_{ur} = compressive strength of the rock, usually lower-bound as a function of depth, α_r = strength reduction factor, b = diameter of the pile, and z_r = depth below the rock surface.

3.15.3.3 *Initial value of soil reaction modulus*
If one were to consider a strip from a beam resting on an elastic, homogeneous, and isotropic solid, the initial reaction modulus K_i (p_i divided by y_i) may be shown to have the following value (using the symbols for rock).

$$K_{ir} \cong k_{ir} E_{ir} \tag{3.86}$$

where E_{ir} = the initial reaction modulus of the rock, and k_{ir} = dimensionless constant.

Equations 3.87 and 3.88 for k_{ir} are derived from experiment and reflect the assumption that the presence of the rock surface will have a similar effect on k_{ir} as was shown for p_{ur} for ultimate resistance.

$$k_{ir} = \left(100 + \frac{400 z_r}{3b}\right) \quad 0 \geq z_r \geq 3b \tag{3.87}$$

$$k_{ir} = 500 \quad z_r \geq 3b \tag{3.88}$$

3.15.3.4 Formulas for family of p-y curves

With guidelines for computing p_{ur} and K_{ir}, the equations for the three branches of the family of p-y curves for rock can be presented. The characteristic shape of the p-y curves is shown in Figure 3.37. The equation for the straight-line, initial portion of the curves is given by Equation 3.89 and for the other branches by Equations 3.90 through 3.92.

$$p = K_{ir} y \quad y \leq y_A \tag{3.89}$$

$$p = \frac{p_{ur}}{2}\left(\frac{y}{y_{rm}}\right)^{0.25} \quad y \geq y_A;\ p \leq p_{ur} \tag{3.90}$$

$$p = p_{ur} \quad y \geq 16 y_{rm} \tag{3.91}$$

$$y_{rm} = k_{rm} b \tag{3.92}$$

where k_{rm} = a constant, ranging from 0.0005 to 0.00005 (see case studies following) that serves to establish the overall stiffness of the curves.

The value of y_A may be found by solving for the intersection of Equations 3.89 and 3.90.

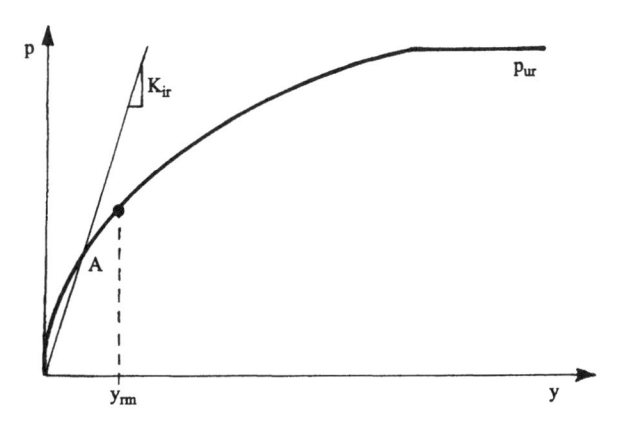

Figure 3.37. Characteristic shape of p-y curves proposed for weak rock.

3.15.4 Comments on equations for predicting p-y curves for rock

The equations predict with reasonable accuracy the behavior of single piles under lateral loading for which experimental data are available. Because of the meager amount of data, the equations should be used with caution. An adequate factor of safety should be employed in all cases; preferably field tests should be undertaken with full-sized piles, with appropriate instrumentation.

If the rock contains joints that are filled with weak soil, the selection of properties of strength and stiffness must be site-specific and will require a comprehensive geotechnical investigation. In those cases, the application of the method presented herein should proceed with even more caution than normal.

3.16 SELECTION OF p-y CURVES

3.16.1 Introduction

In addition to presenting subsidiary information on the formulation of p-y curves, the chapter presents several sets of recommendations and cites the experiments that provided the relevant data. With respect to soil and rock, procedures were presented for the following cases.
- Clay
  - Soft clay with free water
  - Stiff clay with free water
  - Stiff clay with no free water
- Sand above and below the water table
- Soil with both cohesion and a friction angle
- Weak rock

In addition, recommendations were given for dealing with layered soils, a sloping ground surface, the effect of batter, and a shearing force at the base of a pile.

The selection of the p-y curves to be used for a particular analysis is the paramount problem to be solved by the engineer. In most cases the properties of the soils will not be very close to those at the sites where the basic experiments were performed. Therefore, some guidance is given here for selecting p-y curves for use in computations.

3.16.2 Factors to be considered

The quality of the soil investigation and the determination of the parameters describing the soil are of first importance. The recommendations for p-y curves proceed from a knowledge of the undrained shear strength of clays and the friction angle of sands. A later chapter presents some details about obtaining soil properties but, in the absence of a high quality investigation, the engineer selects lower-bound values of shear strenght and makes computations of pile response. Parameters may be varied in computations to provide guidance about performing additional soil studies or perhaps even to recommend full-scale load testing. At the very least, the engineer will be in a position to recommend a desirable value of the factor of safety.

The method of installation of a pile can have an important effect on soil properties and must be taken into account. Chapter 5 deals briefly with that topic. Chapter 6 discusses the

Models for response of soil and weak rock 107

prediction of scour. Some other time-related factors to be taken into account are the variation of the level of the water table, wetting and drying, and future construction.

Plainly, the results from static loadings are more valid because confirmation is provided to some degree by correlation with soil properties. The tests where cyclic loading was employed are relatively few in number

The soil properties and other relevant parameters at the tests used to develop the recommendations can be studied carefully to derive information to guide a given design. For example, data will show that the tests were done for a relatively small range of diameters; therefore, extra care should be used if the pile being designed is much more than one meter in diameter.

3.16.3 *Specific suggestions*

The soft-clay criteria will be used for clays that are normally consolidated or slightly over consolidated. Water is presumed to be at or above the ground surface or the clays would be much over consolidated by desiccation. In the case the clays are normally consolidated or only slightly over consolidated and water is not present above the ground surface, the designer may wish to discount some of the loss of resistance if cyclic loading is to occur. The erosion due to scour was not evident during the tests by Matlock (1970) but the presence of water above the ground surface undoubtedly played a role in the observed loss of resistance.

If the clay is over consolidated and below water, the results of testing show that the effect of cyclic loading was extreme. The site has been re-visited and nothing was found to indicate any significant error in the experimental data. As noted earlier, Wang (1982) built a device to investigate scour that was more discriminating that the pin-hole method of Sherard et al. (1976).

The extreme loss of resistance at the Manor site is believed to be due in some a large degree to cracks that radiated from the pile wall after a few millimeters of that which occurred after lateral deflection. The cracks allowed a passage of water with the result that erosion channels were found at Manor well below the ground surface; thereby reducing lateral resistance. Cracks in the ground surface were observed at Houston where the instrumented test was run in stiff clay with no free water (Welch & Reese 1972). The cracks were evident and, had water been above the ground surface, the erosion would have been enhanced.

In summary for clays, for static loading the results for soft clay and stiff clay appear to have reasonable validation. The cyclic curves for soft clay below water appear reasonable as do the cyclic curves for stiff clay with no water. While there is no reason to doubt the results from the Manor tests for stiff clay under cyclic loading below water, the extreme loss of resistance due to cyclic loading could cause the use a sizeable increase in cost of the foundation; therefore, the designer may wish to perform cyclic-loading tests in stiff clay beneath the water if the size of the job justifies the expense. For sustained loading, lateral consolidation needs to be taken into account for soft clays (see Chapter 6 for a possible experimental technique) but lateral consolidation can be ignored for stiff clays.

With regard to sand, the engineer in most cases will have a considerable degree of confidence in the *p-y* curves for both static and cyclic loading. However, as described later, the driving of the pile will cause sand with a loose or medium density to be densified. Further, cyclic loading could cause further densification with a lowering of the ground surface. The possible changes in the friction angle and other characteristics of the sand need to be considered in design.

Long & Vanneste (1994) studied the influence of cyclic loading on *p-y* curves in sand and analyzed data from a number of tests. One of the conclusions was that one-way cyclic loading had a greater effect than did two-way cyclic loading. The influences of the method of installation and the soil density were also studied.

If the sand is very fine, the engineer must take into account the possible build up of excess pore water pressure during cyclic loading. The determination of the coefficient of permeability of the sand could allow the engineer to decide if rapid drainage would occur due to lateral deflection.

If the sand is very dense or if gravel exists at the site, the driving of a pile might be impossible. Therefore, for the construction procedure that is selected, the effects on the soil of the method of installation must be considered.

In summary, for granular soils the engineer can use the recommended *p-y* curves with considerable confidence, taking into account the possible effects noted above

Very little data exist concerning the response of c-ϕ soils to lateral loading. The engineer may use the suggested *p-y* curves for guidance and perform full-scale field tests where the job justifies the expense.

The *p-y* curves proposed for weak rock will allow the pertinent characteristics of the rock to be considered. While predicting the response of rock to lateral loading is a complex problem, the recommendations presented will allow the engineer to make appropriate designs for many cases of weak rock. For piles that are drilled and grouted firmly into strong rock, the engineer will probably have more concern about the design of the pile where it enters the rock than the lateral resistance of the rock.

Majano et al. (1998) studied the lateral resistance of a friable sandstone by using a specially designed hydraulic ram near the bottom of a 50-m-deep borehole. While the loading arrangement does not simulate the deflection of a laterally loaded pile, finite elements were used to convert the measured data on stress versus deflection into *p-y* curves. The resulting *p-y* curves were softer than would have been computed by the method presented herein. The excavation remained open for several days and O'Neill (2000) suggested the lower lateral resistance was due to stress relief. In an excavation that remains open for some time, stress relief will undoubtedly lead to some loss of lateral resistance, especially if the founding material is below the water table and is a jointed rock or a stiff clay with cracks and fissures.

The reader will be able to gain useful information on the selection of *p-y* curves by reviewing the selection of the soil criteria used for the case studies presented in Chapter 7. A comparison of the results of experiment and the results of analyses where a given set of *p-y* curves were used will allow judgment to be developed concerning the accuracy of the analytical techniques for a fairly wide variety of soils.

CHAPTER 4

Structural characteristics of piles

4.1 INTRODUCTION

The engineer must solve the beam-column equation, presented in Chapter 2, by nondimensional methods or by a computer program. The response of the soil is characterized by equations shown in Chapter 3 or by similar methods. The diameter of a cylindrical pile as a function of depth, or the equivalent diameter of a pile with a non-circular cross section, must be known to solve the soil-response equations. An approach for computing the equivalent diameter of a pile with a non-circular cross section is presented in this chapter.

The solution of the beam-column equation requires values of the bending stiffness $E_p I_p$. If the engineer is interested in small deflections, a constant value for $E_p I_p$ may be employed along the pile. However, in many instances, the engineer is required to find a loading that produces failure, which may be defined as excessive deflection or the formation of a plastic hinge. In the latter case, the value of the ultimate bending moment M_{ult} must be found and, in both instances, the value of $E_p I_p$ at each cross section must be found as a function of the applied loading.

As noted earlier, and as emphasized in this chapter, the nonlinear behavior of soil requires that the load that causes a pile to fail must be found in order to find the safe load that can be applied. In nearly all cases of practical design, the value of M_{ult} is required. The value of $E_p I_p$ in the nonlinear range of the materials of which a pile is constructed is a function of the axial load and the bending moment.

The computation of a nonlinear interaction diagram for values of $E_p I_p$ of a pile made of both steel and reinforced concrete is presented herein. This interaction diagram could be consulted and appropriate values of $E_p I_p$ could be substituted into the difference equations as loadings are incremented in finding the loading at failure. While a knowledge of M_{ult} is essential, experience has shown that very good solutions can be found by using a constant value of $E_p I_p$ in solving the difference equations. Iteration is required because of the nonlinear soil; a second level of iteration with nonlinear properties of structural materials, while potentially beneficial, is tedious and may be unnecessary. The use of nonlinear values of $E_p I_p$ is discussed more fully in Chapters 5 and 6.

4.2 COMPUTATION OF AN EQUIVALENT DIAMETER OF A PILE WITH A NONCIRCULAR CROSS SECTION

The primary experiments, and most of the case studies, have been performed with piles with a circular cross section. However, piles with other shapes are often employed and an

110 *Piles under lateral loading*

equivalent circular diameter for the various shapes is needed in order to employ the recommendations for *p-y* curves given in Chapter 3.

The sketch in Figure 4.1a shows conceptually the stresses from the soil that would act against a pile with a circular cross section when the pile is deflected from left to right. If the ultimate resistance p_u is assumed to have been developed, the arrows on the right side of the section indicate that the soil is in a failure condition. The arrows are drawn to indicate normal stresses and shearing stresses on the front half of the section. The stresses on the back half of the section are reversed in direction, and are indicated to be small, to show that the earth pressures are reduced with deflection.

As a first approximation, one could assume that a pile with a rectangular cross section with the same width as a circular section, one diameter, and with half the depth of a circle, one-half diameter, would behave the same as a pile with a circular cross section.

Then, if the section in Figure 4.1b is considered, which has a width and a depth of one pile diameter, the resistance to deflection would be greater because shearing stresses could act along the back half of the section. With the concepts presented above, the following equation can be written to solve for the equivalent diameter b_{eq} of a rectangular section.

$$b_{eq} = w\left[\frac{p_{uc} + 2\left(d - \frac{w}{2}\right)f_z}{p_{uc}}\right] \qquad (4.1)$$

where w = width of section; d = depth of section; p_{uc} = ultimate resistance of a circular section with a diameter b equal to w; and f_z = shearing resistance along the sides of the rectangular shape at the depth z below the ground surface.

For the undrained strength approach for cohesive soils, the shearing resistance may be computed with the following equation.

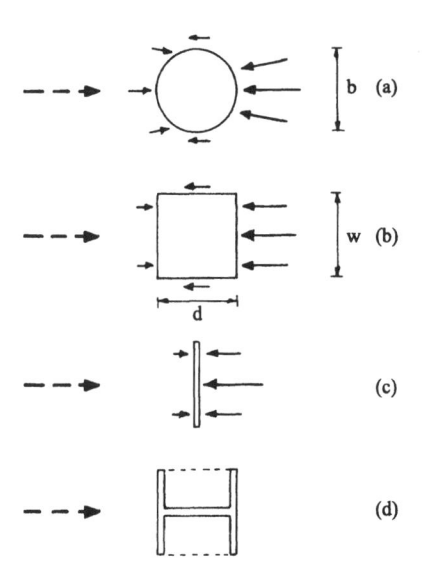

Figure 4.1. Sketches to indicate influence of shape of cross section of pile on p_u

$$f_z = \alpha\, c_u \text{ with } 0.5 \leq \alpha \leq 1 \tag{4.2}$$

where α = shear strength reduction factor; and c_u = undrained shear strength.

The value of c_u will depend on the depth z and on the best estimate of α. In obtaining a value of α, the engineer may wish to use a value derived from equations for the behavior of piles under axial loading.

For cohesionless soils the shearing resistance may be computed with the following equation.

$$f_z = K_z \gamma\, z\, \tan\phi_z \tag{4.3}$$

where K_z = lateral earth pressure coefficient; $\gamma\, z$ = effective vertical soil stress at depth z; and ϕ_z = shear angle (between the soil and the wall of the pile) at the relevant shear strain level.

The value of K_z will be related to the manner in which the pile is installed, and the value of ϕ_z will likely be somewhat lower than the shear angle.

The use of the above equations in computing the equivalent diameters of the shapes shown in Figure 4.1 will be considered. If the dimensions d and w in Figure 4.1b are the same as the diameter b, the equations above will show a somewhat larger equivalent diameter b_{eq}. The flat shape in Figure 4.1c will yield a smaller equivalent diameter. The structural shape in Figure 4.1d, with the direction of loading as shown, can be treated by the equations with the value of α taken as unity, and the value of ϕ_z taken as equal to the shear angle.

The equivalent diameter for a rectangular section, as computed by the above equations, will vary with the shear strength at the site and with the depth being selected. The engineer may wish to select a few depths and compute values of b_{eq}, and then average these values in making a solution.

As an example of the use of the equations, a square section is selected, as shown in Figure 4.1b, with a width and depth of 300 mm. The soil is assumed to be a saturated clay with an undrained shear strength of 50 kPa. The value of α is taken as 0.5; using Equations 4.1 and 4.2, the value of b_{eq} was computed to be 317 mm or an increase of less than 6% over the use of only 300 millimeters.

4.3 MECHANICS FOR COMPUTATION OF M_{ult} AND $E_p I_p$ AS A FUNCTION OF BENDING MOMENT AND AXIAL LOAD

Some manuals include values of the section modulus I for steel members at which a plastic hinge will develop; however, the influence of the axial load is not included. A numerical procedure must be used in the computations.

Equations for the behavior of a slice from a beam or from a beam-column under bending or axial load are formulated. A reinforced-concrete section is assumed in the presentation but the concepts can be applied to a structural-steel shape. The $E_p I_p$ of the concrete member will experience a significant change when cracking occurs. In the procedure described herein, the assumption is made that the tensile strength of concrete is minimal and that cracks will be closely spaced when they appears. Actually, such cracks will initially be spaced at some distance apart and the change in the $E_p I_p$ will not be so drastic. In respect to the cracking of concrete, therefore, the $E_p I_p$ for a beam will change more gradually than is given by the method presented.

112 Piles under lateral loading

Because the nonlinear stress-strain curves for steel and concrete do not indicate a condition for fracture, values of the ultimate strain of these materials are selected to reflect their failure. For concrete, the ultimate value of strain is 0.003; for steel, the ultimate value of strain is 0.015. These values appear to be consistent with those frequently used in practice.

The method shown here can be applied to any member with a symmetrical cross section composed of a combination of concrete and steel. However, for purposes of this document, only standard shapes of sections of reinforced concrete, steel pipe, and structural shapes are considered.

The following derivation adopts the concept that plane sections in a beam or beam-column remain plane after loading. Thus, an axial load and a moment can be applied to a section with the result that the neutral axis will be displaced from the center of gravity of a symmetrical section.

The equations to be solved are as follows:

$$b \int_{-h_2}^{h_1} \sigma \, dy = P_x, \text{ and} \tag{4.4}$$

$$b \int_{-h_2}^{h_1} \sigma y \, dy = M \tag{4.5}$$

A convenient procedure of computation, then, is to: select the angle of rotation for a section; estimate the position of the neutral axis; compute the strain across the section; use numerical methods to solve for the distribution of stresses across the cross section; compute the magnitude of the axial load by summing the forces across the section as indicated in Equation 4.4; modify the position of the neutral axis if the computed value of axial load does not agree with the applied load; repeat the computations until convergence is achieved; solve for the bending moment by numerical methods in implementing Equation 4.5; and obtain the bending stiffness.

The equations for implementing the procedure are shown below; the equations for the mechanics of a beam under pure bending are presented first.

The derivation shown is elementary but is included here for clarity and for a definition of terms. An element from a beam with an unloaded shape of *abcd* is shown by the dashed lines in Figure 4.2. The beam is subjected to pure bending and the element changes in shape as shown by the solid lines. The relative rotation of the sides of the element is given by the small angle d*q* and the radius of curvature of the elastic element is signified by the length *r*. The unit strain ε_x along the length of the beam is given by Equation 4.6.

$$\varepsilon_x = \frac{\Delta}{dx} \tag{4.6}$$

where Δ = deformation at any distance from the neutral axis; and dx = length of the element.

From similar triangles

$$\frac{\rho}{dx} = \frac{\eta}{\Delta} \tag{4.7}$$

where η = distance from neutral axis.

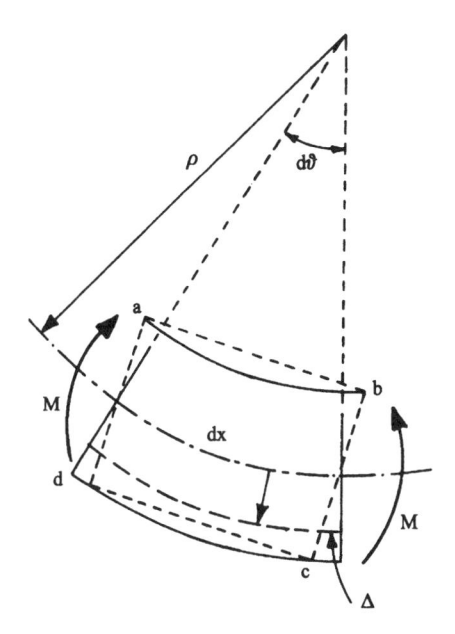

Figure 4.2. Portion of a beam subjected to bending.

Equation 4.8 is obtained from Equations 4.6 and 4.7, as follows:

$$\varepsilon_x = \frac{\eta}{\rho} \tag{4.8}$$

From Hooke's law

$$\varepsilon_x = \frac{\sigma_x}{E} \tag{4.9}$$

where σ_x = unit stress along the length of the beam; and E = Young's modulus.
Therefore,

$$\sigma_x = \frac{E\eta}{\rho} \tag{4.10}$$

From beam theory

$$\sigma_x = \frac{M\eta}{I} \tag{4.11}$$

where M = applied moment; and I = moment of inertia of the section.
From Equations 4.10 and 4.11

$$\frac{M\eta}{I} = \frac{E\eta}{\rho} \tag{4.12}$$

Rewriting Equation 4.12

114 *Piles under lateral loading*

$$\frac{M}{EI} = \frac{1}{\rho} \quad (4.13)$$

Continuing with the derivation, it can be seen that $dx = rdq$ and

$$\frac{1}{\rho} = \frac{d\theta}{dx} \quad (4.14)$$

For convenience, the symbol ϕ_θ is substituted for $d\theta/dx$; therefore, from this substitution and Equations 4.13 and 4.14, the following equation is found, using the subscripts to indicate application to a pile.

$$E_p I_p = \frac{M}{\phi_\theta} \quad (4.15)$$

Also, because $\Delta = h\, d\theta$ and $\varepsilon_x = \Delta/dx$ then,

$$\varepsilon_x = \phi_\theta h . \quad (4.16)$$

The computation for a reinforced-concrete section, or a section consisting partly or entirely of a pile, proceeds by selecting a value of ϕ_θ and estimating the position of the neutral axis. The strain at points along the depth of the beam can be computed by use of Equation 4.16, which in turn will lead to the forces in the concrete and steel. In this step, the assumption is made that the stress-strain curves for concrete and steel are as shown in the following section.

With the magnitude of the forces, both tension and compression, the equilibrium of the section can be checked, taking into account the external compressive loading. If the section is not in equilibrium, a revised position of the neutral axis is selected and iterations proceed until the neutral axis is found.

The bending moment is found from the forces in the concrete and steel by taking moments about the centroidal axis of the section. Thus, the externally-applied axial load does not enter the equations. Then, the value of $E_p I_p$ is found from Equation 4.15. The maximum strain is tabulated and the solution proceeds by incrementing the value of ϕ_θ. The computations continue until the maximum strain selected for failure, in the concrete or in a steel pipe, is reached or exceeded. Thus, the ultimate moment that can be sustained by the section can be found.

4.4 STRESS-STRAIN CURVES FOR NORMAL-WEIGHT CONCRETE AND STRUCTURAL STEEL

Any number of models can be used for the stress-strain curves for normal-weight concrete and structural steel. For the purposes of the computations presented herein, some relatively simple curves are used.

Figure 4.3 shows the nonlinear stress-strain curve for concrete employed herein (Hognestad 1951). Other formulations for the stress-strain curves for concrete have been presented by Eurocode 2 (1992), Comité Euro-International du Béton (1978), and To-

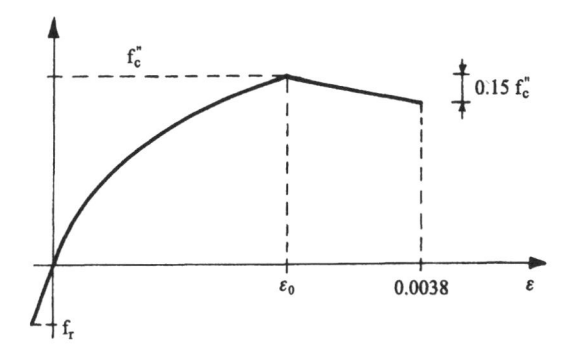

Figure 4.3. Stress-strain curve for concrete.

deschini et al. (1964). The following equations apply to the branches of the Hognestad curve. The value of f'_c is the characteristic value of the compressive strength of the concrete, measured on concrete cylinders 150 mm in diameter and 800 mm in height, and is specified by the engineer. The other symbols are defined below or shown in the figure.

$$f''_c = 0.85 f'_c \tag{4.17}$$

$$f_c = f''_c \left[2\frac{\varepsilon}{\varepsilon_0} - \left(\frac{\varepsilon}{\varepsilon_0}\right)^2 \right] \tag{4.18}$$

$$f_r = 19.7(f'_c)^{0.5} \tag{4.19}$$

$$E_c = 151{,}000(f'_c)^{0.5} \tag{4.20}$$

$$\varepsilon_0 = 1.8\frac{f''_c}{E_c} \tag{4.21}$$

where E_c = initial modulus or tangent modulus presumably of the concrete and the units of E_c, f_r, and f'_c are kPa.

Figure 4.4 shows the idealized elastic-plastic stress-strain (s-e) curve for steel and, as may be seen, there is no limit to the amount of plastic deformation. The curves for tension and compression are identical. The yield strength of the steel f_y is selected according to the material being used, and E is the initial modulus of the steel; the following equations apply.

$$\varepsilon_y = \frac{f_y}{E} \tag{4.22}$$

$$E = 200{,}000 \text{ MPa} \tag{4.23}$$

The models and the equations shown here are employed in the derivations that are shown subsequently.

116 Piles under lateral loading

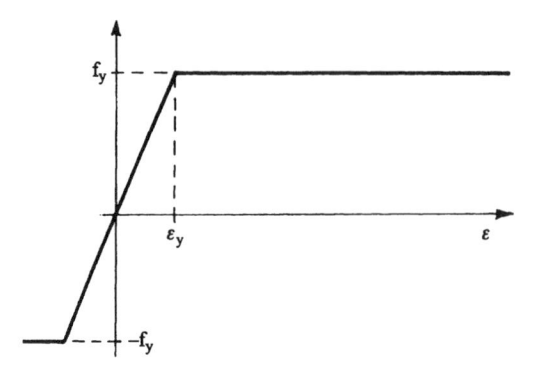

Figure 4.4. Stress-strain curve for steel.

4.5 IMPLEMENTATION OF THE METHOD FOR A STEEL H-SECTION

The assumption can be made without significant error that a constant value of $E_p I_p$ can be used in the computation of the bending stiffness for all ranges of loading. The reduced values of the modulus of the steel after some of the fibers have reached yield will affect the computed value of the moment only slightly. If deflection controls the design, the engineer may wish to modify the equation after the yield stress is reached at the extreme fibers of the structural shape to reflect the nonlinear behavior.

With regard to the ultimate bending moment, handbooks include tabulated values of the plastic modulus. The values are based on the distribution of stresses as shown in Figure 4.5. Therefore, the computation of M_{ult} is done by a simple equation.

$$M_{ult} = f_y Z_x \tag{4.24}$$

where Z_x = tabulated value of plastic modulus (bending is assumed to be about the x-axis).

The relationship in Equation 4.24 holds whenever effects of local buckling do not prevent the development of plastic stress across the entire section, and hence prevent the development of the plastic bending moment.(classes 1 and 2 cross-sections in Eurocode 3, part 1.1).

As an example for the computation of the plastic moment, a section is selected with a depth of 351 mm, a width of 373 mm, a flange thickness of 15.6 mm, and a web thickness of 15.6 mm. The yield strength of the steel is taken as 235 N/mm². The cross-sectional area is 16,900 mm² and the plastic modulus Z_x about the major axis is 2.39×10^6 mm³. The ultimate plastic moment, assuming bending about the major axis, is:

$$M_{ult} = (235)(2.39 \times 10^6) = 5.61 \times 10^8 \text{ Nmm} = 561 \text{ kNm}.$$

Assuming that plastic behavior is developed over the full depth of the section, the value of M_{ult} may be computed approximately as:

$$M_{ult} = (2)(235)[(373)(15.6)(167.7) + (159.9)(15.6)(79.95)] = 5.52 \times 10^8 \text{ Nmm}.$$

The cross-sectional area used in the above computations is slightly less than the total area which accounts for the slightly smaller value by the approximate method.

Horne (1971) presents an equation for solving for the effect of axial loading where the neutral axis is in the web, as follows:

$$a = \frac{P_x}{2t_w f_y} \quad (4.25)$$

$$M_{ult} = f_y Z_x - t_w a^2 f_y, \quad \text{and} \quad (4.26)$$

where t_w = thickness of web of beam.

Horne presented a figure for I-sections that shows the reduction of the ultimate moment due to axial loading (see Fig. 4.6). The abscissa is the applied axial load as a function of the squash load P_u, where P_u is found by Equation 4.27.

$$P_u = f_y A \quad (4.27)$$

where A = cross-sectional area of pile.

No column action is assumed in Equation 4.27; that feature in the design of a pile is discussed in Chapter 6.

As may be seen in Figure 4.6, the reduction in the plastic moment is quite small when the axial load is in the range of 5 to 10% of the squash load. This range is encountered in

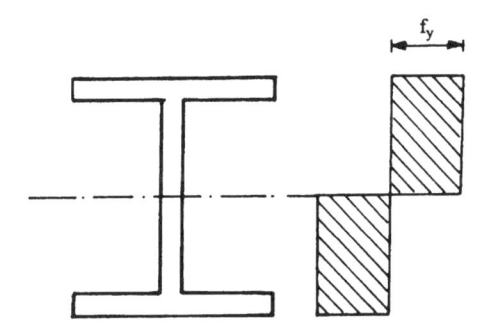

Figure 4.5. Sketch for computing ultimate moment of a structural shape.

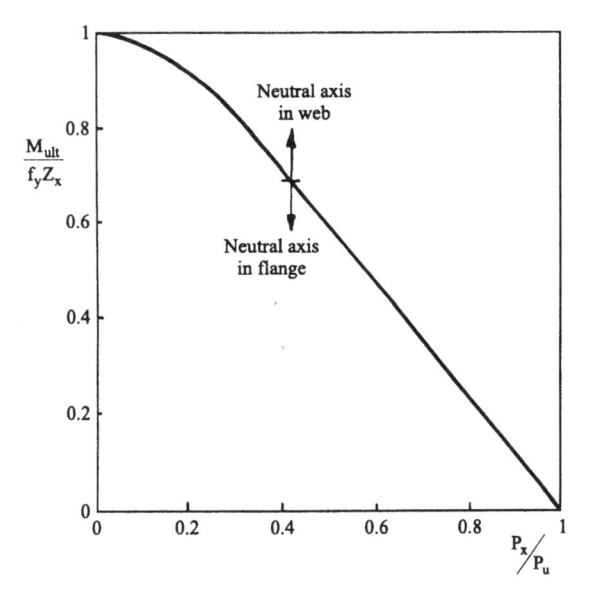

Figure 4.6. Effect of axial loading on ultimate bending moment in I-Sections (after Horne 1971).

118 *Piles under lateral loading*

most designs. When the axial load is relatively large, Figure 4.6 may be used in preliminary design. Alternately, the designer may work out a curve for the particular section by using the equations of mechanics.

For the section selected above, the squash load is

$$P_u = (235,000)(0.01690) = 3970 \text{ kN}.$$

In most cases, a pile that is designed principally to sustain lateral loading would be subjected to an axial load of not more that 5 to 10% of the squash load as computed above; therefore, the reduction of the allowable plastic moment would be small.

4.6 IMPLEMENTATION OF THE METHOD FOR A STEEL PIPE

As for the structural shape, the $E_p I_p$ for elastic behavior may be used without much error in computing the bending moment. The moment of inertia is computed by the familiar equation.

$$I_p = \frac{\pi(d_0^4 - d_i^4)}{64} \tag{4.28}$$

The ultimate bending moment may be computed with simple expressions if the distribution of stresses in the pipe is as shown in Figure 4.5.

$$M_{ult} = f_y Z_p \tag{4.29}$$

where

$$Z_p = \frac{1}{6}(d_0^3 - d_i^3) \tag{4.30}$$

A numerical procedure can be used to investigate the influence of axial loading with results that are similar to those obtained by Horne (1971). The cases that were studied were for a value of f_y of 235,000 kPa, and the ratios of diameter b to wall thickness t ranged from 12 to 48; the results are shown in Figure 4.7 in a plot similar to that in Figure 4.6.

A pipe was selected for study with an outside diameter of 838 mm and an inside diameter of 782 mm ($t = 28$ mm). The yield strength of the steel is assumed to be 235,000 kPa. The following computations were made.

$$A = \pi/4 \, (0.838^2 - 0.782^2) = 7.125 \times 10^{-2} \text{ m}^2$$

$$I_p = \pi/64 \, (0.838^2 - 0.782^4) = 5.85 \times 10^{-3} \text{ m}^4$$

$$E_p I_p = (5.85 \times 10^{-3})(2 \times 10^8) = 1.17 \times 10^6 \text{ kNm}^2$$

$$Z_p = 1/6 \, (0.838^3 - 0.782^3) = 1.838 \times 10^{-2} \text{ m}^3$$

$$P_u = (235,000)(7.125 \times 10^{-2}) = 16,740 \text{ kN}$$

$$M_{ult} = (235,000)(1.838 \times 10^{-2}) = 4320 \text{ kNm (with no axial load)}$$

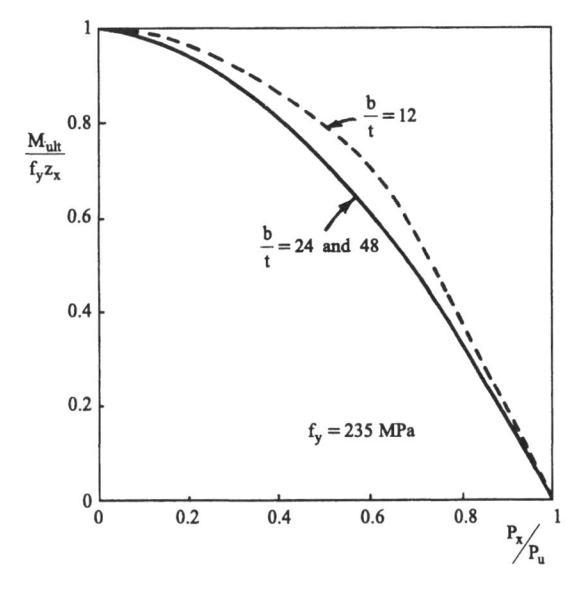

Figure 4.7. Effect of axial loading on ultimate bending moment in pipes.

If it is assumed that an axial load of 250 kN is applied while the pile is being subjected to lateral loading, the value of P_x/P_u is 0.015. Figure 4.7 shows that a negligible correction is needed to M_{ult} to account for the presence of the axial loading.

4.7 IMPLEMENTATION OF THE METHOD FOR A REINFORCED-CONCRETE SECTION

The computation of the necessary parameters for the analysis under lateral load of structural shapes or steel pipes is facilitated by the availability of simple equations. On the other hand, the analysis of a reinforced-concrete section must depend principally on numerical analysis for developing the needed parameters. The variables that must be addressed include geometry of the section, the strengths of the concrete and steel, and the percentage of steel. The number of parameters argue against the presentation of charts or tables that can be entered; rather, the designer should have on hand one of the available computation procedures for the needed values.

The paragraphs that follow present the basics of the computations and further information is given in Chapter 6.

The value of $E_p I_p$ can be taken as that of the gross section. However, the cracking of the concrete will occur early in the loading with a significant reduction in $E_p I_p$. Further reductions occur as the bending moment is increased; therefore, a modification in $E_p I_p$ may be needed for accurate computations, especially if deflection will control the loading.

4.7.1 Example computations for a square shape

Figure 4.8 shows the cross section of a reinforced-concrete pile that is 400 mm square. The strengths of the concrete and steel, the number and size of bars, and the distance from the center of the bars to the outside of the section are indicated.

120 *Piles under lateral loading*

A computer code was employed and Figure 4.9 was prepared, showing the ultimate bending moment as a function of the magnitude of the axial load. The squash load, collapse with only axial loading, was computed to be 5374 kN. Curves such as those shown in the figure may be used to obtain the moment at which a plastic hinge will develop. As shown in the figure, M_{ult} increases with axial loading up to a value of P_x about 2000 kN. If there is some question about the magnitude of the axial loading, the lower value should be selected in order to be conservative.

Figure 4.10 shows computed values of the bending stiffness $E_p I_p$ for a section of the pile as a function of the applied moment. Three values of axial load are assumed. Also plotted in the figure is the value of $E_p I_p$, called gross $E_p I_p$, for the concrete only. Because the computed bending moment is affected only slightly by variations in the value of $E_p I_p$, the gross $E_p I_p$ may be used without much error for the relatively small values of bending moment. Except for the axial load of 2000 kN, the bending stiffness is larger than that of the cracked section for the smaller values of bending moment. This increase is due, of course, to the influence of the steel in the section.

The curves include a section where the $E_p I_p$ changes dramatically for a given value of bending moment. To some extent, the sudden drop in $E_p I_p$ is an artifact of the particular

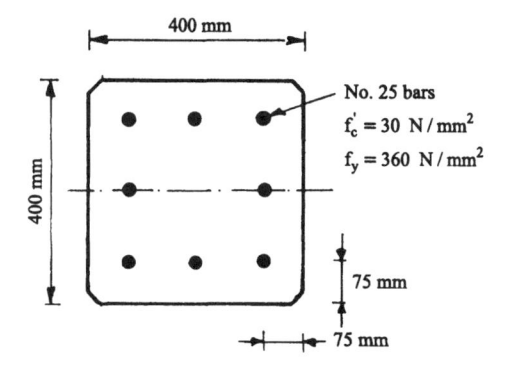

Figure 4.8. Dimensions of square reinforced-concrete pile.

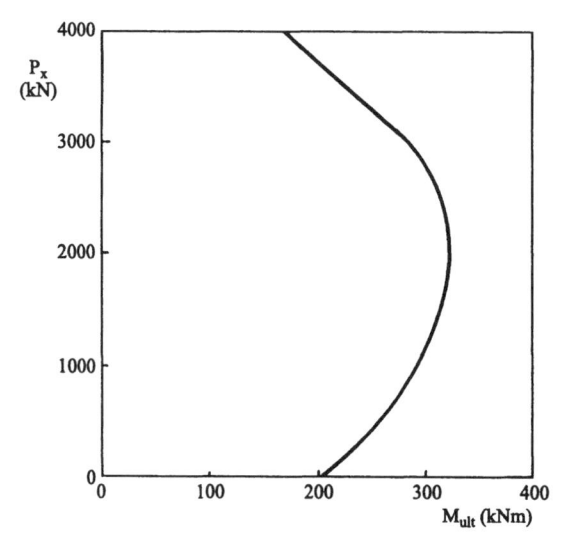

Figure 4.9. Interaction diagram for square reinforced-concrete pile.

Structural characteristics of piles 121

code that is used. The code is written to indicate a sudden loss of bending stiffness with the cracking of the concrete. In practice, the loss of $E_p I_p$ is likely to be much more gradual than indicated in the sketch.

4.7.2 *Example computations for a circular shape*

Similar computations to those for the square shape are given for a reinforced-concrete circular shape, as shown in Figures 4.11, 4.12, and 4.13. The curves are similar in shape to those shown earlier, and a similar discussion applies.

4.8 APPROXIMATION OF MOMENT OF INERTIA FOR A REINFORCED-CONCRETE SECTION

As noted above, the procedures given in the above paragraphs result in a sharp decrease in the value of $E_p I_p$ because the mechanics predicts continuous cracking at a given tensile

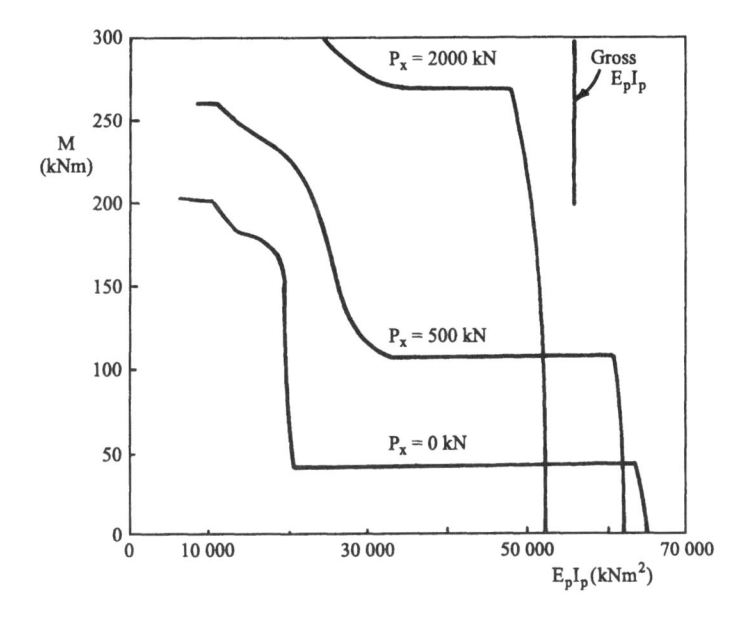

Figure 4.10. $E_p I_p$ as a function of M for pile with rectangular cross-section.

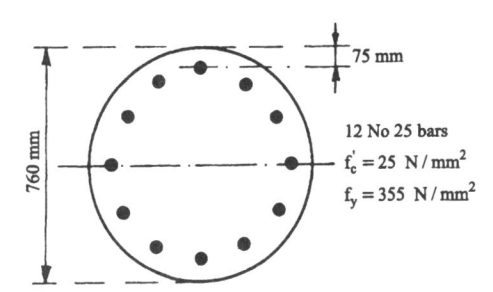

Figure 4.11. Cylindrical reinforced-concrete pile.

760 mm
75 mm
12 No 25 bars
$f'_c = 25$ N/mm²
$f_y = 355$ N/mm²

122 *Piles under lateral loading*

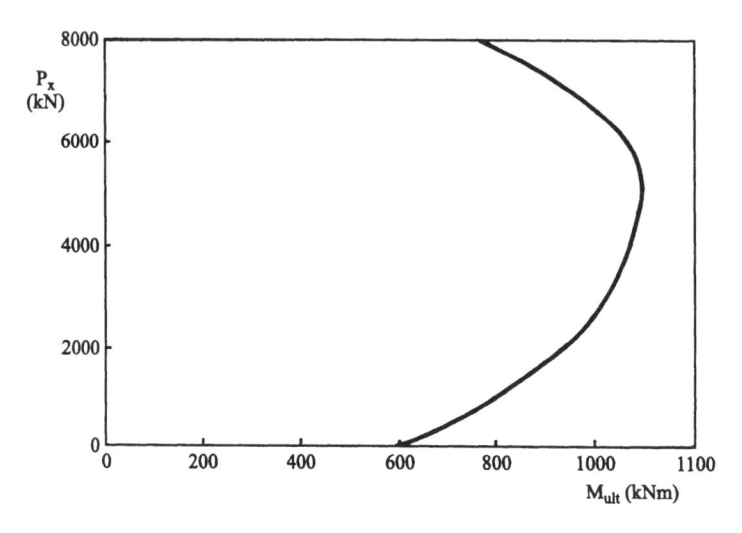

Figure 4.12. Interaction diagram for cylindrical reinforced-concrete pile.

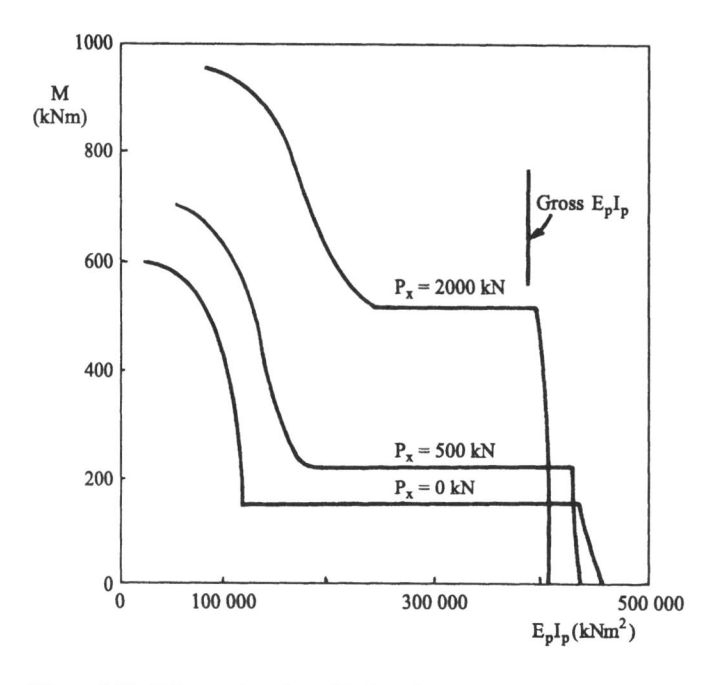

Figure 4.13. $E_p I_p$ as a function of M for pile with cylindrical cross-section.

strain of the concrete. Observations of the behavior of reinforced-concrete sections has yielded an empirical equation that give values of bending stiffness that reduce more gradually, as a function of the applied bending moment, than do the values from mechanics (American Concrete Institute 1989).

$$I_e = \left(\frac{M_{cr}}{M_a}\right)^3 I_g + \left[1 - \left(\frac{M_{cr}}{M_a}\right)^3\right] I_{cr} \qquad (4.31)$$

where

$$M_{cr} = \frac{f_r I_g}{y_c} \qquad (4.32)$$

$$f_r = 7.5\sqrt{f_c'} \quad \text{(for normal weight concrete)} \qquad (4.33)$$

I_e = effective moment of inertia for computation of deflection, I_g = moment of inertia of gross concrete section about centroidal axis, neglecting reinforcement, y_c = distance from the centroidal axis of the gross section, neglecting reinforcement, to the extreme fibers in tension, I_{cr} = moment of inertia of cracked section, and M_a = maximum moment in pile. (Note: Equation 4.33 is units-dependent; therefore, the user should enter the value of f_c' in lbs/in^2, compute the value of f_r in lbs/in^2, and then convert the value of f_r into appropriate units of kPa.)

The value of I_{cr} may be computed by the analytical method, using standard mechanics, presented earlier. In computing bending stiffness, the value of E_p is assumed to remain constant.

The absence of a term for axial load in Equation 4.31 means that the method is limited in scope. However, plotting of the results for no axial load, along with results from the analytical method for no axial load, will reveal a trend that should prove useful in solving a practical problem.

CHAPTER 5

Analysis of groups of piles subjected to inclined and eccentric loading

5.1 INTRODUCTION

The objective of this chapter is to develop a procedure for computing the movement of a cap for a group of piles when subjected to axial load, lateral load, and overturning moment. Several elements of the procedure must be addressed. The mechanics must be addressed, taking into account the nonlinear deflection of each pile head, for both vertical and battered (raked) piles, due to the imposition of an axial load, a lateral load, and a moment.

The mechanics of the problem are discussed first. A brief review is given of some of the relevant literature, and a set of equilibrium equations is presented that can be solved by iteration. A framework is established for the input of the nonlinear response of individual piles.

A brief reference in made to the response of individual piles under lateral loading. The early chapters of the text address lateral loading in some detail. A computer code (subroutine) for individual piles under lateral loading can readily be attached to a global program on pile groups if the influence of close spacing can be explicitly defined.

A review is made of technical literature concerning the deformation of individual piles under axial load. While the objective of the book relates principally to lateral loading, the load versus deflection for axially loaded piles must be addressed. If the cap for a pile group is subjected to an inclined and eccentric load, some of the piles in the group certainly must sustain an axial force. Thus, nonlinear relationships must be selected to define axial load versus deflection for a variety of kinds and sizes of piles.

If piles under lateral loading are spaced close to each other, the piles will influence each other due to pile-soil-pile interaction. The interaction of closely spaced piles under lateral loading is discussed in detail and recommendations are given for modifying the *p-y* curves to account for close spacing.

Close spacing of piles under axial loading must also be addressed as is piles under lateral loading. The problem of pile-soil-pile interaction for axial loading is discussed and recommendations are made for formulating influence coefficients to account for close spacing.

Finally, a comprehensive study is analyzed where a pile group, including batter piles, was subjected to inclined and eccentric loading. The computed values of pile-cap movements were compared with values obtained from experiment. A computer code, based on the computed response of individual piles and on the mechanics noted above, yielded results that agreed well with the experimental values.

5.2 APPROACH TO ANALYSIS OF GROUPS OF PILES

The first four chapters are directed primarily at single or isolated piles under lateral loading; however, most piles are installed in groups and the response of the group to loading is addressed herein. Further, as noted in the title of this chapter, most groups must support loadings that are both lateral and axial. Therefore, the mechanics of a pile under axial loading must be presented, but the design of single piles under purely axial loading is discussed only briefly.

The behavior of a group of piles may be influenced by two forms of interaction: 1. Interaction between piles in close proximity where efficiency is involved; and 2. Interaction by distribution of loading to individual piles from the pile cap. In the first instance the relevant forces are transmitted through the soil, while in the second instance, the forces are transmitted by the superstructure (assumed to be the pile cap in this presentation). If the piles are widely spaced, the pile-soil-pile interaction is insignificant and a solution is made in order to reveal lateral load, axial load, and bending moment to each of the piles in the group.

Equations for the efficiency of closely-spaced piles, under lateral loading as well as under axial loading, are based largely on experimental data. Methods that treat the soil as an elastic, isotropic, and homogeneous material are useful in giving insight into the mechanics of pile-soil-pile interaction but soils behave far differently than the idealized material. Experimental results are not definitive because the various sets of experiments do not isolate each of the many parameters and reveal the influence of each particular parameter. However, a review of a number of relevant studies leads to recommendations for analytical procedures.

The steps in the analysis of a group of piles under generalized loading, axial, lateral, and overturning, are discussed herein: 1. Employ a rational method for computing the movements of the pile cap and the loads to each of the piles in the group, reflecting properly the difference in response of piles that are vertical and those that are battered; 2. Account the reduced *efficiency* of each pile in the group due to close spacing; and 3. Use equations for the stiffness of each pile under axial and lateral loading.

A literature survey is presented, dealing principally with the mechanics of the response of a group of piles. Then, a rational formulation of the mechanics of response is presented which is the basis of the analytical method presented for use in practice. A comprehensive study of the problem of closely-spaced piles is given to provide the basis for degrading the response of some of the piles in the group. The stiffness of each pile under axial and lateral loading is needed in the analytical method and discussed briefly. Finally in the chapter, an example is solved for a case where a group of piles are closely spaced and where the loading comes through the pile cap.

5.3 REVIEW OF THEORIES FOR THE RESPONSE OF GROUPS OF PILES TO INCLINED AND ECCENTRIC LOADS

The development of computational methods has been limited because of lack of knowledge about single-pile behavior. In order to meet the practical needs of designing structures with grouped piles, various computational methods were developed by making assumptions that would permit analysis of the problem.

The simplest way to treat a grouped-pile foundation is to assume that both the structure and the piles are rigid and that only the axial resistance of the piles is considered. Under these assumptions, Culmann (Terzaghi 1956) presented a graphical solution in 1866. The equilibrium state of the resultant external load and the axial reaction of each group of similar piles was obtained by drawing a force polygon. The application of Culmann's method is limited to the case of a foundation with three groups of similar piles. A supplemental method to this graphical solution was proposed in 1930 by Brennecke & Lohmeyer (Terzaghi 1956). The vertical component of the resultant load is distributed in a trapezoidal shape in such a way that the total area equals the magnitude of the vertical component, and its center of gravity lies on the line of the vertical component of the resultant load. The vertical load is distributed to each pile, assuming that the trapezoidal load is separated into independent blocks at the top of the piles, except at the end piles. Unlike Culmann's method, the latter method can handle more than three groups of similar piles. But the method of Brennecke & Lohmeyer is restricted to the case where all of the pile tops are on the same level.

The elastic displacement of pile tops was first taken into consideration by Westergaard in 1917 (Karol 1960). Westergaard assumed linearly elastic displacement of pile tops under a compressive load, but the lateral resistance of the piles was not considered. He developed a method to find a center of rotation of a pile cap. With the center of rotation known, the displacements and forces in each pile could be computed.

Nokkentved (Hansen 1959) presented in 1924 a method similar to that of Westergaard. He defined a point that was dependent only on the geometry of the pile arrangement, so that forces which pass through this point produce only unit vertical and horizontal translations of the pile cap. The method was also pursued by Vetter (Terzaghi 1956) in 1939. Vetter introduced the 'dummy pile' technique to simulate the effect of the lateral restraint and the rotational fixity of pile tops. Dummy piles are properly assumed to be imaginary elastic columns.

Later, in 1953, Vandepitte applied the concept of the elastic center in developing the limit-state-design method, which was further formulated by Hansen (1959). The transitional stage in which some of the piles reach the ultimate bearing capacity, while the remainder of the piles in a foundation are in an elastic range, can be computed by a purely elastic method if the reactions of the piles in the ultimate stages are regarded as constant forces on the cap. The failure of the cap is reached after successive failures of all but the last two piles. Then the cap can rotate around the intersection of the axis of the two elastic piles. Vandepitte resorted to a graphical solution to compute directly the ultimate load on a two-dimensional cap. Hansen extended the method to the three-dimensional case. Although the plastic-design method is unique and rational, the assumptions to simplify the real soil-structure system may need examination. It was assumed that a pile had no lateral resistance, and no rotational restraint of the pile tops on the cap was considered. The axial load versus displacement of each individual pile was represented by a bilinear relationship.

A comprehensive, modern structural treatment was presented by Hrennikoff (1950) for the two-dimensional case. He considered the axial, transverse, and rotational resistance of piles on the cap. The load-displacement relationship of the pile top was assumed to be linearly elastic. One restrictive assumption was that all piles must have the same load-displacement relationship. Hrennikoff substituted a free-standing elastic column for an axially loaded pile. A laterally loaded pile was regarded as an elastic beam on an elastic foundation with uniform stiffness. Even with these crude approximations of pile behavior, the method is significant in the sense that it presents the potential for the analytical treatment of the soil-pile-interaction system. Hrennikoff's method consisted of obtaining influence coefficients for cap displacements by summing the influence coefficients of individual piles in terms of the spring constants which represent the pile-head reactions onto the pile cap. Almost all the subsequent work follows the approach taken by Hrennikoff.

Radosavljevic (1957) also regarded a laterally loaded pile as an elastic beam in an elastic medium with a uniform stiffness. He advocated the use of the results of tests of single piles under axial loading. In this way a designer can choose the most practical spring constant for the axially-loaded-pile head, and nonlinear behavior can also be considered. Radosavljevic showed a slightly different formulation than Hrennikoff in deriving the coefficients of the equations of the equilibrium of forces. Instead of using unit displacement of a cap, he used an arbitrary set of displacements. Still, his structural approach is essentially analogous to Hrennikoff's method. Radosavljevic's method is restricted to the case of identical piles in identical soil conditions.

Turzynski (1960) presented a formulation by the matrix method for the two-dimensional case. Neglecting the lateral resistance of pile and soil, he considered only the axial resistance of piles. Further, he assumed piles as elastic columns pinned at the top and at the tip. He derived a stiffness matrix and inverted it to obtain the flexibility matrix. Except for the matrix method, Turzynski's method does not serve a practical use because of its oversimplification of the soil-pile-interaction system.

Asplund (1956) formulated the matrix method for both two-dimensional and three-dimensional cases. His method also starts out from calculations of a stiffness matrix to obtain a flexibility matrix by inversion. In an attempt to simplify the final flexibility matrix, Asplund defined a pile-group center by which the flexibility matrix is diagonalized. He stressed the importance of the pile arrangement for an economical grouped-pile foundation, and he contended that the pile-group-center method helped to visualize better the effect of the geometrical factors. He employed the elastic-center method for the treatment of laterally loaded piles. Any transverse load through the elastic center causes only transverse displacement of the pile head, and rotational load around the elastic center gives only the rotation of the pile head. In spite of the elaborate structural formulation, there is no particular correlation with the soil-pile system. Laterally loaded piles are merely regarded as elastic beams on an elastic bed with a uniform spring constant.

Francis (1964) computed the two-dimensional case using the influence-coefficient method. The lateral resistance of soil was considered either uniform throughout or increasing in proportion to depth. Assuming a fictitious point of fixity at a certain depth, elastic columns fixed at both ends are substituted for laterally loaded piles. The axial loads on individual piles are assumed to have an effect only on the elastic stability without causing any settlement or uplift at the pile tips.

Aschenbrenner (1967) presented a three-dimensional analysis based on the influence-coefficient method. This analysis is an extension of Hrennikoff's method to the three-dimensional case. Aschenbrenner's method is restricted to pin-connected piles.

Saul (1968) gave the most general formulation of the matrix method for a three-dimensional foundation with rigidly connected piles. He employed the cantilever method to describe the behavior of laterally loaded piles. He left it to the designer to set the soil criteria for determining the settlement of axially loaded piles and the resistance of laterally loaded piles. Saul indicated the possible application of his method to dynamically loaded foundations.

Reese & Matlock (1960, 1966) and Reese et al. (1970) presented a method for coupling the analysis of the grouped-pile foundation with the analysis of laterally loaded piles by the finite-difference method. Their method presumes the use of a digital computer. The formulation of equations giving the movement of the pile cap is done by the influence-coefficient method, similar to Hrennikoff's method. Reese & Matlock devised a convenient way to represent the pile-head moment and lateral reaction by spring forces only in terms of the lateral displacement of a pile top. The effect of pile-head rotation on the pile-head reactions is included implicitly in the force-displacement relationship.

Using Reese & Matlock's method, example problems were worked out by Robertson (1961) and by Parker & Cox (1969). Robertson compared the method with Vetter's method and Hrennikoff's method. Parker and Cox integrated into the method typical soil criteria for laterally loaded piles.

Reese & O'Neill (1967) developed the theory for the general analysis of a three-dimensional group of piles using matrix formulations. Their theory is an extension of the theory of Hrennikoff (1950), in which springs are used to represent the piles. Representation of piles by springs imposes the superposition of two independent modes of deflection of a laterally loaded pile. The spring constants for the lateral reaction and the moment at the pile top must be obtained for a mode of deflection, where a pile head is given only transitional displacement without rotation and also for a mode of deflection where a pile head is given only rotation without translation. While the soil-pile-interaction system has highly nonlinear relationships, the pile material also exhibits nonlinear characteristics when it is loaded near its ultimate strength.

The method of analysis used herein is based on the concept presented by Reese & Matlock (1966), but the solution of the relevant equations is done more conveniently by special techniques (Awoshika & Reese 1971).

5.4 RATIONAL EQUATIONS FOR THE RESPONSE OF A GROUP OF PILES UNDER GENERALIZED LOADING

5.4.1 *Introduction*

A structural theory, principally following the work of Awoshika (1971) and Awoshika and Reese (1971), is formulated herein for computing the behavior of a two-dimensional pile foundation with arbitrarily arranged piles that possess nonlinear force-displacement characteristics. Coupled with the structural theory of a pile cap are the theories of a laterally loaded pile and an axially loaded pile. In this chapter each theory is developed separately. Solutions for all of the theories depend on the use of digital computers for the actual computations.

5.4.1.1 *Basic structural systems*

Figure 5.1a illustrates the general system of a two-dimensional pile foundation. Three piles, with arbitrary spacing and arbitrary inclination, are connected to an arbitrarily shaped pile cap. Such sectional properties of a pile as the diameter, the cross-sectional area, and the moment of inertia can vary, not only from pile to pile, but also along the axis of a pile. The pile material may be different from pile to pile but it is assumed that the same material is used within a pile.

130 *Piles under lateral loading*

The structural system at the pile cap is illustrated in Figure 5.1b. The axial load, lateral load, and bending moment at each pile head must put the pile cap into equilibrium. Also, the individual pile-head loads must be consistent with the movements of each of the pile heads.

There are three conceivable cases of pile connection to the pile cap. Pile 1 in Figures 5.1a & c illustrates a pinned connection. Pile 2 shows a fixed-head pile with its head clamped by the pile cap. And Pile 3 represents an elastically restrained pile, which is the typical case of an offshore structure, shown in Figure 5.2. In Figure 5.2a, the piles are extended and form a part of the superstructure. In Figure 5.2b, the piles rest against knife-edge supports and can deflect freely between these supports. Elastic restraint is provided by the flexural rigidity of the pile itself. The treatment of a laterally loaded pile with an elastically-restrained top, discussed in Chapter 3, gives a useful tool for handling the real foundation.

Piles are frequently embedded into a monolithic, reinforced-concrete pile cap with the assumption that complete fixity of the pile to the pile cap is obtained (Fig. 5.3). However, the elasticity of the reinforced concrete and local failure due to stress concentrations allows the rotation of a pile head within the pile cap. The magnitude of the restraint on the pile from the pile cap is indeterminate, but a range of values may be computed by estimating the p-y curves for concrete and solving for the response of the portion of the pile within the cap by use of the equations for a pile under lateral loading.

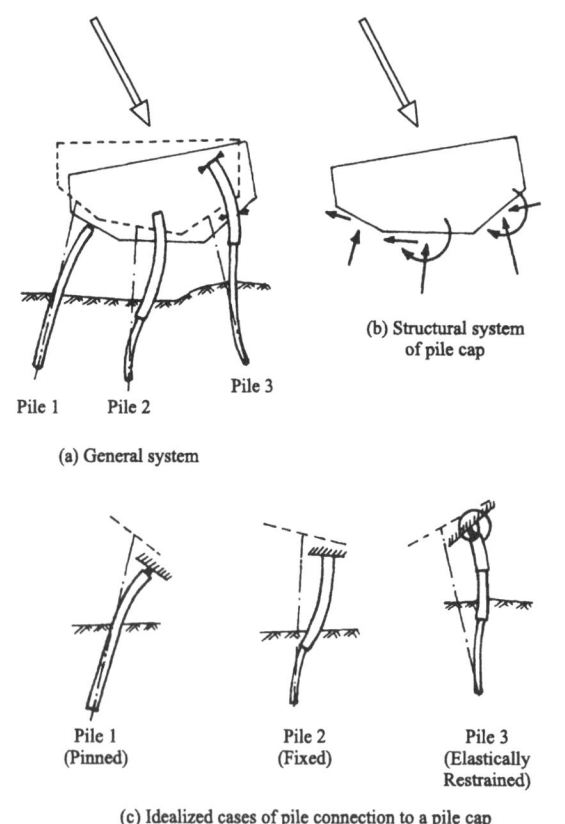

Figure 5.1. Basic structural system A for pile groups.

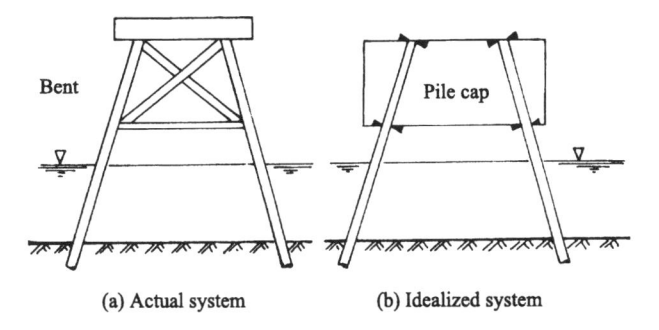

Analysis of groups of piles subjected to inclined and eccentric loading 131

Figure 5.2. Typical offshore structure (after Awoshika 1971).

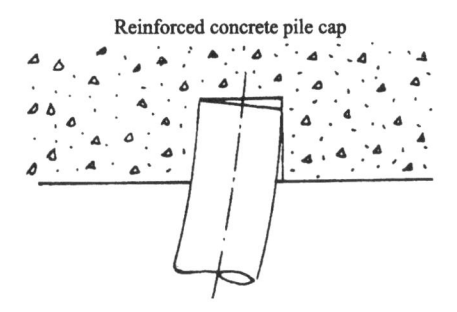

Figure 5.3. Typical pile head in a pile cap (after Awoshika 1971).

The pile cap is subjected to two-dimensional external loads. The line of action of the resultant external load may be inclined and may assume any arbitrary position with respect to the structure (Fig. 5.1a). The external loads cause displacement of the pile cap which results in axial, lateral, and rotational displacements of each individual pile. The displacements of individual piles in turn results in loads on the pile cap (Fig. 5.1b). These pile reactions are highly nonlinear in nature. They are functions of pile properties, soil properties, and the boundary conditions at the pile top.

The structural theory for the pile group uses a numerical method to seek compatible displacement of the pile cap, which satisfies the equilibrium of the applied external loads and the nonlinear pile reactions.

5.4.1.2 *Assumptions*
Some of the basic assumption employed for the treatment of the pile group are presented below.

Two-Dimensionality. The first assumption is the two-dimensional arrangement of the bent cap and the piles. The usual design practice is to arrange piles symmetrically with a plane or planes with loads acting in this plane of symmetry. The assumption of a two-dimensional case reduces considerably the number of variables to be handled. However, there is no essential difference in theory between the two-dimensional case and the three-dimensional case. If the validity of the theory for the former is established, the theory can be extended to the latter by adding more components of forces and displacements with regard to the new dimension (Reese & O'Neill 1967).

132 *Piles under lateral loading*

Nondeformability of Pile Cap. The second major assumption is the nondeformability of the pile cap. A pile head encased in a monolithic pile cap (Pile 2, Fig. 5.1), or supported by a pair of knife edges (Pile 3, Fig. 5.1) can rotate or deflect within the pile cap. But the shape of the pile cap itself is assumed to be always the same for the equations shown herein. That means that the relative positions of the pile tops remain the same for any pile-cap displacement. If the pile cap is deformable, the structural theory of the pile group must include the compatibility condition of the pile cap itself. While no treatment of a foundation with a deformable pile cap is included in this study, the theory could be extended to such a case if the pile cap consists of a structural member such that the analytical computation of the deformation of the pile cap is possible.

Wide Spacing of Piles. The equations developed here are for the case where the individual piles are so widely spaced that there is no influence of one pile on another. However, there are many pile designs where the piles are close enough so that pile-soil-pile interaction does occur, and such interaction is discussed in detail later in this chapter. The effects of pile-soil-interaction can be introduced into the analytical method without difficulty.

Behavior under Lateral Load and under Axial Load are Independent. The assumption is made that there is no interaction between the axial-pile behavior and the lateral-pile behavior. That is, the relationship between axial load and displacement is not affected by the presence of lateral deflection and vice versa. The validity of this assumption is discussed by Parker & Reese (1971). The argument is made, and generally accepted, that the soil near the ground surface principally determines lateral response and the soil at depth principally determine axial response. If overconsolidated clay exists at the ground surface and pile deflection is sizable, the recommendation is made that the soil above the first point where lateral deflection is zero be discounted in computing axial capacity.

5.4.2 *Equations for a two-dimensional group of piles*

To deal with the nonlinearity in the system, the equilibrium of the applied loads and the pile reactions on a pile cap are sought by the successive correction of pile-cap displacements. After each correction of the displacement, the difference between the load and the pile reaction is calculated. The next correction is obtained through the calculation of a new stiffness matrix at the previous pile-cap position. The elements of a stiffness matrix are obtained by giving a small virtual increment to each component of displacement, one at a time. The proper magnitude of the virtual increment may be set at 1×10^{-5} times a unit displacement to attain acceptable accuracy.

5.4.2.1 *Coordinate systems and sign conventions*
Figure 5.4 shows the coordinate systems and sign conventions. The superstructure and the pile cap are referred to the global structural coordinate system (X, Y) where the X and Y axes are vertical and horizontal, respectively. The resultant external forces are acting at the origin O of this global structural coordinate system. The positive directions of the components of the resultant load P_0, Q_0, and M_0 are shown by the arrows. The positive curl of the moment was determined by the usual right-hand rule. The pile head of each individual pile in a group is referred to the local structural coordinate system (x'_i, y'_i), whose origin is the pile head and with axes running parallel to those of the global structural coor-

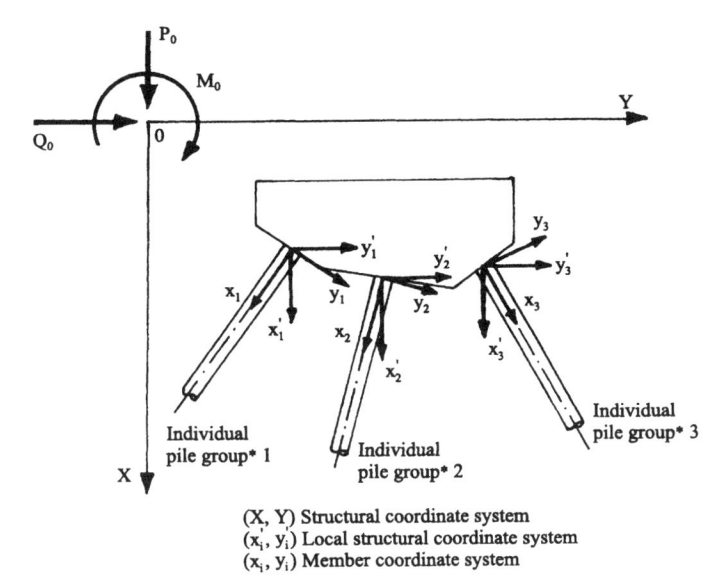

Analysis of groups of piles subjected to inclined and eccentric loading 133

Figure 5.4. Coordinate systems for analysis of pile groups showing positive directions of displacements (after Awoshika & Reese 1971).

dinate system. The member coordinate system (x_i, y_i) is further assigned to each pile. The origin of the member coordinate system is the pile head. Its x_i axis coincides with the pile axis and the y_i axis is perpendicular to the x_i axis. The x_i axis makes an angle λ_i with the vertical. The angle λ_i is positive when it is measured counterclockwise.

Figure 5.5 shows the positive directions of the forces, P_i, Q_i, and M_i exerted from the pile cap onto the top of an individual pile in the ith individual pile group. The forces P_i and Q_i are acting along the x_i and y_i axes, respectively, of the member coordinate system.

5.4.2.2 *Transformation of Coordinates*
Displacement. Figure 5.6a illustrates the pile-head displacement in the structural, the local structural, and the member coordinate systems. Due to the pile-cap displacement from point 0 to point 0' with a rotation α, the ith pile moves from the original position P to the new position P'. The rotation of the pile head depends on the way it is fastened to the cap. The components of pile-cap displacement are expressed by (U, V, α) with regard to the structural coordinate system. The pile-head displacement is denoted by (u'_i, v'_i, α) in the local coordinate system and by (u_i, v_i, α) in the member coordinate system.

The coordinate transformation between the structural and the local structural coordinate system is derived from the simple geometrical consideration:

$$u'_i = U - Y_i \alpha \tag{5.1}$$

$$v'_i = V + X_i \alpha , \tag{5.2}$$

134 *Piles under lateral loading*

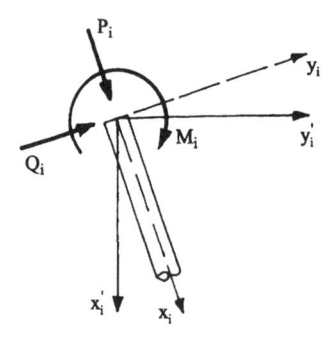

Figure 5.5. Positive direction of forces on single pile in pile group analysis (after Awoshika & Reese 1971).

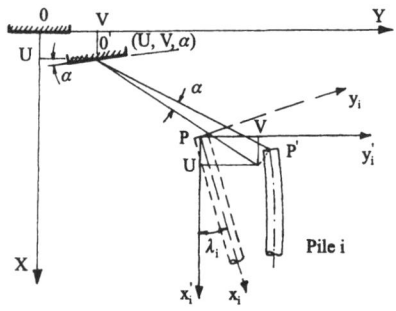

(a) Pile head displacement due to pile cap displacement

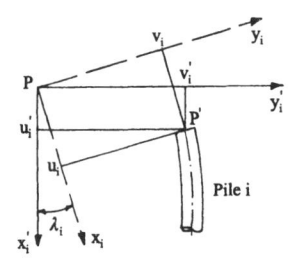

(a) Transformation of displacement between local structural coordinate system and member coordinate system

Figure 5.6. Transformation of displacements (after Awoshika & Reese 1971).

where (X_i, Y_i) = location of *i*th pile head in the structural coordinate system.

The transformation of pile-head displacement from the local structural coordinate system to the member coordinate system is obtained from the geometrical relationship (Fig. 5.6b).

$$u_i = u'_i \cos \lambda_i + v'_i \sin \lambda_i \tag{5.3}$$

$$v_i = v'_i \cos \lambda_i + u'_i \sin \lambda_i \tag{5.4}$$

Substitution of Equations 5.1 and 5.2 into Equations 5.3 and 5.4 yields the transformation relationship between the pile-cap displacement in the structural coordinate system and the corresponding pile-top displacement of the *i*th individual pile group in the member coordinate system.

$$u_i = U\cos\lambda_i + V\sin\lambda_i + \alpha(X_i\sin\lambda_i - Y_i\cos\lambda_i) \qquad (5.5)$$

$$v_i = U\sin\lambda_i + V\cos\lambda_i + \alpha(X_i\cos\lambda_i + Y_i\sin\lambda_i) \qquad (5.6)$$

Equation 5.7 presents in matrix notation the case where the pile head is fixed to the cap so that the cap and the pile head rotate the same amount.

$$\begin{bmatrix} u_i \\ v_i \\ \alpha_i \end{bmatrix} = \begin{bmatrix} \cos\lambda_i & \sin\lambda_i & X_i\sin\lambda_i - Y_i\cos\lambda_i \\ -\sin\lambda_i & \cos\lambda_i & X_i\cos\lambda_i + Y_i\sin\lambda_i \\ 0 & 0 & 1 \end{bmatrix} \begin{bmatrix} U \\ V \\ \alpha \end{bmatrix} \qquad (5.7)$$

The expressions for U_i and V_i will remain the same for the cases where the pile head is free to rotate or is partially restrained, but the expression for α_i must be modified. The matrix expression above is written concisely

$$u_i = T_{D,i} U \qquad (5.8)$$

where u_i = displacement vector of the head of the pile in the ith individual pile group; $T_{D,i}$ = displacement transformation matrix of the pile; and U = displacement vector of the pile cap.

Force. Figure 5.4 illustrates the action of the load on the pile cap and the pile reactions. The load is expressed in three components (P_0, Q_0, M_0) with regard to the structural coordinate system. The reactions in the ith individual pile group are expressed in terms of the member coordinate system (P_i, Q_i, M_i). Decomposition of the reactions of the ith pile with respect to the structural coordinate system gives the transformation of the pile reaction from the member coordinate system to the structural coordinate system.

$$P'_i = P_i\cos\lambda_i - Q_i\sin\lambda_i \qquad (5.9)$$

$$Q'_i = P_i\sin\lambda_i + Q_i\cos\lambda_i \qquad (5.10)$$

$$M'_i = P_i(X_i\sin\lambda_i - Y_i\cos\lambda_i) + Q_i(X_i\cos\lambda_i + Y_i\sin\lambda_i) + M_i \qquad (5.11)$$

Matrix notation expresses the equations above,

$$\begin{bmatrix} P'_i \\ Q'_i \\ M'_i \end{bmatrix} = \begin{bmatrix} \cos\lambda_i & -\sin\lambda_i & 0 \\ \sin\lambda_i & \cos\lambda_i & 0 \\ X_i\sin\lambda_i - Y_i\cos\lambda_i & X_i\cos\lambda_i + Y_i\sin\lambda_i & 1 \end{bmatrix} \begin{bmatrix} P_i \\ Q_i \\ M_i \end{bmatrix} \qquad (5.12)$$

or more concisely,

$$P'_i = T_{F,i} P_i \qquad (5.13)$$

where P'_i = reaction vector of the pile of ith individual pile group in the structural coordinate system; $T_{F,i}$ = force transformation matrix of the pile; and P_i = reaction vector of the pile in the member coordinate system.

It is observed that the force transformation matrix $T_{F,i}$ is obtained by transposing the displacement transformation matrix $T_{D,i}$. Thus,

$$T_{F,i} = T_{D,i}^T \tag{5.14}$$

5.4.2.3 *Solution of equilibrium equations*

With the force and displacement characteristics of each pile at hand, the equilibrium equations for the global structure can now be solved. These are:

$$P_0 + \sum_{i=1}^{n} J_i P_i' = 0 \tag{5.15}$$

$$Q_0 + \sum_{i=1}^{n} J_i P_i' = 0 \tag{5.16}$$

$$M_0 + \sum_{i=1}^{n} J_i M_i' + \sum_{i=1}^{n} J_i Y_i M_i' + \sum_{i=1}^{n} J_i X_i Q_i' = 0 \tag{5.17}$$

where i = values from any 'i-th' individual; J_i = the number of piles in that individual group.

These equilibrium equations are solved by any convenient manner. The stiffness terms (force versus displacement) for the piles are nonlinear; therefore, the equations must be solved by iteration.

Awoshika (1971) performed a comprehensive set of experiments on axially loaded piles, laterally loaded piles, and pile groups with batter piles. The experiments allowed the equations for the distribution of loads to piles in a group to be validated. The experiments will be described and analyzed at the end of this chapter. The intervening material on lateral load, axial load, and the interaction of closely spaced piles will be implemented. Comparisons of values from experiment and from analysis allow the utility method shown above to be evaluated.

5.5 LATERALLY LOADED PILES

5.5.1 *Movement of pile head due to applied loading*

The several earlier chapters address the analysis of single piles to obtain pile-head movements for various boundary conditions at the pile head. The *p-y* method is considered to be the best method currently available for making the required computations. The lateral load and moment at the pile head can be computed readily for the particular lateral deflection that is computed. Thus, the lateral stiffnesses for load and moment can be found as iteration proceeds.

5.5.2 *Effect of batter*

The effect of batter on the behavior of laterally loaded piles was investigated in a test tank (Awoshika & Reese 1971). The lateral, soil-resistance curves of a vertical pile were modified by a constant to express the effect of the pile inclination. The values of the modifying constant as a function of the batter angle were deduced from model tests in sands and also from full-scale, pile-loading tests that are reported in technical literature (Kubo 1965). The criterion is expressed by a solid line in Figure 3.34.

Plotted points in Figure 3.34 show the modification factors for the batter piles tested in the experiments by Awoshika & Reese (1971). The modification factors were obtained for two series of tests independently.

Figure 3.34 indicates that for the out-batter piles, the agreement between the empirical curve and the experiments is good, while the in-batter piles, for the batter angles that were investigated experimentally, did not show any effect of the batter. However, Kubo's experiments show greater batter angles are used in establishing the recommended curve for use.

The values from Figure 3.34 can be used to modify values of p_{ult} which in turn will cause a modification of all of the values of p in the p-y curves. An analytical method may be used to compute the value of p_{ult} by computing the forces on a wedge of soil whose shape is modified to reflect the batter, in or out.

5.6 AXIALLY LOADED PILES

5.6.1 Introduction

The stiffnesses of individual piles under axial loading are needed in order to solve the equation presented in Section 5.4. The topic was discussed in detail by Van Impe in his General Report to the Tenth European Conference on Soil Mechanics and Foundation Engineering (1991). Much of the following material is derived from that paper. While the specific reference is axial loading of the individual pile, the topics presented here will serve to elucidate the discussion presented early on lateral loading and will serve further as pile-soil-pile interaction is dealt with later.

5.6.2 Relevant parameters concerning deformation of soil

A pile under axial load will impose a complex system of stresses in the supporting soil and a corresponding system of deformation will occur in the elements that are affected. Assuming no slip at the interface of the soil and the pile, integration of the deformation of the affected elements in the vertical direction will yield the movement at the pile head. Assuming axisymmetric behavior, mapping of the distribution of stresses in the soil considering the dimensions of the pile, the method of installation, the magnitude of the axial load, and the kinds of soil remains elusive. While such a theoretical method of computing the axial movement of a pile head is currently unfeasible, specific consideration of the stress-strain characteristics of a soil is imperative. Van Impe (1991) refers to Wroth (1972) and writes, *Poor or even dangerous geotechnical design may quite often be blamed to ignoring of the strict interrelationship between soil parameters, their method of determination, the model of the soil strain behavior, the method of analyzing the foundation engineering problem and the corresponding choice of safety factors.* Figure 5.7 illustrates the detailed information that is important in understanding the deformational characteristics of soil.

The importance of in situ methods of determining the deformational properties of soils cannot be overemphasized. The importance is shown graphically in Figure 5.8 (Ward et al. 1959). Van Impe (1991) presented a detailed discussion of the applicability of various in situ methods concerning deformability characteristics of soils.

138 *Piles under lateral loading*

Plate loading test. Application is limited to shallow depths but results show that the average drained Young stiffness can be measured within the depth of influence of the plate. Test is limited in several respects.

Self-boring pressuremeter test. The test has great potential for the measurement of the shear modulus in the horizontal direction and small unloading-reloading cycles may be useful.

Dilatometer test. Empirical correlations will yield values of the tangent constrained modulus of sands and clays.

Cone penetration test. Empirical correlations of deformational characteristics of soil are not generally valid except for normally consolidated sand. Fully drained conditions must be assured.

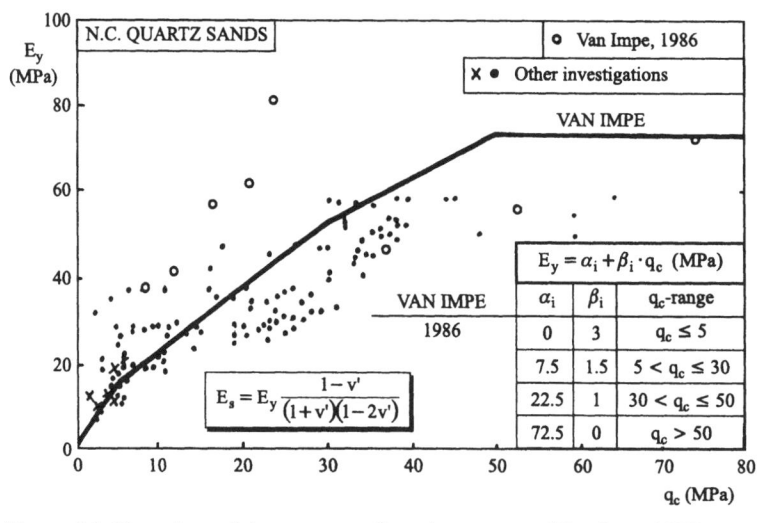

Figure 5.7. Young's modulus versus q_c from the cone test (Van Impe 1988).

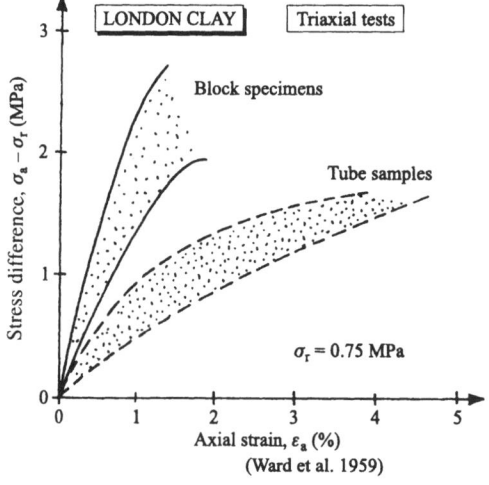

Figure 5.8. Effect of method of sampling on behavior of laboratory specimens (Ward et al.1959) (from Van Impe 1991).

Shear-wave-velocity measurement. The method can yield deformational values of soil under small strain but assumptions must be made concerning the constitutive model, the stress-strain path, and the soil homogeneity.

5.6.3 *Influence of method of installation on soil characteristics*

Driven piles must displace a volume of soil equivalent to the volume of the pile that penetrates. A heave of the ground around the pile normally results for piles in clay. A depression around the pile is sometimes noted when piles are driven into sand because of the densification of the sand due to the vibrations of driving. The excavation for bored piles allows the soil to deform laterally toward the borehole. Continuous-flight-auger piles also cause changes in the characteristics of the soil near the pile.

Excess porewater pressures develop around piles driven into saturated clays. If the piles are driven near each other, the zones of pressure will overlap in a complex manner. The pressures dissipate with time with a consequent decrease in water content and increase in soil strength at the wall of the pile and outward.

No comprehensive attack has been mounted by geotechnical engineers to allow the prediction of the effects of pile installation on soil properties, including deformational characteristics, but some studies have been made that give insight into these effects. Van Weele (1979) got experimental data on the effects of driving a displacement pile (Fig. 5.9). Van Impe (1988) studied the effects of installing a continuous-flight-auger

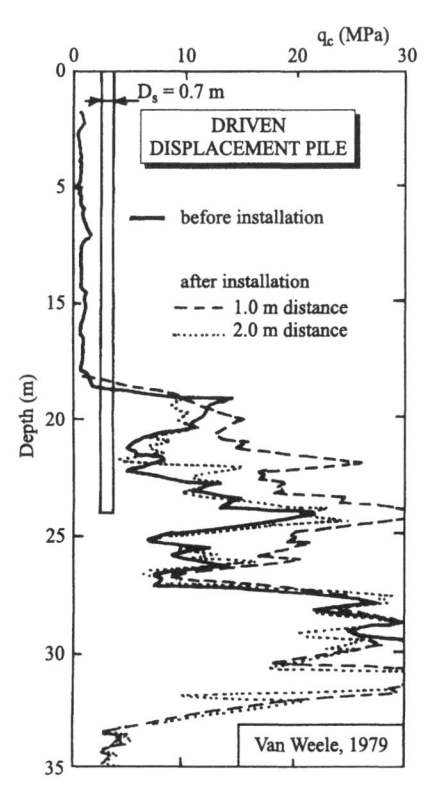

Figure 5.9. Effects of installation of a displacement pile (from Van Impe 1991).

140 *Piles under lateral loading*

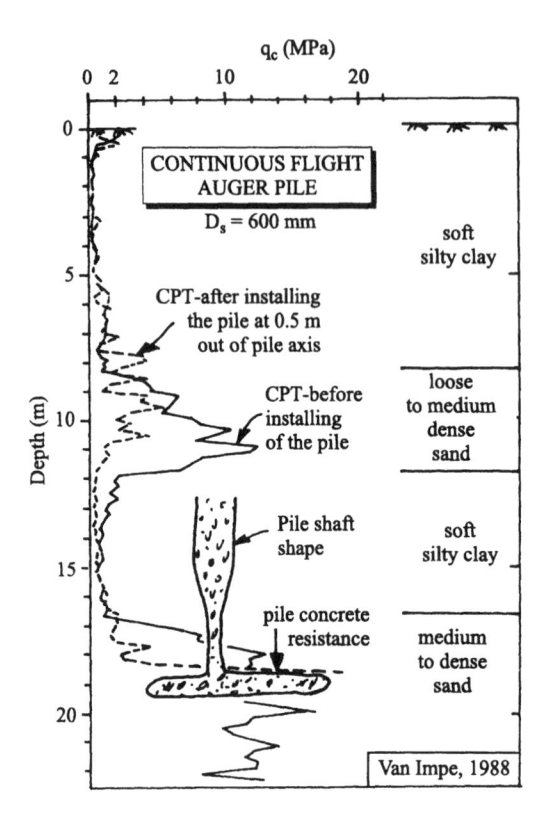

Figure 5.10. Effects of installation of continuous flight auger pile (from Van Impe 1991).

pile (Fig. 5.10). De Beer (1988) presented data on the effects of installing a bored pile (Fig. 5.11). Robinsky & Morrison (1964) present a detailed picture of the effects on the relative density of driving a pile into sand (Fig. 5.12). The data are principally derived from the results of cone penetration tests performed before and after pile driving. The gathering of additional such data, compilation of the data, and detailed analyses will serve to develop methods of predicting the effects of pile installation of soil properties.

5.6.4 *Methods of formulating axial-stiffness curves*

The stiffness is entered as a constant in the equilibrium equations, solutions of the equations are found, the axial movement of each pile is noted, curves giving axial load as a function of movement are entered, and a new stiffness for each pile is computed. Similar procedures are used for the lateral stiffness.

There are basically two analytical methods to compute the load-versus-settlement curve of an axially loaded pile. One method makes use of the theory-of-elasticity. The methods suggested by D'Appolonia & Romualdi (1963), Thurman & D'Appolonia (1965), Poulos & Davis (1968), Poulos & Mattes (1969), Mattes & Poulos (1969), and Poulos & Davis (1980) belong to the theory-of-elasticity method. All of the theories resort to the Mindlin equation, which can be used to find the deformation as a function of a force at any point in the interior of semi-infinite, elastic, and isotropic solid.

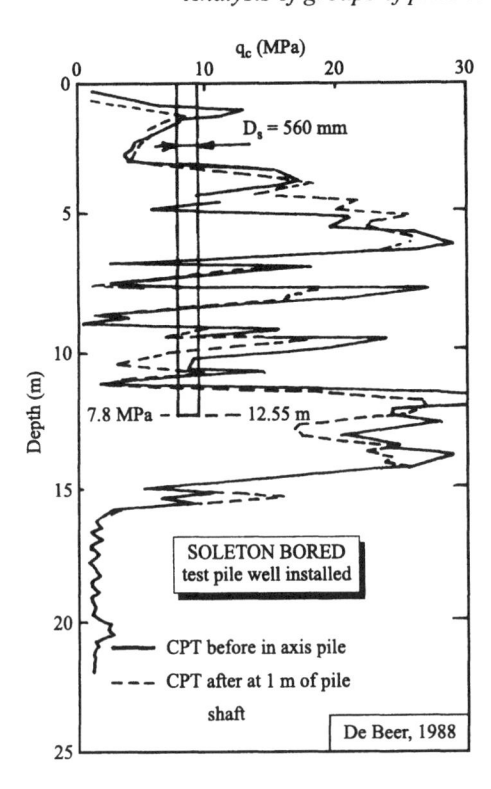

Analysis of groups of piles subjected to inclined and eccentric loading 141

Figure 5.11. Effects of installation of a bored pile (from Van Impe 1991).

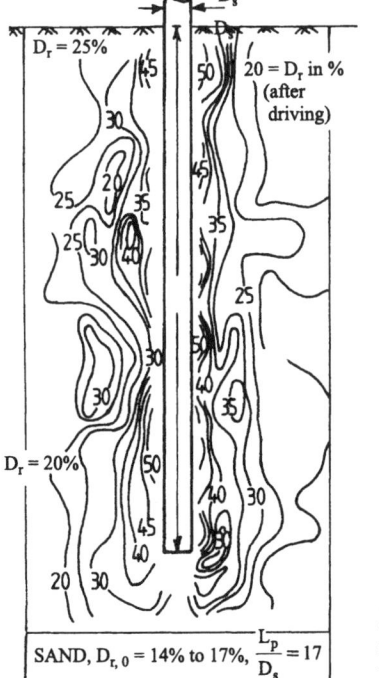

Figure 5.12. Distribution of areas with equal relative density after pile driving (Robinsky & Morrison 1964) (from Van Impe 1991).

142 *Piles under lateral loading*

The displacement of the pile is computed by superimposing the influences of the load transfer (skin friction) along the pile and the pile-tip resistance at the point in the solid. The compatibility of those forces and the displacement of a pile are obtained by solving a set of simultaneous equations. This method takes the stress distribution within the soil into consideration; therefore, the elasticity method presents the possibility of solving for the behavior of a group of closely-spaced piles under axial loadings (Poulos 1968, Poulos & Davis 1980).

The drawback to the elasticity method lies in the basic assumptions which must be made. The actual ground condition rarely if ever satisfies the assumption of uniform and isotropic material. In spite of the highly nonlinear stress-strain characteristics of soils, the only soil properties considered in the elasticity method are the Young's modulus E and the Poisson's ratio ν. The use of only two constants, E and ν, to represent soil characteristics rarely agrees with field conditions.

The other method to compute the load-versus-settlement curve for an axially loaded pile may be called the finite-difference method. Finite-difference equations are employed to achieve compatibility between pile displacement and load transfer along a pile and between displacement and resistance at the tip of the pile. This method was first used by Seed & Reese (1957), other studies are reported by Coyle & Reese (1966), Coyle & Sulaiman (1967), and Kraft et al. (1981).

The finite-difference method assumes that the Winkler concept is valid, which is to say that the load transfer at a certain pile section is independent of the pile displacement elsewhere. Because the curves employed for load transfer as a function of pile movement have been developed principally by experiment, where interaction is explicitly satisfied, the Winkler concept can be used with some confidence.

Close agreement between results from analysis and from experiment for piles in clays has been found (Coyle & Reese 1966), but the results for piles sands show considerable scatter (Coyle & Sulaiman 1967). The effects on the soil of the driving of piles may be more severe in sands than in clays in terms of load-transfer characteristics. In spite of limitations, the finite difference method can deal with any complex composition of soil layers with any nonlinear relationship of displacement versus shear force and can accommodate improvements in soil criteria with no modifications of the basic theory. Improvements in the finite-difference method can be expected as results from additional high quality experiments become available.

The axial-load-settlement curve for a pile is computed by the finite-difference method, described below, by employing curves of load-transfer versus pile movement for points along the sides of a pile and for the end of the pile. The ultimate capacity of a pile is found for some specific movement of the top of the pile. The technical literature is replete with proposals for computing axial capacity and is more voluminous than that for lateral loading. However, the following sections present only specific methods for obtaining the axial stiffness of a driven pile and a bored pile. While the methods have been found to yield results that agree reasonably well with experiment, a comprehensive treatment of axial capacity is not presented.

5.6.5 *Differential equation for solution of finite-difference equation for axially loaded piles*

A graphical model for a pile under axial loading was shown in Figure 1.13. The model can be generalized so that nonlinear load-transfer functions can be input, point by point, along the length of the pile. Also, the stiffness AE of the pile can be nonlinear with length along the pile. Figure 5.13 shows the mechanical system for an axially loaded pile. The pile head is subjected to an axial force P_{top}, and the pile head undergoes a displacement $\bar{z}_t$. The pile-tip displacement is $\bar{z}_{np}$ and the pile displacement at the depth x is $\bar{z}$. Displacement $\bar{z}$ is positive downward and the compressive force P is positive.

Considering an element dx (Fig. 5.13a) the strain in the element due to the axial force P is computed by neglecting the second order term dP.

$$\frac{d\bar{z}}{dx} = -\frac{P}{E_p A_p} \tag{5.18}$$

or

$$P = E_p A_p \left(\frac{d\bar{z}}{dx}\right) \tag{5.19}$$

where P = axial force in the pile (downward positive); E_p = Young's modulus of pile material; and A_p = cross-sectional area of the pile.

The total load transfer through an element dx is expressed by using the modulus m in the load transfer curve (Fig. 5.14a). The maximum load transfer is indicated the symbol f_{max}, a value which must be determined point by point along the length of a pile. The conceptual shape of a curve is shown for load transfer if the pile is subjected to uplift; the curve may or may not differ from the curve for compressive loading.

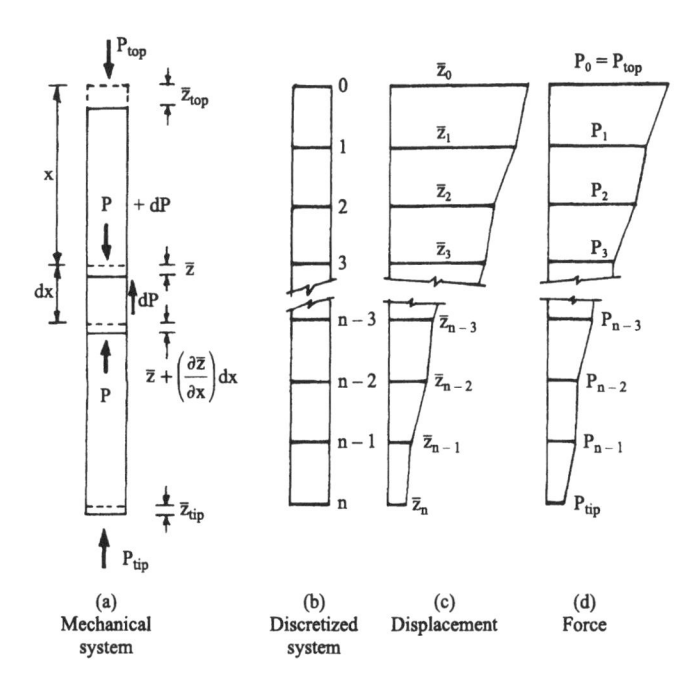

$$dP = -\mu \bar{z} \bar{C} dx \tag{5.20}$$

(a) Mechanical system

(b) Discretized system

(c) Displacement

(d) Force

Figure 5.13. Illustration of mechanics for an axially loaded pile.

144 *Piles under lateral loading*

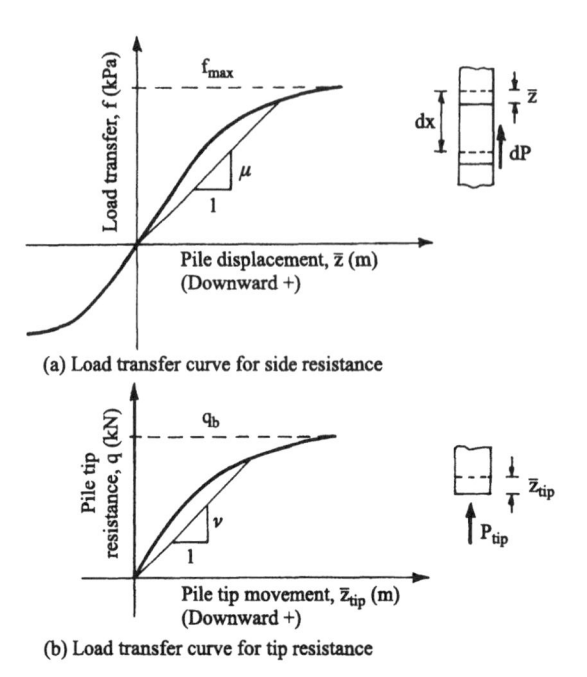

Figure 5.14. Load transfer in side resistance and in tip resistance for a pile under axial loading.

or

$$\frac{dP}{dx} = -\mu \bar{z} \bar{C} \tag{5.21}$$

where $\bar{C}$ = circumference of a cylindrical pile or the perimeter for a pile with a prismatic cross section.

Equation 5.19 is differentiated with respect to x and equated with Equation 5.21 to obtain Equation 5.22.

$$\frac{d}{dx}\left(E_p A_p \left(\frac{d\bar{z}}{dx}\right)\right) = \mu \bar{z} \bar{C} \tag{5.22}$$

The pile-tip resistance is given by the product of a secant modulus v and the pile-tip movement $\bar{z}_{tip}$ (Fig. 5.14b). The maximum load transfer in end bearing is given as q_b, a value that also must be determined point by point along the length of a pile.

$$P_{tip} = v\bar{z}_{tip} \tag{5.23}$$

Equation 5.22 constitutes the basic differential equation which must be solved. Boundary conditions at the tip and at the top of the pile must be established. The boundary condition at the tip of the pile is given by Equation 5.23. At the top of the pile the boundary condition may be either a force or a displacement.

5.6.6 Finite difference equation

Equation 5.24 gives in difference-equation form the differential equation (Eq. 5.22) for solving the axial pile displacement at discrete stations.

$$a_i \bar{z}_{i+1} + b_i \bar{z}_i + c_i \bar{z}_{i-1} = 0 \qquad (5.24)$$

for $0 \leq i \leq n$

$$a_i = \frac{1}{4}(E_p A_p)_{i+1} + (E_p A_p)_i - \frac{1}{4}(E_p A_p)_{i-1} \qquad (5.25)$$

$$b_i = \mu \bar{C} h^2 - 2(E_p A_p)_i \qquad (5.26)$$

$$c_i = \frac{1}{4}(E_p A_p)_{i+1} + (E_p A_p)_i + \frac{1}{4}(E_p A_p)_{i-1} \qquad (5.27)$$

and where h = increment length or dx (Fig. 5.7a and 5.8a).

Equations of the form of 5.24 can be written for every element along the pile, appropriate boundary conditions can be used, and the equations can be solved in any suitable manner. A recursive solution is convenient.

5.6.7 Load-transfer curves

The acquisition of load-transfer curves from a load-test requires that the pile be instrumental internally for the measurement of axial load with depth. The number of such experiments is relatively small and in some cases the data are barely adequate; therefore, the amount of information of use in developing analytical expressions is limited.

There will undoubtedly be additional studies reported in technical literature from time to time. Any improvements that are made in load-transfer curves can be readily incorporated into the analyses.

5.6.7.1 Side resistance in cohesive soil

The curves for f_{max} may be found by a number of methods found in technical literature. The American Petroleum Institute (API) (1993) has proposed a method using Equation 5.28 through Equation 5.30.

$$f_{max} = \alpha_x c_{uz} \qquad (5.28)$$

$$\alpha_x = 0.5 \psi^{-0.5} \quad \psi \leq 1.0$$

$$\alpha_x = 0.5 \psi^{-0.25} \quad \psi > 1.0 \qquad (5.29)$$

$$\psi = \frac{c_{uz}}{\bar{p}} \qquad (5.30)$$

where c_{uz} = the undrained shear strength of the clay at depth z, and $\bar{p}$ = the effective overburden pressure. The constraint in Equation 5.29 is that $\alpha \leq 1.0$. With values of f_{max} the vary from point to point along the length of the pile, the load transfer curves for side resistance can be computed.

Coyle & Reese (1966) examined the results from three, instrumented field tests and the results from rod tests in the laboratory and developed a recommendation for a load-transfer curve. The curve was tested by using results of full-scale experiments with uninstrumented piles. The comparisons of computed load-settlement curves with those from experiments showed agreements that were excellent to fair. Table 5.1 presents the fundamental curve developed by Coyle & Reese.

An examination of the Table 5.1 shows that the movement to develop full load transfer is quite small. Furthermore, the curve is independent of soil properties and pile diameter.

Reese & O'Neill (1987) made a study of the results of several field-load tests of instrumented bored piles and developed the curves shown in Figure 5.15. An examination of Figure 5.15 shows that the maximum load transfer occurred at approximately 0.6% of the diameter of a bored pile. Because the piles tested had diameters of 0.8 to 0.9 m, the movement at full load transfer would be in the order of 5 mm, which is larger than the 2 mm shown in Table 5.1.

Kraft et al. (1981) studied the theory related to the transfer of load in side resistance and noted that pile diameter, axial pile stiffness, pile length, and distribution of soil strength and stiffness along the pile are all factors that influence load-transfer curves. Equations for computing the curves were presented. Vijayvergiya (1977) also presented a method for obtaining load-transfer curves.

5.6.7.2 End bearing in cohesive soil

The work of Skempton (1951) (cited also during discussion of lateral loading) was employed and a method was developed for predicting the load in end bearing of a pile in clay as a function of the movement of the tip of the pile. The laboratory stress-strain curve for the clay at the base of the pile must be obtained by testing, or may be estimated from values given by Skempton for strain, ε_{50}, at one-half of the ultimate compressive strength of the clay. Skempton reported that ε_{50} ranged from 0.005 to 0.02, and further used the theory of elasticity to develop approximate equations for the settlement of a footing (base of a pile). His equations are as follows.

$$q_b = N_c \sigma_f / 2 \tag{5.31}$$

$$w_b / b = 2\varepsilon \tag{5.32}$$

Table 5.1. Load transfer vs pile movement for cohesive soil.

Ratio of load transfer to maximum load transfer	Pile movement in.	Pile movement mm
0.0	0	0
0.18	0.01	0.25
0.38	0.02	0.51
0.79	0.04	1.02
0.97	0.06	1.52
1.00	0.08	2.03
0.97	0.12	3.05
0.93	0.16	4.06
0.93	0.20	5.08
0.93	>0.20	>5.08

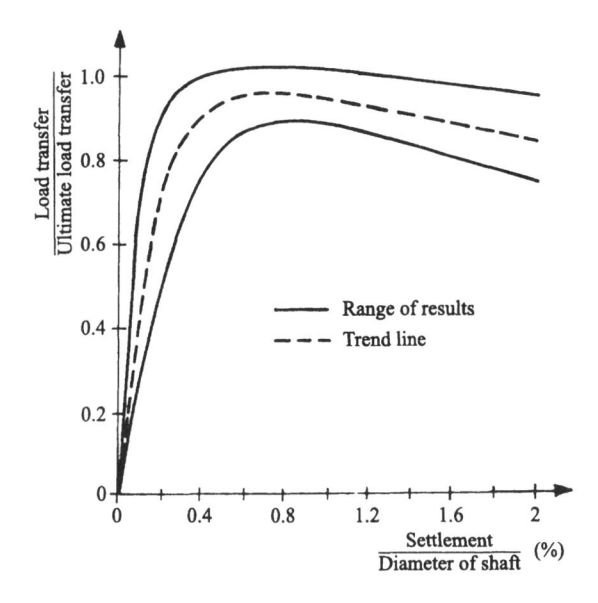

Figure 5.15. Normalized curves showing load transfer in side resistance versus settlement for bored piles in clay (after Reese & O'Neill 1987).

where q_b = failure stress in bearing at base of footing; σ_f = failure compressive stress in the laboratory unconfined-compression or quick triaxial test; N_c = bearing capacity factor (Skempton recommended 9.0); b = diameter of footing or equivalent length of a side for a square or rectangular shape; ε = strain measures from unconfined-compression or quick-triaxial test; and w_b = settlement of footing or base of pile.

The value of ε_{50} can be selected in consideration of whether the clay is brittle or plastic. In the absence of a laboratory stress-strain curve, the shape of the curve can be selected from experience and made to pass through ε_{50}. Equations 5.31 and 5.32 may be used and the load-settlement curve for the tip of the pile can be obtained. The assumption is made that the load will not drop as the tip of the pile penetrates the clay.

Reese & O'Neill (1987) studied the results of a number of tests of bored piles in clay where measurements yielded load in end bearing versus settlement. Figure 5.16 resulted from the studies. Examination of the mean curve shows that a value of about 30 mm will result at the ultimate bearing stress for a pile with a diameter of 0.5 m.

The movement of a pile to cause the full load transfer in end bearing is several times that which is necessary to develop full load transfer in skin friction. The largest strain in skin friction occurs several millimeters away from the wall of a pile when the pile is loaded to failure. End bearing mobilizes the strain on many elements of soil in the zone beneath the tip of a pile. Hence, the movement of a pile to develop the full load transfer in end bearing is a function of the diameter of the pile and can be many times the movement to develop full load transfer in skin friction.

5.6.7.3 *Side resistance in cohesionless soil*
An equation for computing the maximum value side resistance f_{max} in cohesionless soil was suggested the American Petroleum Institute (1993), and presented in Equation 5.33.

148 *Piles under lateral loading*

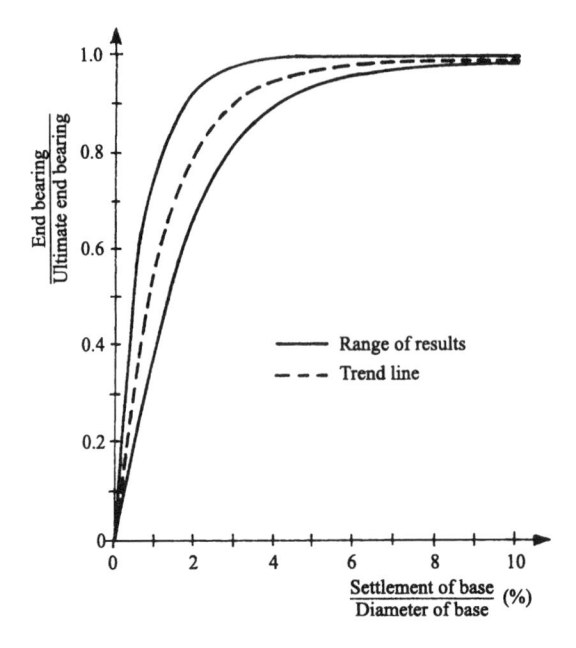

Figure 5.16. Normalized curves showing load transfer in end bearing versus settlement for bored piles in clay (after Reese & O'Neill 1987).

$$f_{max} = K\bar{p} \tan \delta \tag{5.33}$$

where K = a lateral earth-pressure coefficient; $\bar{p}$ = the effective overburden pressure; and δ = the friction angle between the soil and the pile wall. A value of K of 0.8 is recommended for open-ended pipe piles and a value of 1.0 is recommended for full-displacement piles.

The procedure presented in Equation 5.33 is relatively simplistic and can yield good correlations with experimental results if the values of K are selected correctly. Two values are given, indicating that K is a constant with depth, but the value of K is expected to vary with depth and many other factors, including the details of the methods of installation.

Another limitation of Equation 5.33 concerns calcareous or carbonate soils, consisting of soft grains from remains of sea life. Such soils are frequently cemented and open-ended pile piles have been know to penetrate to many meters in such soils under self weight. The soft grains are crushed by the steel shell and cementation prevents the development of lateral pressure with the result that side resistance will be extremely small. The American Petroleum Institute suggests the use of the term *siliceous* in describing soils to which Equation 5.33 is applicable.

Equation 5.33 indicates that f_{max} will increase without limit, but the American Petroleum Institute suggest the limiting values shown in Table 5.2.

Experiments have been reported in literature where tests have been run to solve for the value of δ. A suggestion has been made that the value of δ should be taken as 5 degrees less that the value of ϕ, the friction angle for the granular soil.

Coyle & Sulaiman (1967) studied the load transfer in skin friction of steel piles driven into sand and obtained the curves for piles with diameters ranging from 330 mm to

Table 5.2 Guideline for side friction in siliceous soil (from American Petroleum Institute 1993).

Soil	δ, degrees	Limiting f_{max} kPa
Very loose to medium sand to silt	15	47.8
Loose to dense sand to silt	20	67.0
Medium to dense sand to sand-silt	25	83.1
Dense to very dense sand to sand-silt	30	95.5
Dense to very dense gravel to sand	35	114.8

400 and with a penetration of about 15 m. The water table was near the ground surface and the sand had a friction angle of 32 degrees. An examination of the shape of the curves shows that they can be fitted with the following equation.

$$\frac{f}{f_{max}} = 1.5\left(\frac{z}{B}\right)^{0.15} \quad \text{where} \quad \frac{z}{B} \le 0.07 \tag{5.34}$$

Reese & O'Neill (1987) examined the results of load tests of a number of full-sized bored piles that were instrumented for the measurement of axial load with respect to depth. The results of this study showed that the curves for cohesionless soils were similar to those for cohesive soils and that Figure 5.17 can be used for cohesionless soils.

Mosher (1984) studied the problem of the transfer of load in skin friction (side resistance) of axially loaded piles in sand. He recommended the use of an equation that includes a term for the soil reaction modulus for the sand, a value that will vary with confining pressure and thus is a complex term to evaluate.

5.6.7.4 End bearing curves in cohesionless soil

A procedure for obtaining the value of q_b, the unit end bearing in cohesionless soil, is presented by the American Petroleum Institute.

$$q_b = \overline{p}_t N_q \tag{5.35}$$

where $\overline{p}_t$ = the effective overburden pressure at the tip of the pile, and N_q = a bearing capacity factor. Table 5.3 presents values of N_q and limiting values of q_b.

Vesic (1970) studied the available literature and performed some careful experiments and proposed an equation for computing the load versus tip settlement for piles in sand.

$$w = \frac{C_w q}{(1 + D^2 r) B q_b} \tag{5.36}$$

150 *Piles under lateral loading*

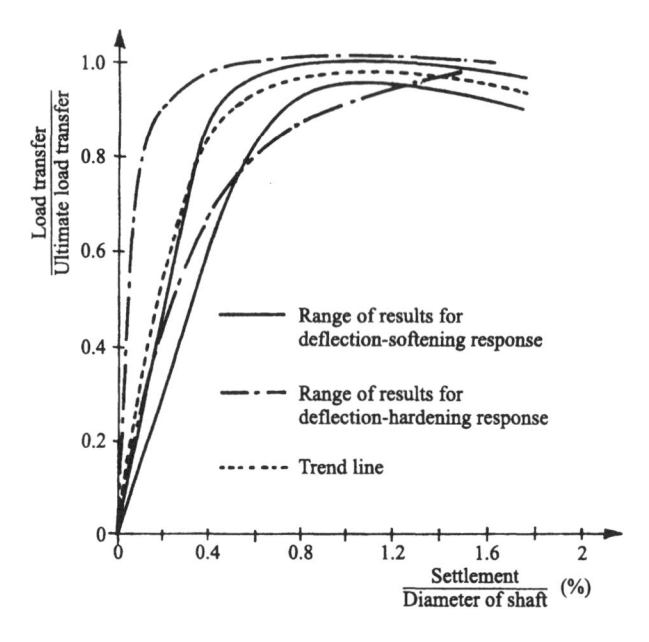

Figure 5.17. Normalized curves showing load transfer in side resistance versus settlement for bored piles in cohesionless soil (after Reese & O'Neill 1987).

Table 5.3 Guideline for tip resistance for siliceous soil (from American Petroleum Institute 1993).

Soil	N_q	Limiting q_b MPa
Very loose to medium sand-silt	8	1.9
Loose to dense sand to silt	12	2.9
Medium to dense sand to sand-silt	20	4.8
Dense to very dense sand to sand-silt	40	9.6
Dense to very dense gravel to sand	50	12.0

where w = settlement, m; q = applied load, kPa; D_r = relative density; B = diameter of tip, m; q_b = ultimate base resistance, kPa and C_w = settlement coefficient (the author found values as follows: 0.00372 for driven piles; 0.00465 for jacked piles; and 0.0167 for buried piles).

(Note: Equation 5.36 is not dimensionally homogeneous so values are dependent on the system of units being used. Values of C_w were recomputed for the SI system.)

Reese & O'Neill (1987) studied the results of experiments with bored piles and developed Figure 5.18. The information in Figure 5.18 was developed from a relatively small

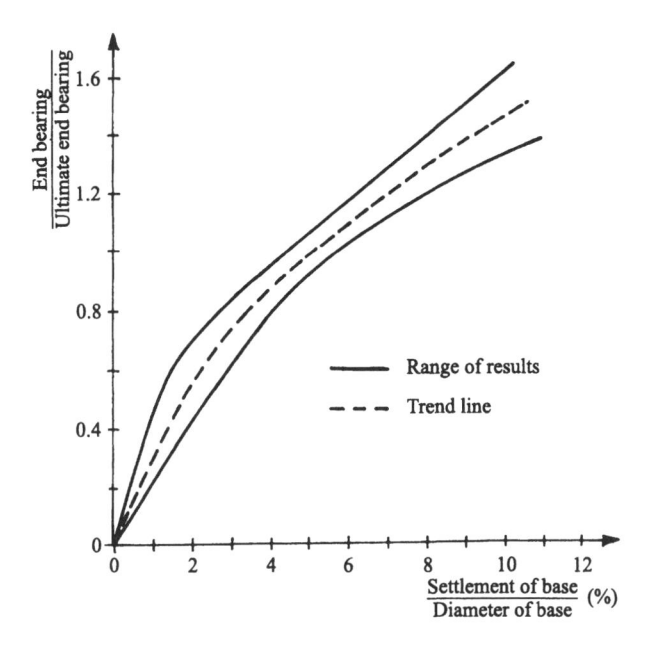

Figure 5.18. Normalized curves showing load transfer in end bearing versus settlement for bored piles in cohesionless soil (after Reese & O'Neill 1987).

amount of data and, as with other methods presented in this chapter, should be used with appropriate discretion.

5.7 CLOSELY-SPACED PILES UNDER LATERAL LOADING

5.7.1 *Modification of load-transfer curves for closely spaced piles*

The method of analysis employed for the behavior of single piles is to employ load-transfer curves, *p-y* curves for lateral loading. The method is extended to the analysis of piles in a group. If the piles are spaced widely apart, the *p-y* curves presented earlier for single piles may be used without modification. As the piles are installed close to each other, their efficiency will decrease and the lateral resistance from the soil will decrease. The decision was made that the most effective way to reflect the loss of efficiency for such piles is to develop procedures for reducing the value of p_{ult} to reflect the close spacing, which in turn will reduce all *p*-values in the *p-y* curves.

The procedure has the distinct advantage on allowing the solution of the nonlinear differential equation for the individual piles in a group when the group is subjected to an inclined and eccentric load. The following sections is this chapter will demonstrate a rational approach, strongly dependent on experimental data, for reducing the values of *p* to reflect close spacing.

152 *Piles under lateral loading*

5.7.2 *Concept of interaction under lateral loading*

The influence of the spacing between piles can be illustrated by referring to Figure 5.19. The assumption is made that all of the piles are fastened to a cap or to a supersturcture and that the lateral deflection of all of the piles will be the same or nearly so. Figure 5.19a shows three closely spaced piles that are in line. It is evident, without resorting to analysis, that the resistance of the soil against Pile 2 is less than that for an isolated pile because of the presence of Piles 1 and 3. Pile 2 may be considered to be in the 'shadow' of Pile 3; the 'shadow-effect' on soil resistance is obviously related to pile spacing.

Similarly, the soil resistance against Pile 2 in Figure 5.19b is influenced by the presence of Piles 1 and 3. The 'edge-effect' on soil resistance is again influenced by pile spacing.

As noted below, some authors (Poulos & Davis 1980, Focht & Koch 1973) have used the theory of elasticity, or a modified version of that theory, to develop influence coefficients that are related to the geometry shown in Figure 5.19c. Those coefficients show that the influence of Pile 1 on Pile 2 can be related to the pile spacing nb and to the angle β. The following sections briefly describe some of the experiments with groups of piles under lateral loading and some of the methods that are proposed for solving of the efficiency of closely-spaced piles.

5.7.3 *Proposals for solving for influence coefficients for closely-spaced piles under lateral loading*

O'Neill (1983) in a prize-winning paper characterizes the problem of closely-spaced piles in a group as one of pile-soil-pile interaction and lists a number of procedures that may be used in predicting the behavior of such groups. He states that none of the procedures should be expected to provide generally accurate predictions of the distribution of loads to piles in a group because none of the models accounts for installation effects. He concluded that there exists a need for more experimental data.

Some of the various proposals are reviewed here in form but not in detail. None of the methods was found to be effective in correlating with data from experiments but do provide a basis for the mostly empirical method that is proposed later to yield reasonable answers to pile-soil-pile interaction.

The theory of elasticity has been employed to take into account the effect of a single pile on others in the group. Solutions have been developed (Poulos 1971, Banerjee & Davies 1979) that assume a linear response of the pile-soil system. While such methods are

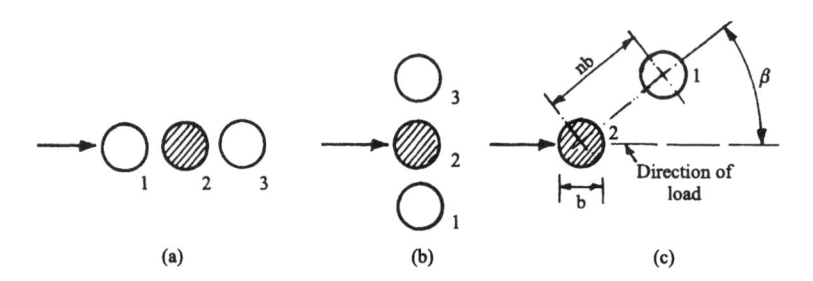

Figure 5.19. Sketches to illustrate influence of pile spacing on pile-soil-pile interaction, (a) in-line piles, (b) sid-by-side piles, (c) piles at an angle with respect to direction of load.

instructive, there is ample evidence to shown that soils cannot generally be characterized as linear, homogeneous, elastic materials.

Focht & Koch (1973) proposed a model that combined the well-documented *p-y* approach of a single pile with the elastic-group effects from Poulos' work. Focht & Koch's modification begins by introducing a term *R* into Poulos' equation as

$$p_k = p_F \sum_{\substack{j=1 \\ j \neq k}}^{m} (H_j \alpha_{pFkj} + RH_k) \qquad (5.37)$$

where *R* = the ratio of the groundline deflection of a single pile computed by the *p-y* curve method to the deflection p_i computed by the Poulos method for elastic soil, H_j = lateral load on pile *j*; α_{pFkj} the coefficient to get the influence of pile *j* on pile *k* in computing the deflection p (curves from the work of Poulos for obtaining the values of the α -coefficients are not shown here), H_k = lateral load on pile *k*; and *m* = number of piles in group (the subscript *F* pertains to the fixed-head case and is used here for convenience; Poulos also presented curves for influence coefficients where shear is applied, α_{pHkj} and where moment is applied, α_{pMkj}).

The above equation can be used to solve for group deflection, Y_g and loads on individual piles. With the known group deflection, Y_g, the *p-y* curves at each depth for a single pile can be multiplied by a factor, termed the '*Y*' factor, to match the pile-head deflection of a single pile with the group deflection, Y_g, by repeated trials. The '*Y*' factor is a constant multiplier employed to increase the deflection values of each point on each *p-y* curves; thus, generating a new set of *p-y* curves that include the group effects. The modification of *p-y* curves, as described above for piles in the group, allows the computation of deflection and bending moment as a function of depth.

From a theoretical viewpoint, group effects for the initial part of *p-y* curves can be obtained from elastic theory. The ultimate resistance of soil on a pile is also affected by the adjacent piles due to the interference of the shear-failure planes, called shadowing effects. Focht & Koch (1973) suggested a *p*-factor may need to be applied to the *p-y* curves in cases where shadowing effects occur. The *p*-factor should be less than one and the magnitude depends on the configuration of piles in a group.

Other approaches regarding the modification of the coefficient of subgrade reaction were also used for pile-group analyses. The Canadian Foundation Engineering Manual (1978) recommends that the coefficient of subgrade reaction for pile groups be equal to that of a single pile if the spacing of the piles in the group is eight diameters. For spacings smaller than eight diameters, the following ratios of the single-pile subgrade reaction were recommended: six diameters, 0.70; four diameters, 0.40; and three diameters, 0.25.

The Japan Road Association (1976) is less conservative. A slight reduction in the coefficient of horizontal subgrade reaction is considered to have no serious effect with regard to bending stress and the use of a factor of safety should be sufficient in design except in the case where the piles get quite close together. When piles are closer together than 2.5 diameters, the following equation is suggested for computing a factor *m* to multiply the coefficient of subgrade reaction for the single pile.

$$m = 1 - 0.2\left(2.5 - \frac{s}{b}\right), \ s < 2.5b \qquad (5.38)$$

154 *Piles under lateral loading*

where s = center-to-center distance between piles; and b = pile diameter.

Bogard & Matlock (1983) present a method in which the *p-y* curve for a single pile is modified to take into account the group effect. Excellent agreement was obtained between their computed results and results from field experiments (Matlock et al. 1980).

As a part of a study where a 3-by-3 group of full-sized piles was loaded laterally, Brown & Reese (1985) reviewed the common approaches for the analysis of pile groups, and concluded that none of the methods was effective in predicting the results that were obtained. The most logical approach appeared to be one that would use the interaction factors for the modification of the *p-y* curves, as was proposed by Focht & Koch (1973). However, the use of interaction factors from the theory of elasticity was unproductive, even for small deflections. The marked difference in the behavior of soils under tension and compression severely limits the application of the theory of elasticity in obtaining the interaction of closely spaced piles under lateral loading.

Modification of *p-y* curves is attractive if some general rules can be found to allow the adjustment of recommendations for *p-y* curves for single piles for the various soils and nature of loading, as presented in Chapter 3. Such modifications would allow loading to each pile in the group to be computed, the effects of cyclic loading to be accommodated, and deflection and bending moment with depth to be computed for each pile in the group. Modifications can be done as shown in Figure 5.20 with *p*-values multiplied by (a_1) and *y*-values multiplied by (a_2). As shown in the material that follows, interaction can be accomplished conveniently by the use of only one of these curve *stretching* parameters.

Scott (1995) performed a comprehensive study of the results of experiments with closely-spaced piles. Experiments were reviewed that were performed both in the field and in the laboratory, and only those tests were analyzed where a single pile was loaded in addition to the pile group. The efficiency is defined as the load on individual piles in the group divided by the load on the single pile *at the same deflection*. Scott noted that the efficiency varied throughout the range of loading; therefore, a reference deflection had to be selected so the results from the various tests could be compared. The reference deflection was selected as 1/50th of the pile diameter.

The loads on the piles corresponding to the reference deflection corresponds approximately to the working load, or the ultimate load divided by a factor of safety ranging from about two to three. A large number of studies were evaluated and the ones which were

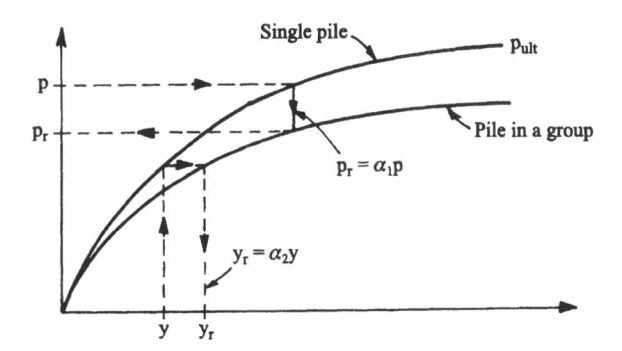

Figure 5.20. Modification of soil resistance for a *p-y* curve for a single pile to for interaction of piles in a group.

judged to have the most complete information are described below. The stipulation was made, of course, that the single pile and the piles in the group had to have the same diameter, had to have been installed in soils with the same characteristics, had to have been installed with the same techniques. and had to have been loaded with the same head conditions, whether fixed-head or free-head. Batter piles were excluded. The meager amount of data did not allow parameters, such as pile diameter and the kind of soil, to be investigated.

The first sets of experiments studied were for piles that were tested in-line or side-by-side. Most of the experiments were performed in the laboratory. The results from the first analyses led to the selection of equations for predicting the efficiency of a pile in a group in terms of the position of a pile in the group and its relative spacing s/b, where s is the center-to-center spacing and b is the pile diameter.

Preliminary analyses were performed and the equations for efficiency were modified by a constant factor, α_g for degrading the p-y curves for pile to pile in the group. Then, with equations on hand for modifying p-y curves for any pile in a group, a series of full-scale experiments were analyzed with the p-y method.

5.7.4 Description and analysis of experiments with closely-spaced piles installed in-line and side-by-side

5.7.4.1 Cox, Dixon and Murphy, 1984

The experiments were performed in the laboratory with steel tubes, 25.4 mm in diameter. The penetrations varied from two to eight diameters because the authors assumed that the near-surface soils have a predominant effect on response to lateral loading. The piles were installed into very soft to soft, plastic clay with a liquid limit of 61. The water content of the soil as installed was 59% and water was kept above the soil surface to prevent drying. The soil was 0.4 m deep in a box that was 0.64-m square and the piles were pushed into place.

The groups consisted of 3 and 5 piles, which were arranged in side-by-side and in-line formations. Single piles were also tested. The penetrations were 2, 4, 6, and 8 in pile diameters. The tests were conducted by applying load at a constant rate of horizontal displacement of 0.97 mm per minute, and most of the tests continued for 7.5 minutes, which gave sufficient displacement to generate the ultimate soil resistance. The heads of the piles were fixed into a frame that prevented rotation, and loads were measured by use of electrical-resistance strain gauges.

The measured efficiencies are shown in Appendix F. Table F.1 presents experimental efficiencies for in-line tests and Table F.2 for side-by-side tests. The authors found that there was no significant difference in the results for different penetrations. Also, the efficiencies of one or two trailing piles did not differ significantly and the efficiencies of side-by-side piles were essentially the same regardless of placement in the group.

5.7.4.2 Franke 1988

Group effects were studied with small-sized piles in the laboratory. The piles were arranged in configurations that were rectangular or approximately rectangular in plan. The center-to-center spacing varied from two to six times the pile diameter. No tests of single piles were reported but Franke asserted that a leading pile acted as a single pile; perhaps single-pile tests were performed to confirm the assertion. The piles were installed into fine sand and electrical-resistance strain gauges were used to measure bending moments. The relative density of the sand ranged from 58% to 88%.

Some of the results from Franke's tests are presented in Table F.3 in Appendix F. On the basis of data such as shown in the table and on other information, Franke reached the following conclusions.
1. The front pile in a group behaves as a single pile.
2. The ratio of the load on a rear pile for in-line piles to that of the load on the front pile can be defined approximately by straight lines on a graph of s (center-to-center spacing) over b (pile diameter) versus efficiency. The efficiency of in-line piles in a group becomes equal to unity for s/b values of 6 or greater.
3. The loads on side by side piles does not vary much with position in the group. The efficiency of side-by-side piles becomes equal to unity for values of s/b of 3 or greater.

5.7.4.3 Prakash 1962

The piles used in the laboratory experiments were aluminum tubes with diameters of 13 mm and with a wall thickness of 0.89 mm. The piles were embedded a maximum of 760 mm in a clean, dry sand, which was placed with relative densities described as loose or dense.

The piles were placed in plan in various rectangular formations. The center-to-center spacings ranged from 2 through 8; single piles were tested. The loads were placed in four increments, and deflections were measured with dial gauges. Readings of deflection were taken after the piles has been in place for 10 minutes. Some of the results reported by Prakash follow.
1. The efficiency of in-line piles in a group becomes equal to unity for s/b values of 8 or greater.
2. The efficiency of side-by-side piles becomes equal to unity for values of s/b of 3 or greater.

5.7.4.4 Schmidt 1981, 1985

An experiment in 1981 was performed in the field with two bored piles 1.2 m in diameter, with a penetration of 28 m, and with a spacing of 2.42 pile diameters. The piles were installed in-line and in very weathered granite. A single pile was tested, and the measured efficiencies are shown by the first line in Table F.4 in Appendix F.

Another experiment in 1981 was also performed with two bored piles 1.2 m in diameter, with a penetration of 16 m, and with a spacing of 1.33 pile diameters. The piles were installed in line and in silt and marl. Again, a single pile was tested, and the measured efficiencies are shown by the second line in Table F.4.

In addition to the results noted above, the 1981 paper reported the measurements of displacement in front of laterally-loaded piles and showed that significant displacements occurred at a distance of 2.0 m in front of piles with diameters of 1.2 m. The displacements were shown to vary significantly with the type of soil.

Schmidt (1985) described the testing of two groups of in-line, bored piles. Each of the groups was installed in-line and with different construction procedures; a variety of types of loading was used. The diameter of all of the piles was 1.2 m and their penetrations were 8.5 m. The piles were installed in uniform, medium dense sand above the water table. Some of the piles were instrumented for the measurement of bending moment as a function of depth. A single, isolated pile was tested in each case so that the efficiency of each of the piles in a group could be found.

One experiment was with two piles with a spacing of 2.0 pile diameters. The measured efficiencies are shown by the third line in Table F.4. Another experiment was with two

Analysis of groups of piles subjected to inclined and eccentric loading 157

piles with a spacing of 3.0-pile diameters and the measured efficiencies are shown in the last line in Table F.4.

Schmidt (1985) presented a remarkable set of data on the efficiencies of in-line piles. Nine experiments are reported, as shown in Figure 5.21, with efficiency of a group being given as a function of the lateral deflection. Diameter, penetration, number of piles, and type of soil are varied. The efficiencies are shown to decrease rapidly with deflection but tend to converge to values between 0.80 and 0.85 at deflections of about 8 mm or more.

Some of the conclusions given by Schmidt in the 1981 and 1985 papers are as follows.
1. The efficiency of groups of piles under lateral loading must be estimated on the basis of individual conditions; such things as soil conditions, spacing, and length of the piles.
2. Test results show that in many cases the group effect is less drastic than some methods show.
3. The group effect vanishes for in-line piles at spacings equal to three diameters or more.
4. With spacing less than three pile diameters for in-line piles, the front pile behaves as a single pile, and group efficiency is governed by the efficiency of the trailing piles. Each of the trailing piles behaves nearly the same.

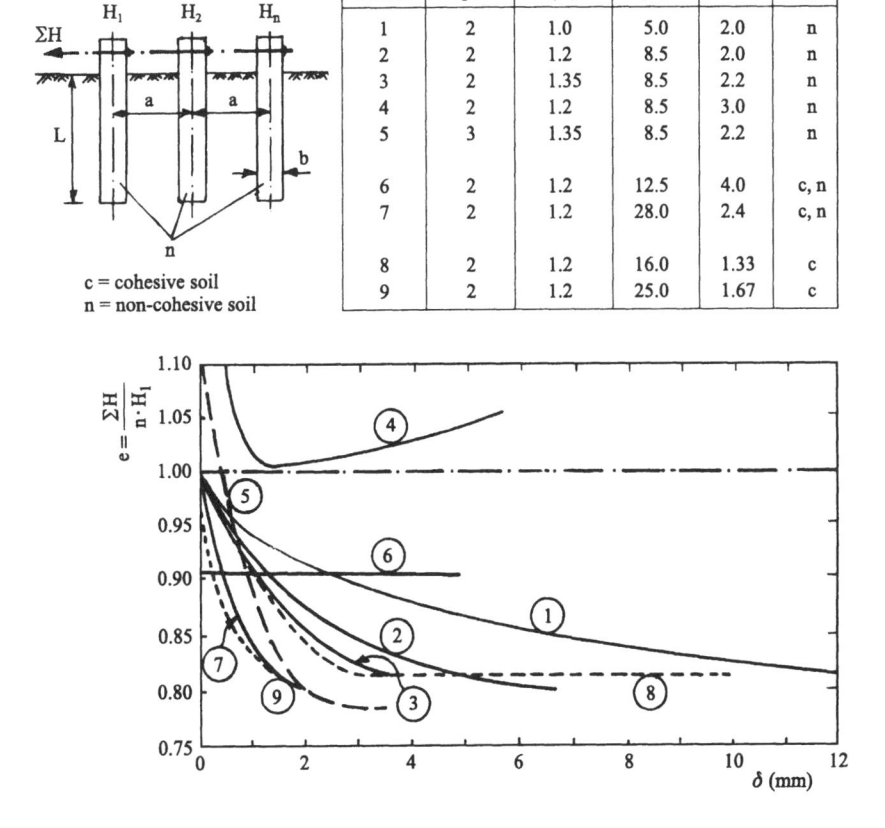

Number of test	Number of piles	Diameter (m)	Length (m)	$\frac{s}{b}$	Soil type
1	2	1.0	5.0	2.0	n
2	2	1.2	8.5	2.0	n
3	2	1.35	8.5	2.2	n
4	2	1.2	8.5	3.0	n
5	3	1.35	8.5	2.2	n
6	2	1.2	12.5	4.0	c, n
7	2	1.2	28.0	2.4	c, n
8	2	1.2	16.0	1.33	c
9	2	1.2	25.0	1.67	c

c = cohesive soil
n = non-cohesive soil

Figure 5.21. Efficiencies of piles installed in-line under lateral load (after Schmidt 1985).

158 *Piles under lateral loading*

5. With pile-head deflections identical due to a rigid cap, bending moments as a function of depth for each pile in a group are approximately the same, regardless of the difference in pile-head loads.

5.7.4.5 *Shibata, Yashimi and Kimura 1989*
Two groups of tests were performed in the laboratory with piles whose lengths were 800 mm. The piles made of aluminum tubing were 20 mm in diameter and piles of chlorized-vinyl tubing were 22 mm in diameter. The loads were placed through a metal plate with joints that allowed unrestrained rotation of the pile heads. Some of the piles were instrumented for the measurement of bending moments.

The in-line and side-by-side groups were tested with spacings of 2, 2.5, and 5 pile diameters. Single piles were tested as well. Several groups were tested with rectangular plans but the results of those tests are not reported here.

The piles were installed in a sand with an average void ratio of 0.76 and a water content of 3.1%. The sand was placed by the boiling method to achieve uniformity and cone-penetrometer tests were performed after the completion of the experiments.

Some of the results from tests by Shibata et al. are presented in Table F.5 in Appendix F. On the basis of data such as shown in the table and on other information, the authors reached the following conclusions.
1. For side-by-side tests, the piles in groups with spacings of three pile diameters approach the behavior of a single pile.
2. For in-line piles, pile-soil-pile interaction was found at spacings of five diameters.

5.7.4.6 *Wang and Reese 1986*
A series of tests were performed on piles that were placed side-by-side. The piles were made of aluminum tubes, 25 mm in diameter, with wall thicknesses of 0.89 mm, and penetrations of 610 mm. The center-to-center spacings in pile diameters were 1.0, 1.25, 1.5, 2.0, 3.0, and 4.0. The piles were instrumented with strain gauges for the measurement of bending moment. A single pile was tested.

The piles were installed into clay with a classification of CH and with a liquid limit of 53. The water content of the clay was 45%. Tests were also performed with piles in sand with a classification of SP. Tests were performed in dense sand with a friction angle of 40 degrees and in loose sand with a friction angle of 34 degrees. The results from the testing are shown in Table F.6 in Appendix F. The conclusions reached from the results of the testing were as follows.
1. For piles in clay and sand, the efficiency with no clear spacing between the piles is approximately 0.5.
2. The efficiency approaches 1.0 for piles in clay and sand when the center-to-center spacing is approximately 3.0 pile diameters. In spite of great care in the testing, some scatter was exhibited in the results.

5.7.5 *Prediction equations for closely-spaced piles installed in-line and side-by-side*

The formulation of the prediction equations was performed by fitting curves to the data representing efficiency versus pile spacing. No differentiation was made with respect to type of soil, diameter of piles, or penetration.

Analysis of groups of piles subjected to inclined and eccentric loading 159

Firstly, a compilation and review of all of the data for side-by-side piles were performed. Logic suggests that the outside piles should carry somewhat more load that the interior piles, but the data did not reveal any significant difference with respect to position in the group. The data show that group efficiency does not become unity until the spacing is nearly four pile diameters.

Secondly, a compilation and review of all of the data for in-line piles were performed. The leading pile was unique in the group but trailing piles, regardless of the number of piles in the group showed no significance difference in efficiency. The leading piles were found to have an efficiency of unity with a spacing of about four diameters and the trailing piles were found to have an efficiency of unity with a spacing of about seven diameters. Equations of the form of $e = a \cdot (s/b)^c$, where a and c are coefficients, was found to fit the data for predicting the loads on side-by-side piles, leading piles, and trailing piles.

Side-by-side piles:

$$e = 0.64 \left(\frac{s}{b}\right)^{0.34} \text{ for } 1 \leq \frac{s}{b} \leq 3.75, e = 1.0, \frac{s}{b} \geq 3.75 \qquad (5.39)$$

Leading piles:

$$e = 0.7 \left(\frac{s}{b}\right)^{0.26} \text{ for } 1 \leq \frac{s}{b} \leq 4.0, e = 1.0, \frac{s}{b} \geq 4.0 \qquad (5.40)$$

Trailing piles:

$$e = 0.48 \left(\frac{s}{b}\right)^{0.38} \text{ for } 1 \leq \frac{s}{b} \leq 7.0, e = 1.0, \frac{s}{b} \geq 7.0 \qquad (5.41)$$

For piles neither directly in-line nor side-by-side, but rather skewed at some angle, the group effects are taken as a modification of the shadow and edge effects. The efficiency is computed from the equation

$$e = (e_i^2 \cos^2 \phi + e_s^2 \sin^2 \phi)^{1/2} \qquad (5.42)$$

where e_i = efficiency of pile where in-line; e_s = efficiency of pile where side-by side; and ϕ = angle between piles (Reese & Wang 1996).

For Pile j in a group with a total of N piles, the group-reduction factor may be found from the following equation.

$$e_j = (e_{1j})(e_{2j})(e_{3j})(e_{4j})....(e_{Ij})....(e_{Nj}), I \neq j \qquad (5.43)$$

In order to implement the power and versatility of the p-y method, the arbitrary, but reasonable, assumption is made that the p-values in the relevant p-y curves can be reduced by implementing Equations 5.39 through 5.43 as follows:

$$-\alpha_j = e_j \text{ (see Fig. 5.20)} \qquad (5.44)$$

and

$$p_{group} = p_{single} \alpha_j \qquad (5.45)$$

160 *Piles under lateral loading*

Equations are now available to compute the response of any pile in a group, regardless of the spacing and the size of the group. The validity of the use of Equations 5.44 and 5.45 will be investigated with the case studies that follow.

The assumption is implicit in the use of the equations that the piles are all of the same size, that the pile-head restraints (either free, fixed, or partially fixed) are identical, and that a pile cap forces all of the piles to deflect the same. When a computer code is employed that models a group in two dimensions, rather than in three dimensions, the efficiency of the piles in each row is averaged in performing the computations. Thus, for a three-by-three group, for example, three values of efficiency are computed, for the front row, the middle row, and the back row.

5.7.6 *Use of modified prediction equations in developing p-y curves for analyzing results of experiments with full-scale groups*

Equations 5.39 through 5.43 will be implemented in the sections that follow to compare predictions for the behavior of groups of piles under lateral loading with results from full-scale experiments with groups of piles. Each of the experiments included a test of a single, isolated pile and results from the single pile were used to obtain experimental values of each pile in the group. As a means of evaluating the validity of Equations 5.39 through 5.43, the first step was to study the results for the single pile. The properties of the soil at the site were adjusted so that the equations for predicting *p-y* curves yielded good agreement between results from analysis and from experiment. Therefore, the behavior of the single pile was used to *calibrate* the site in terms of soil properties. The technique of analysis allowed Equations 5.39 through 5.43 to be evaluated without including the substantial uncertainty about the properties of the soil around the group.

5.7.6.1 *Matlock, Ingram, Kelley and Bogard 1980*
The experiments were performed in the field with steel tubes, 168.3 mm in diameter, 11.9 m in length, and with a wall thickness of 7.1 mm. A pit, about 4.6 m wide by 2.44 m deep, was excavated and braced in order to isolate the piles from a stratum of peat. The piles were pushed into place by use of a template, which also became the loading frame. The moment of inertia of the piles was computed to be 1.17×10^{-5} m^4.

The piles were installed into very soft to soft, dark gray, organic clay. The liquid limit was about 100 and the undrained shear strength, as measured by several methods, ranged rather widely. The analyses that follow are based principally on results from the field vane. The strength at the tops of the piles was taken as 15.4 kPa to about 44.0 kPa at the depth of 14 meters, values that agreed well with results from the field vane. The *p-y* curves for the analyses were formulated from the criteria for soft clay for static loading. As shown below, the selected strength of the soil yielded good agreement between the results from analysis and from experiment for the single pile, the critical first step in evaluating the equations for predicting the behavior of the group.

Two different groups were tested under purely lateral load, a 5-pile group and a 10-pile group, and the axes of the piles formed a circle with a diameter of 0.98 m. The spacing for the 5-pile group was 3.4 diameters and for the 10-pile group was 1.8 diameters. The loading frame was designed for two points of loading, at 0.305 m and at 1.83 m above the ground surface, and load cells were placed at both loading points. Further, the loading apparatus imposed equal deflections at the loading points, where the piles were free to ro-

Analysis of groups of piles subjected to inclined and eccentric loading 161

tate. The top of the piles may then be taken at the lower loading point where the load was known and where the rotational restraint was computed to be 4600 kN-m. A single pile was tested along with the groups.

The piles were tested under both static and cyclic loading; the results for static loading are reported herein. Figure 5.22 shows the numbering of the piles for the 10-pile group; Piles 1, 3, 5, 7, and 9 were in place for the 5-pile group. The curves in Figure 5.23 show the general results from the experiments in terms of the deflection at the pile heads, as a function of the average load on each pile. As may be expected, the efficiency of the piles under lateral loading decreases as the distance between the piles decreases.

According to Matlock et al. (1980), the most important results were those for cyclic loading where the five-pile group behaved like five single piles. However, the cyclic deterioration and gapping for the ten-pile group demonstrated composite-group behavior.

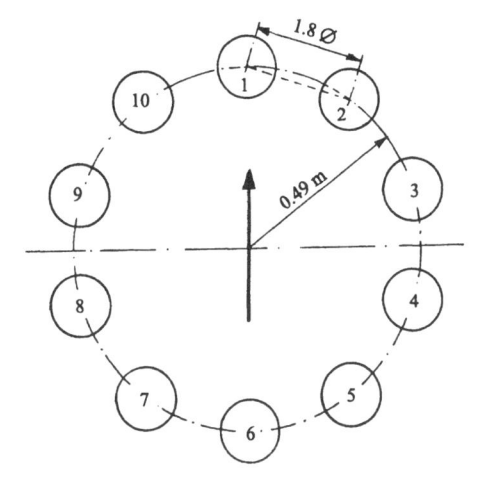

Figure 5.22. Numbering of piles with respect to direction of loading, Matlock test.

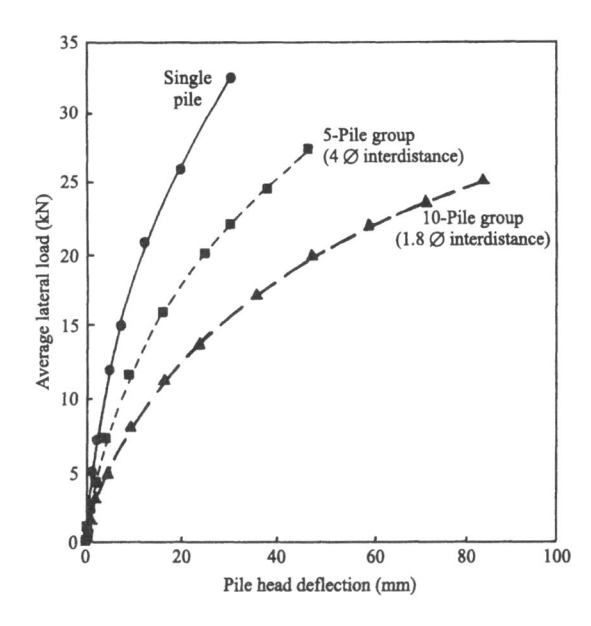

Figure 5.23. Load versus deflection for piles under lateral loading (after Matlock et al. 1980).

162 *Piles under lateral loading*

The analysis of the single pile was done with techniques described in earlier chapters, using the criteria for *p-y* curves for soft clay (Matlock 1970). The details are shown in Figure 5.24 where comparisons are given between results from experiment and from analysis for both deflection and maximum bending moment. The maximum moment occurred at the top of the pile because of rotational restraint and is negative according to the adopted sign convention; but the moment is plotted as positive for convenience in presentation. Agreement is excellent for pile-head deflection for all but the maximum loads and is good for moment where results from analysis are slightly conservative.

The analyses of the 10-pile group proceeded by the use of Equations 5.39 through 5.43. Comparison between the experimental and computed values of pile-head deflection and maximum bending moment are shown in Figure 5.25. Excellent agreement is shown between experimental and computed values of pile-head deflection for the full range of loading. Reasonable agreement is shown between experimental and computed values of maximum bending moment with analytical values being somewhat conservative.

Similar comparisons for the 5-pile group are shown in Figure 5.26. Excellent agreement was obtained between experimental and computed values of bending moment and fair agreement was found for pile-head deflection where analytical values were somewhat unconservative.

According to Matlock et al. (1980), the most important results were those for cyclic loading where the five-pile group behaved like five single piles. However, the cyclic deterioration and gapping for the ten-pile group demonstrated composite-group behavior.

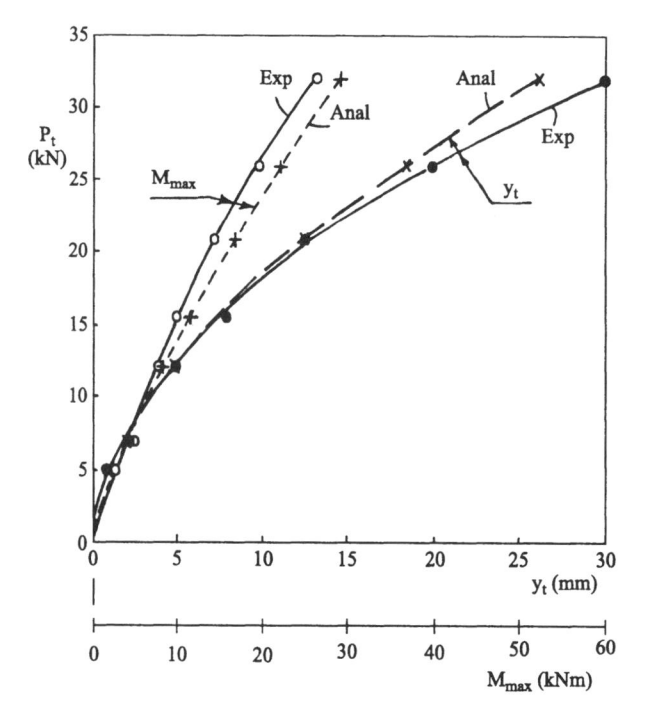

Figure 5.24. Comparison of results from experiment and from analysis for y_t and M_{max} Matlock test of single pile.

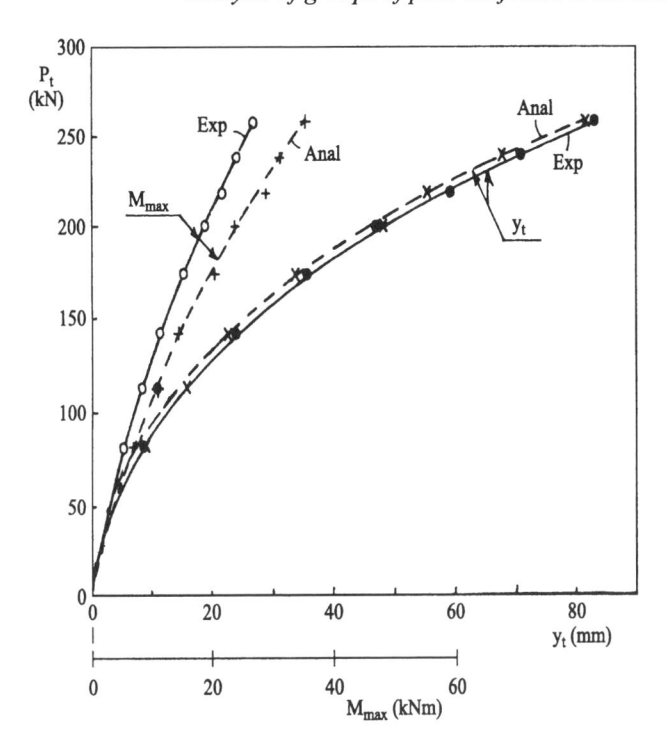

Analysis of groups of piles subjected to inclined and eccentric loading 163

Figure 5.25. Comparison of results from experiment and from analysis for y_t and M_{max} Matlock test of 10-pile group.

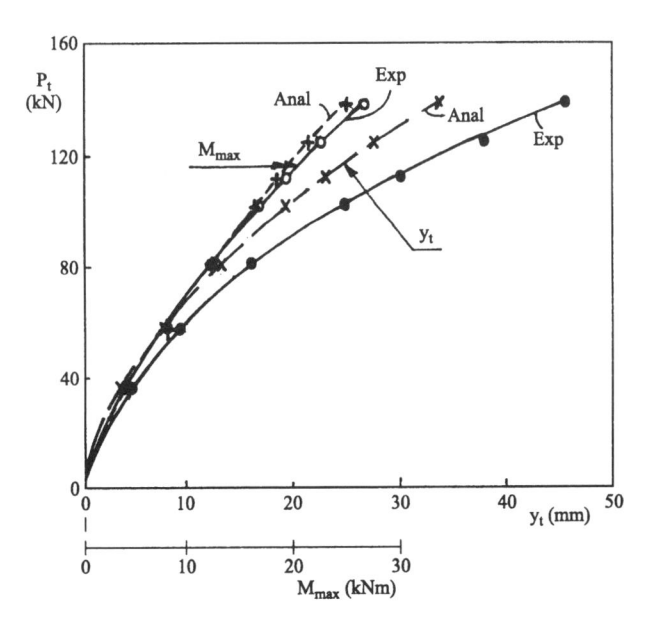

Figure 5.26. Comparison of results from experiment and from analysis for y_t and M_{max} Matlock test of 5-pile group.

164 *Piles under lateral loading*

Table 5.4 shows comparisons between experimental and computed values of efficiency where the experimental values were taken at a deflection of 1/50th of the pile diameter. The displacement selected for the comparison is just an example and is not according to any set of recommendations. The data demonstrate a dramatic increase in the efficiency of the group as the spacing increases. The comparison between experiment and analysis is reasonable.

5.7.6.2 Brown 1985; Brown, Reese and O'Neill 1987
The experiments were performed in the field with steel tubes, 273 mm in diameter, and with a penetration of about 12.2 m. The 3 by 3 group had a spacing of 3-pile diameters center to center. An isolated, single pile was also tested. The group had originally been installed to study interaction under axial loading. For the lateral tests, a shallow pit was excavated around the piles and filled with water to promote saturation. Then, 152-mm-diameter steel tubes, instrumented for measurement of bending moment, were installed along the axis of each of the test piles and grouted into place. After the lateral tests were completed, the soil around the tops of the piles was removed, the piles were reloaded, and calibration curves were obtained, showing readings of electrical-resistance strain gauges as a function of bending moment.

The soil surrounding the piles consisted mainly of stiff, overconsolidated clays. The shear strength was measured by several methods, including triaxial tests, self-boring pressuremeter tests, and quasi-static penetration tests. Values of shear strength that were found to give good agreement between results from analysis and from experiment for the single pile were 55.2 kPa at the ground surface with a linear increase with depth to 138 kPa at a depth of 5.5 m. The values are in reasonable agreement with measurements at the site but somewhat below the average of results from the various tests.

The lateral loads were applied by a frame, allowing a moment-free connection at each pile head. A load cell measured the load to the group and individual cells measured the load to each pile. The readings of deflection transducers, load cells, and bending-moment gauges were recorded by an automatic data-acquisition system. The load was applied in cycles and repeated until equilibrium was achieved, usually at less than 100 cycles. The loading was selected such that the succeeding load was enough larger than the previous load that the first load in the new cycling was judged to be unaffected by the previous cycling. Thus, results were obtained for *static* as well as for *cyclic* loading. Only the results for the static loading will be analyzed herein where static loading is assumed to occur during the first cycle for each of the increments of loads. The increments of load were se-

Table 5.4 Measured and computed efficiencies of piles in the groups tested by Matlock.

5-pile group Pile nos.	Efficiencies Exp.	Anal.	10-pile group Pile nos.	Efficiencies Exp	Anal.
1	0.61	0.84	1	0.43	0.48
3-9	0.595	0.77	2-10	0.52	0.47
5-7	0.735	0.75	3-9	0.35	0.38
			4-8	0.34	0.33
			5-7	0.46	0.33
			6	0.50	0.12
Averages	0.654	0.776		0.388	0.362

lected so that the repeated loads at the previous increment of load was assumed not to affect the results of the loading at the next increment.

The deflection of the pile heads as a function of the average load on each pile is shown in Figure 5.27. The loss of efficiency of the piles in the group is adequately demonstrated. The loss of efficiency is greatest at the higher loads; however, as noted earlier, the experimental efficiency is taken at 1/50th of the pile diameter, rather as a function of load or deflection, which introduces some error in the analytical method.

The results of the analysis of the single pile pile-head deflection and for maximum bending moment are shown in Figure 5.28. The instrumented insert was not installed in the pile but strain gauges for the measurement of bending moment were placed for another study. The moment of inertia of the single pile was reported to be 6.66×10^{-5} m^4 and the moment of inertia of the piles in the group was reported to be 8.62×10^{-5} m^4. In the computation of deflection, the computed values and somewhat less than the experimental ones for the earlier loads. But the overall agreement between computed values and analytical values of deflection and maximum bending moment is good to excellent.

The results of the analysis of the 9-pile group are shown in Figure 5.29. While the curves from analysis show less curvature than the curves from experiment, the overall agreement between experiment and analysis is good.

The efficiency of each pile in the group was found from the measurement of pile-head loads and the averages for front row, middle row, and back row were found to be 0.84, 0.78, and 0.85, respectively. The corresponding values from the analytical method were found to be 0.85, 0.55, and 0.56. In view of the lack of agreement between experimental and computed values for the middle and back rows, the relatively good agreement between deflection and bending moment for the full range of loading is somewhat surprising.

5.7.6.3 *Morrison 1986, Morrison & Reese 1988, Brown et al. 1988*
The tests were performed in the field on the same group tested by Brown & Reese (1985). The clay was excavated to a depth 2.9 m and sand was compacted around the single pile

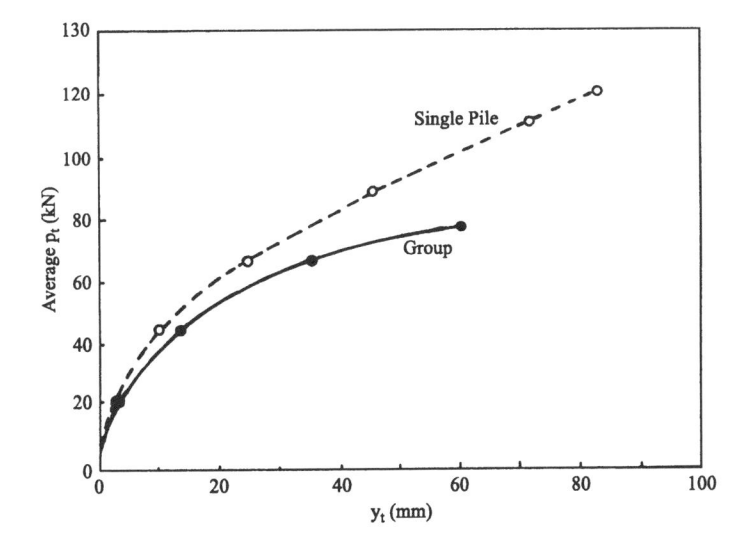

Figure 5.27. Load versus deflection for piles under lateral loading (after Brown 1985).

166 *Piles under lateral loading*

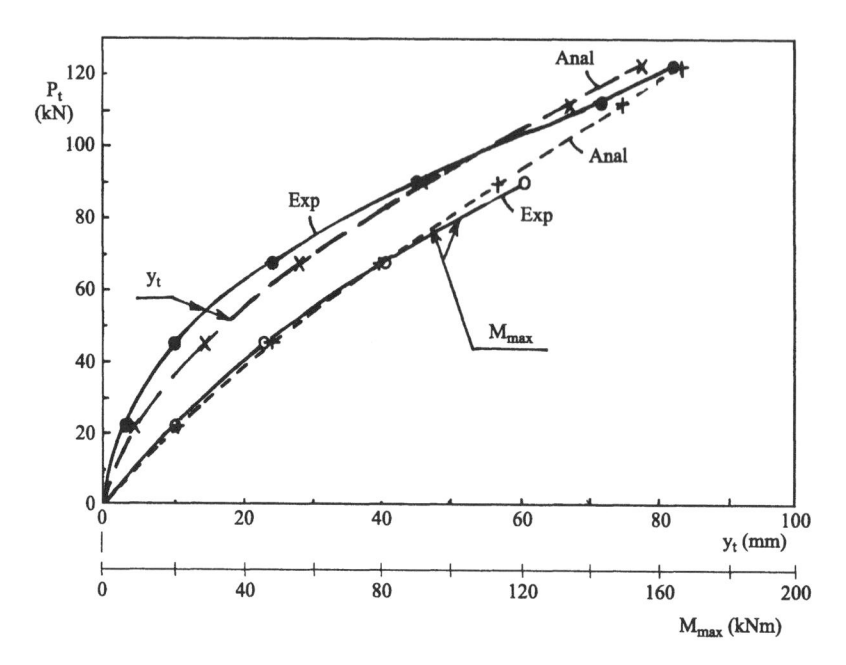

Figure 5.28. Comparison of results from experiment and from analysis for y_t and M_{max} Brown test of single pile.

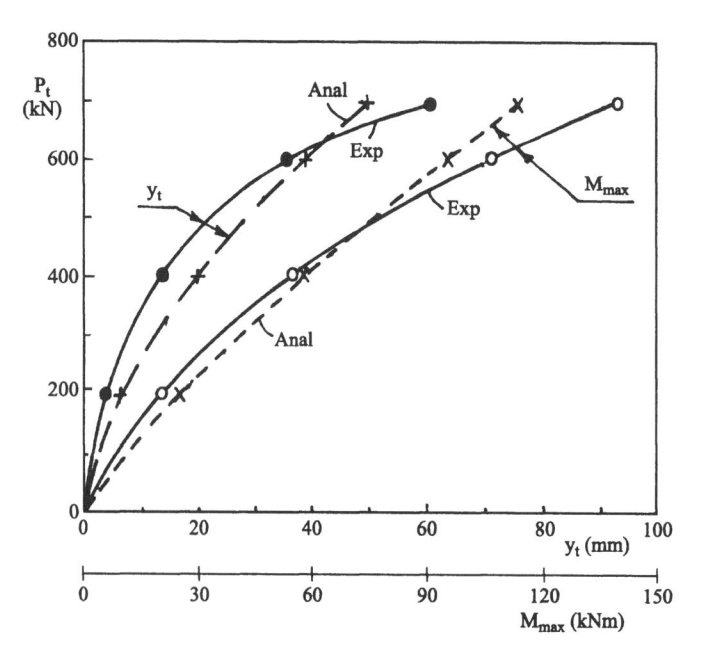

Figure 5.29. Comparison of results from experiment and from analysis for y_t and M_{max} Brown test of group.

Analysis of groups of piles subjected to inclined and eccentric loading 167

and the group of piles by a vibratory-plate compactor in lifts of 200 mm. The sand had a dry density of 5.5 kN/m³ and the friction angle was measured near the ground surface as 38.5°. However, the density and the friction angle were not measured with depth. The sand was saturated by introducing water through perforated pipes at the base of the stratum of sand.

The testing was performed in a manner similar to that used by Brown. The deflection of the pile heads as a function of the average load on each pile is shown in Figure 5.30. As observed with the case for the same group in clay soils, the deflection of the group is significantly greater that that of the single pile.

As for the previous analyses, the first major step was to compute the engineering properties of the of the soil that correlated with the results from the test of the single. Correlation was very poor with the use of the values noted above; the computed deflections were far greater that the experimental values. The friction angle was adjusted until there was agreement between experimental and computed values of deflection and maximum bending moment. A friction angle of 55 degrees was found. A re-evaluation of the procedures at the site revealed that the density shown above was based on the properties of the near-surface soil and a reasonable assumption is that the compaction of the surface layers increased the density of the deeper layers.

In order to obtain an approximate agreement between results from experiments and from analysis, the friction angle of the sand was assigned a value of 55 degrees. The computed values of deflection and maximum bending moment for the single pile are plotted along with the experimental values in Figure 5.31. The agreement of the values of deflection is fair; the analytical values are somewhat higher than the experimental ones. The agreement for the maximum bending moment is excellent for most of the range of loading.

Some comments are in order about the φ-value of the sand. An argument can be made that the higher value may not be unreasonable but of most importance is the aim of the studies described here which is to determine the influence of close spacing on the response of piles in groups where that influence is related to the response of a single pile.

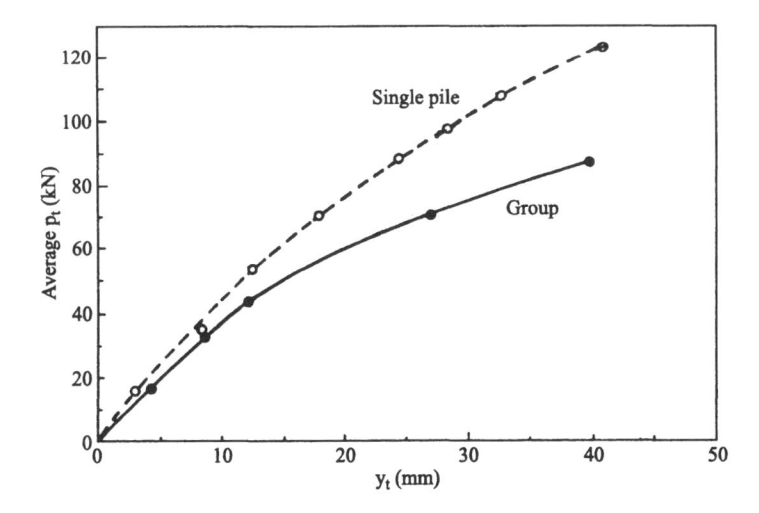

Figure 5.30. Load versus deflection for piles under lateral loading (after Morrison 1986).

168 *Piles under lateral loading*

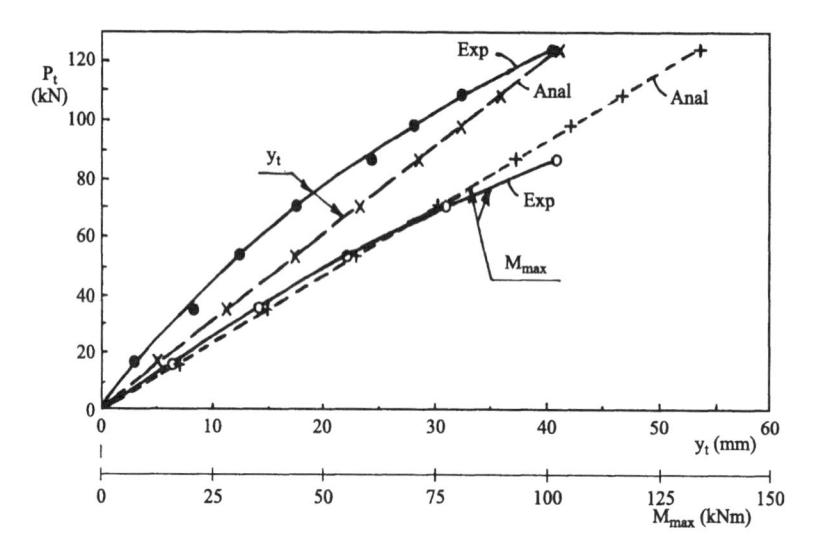

Figure 5.31. Comparison of results from experiment and from analysis for y_t and M_{max} Morrison test of single pile.

Thus, logic suggests that the results for the single pile should agree reasonably well with theory. Chapter 6 presents a comprehensive study of single piles under lateral loading.

Employing the friction angle of 55 degrees, along with other data on the soil, computation was made for the behavior of the group by use of a computer code, such as presented in Appendix D. The results of the comparisons of computed and experimental values of deflections and maximum bending moment are shown in Figure 5.32. The agreement is good for the lower ranges of loading but the analytical method is unconservative for the higher ranges of loading. The analysis shows then the piles behave stiffer that was revealed by experiment. The increased stiffness is also shown in the analysis of the single pile.

The efficiency of each pile in the group was found from the measurement of pile-head loads and the averages for front row, middle row, and back row were found to be 0.68, 0.70, and 0.49, respectively. The corresponding values from the analytical method were found to be 0.85, 0.55, and 0.56. The average of the experimental values is 0.623 and for the analytical values is 0.653; good agreement even though the values for the separate rows vary considerably.

5.7.6.4 *Rollins, Peterson and Weaver 1998.*

The experiments were performed in the field with steel tubes, 273 mm in diameter, with a wall thickness of 9.5 mm, and with a yield strength of 331 MPa. (Peterson & Rollins 1996). The penetration of the piles was 9.1 m. The 3 by 3 group had a spacing of 3-pile diameters center to center. An isolated, single pile was also tested. Inclinometer tubes were installed in the interior of the piles and fixed into place with grout. The nonlinear behavior in bending due to the presence of the grout was ignored and the moment of inertia of the section was computed to be 1.311×10^{-4} m^4.

A granular fill with a thickness of 1.7 m was removed from the site prior to installing the piles. The properties of the soil was measured by the use of various techniques, including the Standard Penetration Test, cone penetrometer test, dilatometer test, push-in pressuremeter test, vane shear test, and laboratory tests. Triaxial tests, consolidation tests and torvane tests were performed on cores from field sampling with a thin-walled tube with an outside diameter of 76.2 mm. Atterberg limits, water contents, and grain-size distribution were obtained from tests of material from the tube samples or disturbed material from the split spoon.

The description of the soil and the shear strength selected for the analysis of the results of the testing of the single pile are: 2.3 m of sandy silt, clayey silt, and lean clay with a shear strength of 40 kPa; 2.5 m of dense, poorly graded sand with a friction angle of 32 degrees; 2.5 m of interbedded layers of sandy silt, silty sand, and lean clay with an shear strength of 40 kPa; 2 m of fine sand with a friction angle of 30 degrees; and underlain with several meters of interbedded sandy silt, silty sand and lean clay with a shear strength of 70 kPa. The values of shear strength are within the range of values reported by the investigator. The water table was at the ground surface after the removal of the fill.

The load was applied by a single-acting hydraulic ram at 0.4 m above the ground surface. Loads were measured with load cells and care was used to remove the effects of friction. The deflection of the pile was measured with dial gauges and with LVDT's. Bending moments along the length of the pile were obtained from interpreting the data from the inclinometers and from strain gauges on the piles.

The deflection of the pile heads as a function of the average load on each pile is shown in Figure 5.33. As expected, the deflection of the group is significantly greater that that of the single pile.

The computed values of deflection and maximum bending moment for the single pile are plotted along with the experimental values in Figure 5.34. The agreement of the values for both deflection and bending moment is good for the early loads but the analytical values are unconservative at the upper loads.

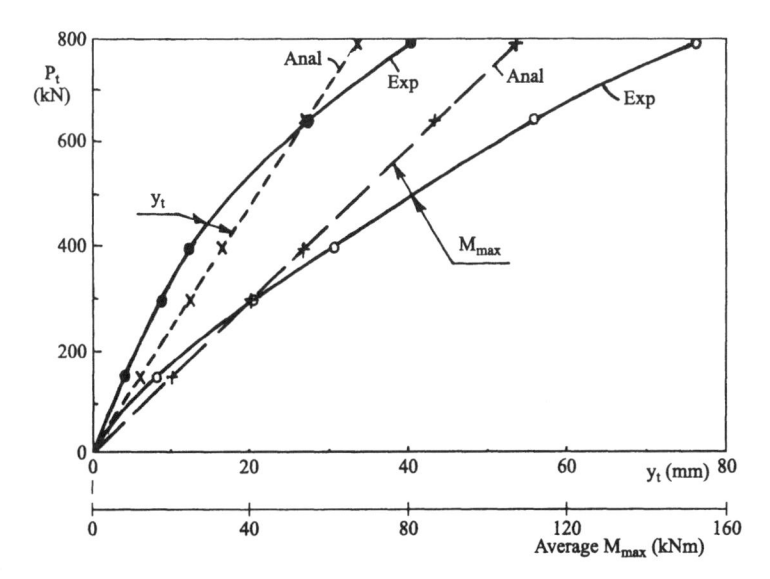

Figure 5.32. Comparison of results from experiment and from analysis for y_t and M_{max} Morrison test of group.

170 Piles under lateral loading

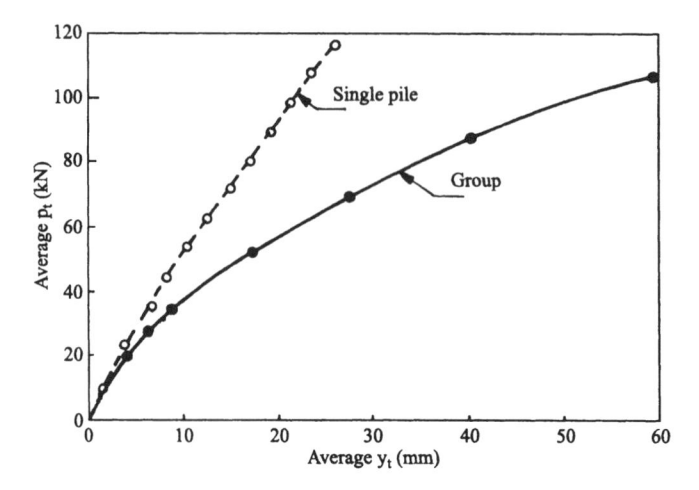

Figure 5.33. Load versus deflection for piles under lateral load (after Peterson 1996).

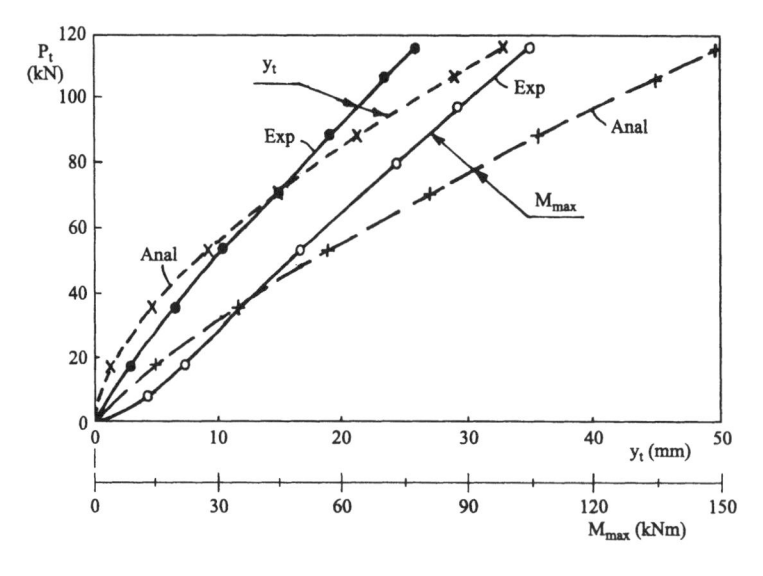

Figure 5.34. Comparison of results from experiment and from analysis for y_t and M_{max} Peterson test of single pile.

Employing the values of the properties of the soil indicated for the analysis of the single pile, Computer Program GROUP (see Appendix D) was used to compute the behavior of the group. The results of the comparisons of computed and experimental values of deflections and maximum bending moment are shown in Figure 5.35. The agreement is for maximum bending moment for the full range of loading but the computed values of deflection are somewhat unconservative.

The efficiency of each pile in the group was found from the measurement of pile-head loads and the averages for front row, middle row, and back row were found to be 0.83,

0.50, and 0.65, respectively. The corresponding values from the analytical method were found to be 0.85, 0.55, and 0.56. The average of the experimental values is 0.660 and for the analytical values is 0.653. The investigators found the efficiencies for the full range of deflection, rather than using a deflection of 1/50th of the pile diameter, as employed in analysis for other tests. There was strong nonlinearity for deflections up to 10 mm (1/30th of the pile diameter) but the efficiencies were relatively constant at 10 mm and at larger deflections; therefore, the efficiencies were determined at a deflection of 10 mm for the Peterson tests.

5.7.6.5 *Ruesta & Townsend 1997a,b*

The tests were done in the field at the site for the construction of a bridge in Florida. A single pile and a 4-pile by 4-pile group were loaded laterally and data taken on deflection and bending moment. The piles were prestressed concrete with square section of 0.762 m by 0.762 m. The center-to center spacing between the piles in the group was 2.29 m. The penetration of the piles below the mudline was 14 m. Several of the piles were instrumented for the measurement of deflection and bending moment along the length of the piles. Tests were performed with the pile heads free to rotate (pinned head) and restrained against rotation. Only the results of the tests of the pinned-head piles will be compared with the analytical results.

The piles were prestressed by using 24 13-mm tension members of ASTM-A-416 steel with a yield strength of 1860 MPa. The concrete had a yield strength of 41 MPa. The analysis of the results of the testing of the single pile was accomplished by the use of a computer code (see Appendix D), which implements the equations for the nonlinear behavior of the reinforced-concrete section, described in Chapter 4.

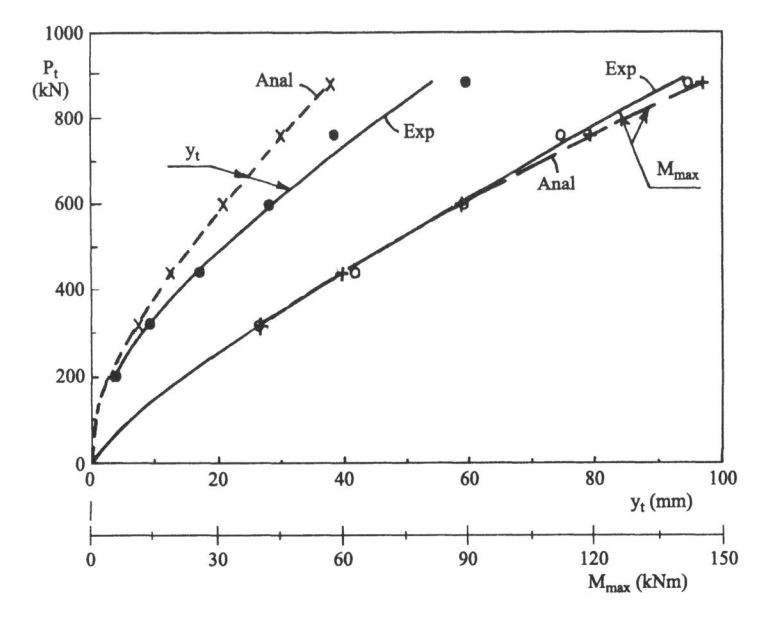

Figure 5.35. Comparison of results from experiment and from analysis for y_t and M_{max} Peterson on test of single group.

172 *Piles under lateral loading*

The properties of the soil was investigated by various methods. The in situ tests performed were the Standard Penetration Test, cone penetration, dilatometer, and pushed-in pressuremeter. Identification of the subsurface materials was done using the disturbed soil from the split spoon. The upper stratum of soil at the site was a sand with a friction angle of 32 degrees and a submerged unit weight of 9.0 kN/m^3. Underlying the sand was a stratum of cemented sand with a friction angle of 42 degrees and a submerged unit weight of 11.5 kN/m^3.

A visual comparison of the deflection of the single with that of the group cannot be presented because the distance from the point of application of the load to the mudline varied. The single pile was loaded at 2.08 m above the mudline and the group was loaded at 1.85 m. The test program was conducted in a river where the water depth was about 1.8 m.

The computed values of deflection and maximum bending moment for the single pile are plotted along with the experimental values in Figure 5.36. The properties of the soil reported by the investigators were employed without modification in the computations. The agreement between experimental and computed values of is good up to a lateral load of about 200 kN. For larger loads, the experimental values of deflection are significantly greater than the computed ones. The agreement between experimental and computed values of bending moment is excellent for the full range of loading.

Employing the values of the properties of the soil indicated for the analysis of the single pile, prediction of the behavior of the group was done with analytical techniques. The results of the comparisons of computed and experimental values of deflections and maximum bending moment are shown in Figure 5.37. The maximum bending moment is only for the front row of the piles where each pile was instrumented for the measurement of bending moment. The agreement of the curves from deflection is good except that the experiment exhibited more curvature than did the analysis. The curves for maximum bending moment for the front row of piles shows that the computation is slightly unconservative.

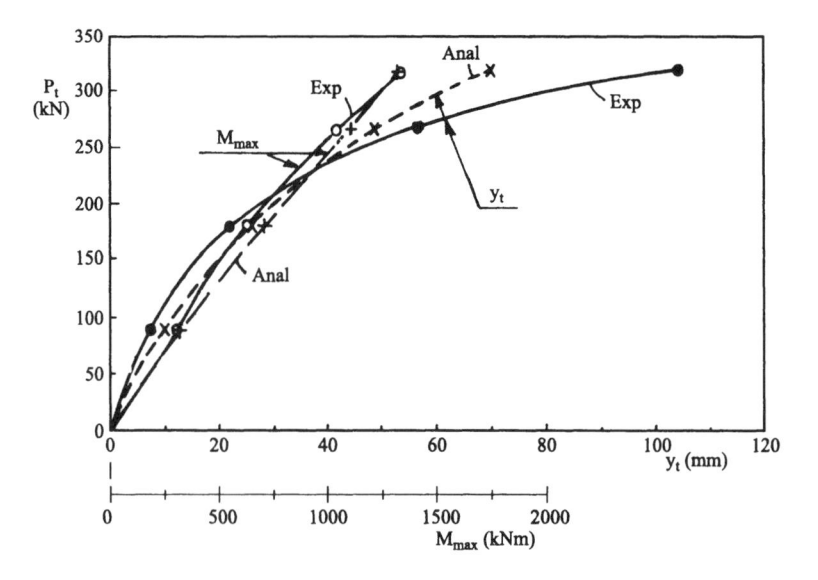

Figure 5.36. Comparison of results from experiment and from analysis for y_t and M_{max}, test of single pile (after Ruesta & Townsend. 1997a).

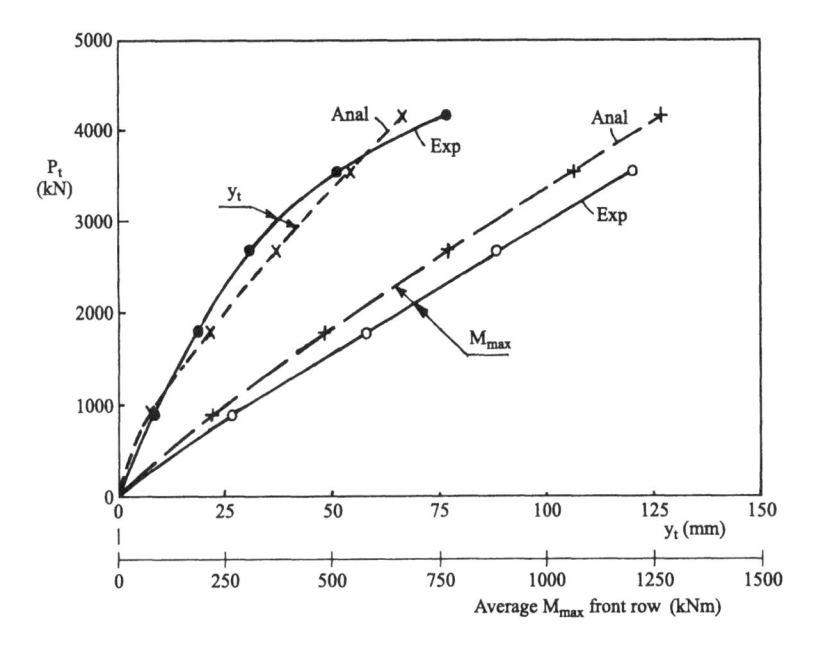

Figure 5.37. Comparison of results from experiment and from analysis for y_t and M_{max} front row of piles Townsend test of group (after Townsend et al. 1996).

The efficiency of each row of piles, from the front row to the back, was given by the investigators as 0.8, 0.7, 0.3 and 0.3, with an average of 0.55. The corresponding values computed by the analytical method are 0.84, 0.53, 0.50, and 0.54, with an average of 0.60.

5.7.7 Discussion of the method of predicting the interaction of closely-spaced piles under lateral loading

While the sample is rather small, perhaps because of the expense of performing full-scale tests of a group of piles under lateral loading, the comparisons of the results of experiment and analysis show that the method of prediction has considerable merit. In no case was the difference between results from experiment and from analysis sufficient to indicate a problem in design if normal factors of safety are employed.

5.8 PROPOSALS FOR SOLVING FOR INFLUENCE COEFFICIENTS FOR CLOSELY-SPACED PILES UNDER AXIAL LOADING

5.8.1 Modification of load-transfer curves for closely spaced piles

In order to maintain a consistent and rational procedure for the analysis of piles in a group, the modification of the load-transfer curves to reflect close spacing under axial load is adopted as was done for piles under lateral load. Thus, the values of f for load transfer in side resistance (Fig. 5.14a) are reduced as are the values of q for load transfer in end bearing. As noted below, neither theory nor experiment can be used to predict pre-

174 Piles under lateral loading

cisely the influence of close spacing under axial load; therefore, where reduction in load transfer is required the values of f and q will be reduced with the same factor. The assumption is made that reduction in the load-transfer values will reflect appropriately the increased settlement due to close spacing.

5.8.2 Concept of interaction under axial loading

The interaction of closely-spaced piles under axial loading is similar in concept to that of lateral loading. Considering a single pile under axial loading, stresses and deformations exist in a zone around the sides of the piles and in a different pattern in the zone around the tip of the pile. Other piles in these zones are influenced, of course; therefore, influence factors must be employed to account for the interaction.

As noted earlier, theories to account for close spacing of piles are limited. The finite-element methods hold promise but formulating constitutive laws for individual elements in three dimensions, accounting for nonlinearities in the natural soil and those due to installation of the piles, awaits a massive effort in research. Therefore, empirical methods must be employed for developing interaction factors.

Experimental results for groups of full-sized, closely-spaced piles are limited, as expected because of the cost of the providing the large axial loads. Evidence from experiments with single piles shows the importance of overburden pressure; therefore, contrary to the testing under lateral loading, the testing of small-sized piles in the laboratory does not provide definitive information. The testing of groups of piles under axial load in the centrifuge is a valid approach but investigating the large number of significant parameters, including influence of pile installation, requires a sustained and concentrated effort that is yet to be implemented.

In view of the above discussion, the use of the proposed factors for accounting for interaction of closely-spaced piles under axial load must be used with caution. Parametric studies are recommended and the full attention of an experienced and knowledgeable engineer must be brought to bear. Fortunately, for practical reasons, piles are usually not installed very close to each other. If installed by driving, the piles will drift and contact previously driven piles, with undesirable results. If installed by drilling, the drilled excavations could contact previously installed piles with undesirable results. Also, equipment is available for installing very large piles by driving or drilling, suggesting that close spacing is unnecessary. However, the redundancy inherit in multiple piles is lost and emphasis is placed on the details of construction and inspection of the large piles.

5.8.3 Review of relevant literature

A review of the technical literature about groups of piles is instructive. Many of the proposed methods are based on the idea of limit equilibrium which is useful because that method is used to analyze the behavior of single piles. The approach presented for groups of piles under lateral load, where a single pile was tested and then a pile group, is plainly absent. Even though experimental data are missing the brief review that follows is instructive with respect to the recommendations that follow.

5.8.3.1 Review by Van Impe
Van Impe (1991) made a comprehensive study of groups under axial load. Piled rafts, where the pile cap is in contact with the soil were considered, but only groups where the

cap was above the soil are considered in this book. Summaries of the study are shown in the following paragraphs.

Equivalent raft method. Prediction of settlement of the group is accomplished by replacing the real foundation with a fictitious equivalent one (Bowles 1988, Tomlinson 1995). A raft is assumed to exist at some depth below the ground surface, depending on the manner the load is distributed from the piles. The imaginary raft will be assumed at the tip of the piles if load is distributed in end bearing. If the load in the piles is distributed only in side resistance, the imaginary raft is frequently assumed to occur at a depth of two-thirds the length of the piles. Other depths for the imaginary raft can be assumed on the basis of the stratigraphy at the site.

In plan the edges of the imaginary raft pass along the exterior of the outside piles of the group. The load on the raft is taken as the total load on the group and is assumed to have uniform distribution on the raft. Uniform distribution is assumed to occur on planes below the imaginary raft where the edges of the lower planes occur at an angle of 26.6 degrees from the vertical (2.0 vertical to 1.0 horizontal) from the imaginary raft. The settlement of the group is then computed by adding that from short-term and long-term loading.

The efficiency of the piles in the group may be computed by taking the ratio of the settlement of the single pile, under the average load of the piles in the group, divided by the computed settlement of the group. As the piles are installed farther and farther apart, the computed settlement of the pile group would become equal to that of the single pile and the efficiency of the group would be unity.

Equivalent pier method. A somewhat similar approach to the equivalent raft method is to consider the soil in between the pile group and the piles as an equivalent continuum defined as an equivalent pier of equivalent dimensions and stiffness. Poulos & Davis (1980) suggested that there are two possibilities for assigning the dimensions to the equivalent.
1. An equivalent pier of the same circumscribed plan area as the group and of some equivalent length L_e (L_e depends both on spacing and L/d, but is virtually independent of the number of piles in the group);
2. An equivalent pier of the same length, L, as the piles, but having an equivalent diameter, de (this approximation is considered more appropriate when piles pass through layered soils or are founded on very different material). Like L_e, d_e is independent of the group's size but it does depend on L/d and pile spacing.

Other recommendations consider the equivalent pier to have the depth of the piles and have a width and breadth to encompass the pile group.

The efficiency of the piles in the group may be assessed by 1. Computing the ultimate capacity of a single pile by appropriate methods; and 2. By computing the ultimate capacity of the equivalent pier by adding the side resistance along the pier to the end bearing of the area of the block at its base. The efficiency (≤ 1.0) is computed by dividing the average capacity of the piles in the group by the capacity of a single pile.

The equivalent pier method is simple, can be applied to any soil profile, is rational, but can not be validated because of a lack (or absence) of experimental results.

Classical form of interaction factors. The method initiated by Poulos in 1968 and based on interaction factors between individual piles in a group, which assumes that the displacement of one pile is influenced by the loading of another pile. has been refined and can account for a large variety of boundary conditions.

An interaction factor, α, describes the additional displacement of a pile in an idealized soil mass due to an equally loaded identical adjacent pile (Poulos 1979, Poulos & Davis 1980). For piles that transmit their loads in side resistance, the interaction factors can take into account the following soil effects: effect of finite layer, effect of enlarged pile base, effect of Poisson's ratio, effect of nonuniform soil modulus, and effect of slip. For end-bearing piles, allowance was made for the effect of finite compressibility of the bearing stratum. Moreover, the interaction between piles of different size was discussed. All these have been studied by Poulos & Davis (1980) and specific design charts were shown for practical purposes (Fleming 1992).

The settlement prediction for pile groups is suggested for the typical cases of either an infinitely rigid or a flexible pile cap. Generally reasonable agreement has been found between theory and field measurements, except for the case of very heavy loads. It seems that there is a tendency for the theory to overpredict the group settlements and to predict a more nonuniform pile-load distribution than actually observed.

Modified interaction factors. Poulos (1988) suggested a modified form of the interaction-factor method allowance being made for a linear variation of the soil modulus near piles and a soil modulus of low strains for the soil between piles in the group The settlement increase due to interaction effects is computed using an average soil modulus E_{sav} which is consistent with the smaller strain levels occurring in the soil between the piles.

The conclusion of this new form is that the settlement due to interaction between piles in a group is reduced. The extent of this reduction depends on the spacing between the piles, the geometry of the group, and relative stiffness of the piles. Significant reduction in group settlement ratio, with respect to the previous method, is shown if the piles are widely spaced.

Single-pile load test coupled with the interaction-factor method. Mandolini & Viggiani (1993) suggested that prediction of settlement for a pile group may be computed starting from the observed settlement of individual test piles. The subsoil is modeled by a succession of horizontal elastic layers of constant thickness and the ratio between the moduli of any two layers has been assumed to be equal to the ratio of the corresponding average values from the cone test.

The value of the Young's modulus is obtained from elasticity theory by taking the settlement of the single pile at a load level corresponding to the average load on the foundation piles. Superposition coefficients are computed in terms of pile spacing with a logarithmic expression plotted for interpolation purposes. The method is limited to prediction of settlement and not to distribution of load to piles in the group.

Additional methods reviewed by Van Impe. A number of other methods were reviewed by Van Impe (1991). In addition to those above, the following procedures are based in some part on the theory of elasticity: Randolph & Wroth (1978), Fleming (1992), Bolton (1994), Van Impe (1991), and Yamashita et al. (1987). Hardin & Drnevich (1972a,b), and Fahey & Carter (1993); made contributions concerning parameters of the soil.

Other methods have been proposed, based on the use of the difference-equation method of solving for the response of single piles and those in a group. In regard to the somewhat more empirical methods of investigating the response of pile groups under axial load, the work of Chow (1986), Chow (1987), and Kraft et al. (1981), were discussed.

5.8.3.2 Review by O'Neill

O'Neill (1983) in a prize-winning paper reviewed 199 articles on group action, many of which included results of tests of groups under axial loading. Some of the factors addressed in the paper dealt with the following points: the state of stress in the soil caused by pile installation (driving, drilling, rotating); the development of excess pore water pressures in fine-grained soils due to installation and subsequent loading and the dissipation of such pressures as a function of time; whether the pile transfers axial load to the soil by side resistance, by end bearing, or both; the character of the soil profile and the differences in the stress-deformation characteristics of coarse-grained and fine-grained soils; downdrag; the influence of the pile cap, whether flexible or rigid, and in contact with soil or not; and the dominant effect of nonlinearity of soils in soil-structure interaction.

Axially loaded piles in sand are usually considered to have efficiencies of unity, regardless of the spacing. Stresses through the soil are certainly transmitted from the piles, along with some deformations. However, the transmitted stresses cause and increase in the effective stress and, hence, a strengthening and stiffening of the sand. The increase in the effective stress due to adjacent piles has been shown to eliminate the loss of efficiency of closely-spaced piles in sand. However, the settlement of the group could be greater that the settlement of a single, isolated pile.

An elementary approach to computing the efficiency of a group of piles in clay is to assume that the soil between the piles will move downward the same amount as do the piles, as assumed in the equivalent-pier method, noted above. A *block* failure can be computed by selecting a prism with the sides just encompassing the exterior of the group and the base equal to the area of the piles and the enclosed soil. The axial capacity of the group can be computed by summing the resistance in skin friction along the sides of the prism and adding the bearing capacity of the base. If the load computed for the prism is less that the sum of the axial capacities of the individual piles, the efficiency of each of the piles is thus less than one.

For many groups of piles in clay soils, the influences of the installation, and the other factors noted in the second paragraph above, are of over-riding importance. Thus, any one of the mathematical models that have been proposed is of limited use in computing efficiency and the capacity of the group to sustain axial load. An example is that the response of a group of piles installed in saturated clay depends strongly on whether or not excess pore water pressures have fully dissipated.

O'Neill (1983) concluded that the efficiencies of piles in a group in insensitive clay are approximately equal to unity if a block failure does not occur (Peck et al. 1974). The group is not to be loaded until excess pore water pressures have dissipated.

5.8.4 Interim recommendations for computing the efficiency of groups of piles under axial load

The following procedure is termed *interim* because research is continuing. If comprehensive results become available from the testing at various sites of single piles and piles in groups and where significant parameters are varied, improved methods of analysis of pile groups under axial load will undoubtedly be proposed. In the meantime, the following procedure is recommended.

178 *Piles under lateral loading*

1. Perform some parametric studies of the particular design being considered to ascertain the importance of the value of the efficiency of the group.
2. The study in (1) shows the efficiency to be an important parameter, make a review of some of the articles noted above.
3. Assume a value of unity for the efficiency of piles installed in sand.
4. Compute the value of efficiency of piles in insensitive clay using the block method proposed by O'Neill. Take steps to ensure that excess pore water pressures have decreased to an acceptable level before placing full load on the group.
5. Perform a field-loading test of a full-sized pile in sensitive clay and in cases where the subsurface profile is complex, preferably testing at various times. Use the results of the testing of the single pile to establish, using techniques noted above, the efficiency of the group.
6. For rare cases, where a project requires a large number of piles and where the efficiency of piles in the group is shown to be critical, perform a field-load test of a pile group.

5.9 ANALYSIS OF AN EXPERIMENT WITH BATTER PILES

5.9.1 *Description of the testing arrangement*

A comprehensive program of the testing of single piles and groups of piles was carried out making use of a test pit (Awoshika 1971). The testing was done in compacted sand and the arrangement is shown in Figures 5.38 and 5.39. The sketches in the figures show that axial load and lateral load could be applied simultaneously to the cap for a pile group.

The test pit was of reinforced-concrete tank, with inside dimensions of 7620 mm long by 3050 mm wide by 3660 mm deep. The number of the single piles installed is shown in Figure 5.39, where a plan view is given. Elevation views of the piles are shown in Fig-

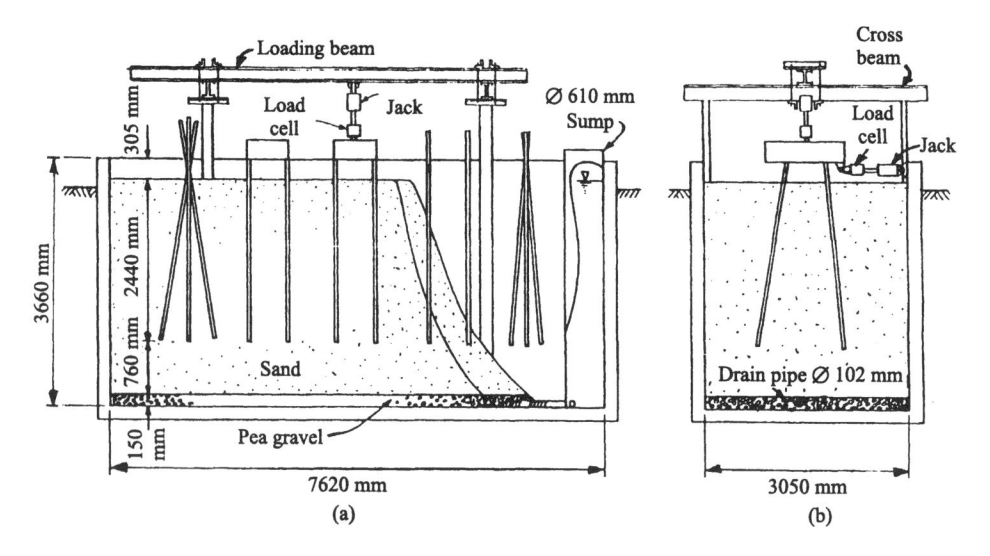

Figure 5.38. Setup for test of single pile and pile groups, (a) elevation view from side, (b) elevation view from end (after Awoshika 1971).

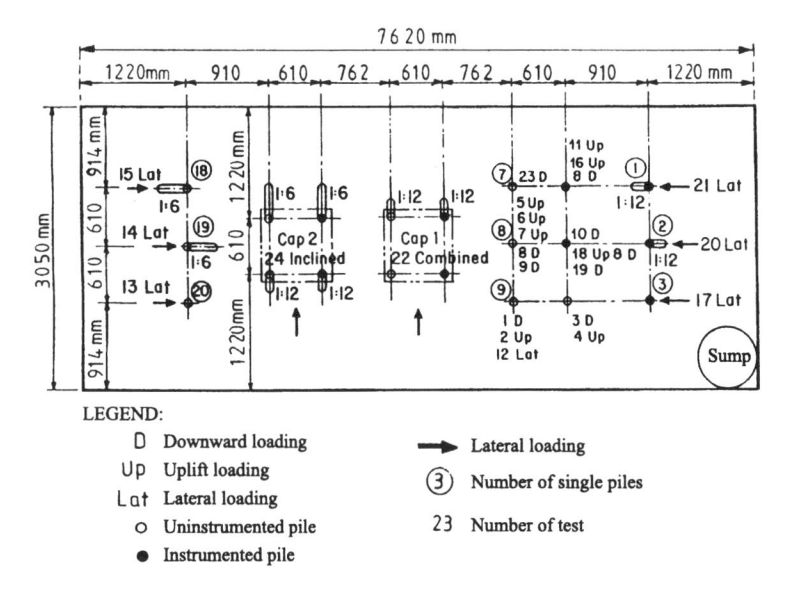

Figure 5.39. Arrangement of test piles and designations of test (after Awoshika 1971).

ures 5.38a and 5.38b. The sketch in Figure 5.38a shows the sand cut away to reveal the placement of the sump. A layer of pea gravel 150 mm thick was placed in the bottom of the tank to facilitate the saturation of the soil by introducing water into the pea gravel at the bottom of the tank.

5.9.2 Properties of the sand

The sand was placed by compaction to be as dense as could be made with hand-operated compaction equipment. After compacting a layer of sand of 760 mm in thickness, the piles were located in the desired positions in the tank while the remainder of the sand was compacted around the piles. The piles penetrated 2440 mm in the sand.

The sand was subangular to slightly subrounded, poorly graded, and fine. The following properties were measured: effective size D_{10} = 0.08 to 0.09 mm; uniformity coefficient = 2.4; specific gravity G_s = 2.679; minimum density γ_{min} = 1.32 gm/mm³; and maximum density γ_{max} = 1.64 gm/mm³. Water was kept at the surface of the sand and the submerged unit weight of the sand was found to be 9.85 kN/m³.

Prior to placing the sand, tests were performed in the laboratory. The results from direct-shear testing are shown in Figure 5.40, and those from triaxial testing are shown in Figure 5.41. From these tests, the friction angle of the sand was 41 degrees.

At the conclusion of the tests of the single piles and the pile groups, undisturbed specimens were trimmed from various depths in the tank. Triaxial tests were performed, and the results are shown in Figure 5.42. As may be seen, excellent agreement was obtained between the results from the nine tests. The friction angle obtained from the undisturbed specimens was 47 degrees. During the testing of the single piles, Awoshika (1971) found from the results of carefully controlled tests that the strength of sand was increasing with time. An examination of the sand around the piles showed some discoloration and

180 *Piles under lateral loading*

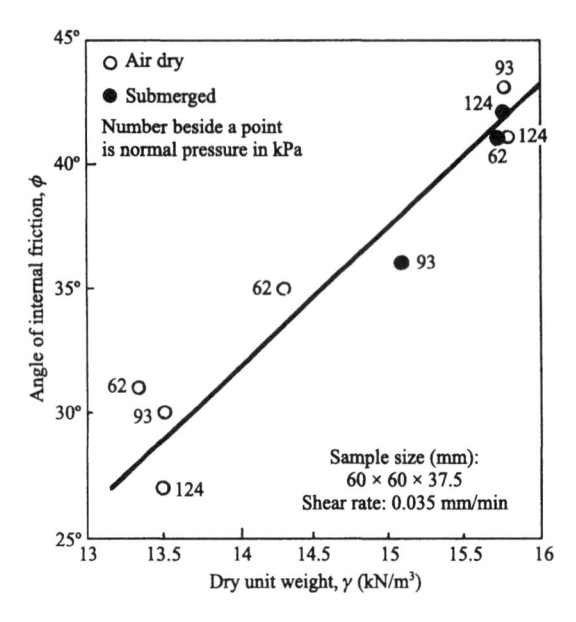

Figure 5.40. Results from direct shear tests (after Awoshika 1971).

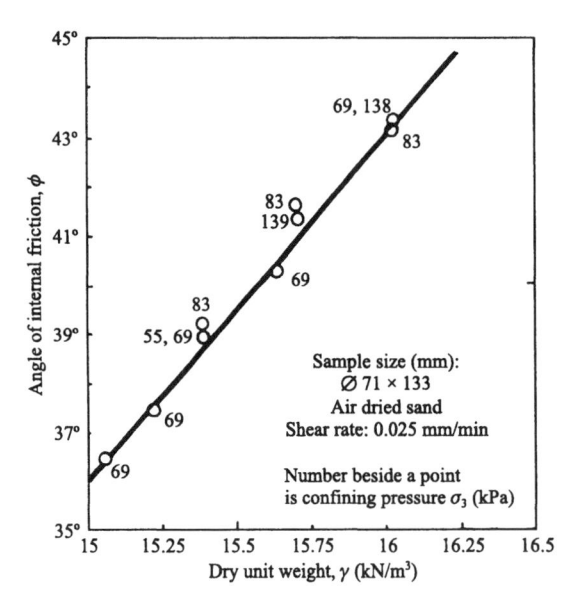

Figure 5.41. Results from compression tests (after Awoshika 1971).

the piles were observed to have corroded. The chemical interaction of the of the products of corrosion with the sand was found to have increased the strength of the sand in the vicinity of the piles. However, the discoloration was not noticed in the undisturbed specimens that were trimmed for testing. The conclusion was reached that the aging of the sand in the tank caused the significant increase in the strength of the sand.

The results from the testing of the piles under axial load, not shown here, agreed well with the friction angle of 47 degrees and that value is used in the comparisons for the group test that follows.

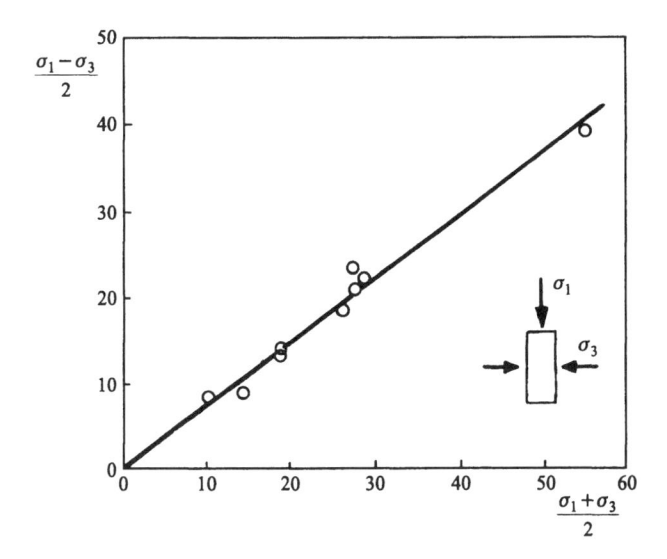

Figure 5.42. Results from compression tests of undisturbed specimens (after Awoshika 1971).

5.9.3 *Properties of the pipe piles*

The steel-pipe piles had an outside diameter of 51 mm and a wall thickness of 1.65 mm. The yield strength of the steel was 448 MPa, the cross-sectional area of the pipe was 2.45×10^{-4} m^2, and the moment of inertia was 7.70×10^{-8} m^4. The value of the $E_p I_p$ was computed to be 15.40 kN-m^2.

Single piles were tested under axial loading and lateral loading. The experimental results from the testing of a single pile under axial loading, with no interaction from adjacent piles, are used in the analysis that follows of the pile group. However, the differential equation using *p-y* curves was solved for the single piles as input to the analysis of the group. The equilibrium equations, described earlier, were solved by iteration to account for the nonlinear behavior of the soil.

5.9.4 *Pile group*

Plan and elevation view of the pile group are shown in Figure 5.43; the group is designated as Cap 2 in Figure 5.39. Piles 1 and 2 were installed with a batter of 1 horizontal to 12 vertical and Piles 3 and 4 had a batter of 1 horizontal to 6 vertical. The tops of the piles were welded to steel angles and then embedded in concrete with the view of fixing the heads of the piles against rotation. The cap weighed 2.22 kN. The loads were applied with an inclination of 12 degrees with the vertical, leaning toward Piles 1 and 2, and with an eccentricity of 10.9 mm, also toward Piles 1 and 2.

The space between the pile cap and the soil was 0.305 m and strain gauges were installed for the measurement of bending moment in the piles. The strain gauges were installed on the piles at 152 mm below the bottom of the cap. The piles had a spacing of 0.622 m at the bottom of the cap. Therefore, the piles were more than 12 diameters apart at the ground surface and no adjustment is necessary for close spacing.

182 *Piles under lateral loading*

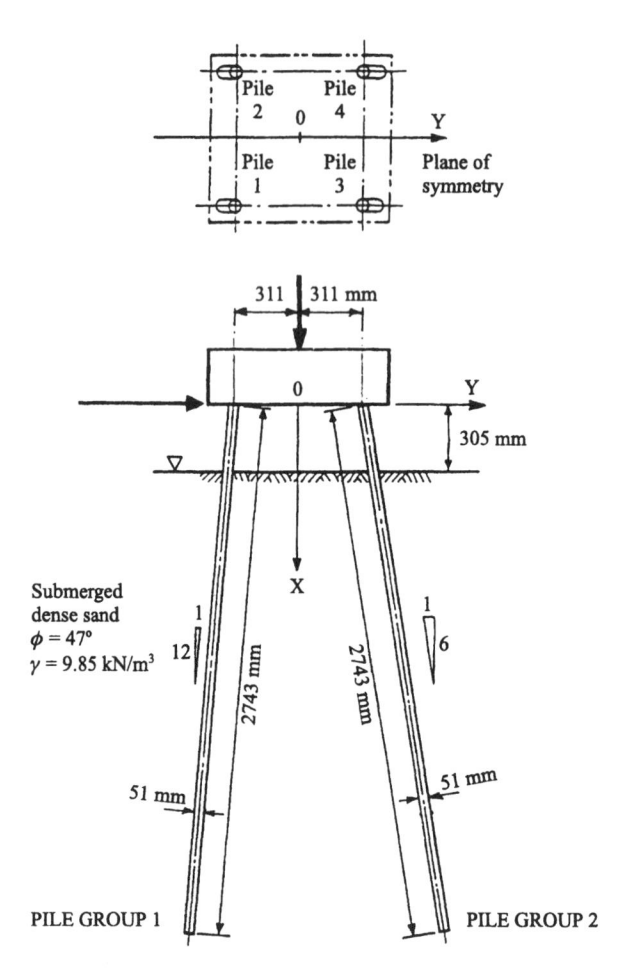

Figure 5.43. Setup for test of pile group (after Awoshika 1971).

The loads were applied with hydraulic rams and acted against beams fastened across the test tank or against the sides of the tank. The displacements were measured with dial gauges. The loads were applied in increments and deflections and rotation were measured for each increment of load. Loading was continued until displacements showed the group to be in a condition of incipient collapse.

5.9.5 Experimental curve of axial load versus settlement for single pile

The curve of axial load versus settlement is shown in Figure 5.44. The settlement at failure is quite small. The small settlement is consistent to the small amount of movement to reach the ultimate load transfer in side resistance and the small amount of movement to develop full load transfer in end bearing because of only a diameter of the pile of 51 mm.

The experimental curves was employed in the computing the response of the group.

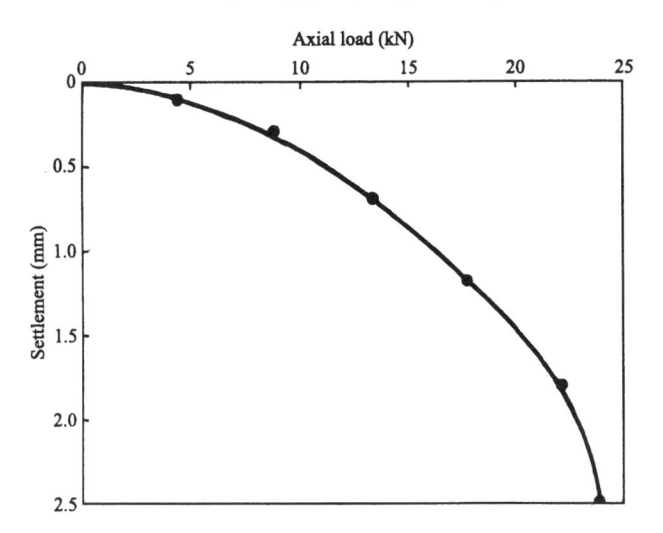

Figure 5.44. Axial load versus settlement for pipe pile in test group (after Awoshika 1971).

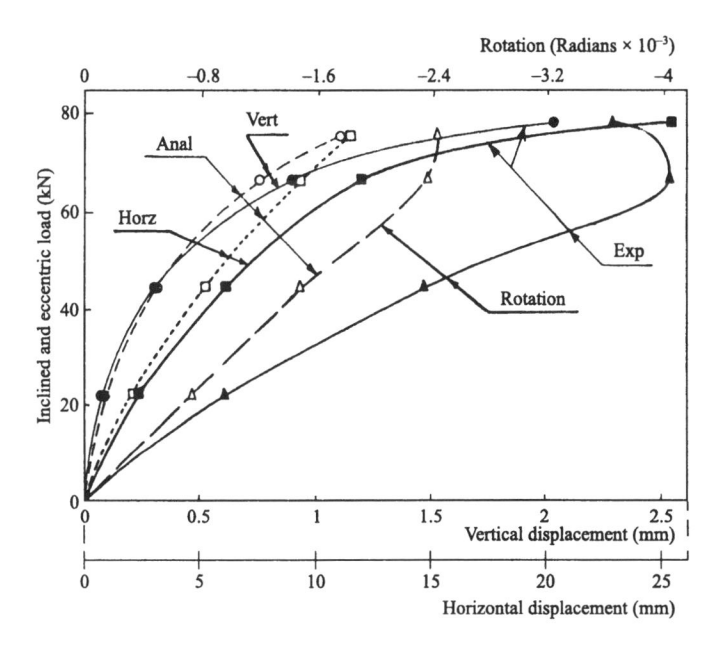

Figure 5.45. Comparison of results from experiment and from analysis for displacements of pile group (after Awoshika 1971).

5.9.6 *Results from experiment and from analysis*

Figures 5.45 through 5.47 present comparisons of the results from experiment and those from analysis. The measured and computed movements of the origin of the coordinate system, shown in Figure 5.43, are shown in Figure 5.45. Good agreement is shown for the

184 *Piles under lateral loading*

vertical and horizontal displacements but agreement between computed and experimental rotation is poor. Rotation is difficult to measure in the field so the error is likely related to the experiment rather than to a deficiency in the analytical procedure.

A comparison of the axial loads on the pile heads from experiment and from analysis is shown in Figure 5.40. The comparison is excellent except for the experimental result for the largest load in Sub-group 2 at which point the piles are at near collapse. As shown in

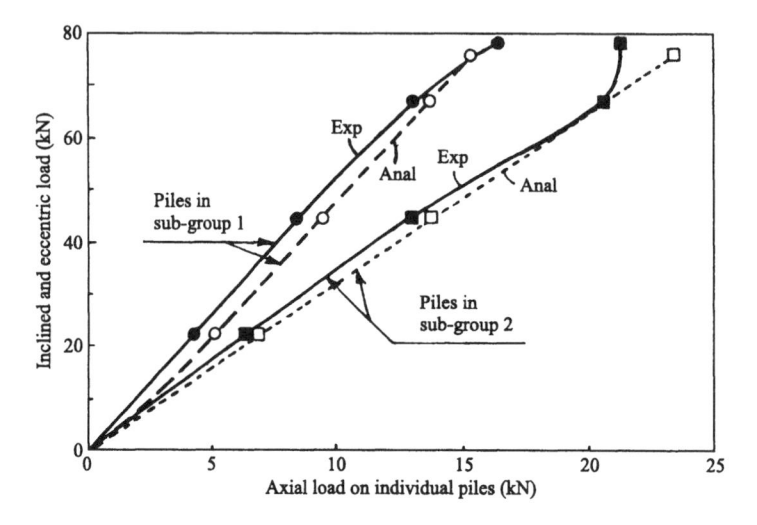

Figure 5.46. Comparison of results from experiment and from analysis for axial load to individual piles (after Awoshika 1971).

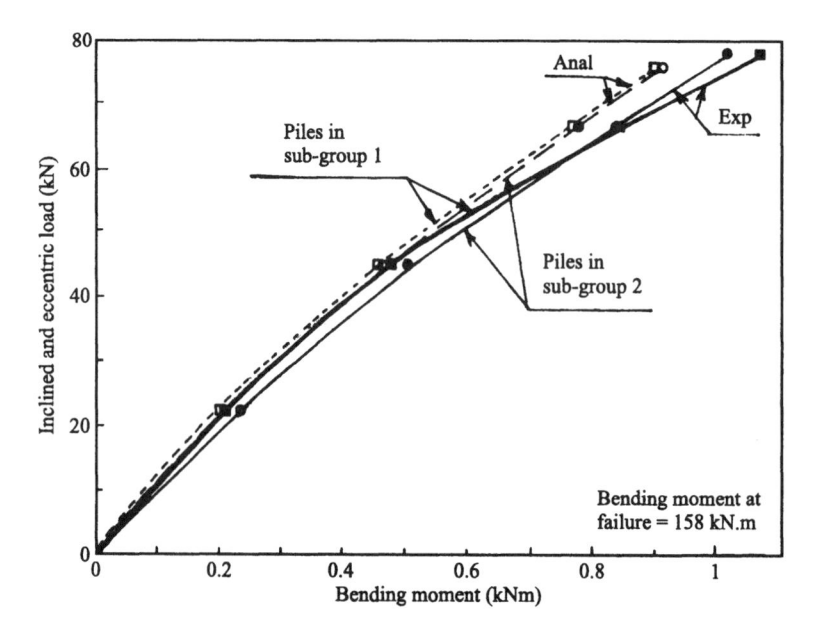

Figure 5.47. Comparison of results from experiment and from analysis for bend at distances of 152 mm below pile heads (after Awoshika 1971).

Figure 5.44, the failure of the piles in axial load occurs at a load of about 24 kN. The capacity of the piles under axial loading controls the amount of loading that could be applied to the group. The applied loading could have been somewhat larger if the piles had had a larger axial capacity.

The bending moments in the piles at a distance of 152 mm below the heads of the piles, for the experiment and for analysis, are shown in Figure 5.47. Agreement is excellent. The maximum bending moment will occur at the pile heads where the piles are fixed against rotation, which will be somewhat larger than the values shown in the figure. However, the computed bending moment at which a plastic hinge will occur is 1.58 kN-m which is larger than the measured values.

5.9.7 Comments on analytical method

The results from the experiments and from the analysis, presented here, show that the analytical method was effective in predicting the behavior of the experimental group. Such information as obtained from proper use of a computer code (see Appendix D), based on the analytical procedure that is presented, can provide valuable guidance to the designer. The use of the method of prediction at the outset of a project can result in the selection of the optimum kind of piles and their characteristics.

CHAPTER 6

Analysis of single piles and groups of piles subjected to active and passive loading

6.1 NATURE OF LATERAL LOADING

Lateral loads on piles can be derived from many sources; a convenient characterization of the sources is to term them *active* or *passive*, as suggested by De Beer (1977). As employed herein, active loading is considered to be time-dependent or *live* loading. Passive loading, on the other hand, is principally time independent or *dead* loading.

Active loading may come from wind, waves, current, ice, traffic, ship impact, and mooring forces. Passive loading is derived principally from earth pressures or potentially moving soil but may also come from dead loading as from an arch bridge. In the following sections, some detail is presented on the nature of the various kinds of active and passive loading.

The solution of a number of examples is presented in this chapter where the piles are subjected either to active loading or to passive loading. The solutions are presented in sufficient detail to provide guidance in addressing similar problems.

6.2 ACTIVE LOADING

6.2.1 *Wind loading*

6.2.1.1 *Introduction*
The first step in the usual practice in the design of a pile-supported structure is to collect data on wind velocities from appropriate sources. The second step is to translate the wind velocities into forces on a structure employing shape factors for the particular geometrical element. The forces may be time-dependent; therefore, a dynamic analysis of the structure may be indicated. The problem is simplified greatly by some agencies. The American National Standards Institute (ANSI 1982) suggests that a minimum static pressure of 0.72 kPa may be used in designing the bracing for a masonry wall.

The velocity may be classified by gusts which are averaged over less than one minute or sustained velocities which are averaged over one minute or longer. The data may be adjusted to a specified distance above the ground or water and may be averaged over a given time.

The spectrum of the fluctuations of the wind velocity about the average may be necessary in some instances. Such data would be useful, for example, in the design of a sign structure that may perform unfavorably to dynamic loading.

188 *Piles under lateral loading*

The data on wind velocities that are collected will have the following characteristics, all as a function of the time of the year: 1. Frequency of occurrence of specified sustained velocities from various directions; 2. The persistence of sustained velocities above a given threshold; 3. Variation of the velocity as a function of the distance above the ground or water; and 4. The probable velocity of gusts associated with the sustained velocities.

With data on the velocity of the wind, forces may be computed by the use of a given shape factor for the structural element. Shape factors may be determined from wind-tunnel tests and take into account factors such as geometry, aspect ratio, roughness, and shielding.

6.2.1.2 *Information of velocities of the wind*

Local information is frequently available on wind velocity (traditionally called *wind speed*) for a given geographical area. For example, the American Society of Civil Engineers (ASCE 1990) presents a map of the United States showing contours of the basic wind speed for the country. The values in the map are for 10 m above the ground surface and adjustments must be made for other heights. Also, other information must be accessed for gusting.

Sachs (1978) presents data, obtained for a period of 5 minutes by a cup anemometer, giving wind speeds at three heights on a mast. At a height of 165 m, the velocity averaged about 95 km/hr but four gusts occurred that averaged about 119 km/hr. At a height of 69 m, the velocity averaged about 80 km/hr but seven gusts occurred that averaged about 93 km/hr. At a height of 13 m, the velocity averaged about 60 km/hr but there was one gust with a velocity of 90 km/hr and nine gusts occurred that averaged about 73 km/hr. The direction of the wind undoubtedly varied some during the five-minute period and, of course, the velocities could have been more severe during other periods of the storm. The gusts did not occur at regular intervals; therefore, exciting the natural frequency of a structure, such as an overhead sign, appears to be unlikely. However, data such as presented by Sachs indicate the difficulty of characterizing the design storm.

The American Association of State Highway Officials (AASHTO 1992) in *Standard Specifications for Highway Bridges* states that a load of 3.6 kPa shall be applied at right angles to the longitudinal axis of trusses and arches and 2.4 kPa to girders and beams. Other information is given in the specifications, such as methods of adjusting for the skew angle of the wind.

6.2.1.3 *Properties of wind*

Equations and procedures are presented, for example by the American Petroleum Institute (API 1993) that may be used in the absence of more specific information on the mean profile, gust factor, turbulence intensity, and wind spectra.

An equation is presented for the mean profile for the velocity of the wind, averaged over one hour at a distance above ground or water in terms of a reference distance. Invariably the wind will gust and equations are available for finding the velocity of a gust in terms of the average velocity of the wind over a period of one hour at some distance above the ground surface. A formulation is available for computing the turbulence intensity, the standard deviation of the velocity of the wind normalized by the velocity of the wind over one hour. The fluctuations of the velocity of the wind can be described by a spectrum and equations are available for computing the spectral density as a function of the standard deviation of the wind velocity.

Wind gusts have spatial scales related to their duration. Three-second gusts are coherent over shorter distances than 15-second gusts. While the shorter-term gusts are appropriate for investigating the forces on individual members, 5-second gusts are appropriate for obtaining the maximum total loads on structures whose horizontal dimension is less that 50 m. Fifteen-second gusts are appropriate for the total static wind load on larger structures. If a dynamic analysis is unnecessary, the one-hour sustained wind is usually appropriate for the total static wind forces on the superstructure of an offshore structure associated with the maximum wave forces.

Force of Wind as a Function of Wind Velocity. The wind force on an object may be computed by use of the following equation (API 1993).

$$F = \frac{w}{2g} V^2 C_g A \qquad (6.1)$$

where F = wind force, N; w = weight density of air, N/m³; g = gravitational acceleration, m/s²; V = wind speed, m/s; C_g = shape coefficient; and A = area of object, m².

In the absence of specific data on shape coefficients, the following values may be used for C_g.

Beam	1.5
Sides of buildings	1.5
Cylindrical sections	0.5
Projected areas of miscellaneous shapes	1.0

Adjustments must be made in the equations for the computation of wind force on surfaces that are not perpendicular to the direction of the wind.

6.2.2 Wave loading

6.2.2.1 Introduction

The following factors must be considered for each offshore structure: wave height and period, marine growth, and hydrodynamic coefficients for computation of wave forces. In addition, the forces from the wind, discussed above, and the current, discussed below, must be taken into account. Fu et al. (1992) presented the flow chart, Figure 6.1, to illustrate the required procedure. For the specific site, the computation procedure is initiated with the current forces and the wave description, wave height and period, usually for the 100-year storm. As shown in the figure, the current forces are modified, according to the structural axis selected for analysis, and considering the factors affecting the blockage of forces against structural elements behind other elements. With wave height and period, a wave-height-reduction factor is selected, depending on the relative direction of the wave. Then, a wave theory is selected, along with the wave kinematics and the wave-kinematics factor, and wave force computations are made. The wave forces take into account the character of the member, the amount of marine growth, and the shielding factor for the particular member.

In the design of pile-supported structures, the loads originating from the wave motion are to be taken into account both in respect of the loading of the individual piles, as well as of the superstructure. The superstructure should be located above the crest of the design wave if possible. Otherwise, large horizontal and vertical loads from the direct wave action can affect the structure, whose determination is not addressed herein. The elevation of the crest of

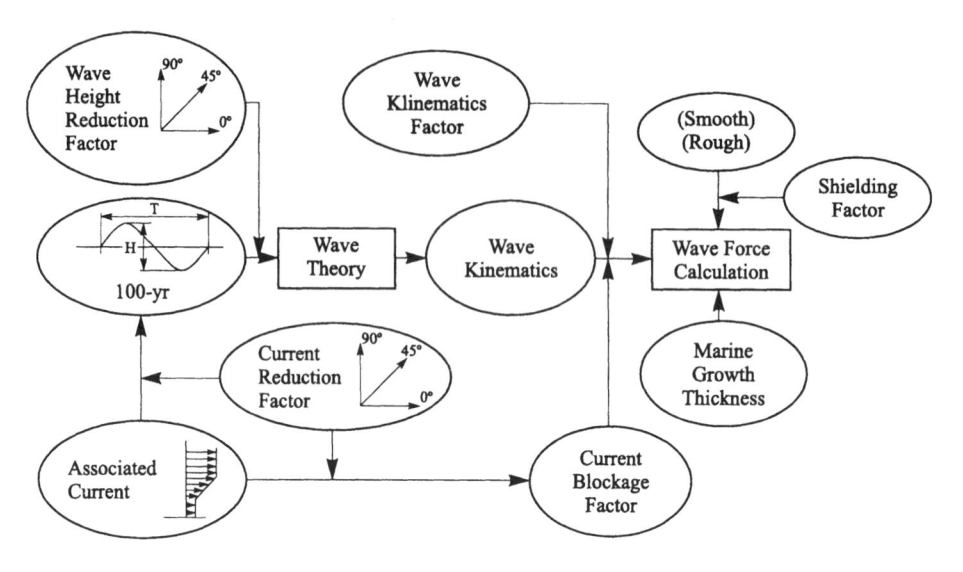

Figure 6.1. Diagram showing steps in the computation of forces from waves and current (after Fu et al. 1992).

the design wave is to be determined in consideration of the simultaneously occurring highest still-water level, as the case may be, taking into account also the wind-raised water level, the influence of the tides, and the raising and steepening of the waves in shallow water.

6.2.2.2 *Example of wave height and period*

Oceanographers have collected data of storm-induced waves in the oceans around the world and have developed predictions for wave heights and periods for given locations. An example of the presentation of such a prediction is presented in Table 6.1. The site is off the coast of Australia, for a water depth of approximately 100 m, and for the 100-year storm. The period of the waves T_z ranges from 5.8 sec and 10.6 sec, the characteristic wave height H_s ranges from 3.6 m to 12.0 m, and the total number of waves is slightly above 18,000. Table 6.1 shows the 35.5 hours of the storm being divided into seven periods as the wave heights build up and decay.

For each of the seven periods of the storm, the distribution of wave height H is assumed to be controlled by a spectrum given by the Rayleigh equation:

$$P_r(H) = \frac{4H}{(H_s)^2} e^{-2\left(\frac{H}{H_s}\right)^2} \tag{6.2}$$

Employing Equation 6.2 and the number of waves for each of the seven periods, the values in Table 6.2 were computed. The relatively small number of waves for the greatest wave heights is of interest, even though the predictions are hypothetical. The equations for predicting the response of the soil to cyclic loading, presented in Chapter 3, are strongly based on experimental results where the piles were cycled under a specific load until stability was achieved. The number of cycles of load necessary to achieve stability was in the order of 50 to 100. Designs would certainly be conservative if only a few cycles of load

Table 6.1 Parameters for a hydrograph of storm waves.

Period (hours)	Duration (hours)	H_s (m)	T_z (sec)	No. of waves N
-21.5 to -11.0	10.5	3.6	5.80	6517
-11.0 to -5.5	5.5	6.0	7.49	2644
-5.5 to -2.0	3.5	11.16	10.22	1233
-20 to 1.5	3.5	12.0	10.60	1189
1.5 to 4.0	2.5	11.16	10.22	881
4.0 to 8.0	4.0	6.0	7.49	1923
8.0 to 14.0	6.0	3.6	5.80	3724
Totals	35.5			18111

Table 6.2 Number of waves of specified height H within each time period during the storm.

Individual wave height H (m)	Period (hours) -21.5 to -11.0	-11.0 to -5.5	-5.5 to -2.0	-2.0 to 1.5	1.5 to 4.0	4.0 to 8.0	8.0 to 14.0
0-1	968	145	20	16	14	105	553
1-2	2132	389	57	48	41	283	1218
2-3	1917	519	90	76	64	377	1095
3-4	1063	521	114	98	81	379	607
4-5	398	429	129	112	92	312	227
5-6	104	301	134	119	96	219	59
6-7	19	183	131	119	93	133	11
7-8	3	97	120	113	86	70	1
8-9	–	45	105	103	75	33	–
9-10		19	88	90	63	13	
10-11		7	71	75	51	5	
11-12		2	54	61	39	2	
12-13		1	40	47	29	–	
13-14		–	29	36	20		
14-15			20	26	14		
15-16			13	18	9		
16-17			8	12	6		
17-18			5	8	4		
18-19			3	5	2		
19-20			2	3	1		
20-21			1	2	1		
21-22			1	1	–		
22-23			–	1			
Total	6604	2658	1235	1189	881	1931	3771

are applied because the difference in the wave forces will be considerable for a wave height of 15 m and compared to a height of 23 m.

6.2.2.3 Kinematics for two-dimensional waves

A convenient approach to the selection of a particular theory for obtaining the velocities and accelerations of water particles as a function of time and position in the wave is given

192 *Piles under lateral loading*

in Figure 6.2 (Barltrop et al. 1990, API 1993). The symbol T in Figure 6.2 is the apparent wave period which is the period seen by an observer who moves with the velocity of the current. The actual period and the apparent period are the same for a zero current; for the usual values of the current, the apparent period is within about 5 to 10% of the actual wave period. The other symbols are: H is wave height, H_b is height of breaking wave, d is mean water depth, and g is the acceleration of gravity. Entering the curve with data for the maximum wave from Tables 6.1 and 6.2, the particle motions may be computed from the Stokes 5 theory or from Stream Function 3. Atkins (1990) presents procedures for computing the particle motions.

6.2.2.4 *Forces from waves on pile-supported structures*
General. In respect of the methods of computation of the forces from waves, two approaches are noted: the method of superposition according to Morison et al. (1950) for slender structural members, and a method based on diffraction theory according to Mac-Camy & Fuchs (1954) for wider structures.

Some detail is presented here on the method of supervision (CERC 1984) which is applicable for non-breaking waves (Fig. 6.2). An approximate method for breaking waves is proposed in a later section.

The method according to Morison gives useful values, if the following condition is met for the individual pile or member (most pile-supported structures meet the condition):

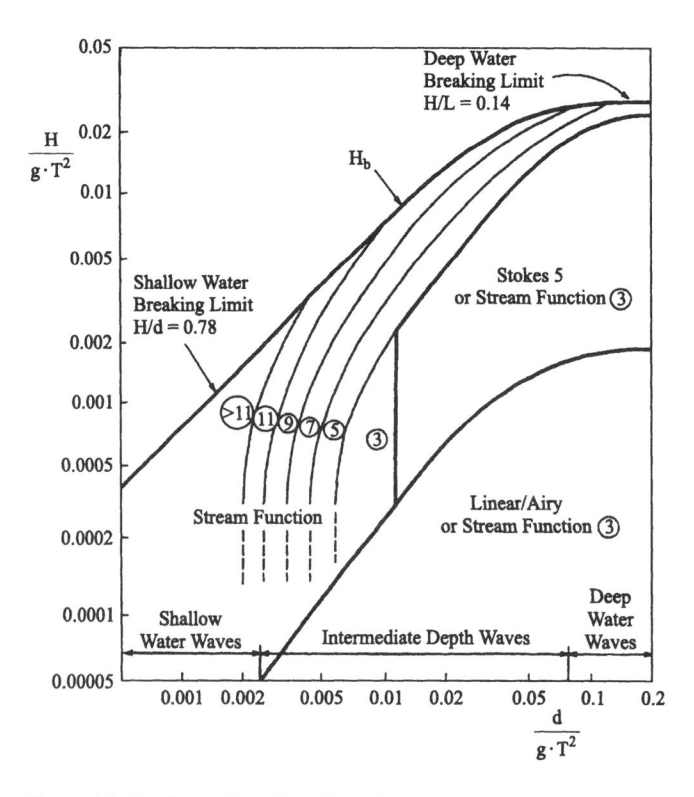

Figure 6.2. Regions of applicability of equations for kinematics of waves (from API 1993, Barltrop et al. 1990).

Analysis of single piles and groups of piles subjected to active and passive loading 193

$$\frac{D}{L} \leq 0.05$$

where D = effective diameter for circular cylindrical member, or width of non-circular member, m; L = length of the design wave, m; and $L = C.T$; C = wave celerity, m/sec; and T = wave period, sec.

Method of computation according to Morison (1950) for non-breaking waves. The computation of the force exerted by waves on a cylindrical object (or another shape) may be computed as the sum of a drag force and an inertia force as follows:

$$p = p_D + p_M = C_D \cdot \frac{1}{2} \cdot \frac{\gamma_w}{g} \cdot D \cdot u \cdot |u| + C_M \cdot \frac{\gamma_w}{g} \cdot A \cdot \frac{\partial u}{\partial t} \qquad (6.3)$$

For a pile with circular cross-section, the equation becomes:

$$p = C_D \cdot \frac{1}{2} \cdot \frac{\gamma_w}{g} \cdot D \cdot u \cdot |u| + C_M \cdot \frac{\gamma_w}{g} \cdot \frac{D^2 \cdot \pi}{4} \cdot \frac{\partial u}{\partial t} \qquad (6.4)$$

where p = hydrodynamic force vector per unit length acting normal to the axis of the member, kN/m; p_D = drag force vector per unit length acting normal to the axis of the member in the plane of the member axis, kN/m; p_M = inertia force vector per unit length acting normal to the axis of the member in the plane of the member axis, kN/m; C_D = drag coefficient; γ_w = unit weight of water, kN/m³; g = acceleration of gravity, m/sec²; D = effective diameter of circular cylindrical member, including marine growth, or width of noncircular member, m; u = component of the velocity vector normal to the axis of the member, m/sec; $|u|$ = absolute value of u, m/sec; C_M = inertia coefficient; A = projected area normal to the axis of the member, m²; and $\partial u/\partial t \approx du/dt$ = horizontal component of the local acceleration vector of the water normal to the axis of the member, m/sec².

A sketch of a pile with forces from drag and acceleration shown at a element dz at a point along the pile is shown in Figure 6.3. Most of the terms in the sketch are defined

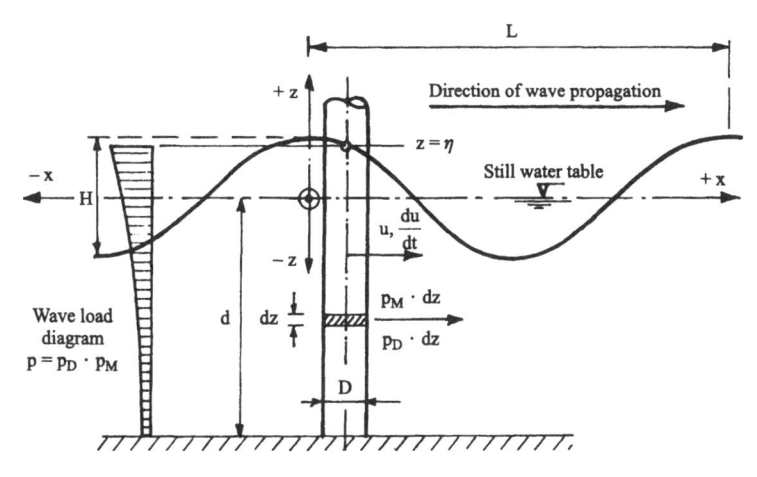

Figure 6.3. Action of a wave on a vertical pile.

194 *Piles under lateral loading*

above; the terms x, z, and η appear in the computation of particle velocity and acceleration with equations noted in Section 6.2.2.3.

Forces from breaking waves. At present, there is no accepted method for computing the forces on a pile from breaking waves; therefore, the Morison formula is employed as an expedient in making computations. The assumption is that the wave acts as a water mass with high velocity but without acceleration. Thus, the inertia coefficient is set to $C_M = 0$, whereas the drag coefficient is increased to $C_D = 1.75$ (CERC 1984).

Safety Requirements. The design of pile structures against wave action is strongly dependent on the selection of the design wave, the wave theory in computations and the selection of the coefficients CD and CM. The selection of the design wave is a matter to be considered by the owner of the structure.

Factors to be considered are the anticipated life of the structure and the chance that a storm of a given magnitude will occur. Statistics are available that give the frequency that a storm of a given magnitude will occur in a specific body of water, such as the Gulf of Mexico. However, data may be quite sparse when a specific site in the Gulf is selected, for example. Therefore, for a particular structure at a particular location in a particular body of water, the selection of the design wave may require a decision from the highest level of management.

6.2.3 Current loading

6.2.3.1 Introduction
While the effects of current must be considered in the design of offshore structures, current plays a major role in the design of foundations for bridges. A guiding concept in the design of offshore platforms is that the platform is relatively austere in the region of the maximum height of the wave, with the deck designed to be above the maximum height of wave. Furthermore, scour is not as severe a problem with offshore structures as for bridges. Therefore, current loading will be treated separately for offshore structures and for bridges.

6.2.3.2 Current loading for offshore structures
Oceanographers have developed data for various geographical areas with respect to currents generated by hurricanes. For example, in most of the Gulf of Mexico, API (1993), the direction of the current in shallow-water (45 m and below) is specified geographically with the maximum current given as 2.1 knots. For deeper water (90 m and above) the maximum current is also specified as 2.1 knots with the direction of the current the same as the direction of the waves. The direction of the maximum wave is specified and coefficients are presented (0.70 to 1.00) are given for modifying the wave and the magnitude of the current as well, depending on direction. For intermediate depths of water, interpolation is used to find the maximum current. The maximum current is specified to occur at the water surface, to remain constant for a considerable depth, and to reduce to 20% of the maximum at the mudline (API 1993).

The storm tide is also specified by API as a function of the depth of water for certain geographic areas of the Gulf of Mexico. Site-specific studies are required for other areas.

In the absence of recommendations by other agencies for areas of the world's oceans, other than the Gulf of Mexico, oceanographers must perform site-specific studies.

6.2.3.3 *Current loading for bridges*

Predicting the maximum flow of a stream at a bridge during the period of recurrence selected for design depends on the availability of statistics for storms in the watershed area. With such statistics, a hydrologic study can be made of the factors that affect the concentration of flow at the bridge and a prediction can be made. Such predictions usually cover a limited period of time because of construction in the watershed. With the height of the stream, along with a prediction of scour (discussed below), computations can be made of the forces on bridge bents and consequently on the piles supporting the bents.

The Standard Specifications of the American Association of State Highway and Transportation Officials (AASHTO 1992) presents a simple, nonhomogeneous equation for the force of stream current on piers.

$$P = KV^2 \tag{6.5}$$

where P = unit force, kPA, V = velocity of stream flow, m/s, K = a constant, depending on the shape of the structural element.

The values of the constant K was converted for English units to SI units with the following results: K is equal to 0.71 for members with a square end, is equal to 0.26 for angle ends where the angle is 30 degrees or less, and 0.34 for circular piers.

6.2.4 *Scour*

Scour is not a condition of loading; however, the scour and erosion of soils at pile-supported foundations, if unanticipated, can create serious instabilities and needs to be addressed along with the various kinds of active loading. The lack of supporting soil along a portion of a pile, perhaps even a small portion, will lead to increased deflections and increased bending moments. Therefore, the amount of scour must be predicted or measures taken to prevent the scour.

The theory of sediment transport can be used to predict the velocity of flow that will move a single-grain particle of soil (silt, sand, gravel cobble, boulder). Boulders are moved downstream by swift-flowing water in the mountains; silt and fine sand are carried by slow-moving river water and deposited in deltas. The theory applies less well to the erosion of clays where cohesion and complex structures exist (Moore & Masch 1962, Gularte et al. 1979). Complexities arise when the water must flow around an obstruction, such as a bridge pier, and velocities increase.

Erosion is prevented by creating a scour-resistant surface. Special blankets, extending well into a river, can be effective to prevent loss of soil at river banks. Reverse filters can be placed, based on the grain-size distribution of the material to be protected (de Sousa Pinto et al. 1959). The layers of the filter, starting from the soil to be protected, and becoming successively larger in size until the size at the mudline is judged to be large enough to resist movement. Each layer that is placed, the filter, should have the following relationships with the layer below, the base. $D_{15filter}/D_{85base} > 5$; $4 < D_{15filter}/D_{15base} < 20$ and $D_{50filter}/D_{50base} < 25$. The subscripts for the D terms in the equations refer to the particular percentage by weight as determined from a grain-size-distribution curve. The specifications are known in the United States as TV grading, because of work by Terzaghi at the Waterways Experiment Station in Vicksburg, Mississippi (Posey 1963, Posey 1971).

With respect to the suggestions noted above, the engineer should be aware the uniformity coefficient of the filter material should be considered, along with the maximum size. A uniform filter material should not be allowed and each layer of the filter should have self-stability.

6.2.4.1 *Scour at offshore structures*
The informal approach of a number of designers of offshore structures is to assume a minimum amount of erosion, perhaps 1.5 to 2 m, and to institute an observational program with the view that anti-scour measures would be instituted if necessary. Such an approach may be risky if the structure is founded in cohesionless soil where bottom currents can occur.

In 1960, a diver-survey was made of scour around an offshore platform in the Gulf of Mexico near Padre Island. The structure was installed in water with a depth of approximately 12 m and some lateral instability was reported even in minor waves. The soils near the mudline consisted of fine sand with a few shells and some stringers of clay. A bowl-shape depression was found to have formed around the structure; the depth of the depression was 2.5 m. The structure was stabilized, using the reverse filters, described above. Three layers of slag from a steel mill were used; the top layer consisted of particles ranging in size from 50 to 200 mm. A survey performed after a hurricane has passed through the area showed that little of the filter material had been lost (Sybert 1963).

Einstein & Wiegel (1970) performed a comprehensive review of technical literature on erosion and deposition of sediment near structures in the ocean. Many aspects of the overall problem were addressed, including sediment, flow condition, and the structure. A total of 415 references were cited but, as could be expected, a general approach was forced on the investigators because of the large number of parameters involved. Furthermore, the absence of cases where all details of observed scour were known prevented the application of specific equations for design.

The American Petroleum Institute (API 1993, p.71) includes the following statement under the topic of Scour in the section on the Hydraulic Instability of Shallow Foundations: *Positive measures should be taken to prevent erosion and undercutting of the soil beneath or near the structure base due to scour. Examples of such measures are 1. Scour skirts penetrating through erodible layers into scour resistant materials or to such depths as to eliminate the scour hazard, or 2. Riprap emplaced around the edges of the foundation. Sediment transport studies may be of value in planning and design.*

6.2.4.2 *Scour at bridges*
Scour has been judged to be the cause of the failure of numerous bridges. Laursen (1970) presents photographs of 12 failed bridges and noted that many other examples could be added. While special mats may be used to prevent erosion at the banks of a river, the erosion of the soil at the stream bed is usually unavoidable. The use of riprap or other techniques for a relatively short stretch of a river, in the vicinity of a bridge, is usually ineffective because the scour will start at the upstream edge of the protection and proceed to erode the entire zone of protection. The number of factors that influence the depth of scour are so numerous, and frequently time-dependent, that predictions of the depth may vary widely. At a proposed bridge for a major city in the United States, officials predicted that a flood would erode all of the alluvial soil down to the bedrock. A portion, if not all,

of the stream bed is re-deposited with the abeyance of the flood so that measurements of the depth of scour must be made during the flood, which could be difficult.

The Standard Specifications for Highway Bridges, American Associations of State Highway and Transportation Officials (AASHTO 1992, p. 92) requires that the probable depth of scour be determined by subsurface exploration and hydraulic studies. The American Association of State Highway and Transportation Officials has issued guidelines for the performance of the hydraulic studies (AASHTO 1992).

6.2.5 Ice loading

The lateral loads from sheets or masses of ice may be a critical factor in the design of some structures. Locks and dams along the northern sections Mississippi River in the United States, and along other rivers as well, must be designed to withstand forces from ice that can occur in various forms and with various characteristics. The ice may pass through the locks or lodge against the dams and apply loads as a function of numerous factors, including the current if the river is in the flood stage.

With the discovery of deposits of petroleum in Cook Inlet on the southern coast of Alaska, emphasis was given to the geometry of the superstructure and the design of piles to withstand the loading from sheets of ice that commonly flowed down the waterway in winter time. Experience had shown the vulnerability of waterfront structures to the forces from the ice. Designs were developed that presented only an unbraced vertical column at the depth of the ice. Some dramatic movies of the performance of the structures showed the fracturing of the sheets of ice in a ratcheting manner. As the ice moved against the structure, the lateral force would increase until the ice fractured, and the force would then decrease to build up again.

Numerous factors are associated with the lateral force of ice on a structure, including the strength of the ice, the geometry of the moving mass, the velocity of approach, and the nonlinear force-deflection characteristics. Such factors are investigated on a site-specific basis. However, AASHTO (1992) suggest a simple equation for the 'horizontal forces resulting from the pressure of moving ice' (pg. 26).

$$F = C_n p t w \qquad (6.6)$$

where F = horizontal ice force on pier, kN; p = effective ice strength, kPa; t = thickness of ice in contact with pier, m; w = width of pier or diameter of circular-shaft pier at the level of the ice action, m; C_n = coefficient from the following table.

Inclination of nose to vertical	C_n
0° to 15°	1.00
15° to 30°	0.75
30° to 45°	0.50

AASHTO recommends a careful evaluation of the local conditions governing the parameters in Equation 6.6 before making using the equation for design The strength of the ice, for example, is expected to range from about 700 kPa to 3000 kPa, depending on whether the ice is breaking up or is at a temperature significantly below the melting point.

6.2.6 Ship impact

An important feature in the design of a bridge or other structures along a navigable waterways is to protect the structure against severe damage if a ship loses control. In mid-December 1996, the *Bright Field*, a 234-m-long freighter, loaded with 51,000 tons of grain, lost steerage, crashed into the Riverwalk, New Orleans, Louisiana, destroyed commercial facilities, and injured 116 people (*Austin American-Statesman 1996*). The use of rock-filled barriers, as is done for some bridge piers, is not possible at New Orleans because of the many docks for river boats and tourist ships. Piles under lateral loading, however, have found a useful role as breasting dolphins in the protection of marine facilities when controlled docking is possible.

A single pile, driven a sufficient distance below the mudline, and with several meters of unbraced length above the mudline, can deflect a considerable distance without damage. The relative flexibility of such a pile is an advantage when the pile is used as a breasting dolphin.

Some breasting dolphins have been constructed of timber piles, which are driven in a circular pattern with a slight batter. The pile heads are lashed together with steel cables. Such a design may resist a relatively large lateral force but deflection is relatively small before damage occurs. A series of such timber-pile dolphins were in place at a dock in the Gulf of Mexico when a large tanker was docked for the first time. Even though the velocity of the tanker had been reduced by tugs to a fraction of a meter per second, many of the timber dolphins broke with loud popping sounds at the ship came against the dolphins.

A ship being berthed approaches the dock with a given velocity and attitude, such that one or more dolphins are contacted. Harbor and other authorities may establish rules that must be followed by the operators of the ship. Factors that are considered are the nature of the harbor, the weather at docking, the displacement of the vessel, and perhaps the nature of the soils near the mudline. Table 6.3 shows stipulated berthing velocities for the design of facilities for various ports.

A perusal of Table 6.3 is of interest. The average of the stipulated berthing velocities is about 0.20 m/sec; the lowest value is 0.08 m/sec and the highest value is 0.30 m/sec. The data suggest that a value is to be set for each dock but the reasonable range shown in the table indicates berthing practice over a wide geographical area.

A single dolphin may be designed by techniques presented herein to sustain an amount of energy. In addition to breasting dolphins, there are many auxiliary devices for absorbing the energy of a docking vessel. The example computation, presented in this chapter, is not intended to suggest a comprehensive approach to the design of a docking system.

6.2.7 Loads from miscellaneous sources

The sections above present the principal sources of active loads. There can be a number of other sources and the designer can use creativity to ensure that all of the active loads are taken into account. Temperature effects can cause members to shrink and expand with resulting lateral loads on piles. Traffic on bridges can cause lateral loads on curved roadways and may cause lateral loads by sudden braking. Of course, for piles that are installed on a batter, a component of the vertical loads will cause lateral deflection of a pile. Therefore, a careful evaluation of the vertical loading is usually necessary to design properly piles under lateral loading.

Table 6.3 Approach Velocities Assumed for Design of Dock-and-Harbor Facilities.

Harbor	Displacement of design vessel (tons)	Berthing velocity normal to dock (m/sec)	Remarks
Kitimat	24,000	0.15	Report in 1960
Rotterdam	45,000	0.25	Committee Report Dec. 1958
Thames Haven	60,000	0.30	Fendering system designed for 25% of maximum kinetic energy of vessel. Committee Report Aug. 1959
Finnart Oil	100,000	0.18	Committee Report Sept. 1959
Terminal	65,000	0.22	
Amsterdam	60,000	0.15	Report in 1954
Point Jetty, Davenport	40,000	0.15	One-half the weight of vessel is used in computing energy to be absorbed. Report in 1955
Singapore	20,000	0.30	Velocities considered upper limits. Approach velocity of 0.12 m per second is usual. Report in 1955
Sumatra Parking Terminal	50,000	0.08	Report in 1960
Isle of Kent	32,000	0.24	Committee Report July 1954

6.3 SINGLE PILES OR GROUPS OF PILES SUBJECTED TO ACTIVE LOADING

6.3.1 *Overhead sign*

6.3.1.1 *Introduction*

Large numbers of overhead signs are constructed for advertising, where allowed, and for providing information to highway drivers. The larger ones have two columns that support a structure for holding the sign. A smaller sign can be supported by a single column. The foundations for the columns can be supported by a single pile or by multiple piles. In the latter case, the piles can be analyzed as a group.

200 Piles under lateral loading

6.3.1.2 Example for solution
The example is a sign with a single column and with a foundation consisting of a single steel-pipe pile. The problems to be solved are: The diameter and bending stiffness of the pile; the required penetration of the pile; and the expected deflection of the sign during the storm. The sign is patterned after advertising signs along some highways: dimensions of sign, 4 m wide by 3 m high; height of midpoint of sign, 8.5 m; velocity of wind at sign, 32 m/sec; soil at site, overconsolidated clay; undrained shear strength of clay, 70 kPa; water table, 10 m below ground surface; total unit weight, 19 kN/m^3; and number of cycles of loading, 1000. The value of ε_{50} was selected as 0.007 from Table 3.5

The forces from the wind against the sign are computed from Equation 6.1, with C_s selected as 1.5 and w as 11.88 N/m^3. The force is computed to be 11.15 kN.

6.3.1.3 Step-by-step solution
The first step is to arrive at and approximate diameter and stiffness of the pile near the ground surface. The assumption is made that the building authority for the region has specified a factor of safety of 2.2 for the design. Further, the factor of safety is used to upgrade the load because of the nonlinear nature of the response of the soil. Therefore, the ultimate bending moment can now be estimated. Experience has shown that the maximum moment will occur near the ground surface when the loading is dominated by an applied moment. Thus, the moment arm is selected as 8.5 m plus 1.0 additional m, leading to an approximate value of M_{ult} of (9.5)(11.15)(2.2) or 233.0 m-kN. For the purposes of these computations, the axial load is assumed to be negligible until a final check of stresses is made.

The value of the strength of the steel in the pipe is selected as 250 MN/m^2 and Equation 4.29, $M_{ult} = f Z_p$, is used to solve for the value of Z_p as of 9.32 × 10^{-4} m^3. Equation 4.30, $Z_p = 1/6(d_0^3 - d_i^3)$, can now be used to find a pipe section that will satisfy the requirement of the maximum moment. However, the designer will normally query the local suppliers of steel pipe to learn the sizes that are readily available. Proceeding with Equation 4.30, the selection of a steel pipe with an outside diameter of 300 mm leads to a wall thickness of 11.17 mm. A wall thickness of 12 mm is selected without regard to the availability of such a size. Using the equations noted above, the ultimate bending moment for the pipe section is 249.0 m-kN, somewhat higher than the value required.

With the selection of a trial pile, the next step is to make solutions with the professional version of Computer Program LPILE (see Appendix D). The stiffness of the selected steel pipe is 22.55 MN-m^2. Assuming the properties of the soil as given above and that the loading will be cyclic, the pile was modeled as a single member. The extension above the ground line is 8.5 m and the penetration is assumed to be 20 m to ensure that *long-pile* behavior will hold.

The lateral load is applied in increments at the midheight of the sign to a magnitude greater than the factored load of (11.15)(2.2) or 24.53 kN. Figure 6.4 presents a plot of the maximum bending moment, deflection at the ground line, and deflection at the midheight of the sign, all as a function of the applied lateral load. Using the value of M_{ult} of 249.0 m-kN, the value of P_t that would cause a plastic hinge was found to be 29.1 kN, as shown in the figure. The factor of safety was then computed to be 29.1/11.15 or 2.61.

Referring to Figure 6.4 and assuming linear behavior between load increments, the lateral deflection at the midheight of the sign at the ultimate load was computed to be 453 mm and the deflection at the groundline was 16 mm. At the computed wind load of

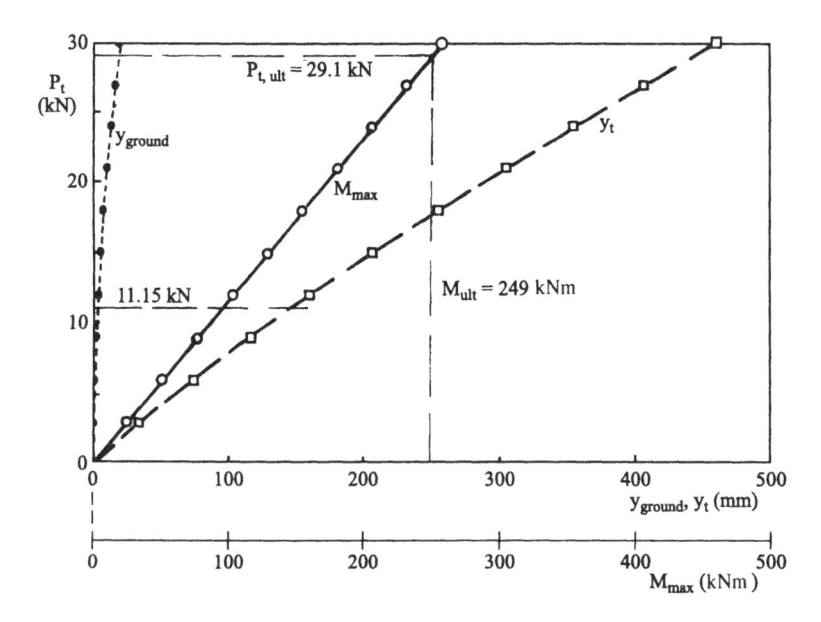

Figure 6.4. Computed response of an overhead sign to wind loading.

11.15 kN, the sign was computed to deflect 149 mm, the ground-line deflection was 3 mm, and the ultimate bending moment was 95.3 m-kN.

The next step in the solution is to compute the necessary penetration of the steel pile. The required procedure is to gradually reduce the penetration and to compute the deflection. When the penetration becomes inadequate, the deflection will increase, which indicates that the tip of the pile is deflecting. If there are two or more points of zero deflection along the pile, the tip of the pile will not deflect and the deflection will be almost unchanged. The load used in the computation was 29.1 kN, which included the factor of safety of 2.61. The results of the computations with the reduced penetration of the pile are shown in Table 6.4 and Figure 6.5.

Figure 6.5 shows that the 'rule-of-thumb' of two points of zero deflection as necessary to establish the 'critical' penetration is apparently valid, because the deflection at the sign remains constant beyond a penetration of 4.5 m. However, the value selected or computed for ε_{50} influences strongly the computed values of deflection. For example, computations with the use of a value of ε_{50} of 0.02, not shown here, led to a suggested penetration of 6 m. Therefore, a prudent designer will perform parametric studies using a range of values of ε_{50} and make a selection for the penetration on the basis of the results. For this particular case, a penetration of 6 m is suggested, even though factored loads were used in the computations.

6.3.1.4 Discussion of results

The results of the computations showed that the maximum bending moment did, in fact, occur very close to the ground line, and the approximate method of finding the initial dimensions of the cross section of the pile was valid. However, as the moment arm for a particular design becomes relatively small, initial computations with the computer program may be necessary to obtain a trial size. Because of the nonlinear nature of the re-

202 Piles under lateral loading

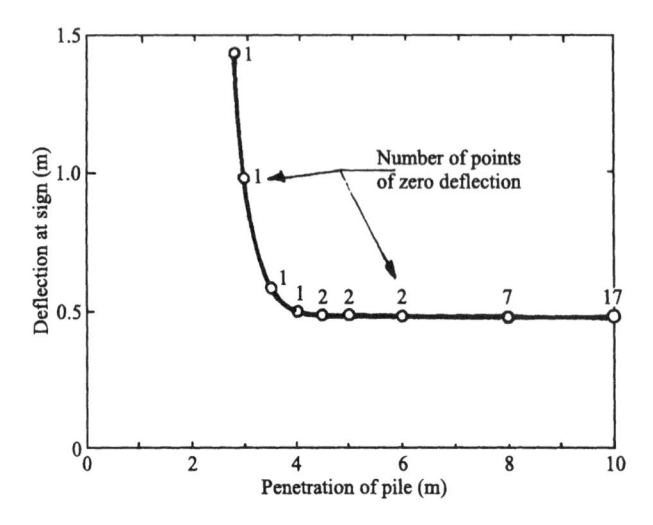

Figure 6.5. Influence of penetration of pile on deflection of overhead sign at a constant load.

Table 6.4. Influence of Penetration on the Deflection at Sign.

Penetration	Deflection at sign	number of points of zero deflection
m	m	
15	0.4796	42
10	0.4796	17
8	0.4796	7
6	0.4798	2
5	0.4805	2
4.5	0.4803	2
4	0.4920	1
3.5	0.5773	1
3	0.9802	1
2.8	1.429	1

sponse of the soil, no alternative method can be suggested to obtain the necessary penetration of the pile other that shown in Figure 6.5, except that a range of values of ε_{50} must be investigated, depending on the conditions at the site.

The engineer must ascertain that the design shown above will meet the standards and specifications of the governmental body that has cognizance over the area where the construction is to be done. Additional computations could reduce the cost of the foundation by either reducing the size of the steel-pipe pile or by finding a structural member that would satisfy the loadings. However, the savings to be gained may be less than cost of the computations.

The criteria for cyclic loading was used in characterizing the response of the soil. However, the engineer may wish to examine the ground line deflection with respect to the soil response to gain information on whether or not continued repeated loading could cause a significant reduction of soil resistance. At the expected lateral load at the sign of 11.15 kN, the deflection at the ground line is about 3 mm. A small gap could open in the stiff clay and, if there is precipitation, there could be some loss of resistance because of soil erosion. Therefore, some improvement of the ground surface would be advisable. For example, the response of the soil can be improved by casting a concrete slab at the ground line; however, analyses would still be necessary with a proper set of p-y curves assigned to the concrete.

Data on the gusting of the wind fails to show that the gusts will come at regular intervals; therefore, excitation to develop resonance in the system is not likely. Further, the computation of the natural frequency of the system is complex because a single-degree-of-freedom is unlikely to be approximately correct and because of the nonlinear response of the soil. The engineer could take advantage of any available, empirical evidence of the behavior of similar structures in storms.

6.3.2 Breasting dolphin

6.3.2.1 Introduction
In the presentation of information on the loading of breasting dolphins (Art. 6.2.6), the large number of parameters that influence the design is indicated. Therefore, the engineer will probably be required to undertake an extensive study to select the factors that will guide the design. Such factors may be well established by the harbor authority or a comprehensive study may be required. The example that follows is based on the premise that the harbor authority has set guidelines or that a thorough study has been done to establish design parameters.

6.3.2.2 Example for solution
The vessel displacement is stated as 40,000 tons, the velocity at impact is given as 0.152 m/sec, and the vessel is assumed to contact two dolphins simultaneously. Using the equation for the energy of the vessel as the breasting dolphin is contacted, $E_N = (W/2g)v^2$, the energy to be absorbed by a single breasting dolphin is 231 kN-m, using kN for convenience in the computations. The value of P_t will be factored upward so that the bending moment to cause first yield has a sufficient margin of safety.

Soil borings revealed a variety of soils at the site and study led to the selection of the following properties for the analysis reported herein.

Depth m	Undrained shear strength kPa	Total unit weight kN/m³	ε_{50}*	Material
0				
	0	0		Water
11.3				
	192	15.7	0.005	Clay
15.2				
	77	15.7	0.01	Clay
50				

*Assumed to have been obtained from soil tests

204 *Piles under lateral loading*

The clay in the first 4 m below the mudline exhibited some fissuring. A careful study of all available information suggested that the data in the above table can be used without change in developing the *p-y* curves. The soft-clay criteria with cyclic loading is considered appropriate; however, the clay will behave statically during the first few loadings. The factor of safety employed in design is assumed to be adequate to prevent a problem. Prudence suggests that a second set of computations, using static criteria, would be in order. Such computations will follow the procedures set down herein and are not shown in the interest of brevity.

The tide at the location of the pier on the Gulf of Mexico is almost negligible and the assumption is made that the docking vessel with strike the dolphin at 14.6 m above the mudline (3.3 m above the water surface). The design of a breasting dolphin becomes complicated if the depth of water varies significantly with time. For example, a design was made for the docking of vessels along the Mississippi River, where the water surface can vary several meters during the year.

The energy that can be developed by the breasting dolphin is related directly to the load-deflection curve, but the deflection is a function of the bending stiffness of the dolphin (pile). Because the ultimate bending stress and maximum allowable deflection are functions of the strength of the steel, a high-strength steel is nearly always preferable. For the example shown here, the strength of the steel was selected as 345,000 kPa. An important further consideration is that the pile should be tapered by reducing the wall thickness in zones of lower bending stress. The deflection will be increased to increase the developed energy without causing excessive stress in the steel.

Rough computations are difficult because finding the lateral load P_t, factored upward for safety, involves too many parameters. Therefore, a pile with an outside diameter of 1.4 m and with a wall thickness of 60 mm in the region of maximum stress was selected with only some computations on an earlier design for guidance. The adequacy of the selection, considering the tapering to yield the maximum amount of deflection at the point of application of the load, is to be determined by repeated trials with Computer Program LPILE (See Appendix D).

6.3.2.3 *Step-by-step solution*
The first step is to make a trial solution with the following pile: length = 50 m; outside diameter = 1.4 m; and wall thickness = 60 mm. The *p-y* curves for the soil specified above were modeled by the soft-clay (plastic) criteria assuming cyclic loading.

The curve of load versus deflection at the pile head (14.6 m above the mudline) is shown in Table 6.5. A preliminary integration of the area under the load-deflection curve, not shown here, was necessary to ensure that sufficient load was applied to yield an amount of energy to balance that of the ship at touching (231 kN-m). Numerical integration of the data in Table 6.5 for an energy of 231 kN-m yielded a lateral load of 1097 kN and a deflection at the point of application of the load of 373 mm.

The bending moment for the load of 1097 kN was computed with the computer program and the maximum moment was found to be 0.1733 kN-m. Using the I_p of the 60-mm section, 0.0568062 m^4, the maximum bending stress was computed to be 213,600 kPa. If the strength of the steel is 345,000, the factor of safety at first yield of the steel is 1.62. However, the factor of safety will increase if the pile is tapered, that is, if the wall thickness is reduced in regions where the bending moment is less than the maximum. The reduced bending stiffness along the dolphin will result in an increase in the deflection at the head of the dolphin and, hence, an increase in the energy that can be offset. The bending-moment curve for a lateral load of 1097 kN was examined, and wall thicknesses were selected so that in no point along the pile would the bending moment be less than 1.62. The results of the analysis are shown in Table 6.6.

The values shown in Table 6.6 and then used to obtain data for finding the lateral load and corresponding deflection that will yield an energy of 231 kN-m.

With a revised schedule of pile stiffness, a new curve of load versus deflection at the pile head can be computed. The results are shown in Table 6.7. Numerical integration of the data in Table 6.7 for an energy of 231 kN-m yielded a lateral load of 1029 kN and a deflection at the point of application of the load of 391 mm.

The next step is to solve for a bending-moment curve for a lateral load of 1029 kN and to check the computed bending stresses against the allowable bending stresses for the determination of the factor of safety along the dolphin. The results of the computations are shown in Figure 6.6 and Table 6.8. In the table the values of the computed bending moment, using Computer Program LPILE, at the changes in wall thickness are tabulated along with the computed values of allowable bending moment at first yield and allowable bending moment assuming a plastic hinge. The computed factors of safety are also shown in Table 6.8.

Table 6.5. Trial computations for energy developed by dolphin assuming constant wall thickness of 60 mm.

Lateral load	Deflection at point of application of load
kN	m
0	0
100	0.0195
200	0.0441
300	0.0721
400	0.1027
500	0.1359
600	0.1711
700	0.2083
800	0.2475
900	0.2883
1000	0.3307
1100	0.3749
1200	0.4203

Table 6.6. Schedule of wall thickness and moment of inertia along the length of the dolphin to yield an increased value of computed energy.

Depth	Wall Thickness	Moment of Inertia
m	mm	m^4
0 to 7	25	0.0255300
7 to 11.5	42	0.0413451
11.5 to 22	60	0.0568062
22 to 25	42	0.0413451
25 to 50	25	0.0255300

206 Piles under lateral loading

Table 6.7. Computations for energy developed by dolphin using step-tapered wall thickness.

Lateral load	Deflection at point of application of load
kN	m
0	0
100	0.0219
200	0.0490
300	0.0799
400	0.1140
500	0.1511
600	0.1911
700	0.2337
800	0.2787
900	0.3264
1000	0.3764
1100	0.4285

Table 6.8. Computed bending moments and allowable bending moments for first yield of steel and for a plastic hinge, along with the respective factors of safety.

Depth	Computed bending moment	Allowable bending moment at first yield	Factor of safety	Allowable bending moment or plastic hinge	Factor of safety
m	kN-m	kN-m		kN-m	
7.0	7200	12,580	1.75	16,310	2.27
11.5	11,800	20,380	1.73	26,730	2.27
17.0	16,200	28,000	1.73	37,190	2.30*
22.0	10,600	20,380	1.92	26,730	2.52
25.0	5310	12,580	2.37	16,310	3.07

*Maximum

The final step in the analysis, assuming the factors of safety are satisfactory, is to use computer program and gradually reduce the penetration of the dolphin to find the critical penetration. Excessive deflection was found to occur at a length of the pile of between 33 and 34 m. The deflection for the large-diameter pile was found to be sensitive to small changes in penetration; therefore, a pile with a total length of 40 m is warranted, giving a penetration below the mudline of 25.4 meters.

6.3.2.4 Discussion of results

Whether or not the dolphin with a diameter of 1.4 m is satisfactory with the factors of safety shown in Table 6.8 is a matter for discussion by the engineer in consultation with the owner of the project. Another steel-pipe section could be selected, the strength of the steel could be varied according to locally-available material, and consideration given to the use of auxiliary material for absorbing energy, and the computations repeated to provide more information on which to make a decision. A number of materials or devices are available for the use as auxiliary energy absorbers. As an example, about 80 kN-m of energy can be absorbed by

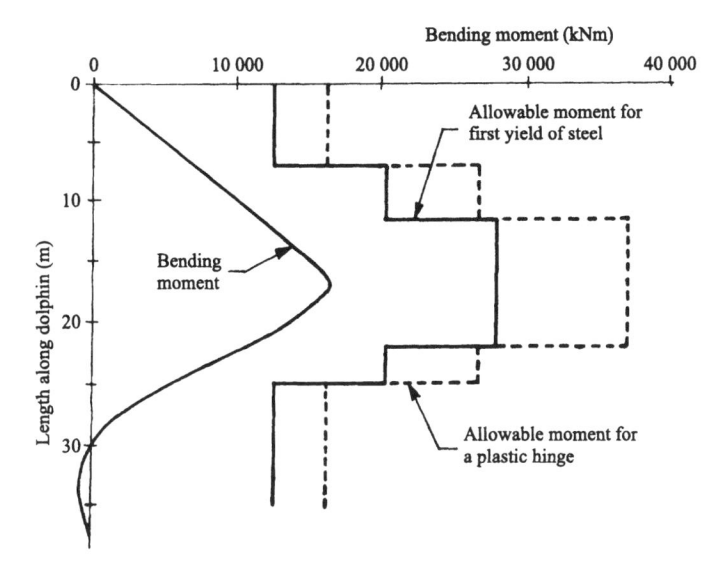

Analysis of single piles and groups of piles subjected to active and passive loading 207

Figure 6.6. Plots of bending-moment curve for an applied load of 1.029 kN and allowable bending moment for breasting dolphin.

using five m of cylinder bumpers, 0.5 m by 0.25 m, so that they are loaded radially (Quinn 1961, p. 290). The amount of computational time could be considerable to work out a variety of solutions, considering the number of important parameters that are involved.

As noted earlier, the dolphin will behave with more stiffness under the initial lateral loads than indicated by the computations because of the use of the recommendations for the response of the soil under cyclic loading. Therefore, some computations assuming the response of the soil for static loading might be useful.

A breasting dolphin can be readily tested experimentally by applying increments of lateral load at the head of the dolphin and measuring the deflection of the dolphin for each of the individual loads. If there is concern about scour or about deposition of soil near the dolphin, the influence the these factors can be studied by modifying the soil at the mudline.

While the computations presented herein are critical to the proper design of a breasting dolphin, a number of other factors must be considered. For example, consideration must be given to the use of 'rubbing blocks' to distribute the loading against the side of docking vessel to prevent damage of the vessel itself.

The designer must consider the response of the vessel and the breasting dolphins after the dolphins have been deflected by the moving vessel. The energy in the deflected dolphins would be expended by pushing the vessel away from the dock. However, lines are tied to mooring dolphins and the damping of the energy in the dolphins would occur rapidly.

6.3.3 Pile for anchoring a ship in soft soil

6.3.3.1 Introduction
One or more ship's anchors are commonly used to allow the ship to remain in relative the same position in the ocean. Flukes will cause the anchor to sink below the mudline, de-

pending on the strength of the soil, and resistance (usually termed *holding power*) is derived from the soil as the anchor acts as a kind of a plow. If the soil at the mudline is weak, the resistance is low and a number of anchors may be necessary, depending, of course, on the lateral forces that will be applied to the ship.

A drilling ship is sometimes used in offshore operations and the ship must remain in virtually the same horizontal position for a considerable period to time. A substitute anchor, consisting of a driven pile, can find a useful role if the near surface soil at the mudline is quite soft. As shown in Figure 6.7, the pile can be driven with a follower so that its top is a desirable distance below the mudline. The anchor chain is attached to a bracket at some point along the pile and a tensile force will cause the chain to depart at some angle from the pile and assume a curved shape to the mudline. The components of the tensile force in the chain at the pile will cause a lateral force, an upward force, and possibly a moment, depending on the details of the bracket. The pile should have an appropriate factor of safety against being pulled upward and against the development of a plastic hinge.

Before the design of the pile, equations must be developed to predict the configuration of the chain. The following section presents the relevant equations, and the solution of an example problem.

6.3.3.2 *Configuration of anchor chain*
The concepts presented for the computation of the position of an anchor chain as its lower end moves from the vertical, along the driven pile, to a position of equilibrium can be applied to any soil. However, the equations presented here are applicable only to cohesive soil (Reese 1973). The principal assumptions in the development of the equations are:
1. After being subjected to a tensile load, the chain lies in a vertical plane; 2. The position of the chain is a curve formed by a succession of arcs of a circle; 3. The soil surrounding the chain reaches a limiting state of stress where the ultimate resistance presents further

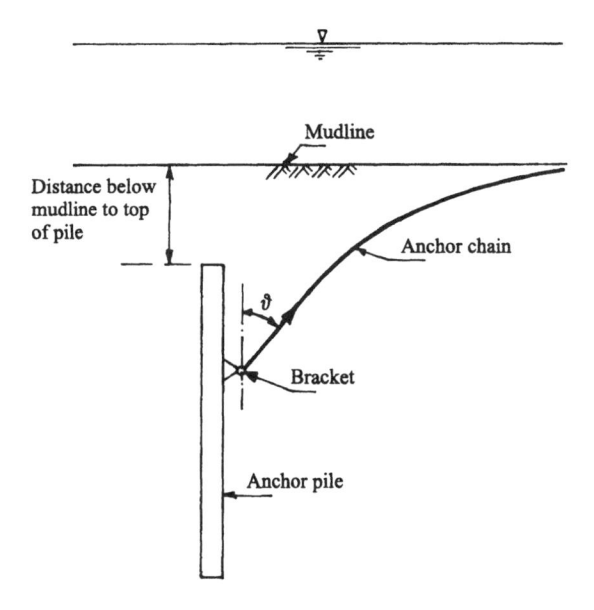

Figure 6.7. Anchor pile at an offshore location.

Analysis of single piles and groups of piles subjected to active and passive loading 209

movement of the chain; 4. The chain become horizontal at the mudline; 5. The undrained strength of the clay is constant along each of the arcs of a circle; and 6. The tensile force in the chain remains constant.

With respect to (6) above, the assumption is made that the horizontal loading on the drilling ship is repeated, and increasing to the final value of tension. With the alternate increase in load and then the relaxing, the axial movement of the chain is expected to be small at the final load. Thus, the tension is not expected to be reduced with distance below the mudline. This assumption is undoubtedly conservative. The analysis can be extended to the case of a decrease in tension with distance from the anchor (Bang & Taylor 1994). Because the tension will depend on the relative axial movement of the chain with respect to the soil, the change in the tension is a matter of some question.

An element from the chain is shown in Figure 6.8 and the equilibrium of the element is satisfied by solution of the following equations. An angle α is chosen to define an arc of a circle and an expression is derived to compute the value of R, the radius of the circle.

$$2T \sin \frac{\alpha}{2} = F = p_u R \alpha \qquad (6.5)$$

$$p_u = Kcd \qquad (6.6)$$

where K = bearing capacity factor for the embedded chain in undrained clay, ranging generally from 9 to 11, c = the undrained strength of the clay, and d = equivalent diameter or width of the chain.

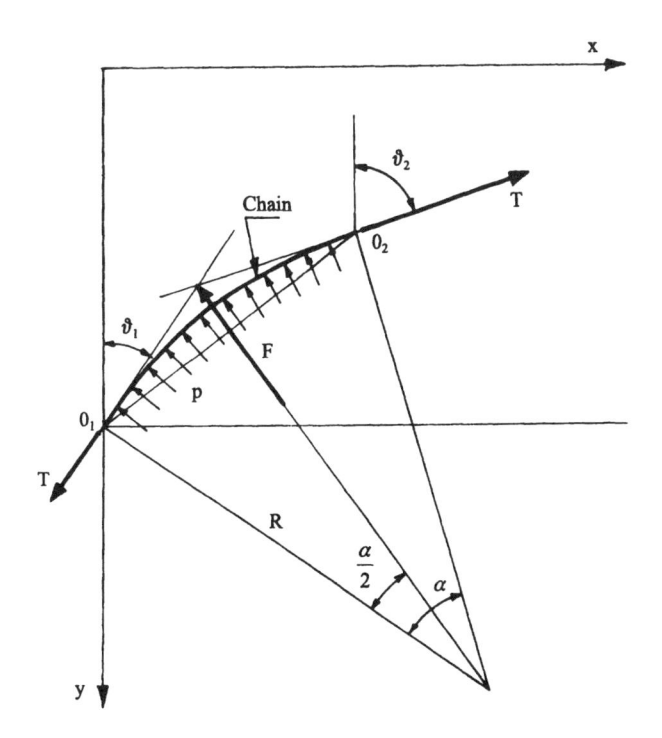

Figure 6.8. Segment of an anchor chain showing geometry and applied forces.

210 Piles under lateral loading

For small angles, $\sin(\alpha/2)$ is approximately equal to $\alpha/2$ and Equation 6.5 becomes $2T(\alpha/2) = Kcd R\alpha$. Thus,

$$R = \frac{T}{Kcd} \tag{6.6}$$

The length of the chord, χ, is needed and may be used to compute the forward coordinates.

$$\chi = 2R\sin\frac{\alpha}{2} \tag{6.7}$$

$$x_{i+1} = x_i + \chi \sin\beta \tag{6.8}$$

$$y_{i+1} = y_i - \chi \sin\beta \tag{6.9}$$

where $\beta = 90° - (\theta_I + \alpha/2)$.

A solution proceeds by selecting a value of tension to be applied to the anchor chain, establishing a trial position where the chain is to be attached to the anchor pile, and tabulating the undrained shear strength with depth. The computer code is written to start with a given angle of theta at the anchor pile. The equations are solved for the curved position of the anchor chain, point-by-point, and convergence is checked. Convergence requires that the chain becomes horizontal at the mudline, within the tolerances set for angle and vertical position. If convergence is not achieved, another value of the angle theta is selected and the solution is repeated. The equations are not complex and convergence occurs with little computer time.

6.3.3.3 Example for solution

The problem of an anchor pile that is to be addressed is shown in Figure 6.9. A drill ship is to operate at a location where the near-surface soil is soft clay. A pile with a diameter of 1.52 m, a wall thickness of 25 mm and a length of 30 m is driven with a follower to 6.3 m below the mudline. The properties of the soft clay are given in Table 6.9. The chain was constructed of 3.25-inch rod (82.6 mm) and the average width was 282 mm. The chain had a breaking strength of 5382 kN and a proof-test strength of 3577 kN.

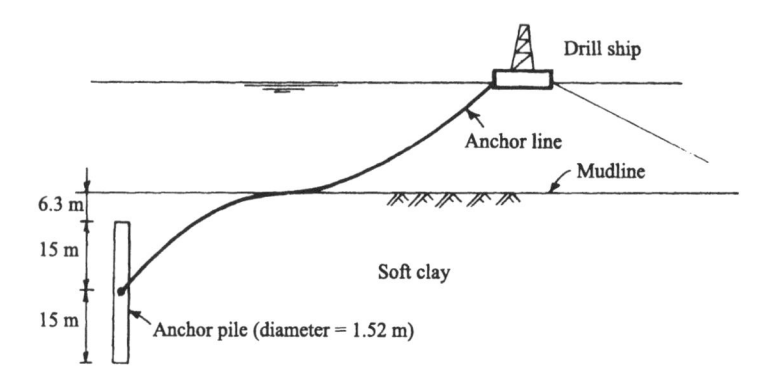

Figure 6.9. Sketch showing layout of problem of design of an anchor pile.

6.3.3.4 Step-by-step solution

The bearing capacity factor K for the chain was selected as 9.0, giving a value of Kd of 2.538 m. The equations for the position of the chain were solved, using the soil properties shown in Table 6.9, and the computed values of theta for various amounts of tension in the chain are shown in Table 6.10.

For an example computation of the anchor pile, the tension in the chain was assumed to equal to the proof-test strength, giving a factor of safety of 1.50 with respect to the ultimate capacity of the chain. The computed position of the chain for a tensile load of 3580 kN is shown in Figure 6.10.

With a value of tension of 3580 kN and a value of theta of 55 degrees, the component of axial load was 2053 kN and the component of lateral load was 2933 kN. Using the values of shear strength in Table 6.9 and the dimension of the anchor pile, the resistance to uplift was computed to be 2612 kN, yielding a factor of safety against pullout of 1.27. Computer Program LPILE (see Appendix D) was used and the resulting curves of deflection and bending moment for a lateral load of 2933 kN at the midheight of the anchor pile are shown in Figure 6.11. The maximum deflection was computed to be about 40 mm which should be acceptable. The maximum bending moment was computed to be 5695 m-kN. Using the dimensions of the cross section of the pile and assuming a strength of steel of 250,000 kN/m^2, the bending moment at first yield of the steel was computed to be 10,800 m-kN and the bending moment to develop a full plastic hinge was computed to be 14,000 m-kN. Thus, the factor of safety in bending at first yield of the steel was 1.90.

6.3.3.5 Discussion of results

The solution of the problem of a pile used as an anchor for a floating vessel is performed in a straightforward manner. The technology for the analysis of a pile under lateral load-

Table 6.9. Properties of soft clay at site of anchor pile.

Depth	Undrained shear strength	Total unit weight	Strain at 50% of ultimate strength
m	kN/m^2	kN/m^3	
0	2.90	10.9	0.01
9.8	12.48	11.8	0.01
10.2	17.23	12.6	0.01
20.1	17.23	12.6	0.01
30.5	22.1	14.1	0.01
50	22.1	14.1	0.01

Table 6.10. Computed values of theta as a function of tension in the anchor chain.

Tension, T kN	Theta degrees
500	13
1500	35
2000	43
2220	46
2670	50
3120	54
3580	55

212 *Piles under lateral loading*

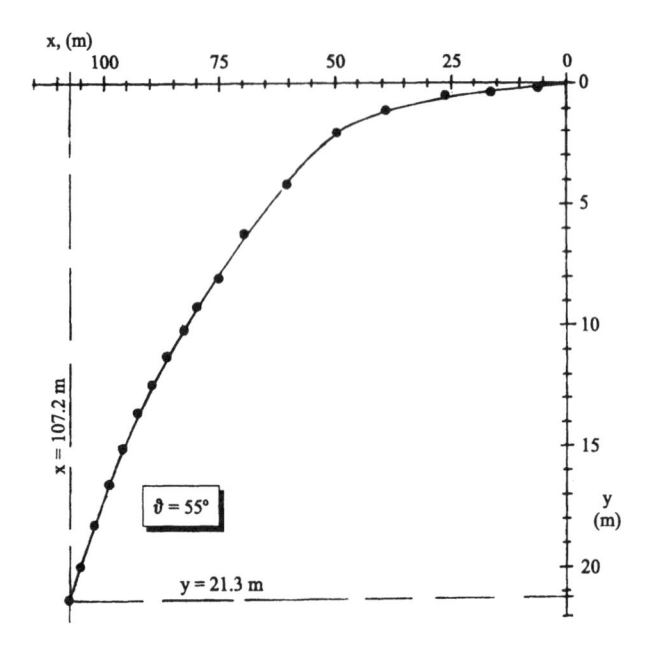

Figure 6.10. Computed position of anchor chain for a tensile load of 3.580 kN.

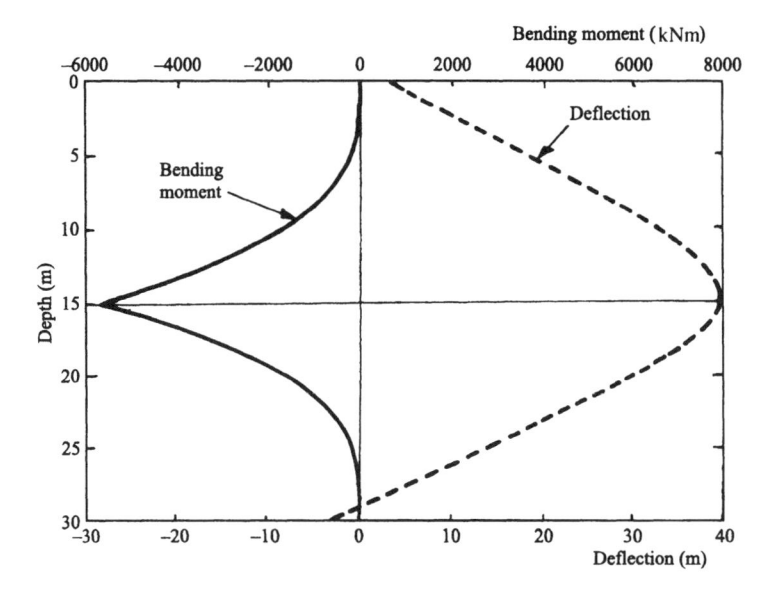

Figure 6.11. Curves showing bending moment and deflection of anchor pile.

ing was applied to find the deflection and bending moment in the anchor pile with no particular difficulty.

As noted earlier, the principal uncertainty in the solution resides with the magnitude of the tensile load at the wall of the pile. The assumption that no reduction in tension is ob-

Analysis of single piles and groups of piles subjected to active and passive loading 213

served along the anchor line may be excessively conservative. If design of such anchor piles is a frequent occurrence, a useful procedure would be to install remote-reading load cells in the anchor line at appropriate intervals. Attention should be given not only to the amount or reduction in the tensile load from point to point along the line but also to the load-deflection characteristics of the unit load transfer.

While the results show the pile that was selected to be satisfactory, the expenditure of additional time by the engineer would be useful. The parameters that should be investigated are: the position of the top of the anchor pile, the diameter and wall thickness as a function of depth, and the optimum position of attaching the anchor line to the pile. Attention should also be given to appropriate techniques for recovering the anchor piles after they are no longer needed.

6.3.4 *Offshore platform*

6.3.4.1 *Geometry of platform and method of construction*
The upper deck of the platform is square, with a dimension of 19.81 m. The platform is also square at the level of the bottom panel point, with a dimension of 24.49 m. There are four main piles, one at each corner of the deck, that are driven on an outward batter. Twelve conductor pipes are driven to provide the initial casings for drilling the wells. The conductor pile are driven through close-fitting openings (slots) so as to move laterally with the deflection of the platform. Therefore, the conductors can provide lateral resistance but do not take any of the axial load from the platform proper. However, each conductor pipe is designed to sustain an axial load of 1112 kN that could arise from drilling operations.

The marine contractor endeavors to position an offshore platform so that its principal axes are parallel and perpendicular to the expected direction of the maximum storm. The analysis that follows is based on the assumption that the maximum forces from the storm hits the platform on one of these axes. Plainly, the maximum waves and winds could produce the maximum force from some oblique direction, in which case the platform could twist about its vertical axis. Thus, the two-dimensional equations for analyzing a group of piles would be inappropriate. The two-dimensional equations have been extended to the three-dimensional case, but the assumption is made for the example shown here that three-dimensional analysis is unnecessary.

An elevation view of one of the two bents in the platform is shown in Figure 6.12. The sketch defines one of two bents that are assumed to behave in an identical manner. Thus, six of the conductor pipes can resist lateral loading. The loadings on the bent are shown by horizontal or vertical arrows and are applied at panel points. The horizontal or lateral loads are derived from waves, currents, and wind as discussed in a previous section. In normal practice, the horizontal loads derive from the storm that is assumed, frequently the so-called 100-year storm. The vertical forces are due to the loads from equipment and supplies on the deck of the platform. The resulting loads and moment are shown by the heavy arrows and are used to analyze the foundation.

Construction is accomplished by placing a jacket, or template, on the ocean floor with a height of 17.06 m, a few meters greater than the water depth of 14.33 m. A derrick barge is used to drive the piles and conductor pipe. The tops of the piles are welded to the top of the jacket. A deck section with a height of 10.98 m, fabricated to close tolerances is lifted and its legs are stabbed into the tops of the piles. The legs of the deck section are welded

214 *Piles under lateral loading*

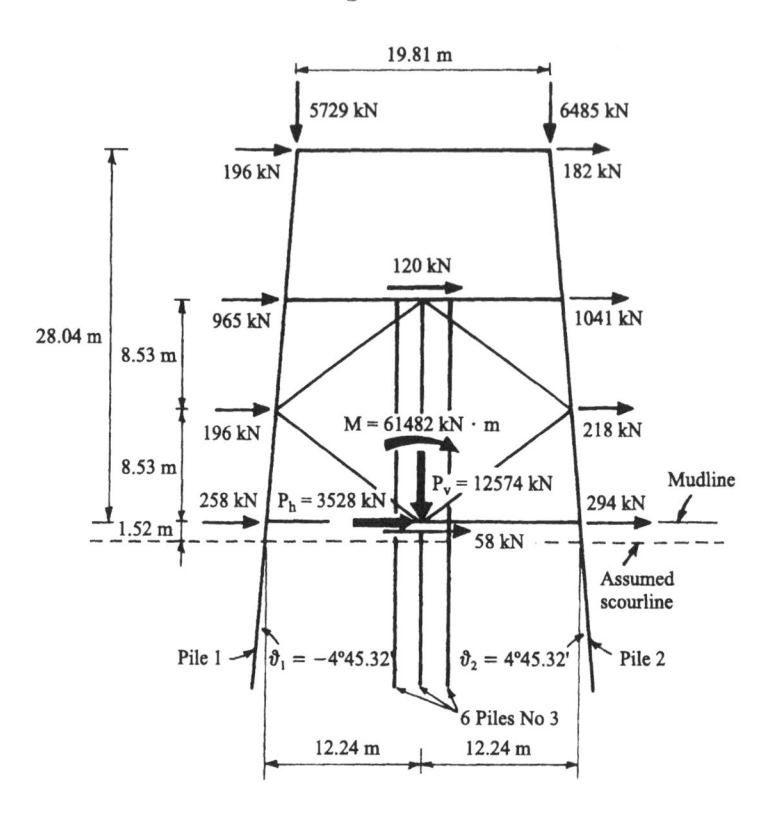

Figure 6.12. Elevation view of one bent in an offshore platform, showing loadings and dimensions.

into place and construction proceeds by placing the drilling equipment and supplies on the deck. The sketch in Figure 6.12 shows that the geometry of the platform is relative austere above the jacket with the view that the minimal area of the structure will limit the lateral forces from waves. The maximum unit forces from waves occur at the wave crest. Oceanographers make careful studies of all of the factors affecting the maximum wave height to be expected to ensure that the deck itself will be above the height of the maximum wave.

6.3.4.2 *Factors of safety*
The three agencies that have been most active in establishing criteria for the design of offshore platforms are: the American Petroleum Institute, Det Norske Veritas, and Lloyd's Register. Information from the API (1993) will be presented here as a example of the guidelines that have been established. Table 6.11 gives the factors of safety that are recommended for the penetration of piles.

The factors of safety with respect to the response of the piles under lateral loading are not presented in such a formal manner but the above factors provide guidelines for judging the adequacy of a design.

6.3.4.3 *Interaction of piles with superstructure*
The designer of the platform has several options with respect to the manner in which the piles will behave under lateral loading. The first option is to decide if the jacket legs are to

Analysis of single piles and groups of piles subjected to active and passive loading 215

Table 6.11. Values of factors of safety for pile penetration (API 1993).

Load condition	Factors of safety
1. Design environmental conditions with appropriate drilling loads	1.5
2. Operating environmental conditions during drilling operations	2.0
3. Design environmental conditions with appropriate producing loads	1.5
4. Operating environmental conditions during producing operations	2.0
5. Design environmental conditions with minimum loads (for pullout)	1.5

be extended and, if so, the amount of the extension. If soft clay exists at the mudline, the weight of the platform is usually sufficient to cause the jacket legs to penetrate a few meters. Mud mats are frequently placed under the bottom level of braces to ensure that the entire jacket does not penetrate into the soft clay. In the current case, as shown in Figure 6.12, the jacket legs were designed to penetrate a distance of 1.52 m, the same depth at which scour is expected to occur after some time.

Two options are available with the extended jacket leg: 1. A gasket may be placed at the bottom of the jacket leg and the entire annular space between the outside of the pile and the inside of the jacket leg can be filled with grout, as is planned for the current design; or 2. A permanent shim or spacer can be fabricated into the bottom of the jacket leg to ensure close contact between the pile and the jacket; shims are also placed at each panel point in the jacket. With either of the two options noted above, some of the bending moment near at the top of the pile is transferred into the jacket. Some designers, however, prefer not to have a jacket-leg extension or to grout the pile to the jacket in order to minimize the bending stress at the joint at the lower panel point. That joint is vulnerable to cracking due to repeated loading. A design with no grouting and with shims or spacers at each of the panel points will cause the pile itself to sustain all of the bending stresses. Such a procedure, not employed for the present design, is more costly with respect to the volume of steel but the structure is less vulnerable to fatigue cracking.

6.3.4.4 *Pile-head conditions*
Elementary mechanics can be used to obtain pile-head conditions (boundary conditions) that are expected to be close to the values one would obtain by more sophisticated analysis. Figure 6.13 shows models of the main piles and the conductor pipe with the assumption that the braces at the panel points provide only support and do no affect the bending moments. The relationships between the pile-head moment (M_t) and the slope at the pile head (S_t) may be found as shown in the figure by assuming that the pile, above the mudline, will behave as a continuous beam. The point should be made, however, that the slope of the piles at the bottom panel point (mudline usually) cannot be found directly

216 Piles under lateral loading

from the equations in Figure 6.13 but must be modified to reflect the rotation of the superstructure under the combined loadings.

Because of the nonlinearity of the soils, and sometimes of the material in the piles, a more accurate solution for the pile-head conditions requires iteration between the foundations and the superstructure for each set of loadings. For the present problem, the equations shown in Figure 6.13, modified to reflect the rotation of the structure, can be used for the initial analyses with the further assumption that the pile heads remain the same plane after loading as before. The resulting loads and moments at each pile head can be applied to the superstructure to allow an analysis of the superstructure. Then the positions and rotations at each pile head can be compared to the ones from the foundation analysis. If necessary, modifications are made in the pile-head-boundary conditions and the cycle repeated. Only a few cycles are usually necessary to converge to a correct solution for a given set of loadings. Such a procedure may or not be necessary, depending on the particular case.

The equations in Figure 6.13 are used as initial boundary conditions for a computer code and the program automatically modifies the boundary conditions for each pile by employing the computed slope of the superstructure. That procedure was employed to obtain the results shown below.

6.3.4.5 Soil conditions at the site
A comprehensive investigation of the soils at the site was undertaken. The exploration equipment was mounted on a barge or a boat and the sampling employed the wire-line

(a) For piles 1 and 2 (main piles)

- 914 m
- 711 m
- 711 m
- 914 × 32 mm, $I_p = 8.5848 \times 10^{-3}$ m^4
- Jacket 991 × 10 mm, Pile 914 × 41 mm, $I_p = 1.43454 \times 10^{-2}$ m^4
- $\dfrac{M_t}{S_t} = 11991 \times 10^{-6}$ kN·m

(b) For pile 3 (conductors)

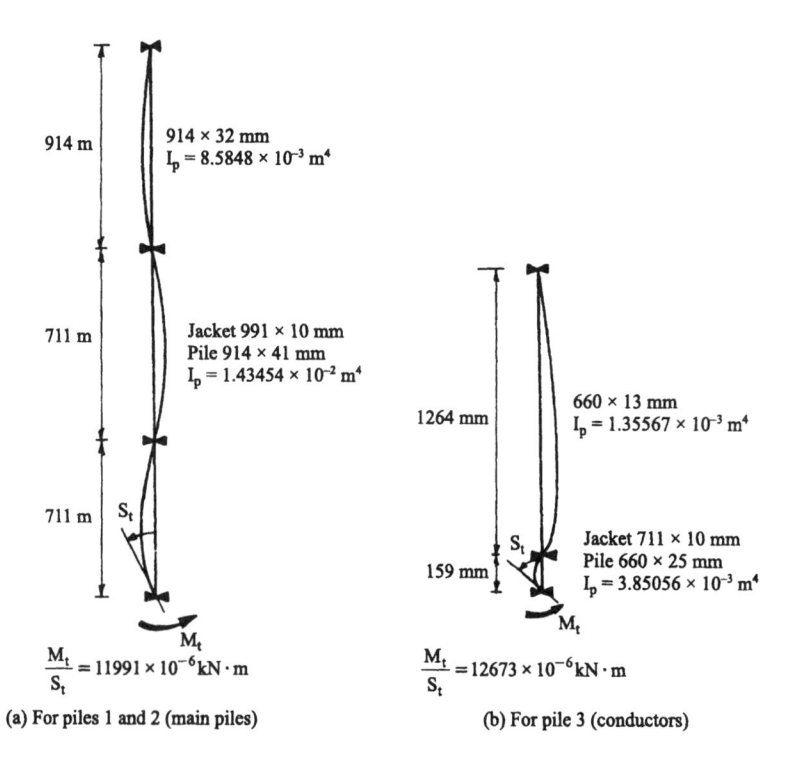

- 1264 mm
- 159 mm
- 660 × 13 mm, $I_p = 1.35567 \times 10^{-3}$ m^4
- Jacket 711 × 10 mm, Pile 660 × 25 mm, $I_p = 3.85056 \times 10^{-3}$ m^4
- $\dfrac{M_t}{S_t} = 12673 \times 10^{-6}$ kN·m

Figure 6.13. Sketches of portions of superstructure of offshore platform to allow computation of rotational restraint at pile heads.

Analysis of single piles and groups of piles subjected to active and passive loading 217

technique. A oil-field-supply boat of the 40 to 50 m class is frequently used at the drilling platform (Emrich 1971). The boat can include all of the facilities that allow an efficient marine operation. A center well is installed of less than one meter in diameter and the drilling derrick is set above the well. A hole with a diameter of about 180 mm is advanced with rotary-drilling tools, employing drilling mud as necessary. The interior of the drill pile is flush-jointed and with a diameter of 89 mm, allowing the deployment of a 55-mm O.D. sampling tube. When the hole has been drilled to the desired depth, the drill pipe is lifted a few meters and held at the rotary table with slips. The derrick and a thin wire line is then used to lower the sampling tube to the bottom of the drilled hole, with depth being monitored with a wire-line revolution counter. The wire line is used to lift a sliding, center-hole weight about 1.5 m.; then the weight is dropped a sufficient number of times to achieve a penetration of 0.6 m. The number of blows is counted to drive the sampler. The sampler with soil is retrieved for some testing at the site. Portions of the samples are protected against loss of moisture and taken to the laboratory for further testing. The wire-line procedure allows the completion of the sampling at a particular borehole without retrieving the drill pile, leading to a practical procedure.

The results of the tests at the site of the offshore platform are shown in four tables. Table 6.12 shows descriptions of the soils encountered and the results of Atterberg Limit

Table 6.12. Summary of test results, soil description and atterberg limits water depth 14.33m.

Depth, m	Soil description	Atterberg limits	
		PL	LL
14.94	Soft gray sandy clay with some decaying wood fragments	17	38
17.37	Soft clay, silty with silt layers, trace of tan clay	23	68
20.42	Firm gray clay with silt layers	30	88
23.47	Firm gray clay with silt lenses and silt pockets	23	71
26.37	Firm to stiff gray clay with some seams of silt and fine sand	27	78
29.57	Gray clay with occasional silt lenses below 29m	26	81
38.71		27	89
43.59	Stiff gray clay		
47.24	Stiff gray clay with sandy silt layers		
49.68	Stiff gray clay with sandy silt layers		
57.30		26	79
61.87	Red clay cuttings noted		
62.79	Layered stiff clay and sand		
64.62			

218 *Piles under lateral loading*

tests; Table 6.13 shows the results of unconfined compression tests of both undisturbed and remolded soil; Table 6.14 shows the results of consolidated-undrained triaxial tests of both undisturbed and remolded soil; and Table 6.15 shows the results from vane tests performed to a limited depth at the site.

Emrich (1971) reported on studies aimed at arriving at the degree of disturbance caused by driving the sampling tube as opposed to the recommended procedure of seating the sampler with a steady push (Hvorslev 1949)[*]. Three borings were drilled to a depth of

Table 6.13. Summary of test results, unconfined compression tests Water depth 14.33 m.

Depth m	Undisturbed				Remolded			
	w %	γ_d kn/m^3	q_u kPa	ε_f	w %	γ_d kn/m^3	q_u kPa	ε_f
14.93	27.3	15.05	38.1	0.092	27.4	15.19	34.8	0.02
17.37	42.4	12.21	36.6	0.141	40.0	12.39.	30.4	0.132
20.42	55.8	10.34	65.9	0.075	51.0	10.89	31.9	0.107
23.47	67.5	11.55	76.4	0.079	43.7	11.88	30.3	0.127
26.37	47.4	11.67	94.4	0.071	44.3	11.78	38.6	0.152
29.57	53.9	11.01	92.9	0.045	42.5	12.11	38.7	0.20
38.71	47.7	11.06	75.8	0.063	43.8	11.83	48.7	0.149
48.16	28.1							
57.30	42.8	11.69	42.6	0.129	39.6	12.52	61.8	0.150
71.32	21.1							

Table 6.14. Summary of test results, consolidated-undrained triaxial tests, offshore site water depth 14.33 m.

Depth	Undisturbed					Conf. Pressure		Remolded				
	w %	%	γ_d kN/m^3	q_c kPa	ε_f	ΔV %	kPa	w %	%	γ_d kN/m^3	q_r kPa	ε_f	ΔV %
14.78	31.8	30.5	14.06	80.4	0.080	4.0	3.8	31.1	28.5	14.22	57.4	0.200	3.8
17.22	46.5	46.0	11.40	73.1	0.093	1.4	21.1	46.3	45.1	11.66	44.0	0.150	3.0
20.27	54.9	54.2	10.23	112.3	0.047	2.6	41.2	55.4	52.1	10.34	47.3	0.119	4.6
23.32	44.6	43.5	11.55	117.7	0.060	3.5	62.2	42.0	38.5	12.14	62.3	0.086	4.5
26.21	39.9	38.5	12.08	118.5	0.040	4.9	82.3	40.3	33.0	12.50	67.3	0.092	8.6
29.41	46.5	45.0	11.31	169.4	0.047	2.6	103.4	47.1	41.8	11.39	82.3	0.151	9.0
38.56	49.2	48.2	10.92	152.1	0.040	2.7	123.5	47.1	43.9	11.31	94.1	0.080	5.3
57.45	40.2	35.1	11.78	163.7	0.150	6.1	289.2	40.7	34.8	12.35	161.5	0.120	5.3

Table 6.15. Summary of test results, strength for vane tests water depth 14.33 m.

Depth m kPa	Maximum strength kPa	Minimum strength
14.9	42.6	15.2
17.4	54.9	22.8
20.4	65.5	24.3
23.5	109.6	33.5

[*] The comprehensive report by the late Dr. Hvorslev, while published some time ago, contains much information of lasting value.

91 m at an onshore site near the Mississippi River. The soils were typical of those that occur over a considerable area of the Gulf of Mexico. Samples were taken with the wire-line sampler described above, with a wire-line sampler with a diameter of 76 mm, with an open push-sampler with a diameter of 76 mm, and with fixed-piston sampler with a diameter of 76 mm. Unconfined compression tests were performed on specimens from each of the methods of sampling. In addition, field-vane tests were performed and miniature-vane tests were performed at the ends of the tube samples. Emrich reported that the strengths from the various unconfined-compression tests and the vane tests were plotted on the same graph with the results from the fixed-piston sampler assumed to yield correct results. Compared to results from the fixed-piston sampler, the following results were obtained: 57-mm wire-line sampler, 64%: 76-mm wire-line sampler, 71%; and the open-push sampler, 95%. Scattered results were obtained from the vane tests but the results were generally higher that the results from the fixed-piston tests.

An exhaustive study of the data that are presented on the strength of the clay at the sites is unwarranted for several reasons. With respect to Table 6.14, the investigators presumably expected that subjecting specimens to a confining pressure equal to the overburden pressure would reflect a gain in strength that offsets the strength loss in sampling. However, using data from Leonards (1971), the value of $(c/p)_n$ for the undrained shear strength of normally consolidated clay c as a function of the effective overburden pressure p is 0.299. Thus, the clay at the site is substantially *under consolidated* and consolidated-undrained tests, shown in Table 6.14 would be expected to reflect shear strength considerably more than the actual. Such an interpretation, however, is not confirmed by the moderate loss of water content during the consolidation phase of the tests, as shown in Table 6.14. Nevertheless, the logic of using the results from the consolidated-undrained tests is unsubstantiated.

Strong dependence could be placed on the results from the vane tests, as shown in Table 6.15, which show strengths that are substantially higher than those from the unconfined-compression tests. However. nearly all of the experimental data on the response of piles to axial and lateral loading have depended strongly on results from unconfined-compression tests rather than results from vane tests.

Even though the work of Emrich, reviewed above, suggests that the strength of the unconfined-compression tests of wire-line samples from a sampling tube with a diameter of 57 mm (presumable the size used for the results shown in Table 6.13) can be multiplied by a factor of 1.5 to achieve a strength that would be obtained from a fixed-piston sampler. Two points argue against such a procedure. Firstly, unless much more data can be obtained on the comparative results of unconfined compressive strength from various sampling techniques, the application of Emrich's results to all sites is unwise. Secondly, the driving of a pile into soft, saturated clay cause remolding and excess pore-water pressures at the wall of the pile. Some attempts have been made to quantify the initial loss of strength of the clay and the subsequent regain of strength (Seed & Reese 1957, Reese 1990), but no method has been accepted by the geotechnical engineering community. Therefore, to reduce the possibility that a failure could occur if a pile is loaded axially too soon after installation, a lower value of undrained shear strength is selected rather than a substantially higher value.

The result from the above discussion is that the undrained shear strength from the unconfined compression tests, shown in Table 6.13, is accepted. A review of the results in the table indicates two anomalies: 1. The remolded strength at a depth of 57.3 m is greater

220 *Piles under lateral loading*

Table 6.16. Values of shear strength accepted for analysis (depth measured from mudline).

Depth m	Undrained shear strength kPa	Angle of internal friction deg
0	0	
1.52	0	
1.52	22.1	
13.10	48.3	
22.86	40.0	
50.29	40.0	
50.29		36
62.00		36

than the undisturbed strength; and 2. The data on compression tests is very limited from a depth of 38.7 m to 64.6 m where the sand was encountered. No explanation was given for the scarcity of results in the lower part of the boring. Table 6.16 shows the values of undrained shear strength that are employed in the analyses that follow. The clay will probably be considerably stronger than in the table after the piles have been in place for several weeks.

In view of the factors noted above about the testing of the soil, the value of ε_{50} used in the analyses was selected as 0.02 by referring to Table 3.3. The argument could be made that a somewhat higher value could be used for the deeper soils; however, 1. The properties of the soils near the mudline will dominate the results, and 2. The higher value of ε_{50} will be conservative with respect to deflection. The computation of the maximum bending moment, of principal interest in the present analyses, is hardly affected by the value employed for ε_{50}.

6.3.4.6 *Preliminary dimensions of piles and axial capacities*

The lengths of the various segments of the main piles (Piles 1 and 2) and the conductor pipe (Pile 3) were selected on the basis of some trial computations and previous experience (see Fig. 6.14). Also, considered was the necessary mass of the pile for driving. The conductor pile with lengths of 30.48 m were in the relatively soft clay for their entire length and offered no particular problem in installation. However, the main piles, with lengths of 54.86 m were designed to penetrate 4.57 m into the sand. If the piles could not be driven to design depth in the sand, the thick-walled sections necessary for sustaining bending moment would be out of place. Therefore, the contractor made some preliminary analyses to obtain reasonable assurance the main piles could be driven to the penetration of 54.86 m. Downward capacity can usually be obtained by end bearing but the resistance to the expected uplift can be a problem. If a pile fails to penetrate the proper distance into the sand, the use of jetting to loosen the soil is unacceptable. Many offshore designs have resorted to drilling and grouting in order to place the piles to the required depth.

Using the data presented above, load-settlement curves were computed for the main piles with the following results: downward capacity, 15,970 kN, settlement, 51 mm; upward capacity, 6430 kN, uplift, 20 mm. Actually the downward capacity is expected to increase with increasing depth, but the settlement of 51 mm was judged to be a limiting value for the present computations. The conductor pipe were computed to have a capacity of 2360 kN fully adequate to sustain the load from drilling operations of 1112 kN, a load

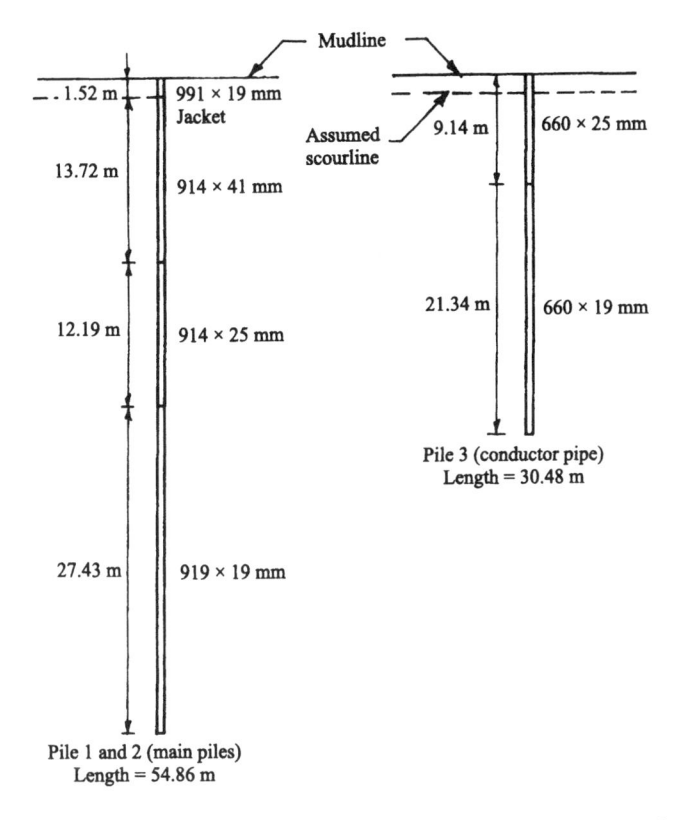

Figure 6.14. Trial dimensions and bending stiffness of main piles and conductor pipe for offshore platform.

that would be on only one conductor at a time. Settlement was not a problem because the conductor pipe would not be fastened rigidly into the jacket.

6.3.4.7 Results of analyses

The loads and other data outlined above was entered into Computer Program GROUP, using the technology described in Chapter 5. The assumption was made that the conductor pipe were far enough apart that no reduction was necessary for pile-soil-pile interaction. The following results, from the 100-year storm without factoring, for the movements of the origin on the coordinate system (point where loads were applied): downward movement, 15.31 mm; horizontal movement, 45.12 mm; and rotation, 4.377×10^{-4} radians. The loads and movements at each of the pile heads are shown in Table 6.17.

A brief examination of the results in Table 6.17 reveals some interesting facts. The six conductor pipes with a smaller diameter than the main piles sustain the major portion of the lateral load, illustrating the importance of construction details. While the diameter of the conductor piles is smaller, the rotational restraint is significant because of the short distance between shims above the top of the pile. Further, if the lateral loads are summed and compared with the applied lateral load, the influence of the batter of the main piles can be seen. That is, the lateral component of the axial load of Pile 2 is important in resisting the lateral load on the platform.

222 *Piles under lateral loading*

Whether Pile 1 or Pile 2 has the greatest stress is not immediately apparent from the data in Table 6.17. The axial load on Pile 2 of 9370 kN, almost three times that of Pile 1, but still substantially less than the ultimate load of 15,970 kN. The results for Pile 2 will be analyzed to compare computed stresses with allowable values.

Figure 6.15 shows a plot of computed values of bending moment and combined stress as a function of length of Pile 2, along with the stress that can be sustained at first yield of the steel. The axial load is assumed to remain unchanged in computing the combined stress. The influence on the results of the change in wall thickness of the pile is evident. The most critical stress is below 25.5 m where the stress is altogether from the axial load. The assumption that the axial load is unchanged with depth is obviously incorrect; therefore, axial stress will not control. Also, the same argument holds for the large stress at a point just below 15 m. A further examination of the values of combined stress in Figure 6.15 suggests that the wall thickness on the pile is sized rather well over its length, and that the most critical stress will occur at or near the top of the pile.

Additional runs were made with a computer code by increasing the loads at the origin of the global coordinate system by an equal percentage. The results for the case where the global

Table 6.17. Computed movements and loads at each pile head (first loading).

	x_t mm	y_t mm	α rad	P_t kN	M_t m-kN	P_x kN
Pile 1	6.2	45.8	0.255×10^{-2}	564	−2530	3240
Pile 2	24.3	43.3	0.246×10^{-2}	524	−2420	9370
Pile 3	–	45.1	0.140×10^{-2}	322	−1210	–

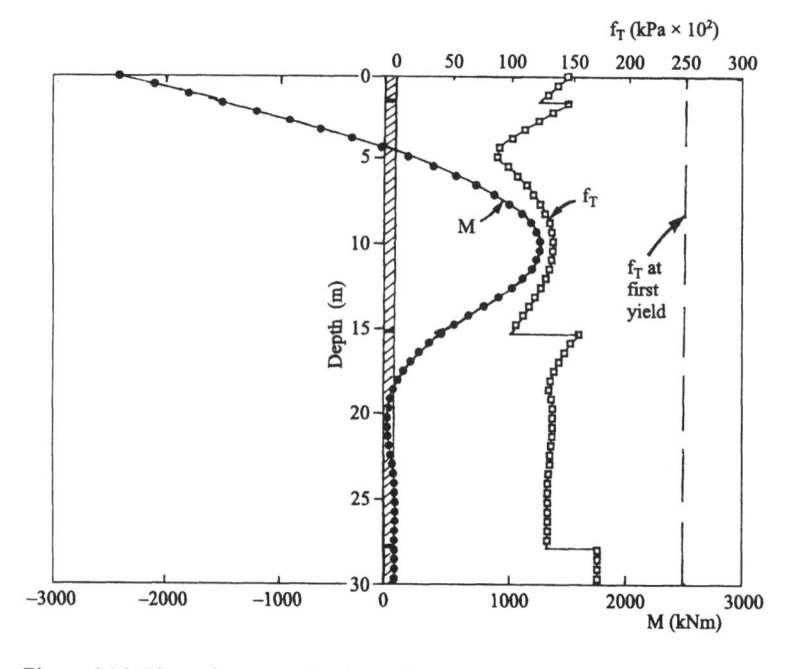

Figure 6.15. Plots of computed values of bending moment and combined stress along the length of Pile 2 of offshore platform.

loads were factored upward by 1.50 are presented. If one can say that the factor of safety should reside in the factoring of the loadings, the use of 1.50 is consistent with the API factors for pile penetration. The movements of the origin on the coordinate system were: downward movement, 57.00 mm; horizontal movement, 104.0 mm; and rotation, 0.3194×10^{-2} radians. The loads and movements at each of the pile heads are shown in Table 6.18.

With the factoring of the global loads by 1.50, the nonlinear response of the system, due to the nonlinearity of the soil, is apparent. The origin of the coordinates moved vertically by a factor of 3.72 and moved horizontally by a factor of 2.3. Referring to Table 6.18, a principal factor in the nonlinear movement is that Pile 2 is in the nonlinear range of axial response. The pile moved downward 104 mm with an axial load of 14,100 kN. The axial load is less than the ultimate axial load of 15,970 kN; however, because of the lack of stiffness in the load-settlement curve, little additional load can be applied to the group before collapse is indicated. The maximum combined stress that was computed for the factored loads was 219,000 m-kN, less that the value of 250,000 m-kN that can be sustained at first yield.

6.3.4.8 Discussion of results

The brief presentation of the analysis of piles for the foundations of an offshore platform exposes a number of important problems that must be addressed. The methods of analyses of piles under axial and lateral loading, and of pile groups, appear to be well suited to address design. Plainly, many other factors are of importance; in particular, the determination of the in situ properties of the subsurface soils and the estimation of the change in those properties due to pile installation and subsequent loading.

In the present analysis, the weakness in the system lies in the lack of capacity and stiffness of the main piles under axial loading. The increase in those characteristics presents a practical problem at the present location. The main piles will need to be driven more deeply into the stratum of sand so that more of the supporting load will be derived from side resistance, where the load transfer-movement mechanisms are stiffer that for end bearing. Not only will the determination of the precise characteristics of the deep layer of granular soil be difficult, but the preliminary analyses leading to methods of driving the piles will be uncertain. As noted earlier, the use of other methods of placing the supporting piles may be considered.

6.4 PASSIVE LOADING

6.4.1 Earth pressures

The solutions presented in the following sections emphasize loadings from moving soil or from earth pressure; therefore, a brief discussion is presented of earth pressure, particularly related to the variation of the pressure with soil movement.

Table 6.18. Computed movements and loads at each pile head (global loads factored by 1.5).

	x_t mm	y_t mm	α rad	P_t kN	M_t m-kN	P_x kN
Pile 1	9.2	105.0	0.643×10^{-2}	835	−3880	8350
Pile 2	104.0	95.6	0.617×10^{-2}	718	−3570	14,100
Pile 3	–	104.0	0.473×10^{-2}	496	−1950	–

224 Piles under lateral loading

As shown in Figure 6.16, one can imagine an infinitely thin, infinitely stiff membrane placed vertically in a granular deposit. The soil on one side of the membrane is assumed to be removed without lateral movement and the pressure at an element along the membrane is examined. The vertical stress is equal to the unit weight of the soil γ times the depth z. The coefficient of lateral earth-pressure-at-rest K_o is multiplied by γz to find the lateral earth pressure at the element.

The lateral earth pressure decreases if the rigid wall is moved away from the soil. The zone of reduced pressure as the wall moves away is called *active*. When the lateral earth pressure achieves the lowest value, the coefficient is designated K_a and is called the *minimum coefficient of active earth pressure*. Conversely, if the rigid wall is moved toward the soil, the zone of increased earth pressure is called *passive* and, when the lateral earth pressure achieves the highest value, the coefficient is designated K_p and is called the *maximum coefficient of passive earth pressure*. For idealized soils with favorable geometry's of the earth mass, the values of K_a, K_p, and perhaps K_0, may be computed analytically. However, from the point of view of the analyses presented herein, no techniques are available for computing the movements required to develop the limiting coefficients for idealized soils, much less for the heterogeneous, anisotropic, nonlinear, and complicated soils normally found in nature. Therefore, for the analyses that follow, the loadings from earth pressures are based on limiting values or on results from experiments.

6.4.2 Moving soil

Moving soil is encountered in practice when piles are placed in an unstable slope. The design of such piles may be based on the assumption that forces from moving soil will act against the piles. The forces, acting against the moving soil, are sufficient to put the slope into equilibrium. An appropriate factor of safety is build into the analysis. As discussed by Stewart et al. (1994) moving soil is also encountered at bridge abutments. Results are presented from centrifuge tests and approaches are described for designing pile-supported bridge abutments.

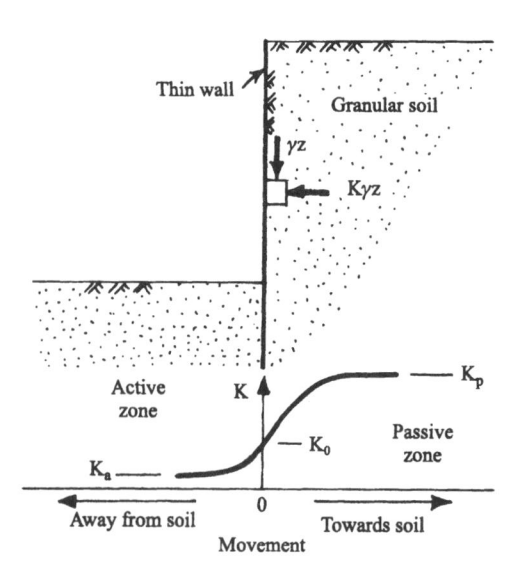

Figure 6.16. Concepts of active and passive earth pressures related to movement of a rigid wall.

Another case of moving soil is when piles support axial loads and lateral loads at their tops, are in a deep fill in a valley, and the fill will move vertically and laterally with time (Reese et al. 1989). Analyses can be done conveniently with *p-y* curves with the origin of the curves displaced with depth to reflect the computed movement of the soil.

A third case of moving soil is where piles are in place prior to the construction of an excavation or an embankment. Poulos & Chen (1997) did a two stage analysis by use of the finite-element method and the boundary-element method. Design charts were presented and comparisons were given between measured behavior of soil and predicted behavior.

The mechanics of the case of soil moving against a group of piles is similar to the case of piles moving against the soil, as discussed in Chapter 5. The method presented there of modifying *p-y* curves to account for close spacing is employed for the analyses presented herein. However, a study of other methods of dealing with closely spaced piles in moving soil has contributed to the development of the recommendations for the interaction coefficients. A brief review of some of those methods is presented here.

Wang (1986) did a comprehensive study of the various analytical methods that can be employed in analyzing a closely-spaced group of piles in moving soil. Desai & Appel (1976) developed a three-dimensional finite-element code for the analysis of piles in a group. Linear analysis was used for the piles and nonlinear analysis for the soil. Shie & Brown (1991) and Brown & Shie (1992) did further work with finite elements and produced some useful results. Bransby (1996) used two-dimensional finite-element analysis and differentiated between *p-y* curves for active loading and *p-δ* curves for passive loading. The method was described in Chapter 3 and difficulties in its use in obtaining *p-y* curves were discussed. The problems of solving of the response of a group of closely-spaced piles are even more formidable than for the single pile; therefore, the benefits from the promising FEM are yet to be realized.

The continuum model described by Poulos (1971) and modified by Focht & Koch (1973), also described in Chapter 5, was discussed by Wang. A further modification by Ha & O'Neill (1981) of the continuum model resulted in a computer code, PILGP1. The continuum model is deficient in that the soil is assumed to sustain tension. Experimental results suggest that elastic theory overestimates the distance from a displaced pile that a significant amount of soil movement occurs.

An elementary concept for the response of a group of closely-spaced piles is to assume that the soil within the group will move with the piles; therefore, the group can be treated as a single pile of large diameter. The concept was used for groups under axial load many years ago by Terzaghi & Peck (1948) and extended by Reese (1984) to piles under lateral load. Bogard & Matlock (1983) analyzed experiments with a group of piles in a circular pattern under lateral load and used a variation of the single-pile concept in developing predictive equations.

Wang (1986) extended the wedge model for the ultimate lateral resistance against single piles, described in Chapter 3, to the case of closely-spaced piles in a side-by-side row. For piles in sand, he found that, for a zero clear space, the ultimate resistance is about one-half of that for a single pile. Further, he found that the ultimate resistance for the piles in the group with a center-to-center spacing of about three diameters was identical to that of the single pile. Wang studied the influence of spacing on the ultimate lateral resistance of piles in a group in cohesive soil. For a zero clear space, the ultimate resistance was again about one-half of that for a single pile. For other spacings, the ultimate lateral resistance was a function of the strength of the clay and the depth of the wedge, but no significant interaction was computed if the clear spacing was two to three clear diameters.

The theory of plastic flow of rigid, perfectly plastic materials under plane-strain conditions is well established in technical literature (Sokolovskii 1965, Hill 1950). Broms (1964) used slip-line theory to compute the ultimate lateral resistance of various shapes as a function of the cohesion c. He computed values, ranging from 8.28 to 12.56 times c, for smooth and rough plates, a smooth cylinder, and a smooth and rough square shape. Broms also assumed plane strain, as indicated by the values he computed. If three-dimensional behavior could have been modeled, the values shown above would be reduced to zero or to a relatively small value for the ultimate lateral resistance at the ground surface. Experience has shown that the behavior of laterally-loaded piles is largely controlled by the soil response that is influenced by the presence of the ground surface.

Various investigators have proposed methods of predicting the lateral forces against piles in a group from moving soil. While the methods can give insight into the problem of interaction between closely-spaced piles, all fall short of dealing properly with the three-dimensional manner of soil response. An exception is the finite-element method, but the difficulties of obtaining results that may be used in practice with FEM are formidable, as discussed above. Therefore, the method employed herein is to use modified p-y curves, as described in Chapter 5, where the modifications are based largely on experimental results.

6.4.3 Thrusts from dead loading of structures

While earth pressures and moving soil constitute mainly to the passive loading of piles, the lateral loads from an arch bridge or from some trusses of buildings can also cause passive loading. One of the designs presented below, for example, deals with loadings that are derived from the dead load on a frame used in the construction of a building in the shape of a pyramid.

6.5 SINGLE PILES OR GROUPS OF PILES SUBJECTED TO PASSIVE LOADING

6.5.1 Pile-supported retaining wall

6.5.1.1 Introduction
A common solution is to employ piles to support a retaining wall to ensure that failure of the wall does not occur by rotation or sliding. The solution, using the technology employed herein, is straightforward except that, as noted earlier in this chapter, the earth pressure depends on the movement of the wall. Furthermore, the magnitude and distribution of the earth pressure after construction is largely indeterminate because the methods of placing the backfill can vary widely. The recommendations of Terzaghi et al. (1996), embodied in charts, are based on the use of the equivalent-fluid method and a perusal of the charts shows the importance of the selection of an appropriate material for the backfill. A comparison of values from theory with values from the charts shows that the charts are close to those from Rankine theory for active earth pressure. Therefore, the assumption is implicit in the Terzaghi charts that the wall is capable of some movement without distress if the pressures from the backfill were greater than the chart values (see Section 6.4.1).

6.5.1.2 Example for solution

The wall selected for analysis is shown in Figure 6.17. The piles are 356 mm in outside diameter, 2.44 m apart along the wall, and penetrated below the base of the wall to a depth of 12.2 m. The soil at the site consisted of silty clay with the water table at a depth of 5 m. The water content averaged 20% in the top 5.0 m and 44% below that depth. The undrained shear strength of the silty clay varied considerable with depth; a value of 144 kN/m^2 was the average in the region of the significant deflection of the piles. The unit weight of the soil was 7.2 kN/m^3 below the water table and 17.0 kN/m^3 above the water table. The value of ε_{50} was assumed to be equal to 0.005, a value consistent with values shown in Tables 3.3 and 3.5.

A field-loading test was performed at the site and the results are shown in Table 6.19. The test was performed on a pile with identical geometry and penetration of those to be used in construction. The soil conditions were relatively consistent across the site. The data in Table 6.19 are plotted in Figure 6.18 and, as may be seen, the pile experienced a failure by plunging.

The reinforcing for the 356-mm diameter pile consisted of four No. 20 reinforcing bars. The concrete cover from the center of the bars to the face of the pile was 0.0765 mm. The steel and concrete in the pile had ultimate strengths of 415,000 kPa and 18,400 kPa, respectively. A computer code was used to compute the ultimate bending moment M_{ult} and

Table 6.19. Results of field loading of test of a pile identical to those in retaining wall.

Load, kN	Settlement, mm
0	0
309	1.1
605	2.3
670	3.1
703	3.9
715	5.0
739	7.1

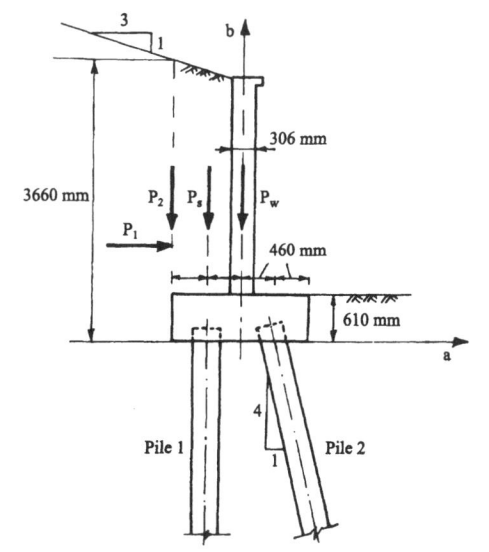

Figure 6.17. Sketch of a pile-supported retaining wall.

228 *Piles under lateral loading*

the bending stiffness $E_p I_p$. The axial load at failure with no bending moment, the 'squash' load, was computed to be 3036 kN. Computer runs were made with the following axial loads in kN: 0, 200, 400, 500, 600, 800, 1000, 1200, and 1550. The interaction diagram for M_{ult} as a function of the axial load is shown in Figure 6.19; the curve will be used to define the failure of a pile in bending. As shown in Figure 6.20, values of $E_p I_p$ are plotted as a function of bending moment and axial load. The wide range in the value of $E_p I_p$ is apparent and appropriate values will be interpolated and employed in the analyses that follows of the pile group. Experience has shown that computed values of bending moment are not very sensitive to the values of $E_p I_p$ that are employed; however, interpolation will be done carefully in getting input for computer runs.

6.5.1.3 *Step-by-step solution*

The backfill is assumed to consist of a free-draining, granular soil without any fine particles. The surface of the backfill is assumed to be treated to prevent inflow of water

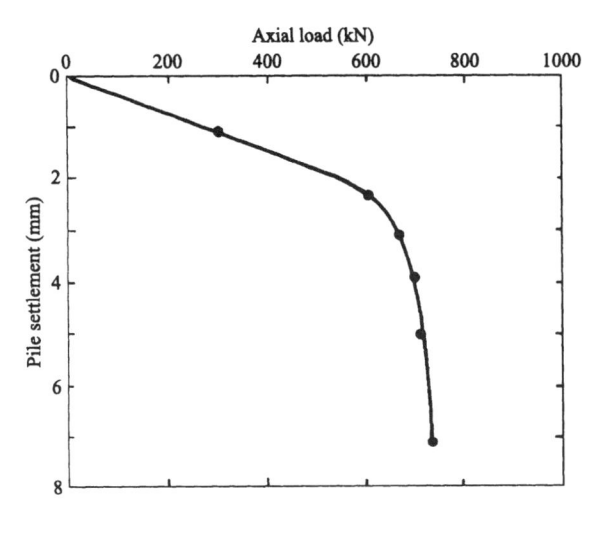

Figure 6.18. Curve of axial load versus settlement for reinforced-concrete pile supporting retaining wall.

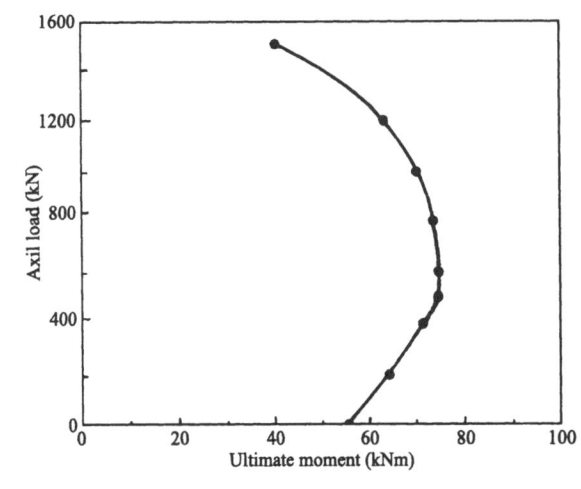

Figure 6.19. Interaction diagram for reinforced-concrete pile used to support retaining wall.

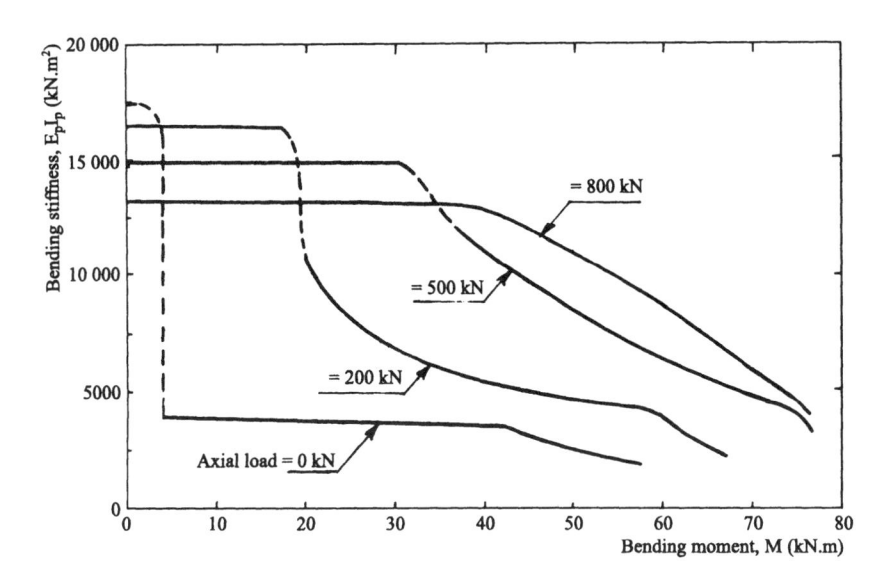

Figure 6.20. Bending stiffness as a function of bending moment and axial load for concrete pile supporting retaining wall.

and weep holes are to be provided at the base of the wall to prevent the collection of water.

Employing the charts by Terzaghi et al. (1996), the forces P_1 and P_2 were computed to be 85.1 and 26.3 kN, respectively, for the 2.44-m length of wall. Employing the unit weights of soil and concrete, the forces P_s and P_w were computed to be 101.4 and 112.1 kN, respectively. The forces P_s and P_w are not likely to be significantly in error because the unit weights of the materials are 'standard' values and should not change in time; therefore, the load factors will be applied to the forces P_1 and P_2 which are dependent on earth pressures. Sometime in the future a surcharge or a line load could be applied at the back of the wall, the backfill may not consist or the select materials that are assumed, or the drainage may be plugged.

The unfactored, individual loads are resolved at the origin of the a-b coordinate system as follows: P_v = 112.1 + 101.4 + 26.3 = 239.8 kN; P_h = 85.1 kN; M = (85.1)(3.66/3) – (101.4)(0.38375) – (26.3)(0.920 = 40.7 m-kN. Rather than using the bending stiffness $E_p I_p$ from Figure 6.20, the presence of the steel bars was ignored and the $E_p I_p$ for the gross section (15,930 kN-m^2)was used for the initial computations. Referring to Figure 6.20, the value of the $E_p I_p$ for the gross section is close to the average value of $E_p I_p$ for moderate values of axial loads and for the lower values of bending moment.

In considering the rotational restraint provided to the pile head by the base of the wall, Figure 6.17 shows only a moderate penetration of the piles into the concrete. Therefore, the pile heads are assumed to be free to rotate. However, reinforcement should be employed for Pile 1 in particular to ensure that the pile can take tension; such reinforcement could contribute to the restraint against rotation. Such a factor should be considered in the 'fine tuning' of the design.

Runs were now made with the computer code. Significant results from the computations are shown in Figure 6.21. The loads were factored by increasing P_1 and P_2 by the

230 *Piles under lateral loading*

multipliers indicated. The reasonable assumption is made that P_s and P_w will not change if drainage is assured. The curves in the figure show nonlinear response for both the lateral deflection of the wall Δ_h and for the maximum bending moment M_{max}. The computed values of the maximum bending moment for the two piles was very close and the average value is plotted. At a load factor of 3.5, the maximum bending moment reached the ultimate value, as found in Figure 6.19, and another run was made by adjusting the $E_p I_p$ values of Pile 1 and Pile 2 by interpolation from Figure 6.20. The results are shown in Figure 6.21. There was a small increase in the maximum bending moment and a large increase in the computed deflection.

The wall was far from failure in respect to axial loading on the piles. For the load factor the computed axial load on Pile 1 was only 23 kN and on Pile 2 was 125 kN, far less than the capacity of the piles as shown by Figure 6.18. It is of interest to note that the computed axial deflection of the wall was just 2 mm for the load factor of 3.5 and that the rotation of the wall was actually negative by one-tenth of a degree. These results are explained by noting that the top of the raked pile actually moves up as lateral deflection occurs.

6.5.1.4 *Discussion of results*
The factor of safety of the piles is much greater with respect to axial load than with respect to bending moment. Consideration could be given to increasing the amount of reinforcing steel in the upper few meters of each pile in order to achieve a more balanced design.

The 21 mm of horizontal deflection of the wall, divided by the height of the wall of 3660 mm yields a value of Y/H of 0.0057. Figure 6.22 gives values of rotation (or lateral deflection) necessary to develop the minimum amount of active pressure. The value of Y/H that was obtained appears to be adequate except in unusual circumstances.

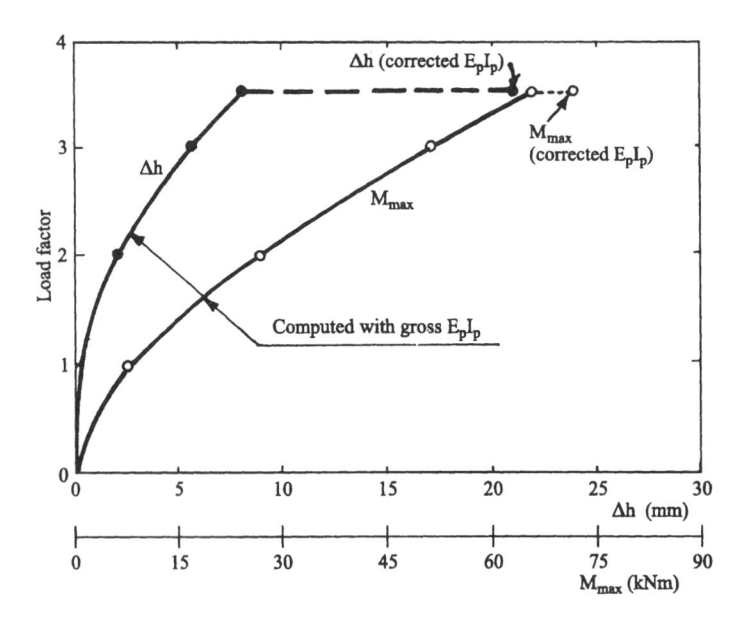

Figure 6.21. Response of pile-supported retaining wall to lateral loading.

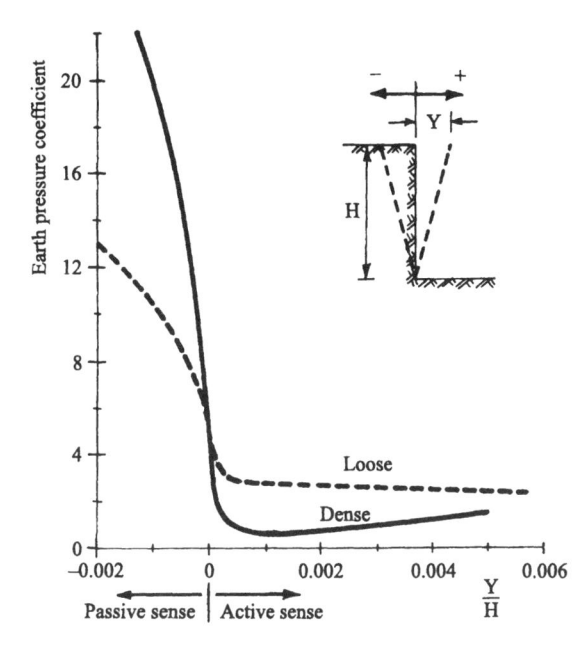

Figure 6.22. Relationship between the coefficient of lateral earth pressure and rotation of a rigid wall as a function of the density of sand.

Other runs were made with the computer code for the case were the raked pile was converted to vertical. Failure in bending of both of the piles was computed to have occurred with a load factor of 3.0 rather than 3.5. However, the lateral deflection at the load factor of 3.0 was computed to be 38 mm; thus, a considerable increase in the flexibility of the wall was achieved with a small decrease in the load factor. The axial load on Pile 2 with the load factor of 3.0 was computed to be 125 kN, the same as for the raked pile.

A series of computations of the sort noted above should lead to an optimum design. Factors to be considered are the geometry and strength of the piles, the penetration and spacings of the piles, the way the piles are fastened to the base of the wall, and the desirable flexibility of the foundation system. The amount of engineering effort that is desirable to invest depends, of course, on the length of the wall and on the consequences of a collapse.

6.5.2 Anchored bulkhead

6.5.2.1 Introduction

The differential equation for a pile under lateral loading can be applied directly for the solution of the response of an anchored bulkhead. The equation includes the following features that facilitate the analysis: 1. Distributed loads along a length of the pile; 2. Discrete, numerical p-y curves that will simulate the anchor; and 3. Normal p-y curves that will reflect the response of the soil below the dredge line. Each of these features will be illustrated in the example that is presented.

6.5.2.2 Example for solution

The wall to be analyzed is presented in Figure 6.23, along with the earth pressures that act. The earth pressures were computed by assuming the wall has deflected to allow the

232 Piles under lateral loading

mobilization of the minimum active pressures, computed by the Rankine equations. As shown, the earth pressures include the effect of a surcharge loading and the effect of a sudden drawdown. The needed values are the maximum bending moment in the sheet pile wall, the load and deflection of the anchor rod, and the factor of safety against a deep-seated failure. The anchor rods are spaced 2.44 m apart and wales will be used to transfer the load from the wall to the anchors. Wales will not be analyzed here.

The data from Figure 6.23 are from Peck et al. (1974) except that conversion has been made from English to SI units. The authors did not give a value of ε_{50}; therefore, a value of 0.01 was selected following suggestions presented in Chapter 3.

The sheet pile that was selected for analysis is a PDA27, a classification available in the United States. The dimensions of the cross section are given in Figure 6.24 in English units. The steel is ASTM A-572 GR 50 with a minimum yield stress of 50,000 psi (345,000 kPa).

Some anchored bulkheads have traditionally been analyzed by statics with the assumptions of either *free earth support* or *fixed earth support*. In the first instance, the penetra-

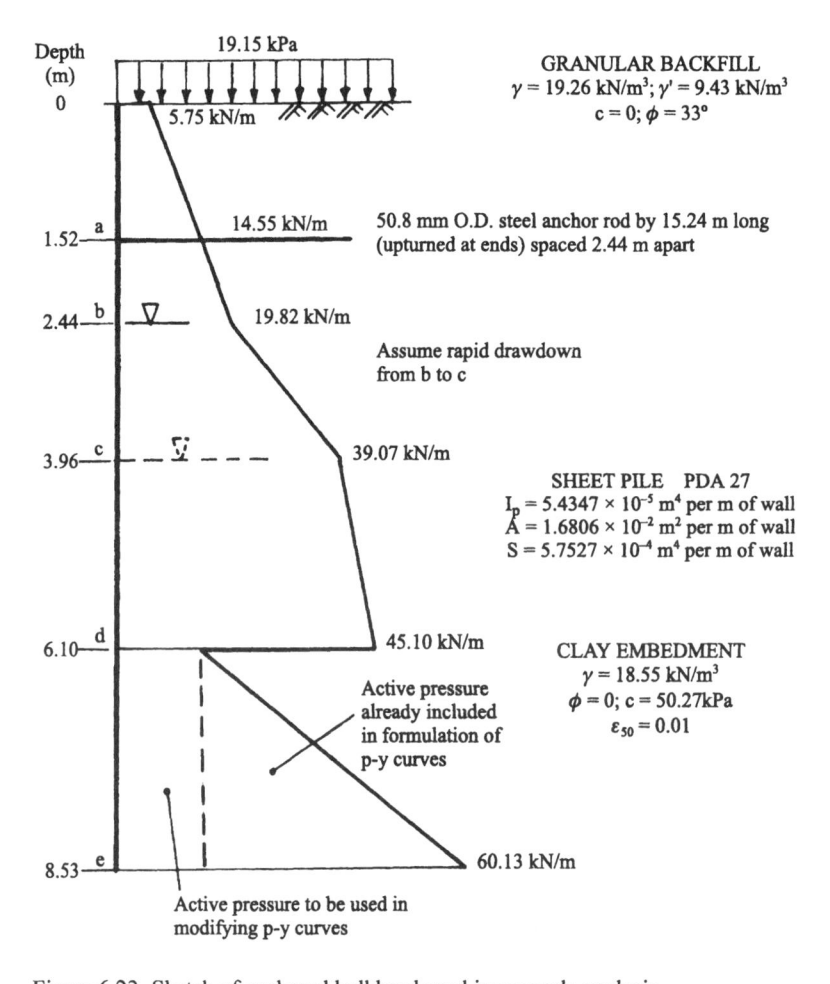

Figure 6.23. Sketch of anchored bulkhead used in example analysis.

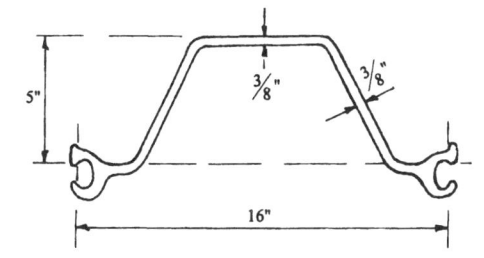

Figure 6.24. Dimensions of cross section of the steel sheet pile used in analysis of an anchored bulkhead (English units).

tion of the wall is just sufficient to provide stability against the lateral failure of the base of the wall. In the second instance the wall penetrates farther and is assumed to be fixed against rotation at some distance below the dredge line. The analytical approach presented here deals automatically with either case and no assumption is needed about the nature of the support along the wall below the dredge line.

6.5.2.3 *Step-by-step solution*
Following the procedure suggested by Peck et al. (1974) a deep seated failure will occur if the vertical pressure behind the wall at the level of the dredge line is greater than the bearing capacity of the clay inside the wall. The vertical pressure behind the wall is equal to $19.15 + (6.10)(19.26) - (6.10 - 3.96)(9.80) = 115.66$ kN/m^2. The bearing capacity of the clay may be taken at 5c or $(5)(50.27) = 251.35$ kN/m^2. The factor of safety against a deep-seated failure, then, is $(251.35)/(115.66) = 2.17$. Thus, the computations show the wall to be safe against a deep-seated failure.

Attention is now directed to the initial earth pressures and to the construction of the *p-y* curves for the portion of wall below the dredge line. If wall friction is neglected, a possible solution is to formulate two sets of *p-y* curves, one set for behind the wall where the overburden exists and the other in front of the wall below the dredge line. Haliburton (1968) presented such an approach; however, experimental evidence is lacking on the nonlinear variation of these separate sets of curves as a function of the displacement of the wall. Therefore, simplifying assumptions are given below that allow the presently available *p-y* curves to be modified in a rational manner.

The following assumptions are made: 1. The active-pressure condition will exist in the soil behind the wall above the dredge line; 2. The *p-y* curves for a continuous wall are the same as for a widely-spaced pile except that the *p*-values are reduced by one-half; and 3. The *p-y* curves for the wall below the dredge line may be constructed initially by using the recommended values on the passive-pressure side (inside the wall) and subtracting from the ultimate *p*-values the portion of the value of the active pressure (outside the wall) due to the overburden pressure. The *p-y* curves already include the effect of active pressure for ground-level conditions. As discussed below, Assumption (3) may be modified as noted below.

Assumption (1) merely states that the deflection of the wall toward the excavation will be sufficient to allow the development of the minimum active pressure even in the vicinity of the anchor rods. Wang (1986) did experimental and analytical studies to show that Assumption (2) is essentially correct. Assumption (3) will require no modification if computations show that the wall has an inward deflection over its full length, including below the dredge line. However, if there is a point of zero deflection, the *p-y*

234 *Piles under lateral loading*

curves above the zero deflection will remain unchanged but an active-pressure component will be added to the *p-y* curves below the point of zero deflection. Thus, trial solutions will be necessary.

The *p-y* curves for the clay below the dredge line is computed by use of Equations 3.22 through 3.25 with the following modifications: The ultimate *p*-values for the clay, with the inward movement of the wall, are reduced by the active pressure behind the wall, due only to the overburden above the dredge line. The ultimate values of *p* were not reduced initially for the wall as related to the pile with the view that checks will be made after reviewing the results. The resulting *p-y* curves are shown in Table 6.20.

The anchor rods have a diameter of 50.8 mm, have a length of 15.24 mm and are spaced at 2.44 m apart along the length of the wall. The resistance of an anchor rod to deflection is modeled as a stiff *p-y* curve, which acts over one increment of length along the wall. If the wall is assumed to be divided into 100 increments in the solution, the length of an increment is 0.0853 m. If the area of the rod is 0.002027 m^2, the length of the rod is 15.24 m, and the modulus of elasticity of steel is 200,000,000 kN/m^2, the stiffness of the anchor may be computed as follows: $(p/y)_{anchor}$ = (.002027)(200,000,000)/(15.24)(.0853) = 311,900 kN/m^2. Because the anchors are spaced 2.44 m apart along the wall, the stiffness needs to be reduced to 126,800 kN/m^2 in the analysis of a one meter strip of the wall.

The next step is to employ a computer code to analyze the wall. The solution with the loading shown in Figure 6.23 resulted in a maximum moment in the sheet-pile wall of 104.3 kN-m, yielding a bending stress of 103.4/0.00057527 = 179,700 kPa which is substantially less than the 345,000 kPa at minimum yield. The corresponding solution employing statics give a bending stress of 171,700 kPa. The computed deflection at the anchor rod of 0.00803 m yielded a force of (0.00803)(126,800)(0.0853)=86.85 kN for the assumed one meter length of the wall. The force in the anchor rod for a spacing of 2.44 m was 211.9 kN, yielding a tensile stress of 104,600 kPa, a nominal value. The corresponding force computed by statics was 277 kN.

The difference in the solutions from statics and from the differential equation was most pronounced with respect to the response of the soil. In using statics, the passive pressure is assumed to be distributed over the entire length of the wall below the dredge line, while the computer solution yields soil resistance as a function of the deflection of the sheet-pile wall. The results from the computer showed a point of zero deflection at about 0.62 m above the tip of the piles. The maximum soil resistance was computed to be 67.7 kN/m at a depth of 0.20 m below the dredge line, where the interpolated value of ultimate resistance is 154 kN/m. The maximum soil resistance was computed to be 52 kN/m at the tip

Table 6.20. Computed values of *p-y* curves for clay below the dredge line, anchored bulkhead.

	Depth	
	0 m	2.438 m
y, m	*p*, kN/m	
0	0	0
0.0125	24.0	56.9
0.025	30.2	71.7
0.05	38.1	90.4
0.1	48.0	113.9
0.2	60.5	143.4
10	60.5	143.4

of the piles, where the value of ultimate soil resistance is 218.5 kN/m. Thus, the soil is well below the failure condition.

The soil pressures against the wall were increased by factors to find the loading that would cause a failure of the wall. A multiplier of 1.50 was used and the bending moment in the sheet pile was found to be close to yielding. The examination of the results of the computation with the modest load factor gives insight into the functioning of the wall. Accordingly, curves for deflection, bending moment, shear, and soil reaction are shown in Figures 6.25 through 6.28, respectively, as a function of depth.

The curve in Figure 6.25 shows that the maximum deflection occurred at a depth of 4.5 m and was 82 mm. A negative deflection of 37 mm (away from the excavation) was computed at the top of the piles and the deflection at the dredge line was 64 mm. The negative deflection at the tip of the piles was computed to be 4 mm. The significant influence of the stiffness of the anchor rod on the deflection is apparent.

The deflection at the anchor rod at a depth of 1.52 m was computed to be 0.0129 m, yielding a soil reaction of 1640 kN/m, as shown in Figure 6.28. The anchor load was computed to be 343 kN, giving a tensile stress of 169,000 kPa, well below the failure stress. The influence of the anchor is illustrated in the upper portion of the shear diagram in Figure 6.27.

The maximum bending moment was 175 kN-m, as shown in Figure 6.26, yielding a computed bending stress of 304,200 kPa, a value that is slightly below the ultimate stress of 345,000 kPa.

Figure 6.28 shows that the soil reaction is almost uniform for about one and one-half meters below the dredge line, with a maximum value of 56 kN/m which is well below the interpolated value of ultimate resistance. The maximum soil resistance at the tip of the

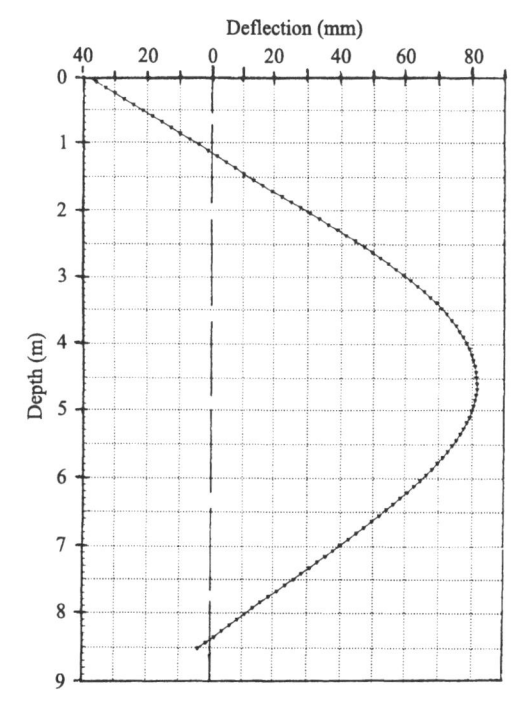

Figure 6.25. Deflection along anchored bulkhead with an assumed load factor of 1.5.

236 *Piles under lateral loading*

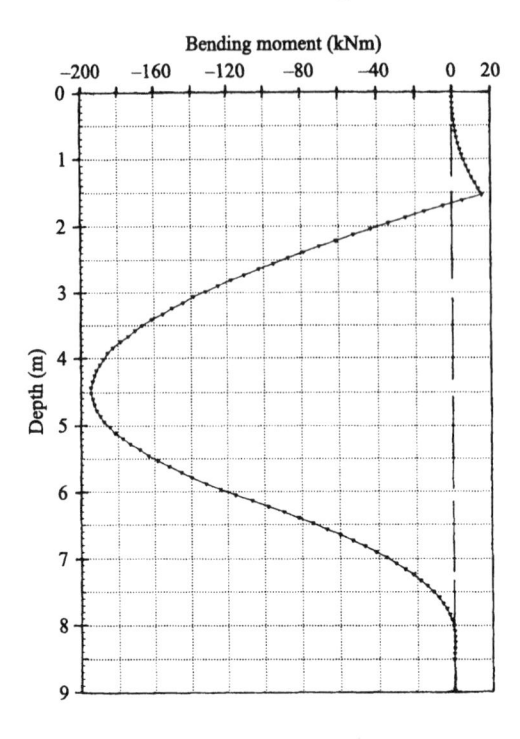

Figure 6.26. Bending moment along anchored bulkhead with an assumed load factor of 1.5.

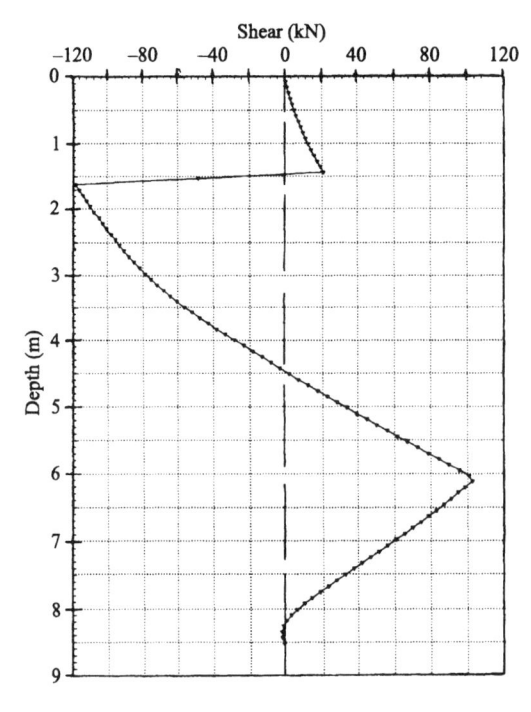

Figure 6.27. Shear along anchored bulkhead with an assumed load factor of 1.5.

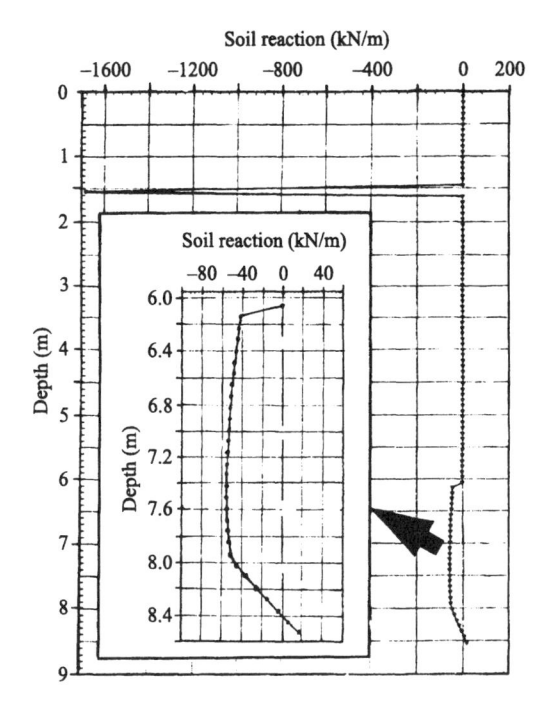

Figure 6.28. Soil reaction along anchored bulkhead with an assumed load.

piles is 20 kN/m which is well below the ultimate value of 218 kN/m. Thus, with the load factor of 1.50, the soil is not close to the failure condition, even if the reduction of 50% had been made in the ultimate values of p. Figures 6.25 and 6.28 show that the piles have a point of zero deflection at about a half meter above the tip.

6.5.2.4 Discussion of results
The input for the computer code can be readily coded to solve the problem of the anchored bulkhead. While the equations of static equilibrium can be employed to obtain values that can be used to make a reasonable design for the case of a single anchor rod but a solution using statics for multiple anchors would be unsatisfactory. The solution of the differential equation, as illustrated in Figures 6.25 through 6.28, give the designer considerable insight into the actual response of the wall to earth pressure.

With respect to the earth pressure, active pressure is inappropriate at the top of the wall. Figure 6.22, and other such information, can be consulted to allow the designer to decide if the wall is safe against a bending-moment failure, considering the modest load factor that produced a yield moment. Critical factors are the earth pressure developed during construction and many other details of the construction process.

6.5.3 Pile-supported mat at the Pyramid Building

6.5.3.1 Introduction
Passive loading, for the example shown here, is derived from the dead load of a structure, with the live load constituting only a small portion of the total load. Thus, the loading of an arch bridge may be considered as passive. Loading for some other structures is derived

238 Piles under lateral loading

principally from dead load; the loading can be considered as passive. The building in the example employs pile-supported mats with two-dimensional loading; the piles are raked to minimize the lateral deflection.

6.5.3.2 Example for solution

A sports arena and general-service building in the shape of a pyramid (Fig. 6.29) was constructed near the banks of the Mississippi River at Memphis, Tennessee. The structure has dimensions in plan of 140m by 140 m and the ribs supporting the walls (roof) were supported by mats at the corners and at intervals of 17 m along the four sides. The alluvial soil varies considerably from point to point at the site and separate soil profiles were developed for the design of each pile-supported mat or for a group of mats. The movement of the mats under the combined loadings, lateral, axial, and overturning, was restricted in order to prevent wrinkling or other distress of the medal cladding of the building (Griffis 1990).

A corner mat that was analyzed is shown in Figure 6.30. The piles had a nominal diameter of 406 mm but measurements indicated the as-built diameter to be 427 mm. Construction of the piles was done by drilling to the projected depth with a continuous-flight auger, placing a single rebar in the center- hole of the auger, and pumping grout through the center-hole as the augur was lifted. After the auger had been withdrawn, rebar cages consisting of six bars with diameters of 22.2 mm was vibrated into the fluid grout to a depth of 9 meters. The compressive strength of the grout after 28 days was found by testing to be 41.4 MPa and the strength of the steel was specified to be 350 MPa.

The soil profile is shown in Figure 6.31. The water table was shallow and at a depth of 3.3 m below the ground surface. The miscellaneous fill at the ground surface was removed

Figure 6.29. Photograph of Pyramid building.

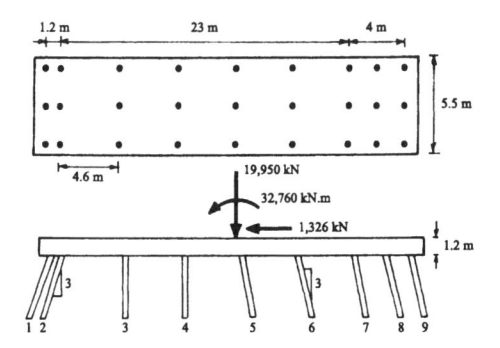

Figure 6.30. Plan and elevation of a corner raft, Pyramid Building.

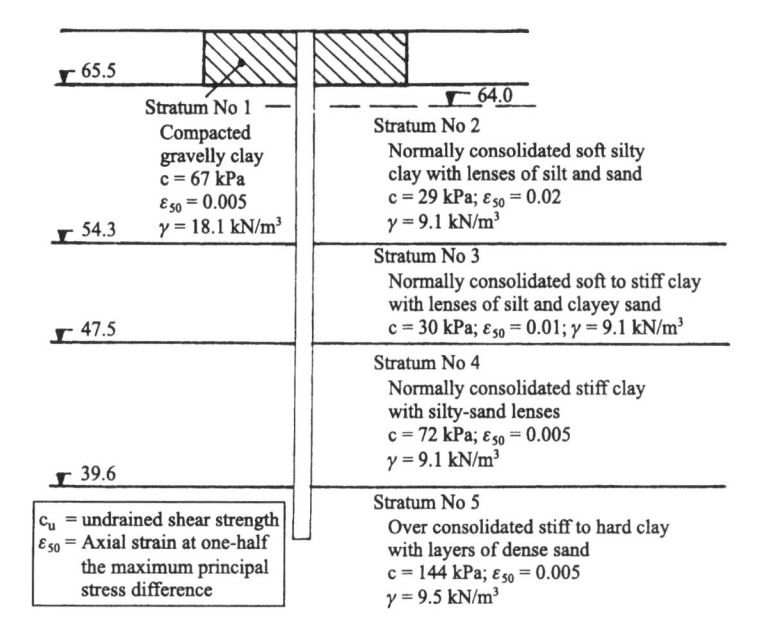

Figure 6.31. Soil profile at corner raft, Pyramid Building.

and replaced with a compacted gravelly clay to a depth of 1.8 m; the selected fill was compacted to 95% of Proctor density. The soil properties were determined from tube tests and from the Standard Penetration test, where applicable. The values of ε_{50} shown in Figure 6.31 generally follow the recommendations shown in Chapter 3 and are modified slightly on the basis of data from soil reports. Consolidation tests were performed in order to allow investigation of the possibility of load-shedding due to consolidation of the upper sediments.

6.5.3.3 Step-by-step solution

The design of the pile-supported mats was initiated by the testing of individual piles. The tests under axial loading were performed with piles with two penetrations: 21.8 and 30.9 m (Reese et al. 1993) and analyses with the *t-z* method showed that the results were consistent with predictions. Results for the longer piles (Fig. 6.32) were found to be suit-

240 Piles under lateral loading

able for use in the design of the corner mat after analyses of the effects of load-shedding showed the curve to be suitable for long-term loading. The tests were performed in accordance with procedures of the American Society for Testing and Materials (ASTM 1992) with the results shown in Table 6.21 and Figure 6.32.

Solution of the nonlinear differential equation for the model of the axially-loaded pile, not shown here, gave close agreement with the results from the experiment but it was decided to use the direct experimental results (Table 6.21) of the computer code to characterize the axial stiffness of the piles. The three piles in each of Sub-Groups 1 and 2 are relatively close together and consideration was given to degrading the axial response of those two sub-groups because of the close spacing. Reference to the loadings in Figure 6.30 shows that these sub-groups will receive the greatest axial load. However, the spacings are large except in one direction and the decision was made to assume the piles in Sub-Groups 1 and 2 responded under axial loading as did all of the other piles. This decision is reinforced because of the redundancy in the system. That is, if the piles in these two groups became overloaded no failure would occur because the load-settlement relationships do not show movement-softening. Rather, rotation and axial deflection of the pile cap would occur and load would be transferred to other piles in the group. The

Table 6.21. Results of axial-load testing of a pile at the pyramid building with penetration of 30.9 m.

Load, kN	Settlement, mm
0	0
331	1.25
691	2.5
939	3.9
1166	5.3
1409	7.7
1597	10.2
1768	12.8
1994	17.8
2221	25.4

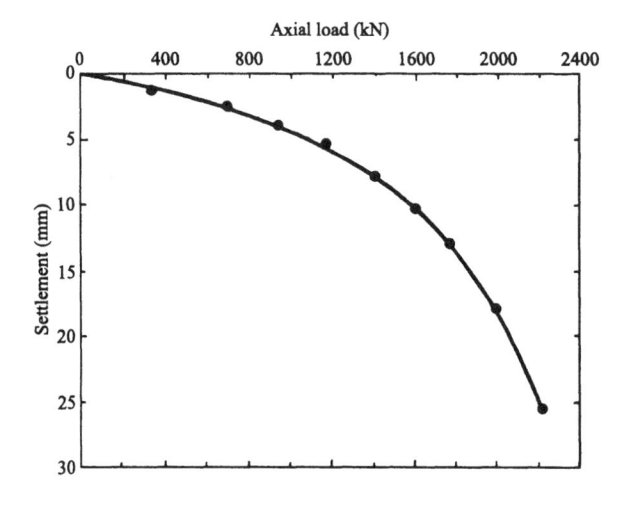

Figure 6.32. Axial load versus settlement for a pile with a penetration of 30.9 m, Pyramid Building.

Analysis of single piles and groups of piles subjected to active and passive loading 241

lighter-loaded piles in the remainder of the group could well sustain the additional axial loading if Sub-Groups 1 and 2 settled excessively.

Piles were also tested at the site under short-term, lateral loading (Reuss et al. 1992) after placement of the compacted gravely clay at the ground surface. The results of the testing are shown in Table 6.22 and in Figure 6.33. The pile under lateral loading was analyzed using a computer code with the subroutines for p-y curves as presented in the program. The results from analyses are somewhat conservative but the agreement between the results is certainly acceptable. The analytical method was employed for modeling the piles under lateral loading in the computer code.

Tests of a pile under lateral loading were also conducted before the gravelly clay was in place. The primary aim of the testing was at determining the response of the soft silty clay to sustained loading. Excess porewater pressure around the piles would occur on the application of the lateral loading and investigation was undertaken to ensure that the dissipation of the excess pressures would not lead to unacceptable lateral deflection. Two loads were used and the results are shown in Table 6.23. Plotting the data from Table 6.23 with a natural scale showed that the deflection increased rapidly with time. The plotting of the data from sustained loading in Figure 6.34 shows time with a logarithmic scale and deflection with a natural scale. While an appropriate theoretical solution cannot be applied, the expectation is that the rate of increase in deflection can be patterned by the theory of one-dimensional consolidation. Therefore, the time on the abscissa could be expressed as

Table 6.22. Results of lateral-load testing of a pile at the pyramid building.

Load, kN	Deflection, mm
0	0
22.2	1.8
44.5	3.0
66.7	5.6
90.0	11.7
111.2	17.5
133.4	27.7

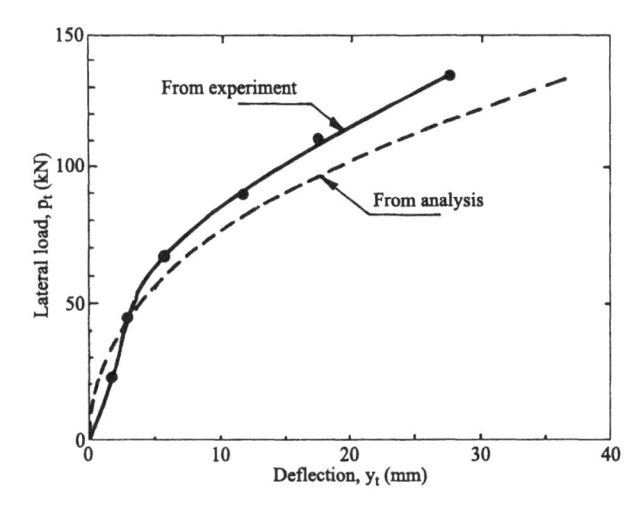

Figure 6.33. Results of testing CFA pile under short-term lateral loading, Pyramid Building.

242 Piles under lateral loading

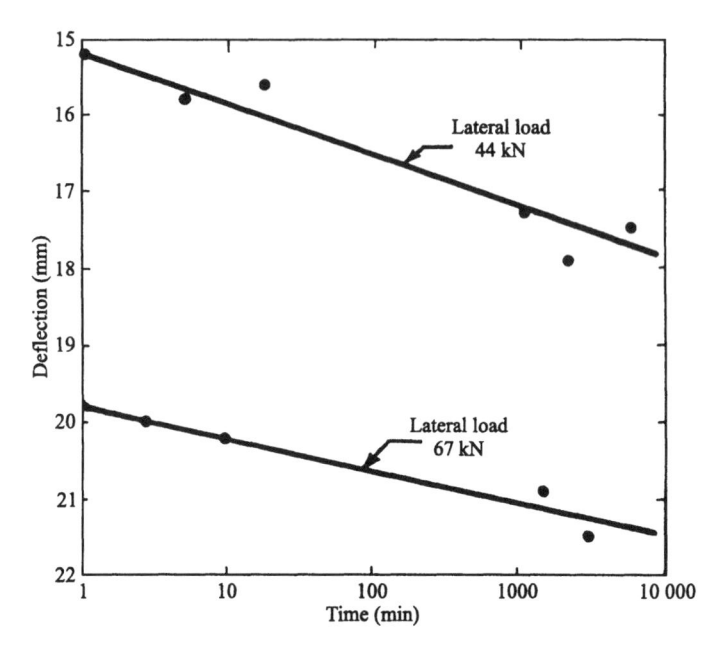

Figure 6.34. Results of testing CFA pile under sustained lateral loading, Pyramid Building.

Table 6.23. Results of testing of a pile at the pyramid building under sustained lateral load.

Time, min	Load, 44 kN Deflection, mm	Load, 66 kN Deflection, mm
1.0	15.2	19.8
2.8		20.0
5.5	15.8	
10	15.6	20.2
1100	17.3	
1500		20.9
2200	17.9	
3100		21.5
6000	17.5	

$$t = T_3 \frac{H_a^2}{c_v} \tag{6.10}$$

where t = time, T_3 = time factor for the three-dimensional distribution of excess pore pressure from the axial loading of a pile, H_a = average length of the drainage path, and c_v = coefficient of consolidation.

The assumption can be made that H_a and c_v are constants for the particular case; therefore, the time will vary as T_3. In the absence of a theory, the assumption can be made that T_3 will decrease rapidly with time. Thus, the deflections should approach a limiting value as time increases. The dial-gauge reading at one minute for the load of 44 kN was

15.2 mm; stretching the readings by a factor of 1.2 yields a reading of 18.2, which corresponds to a time of 25 days. For the load of 67 kN, the dial-gauge reading for one minute was 19.8 mm; stretching the readings by a factor of 1.2 yields a reading of 23.8 mm, which corresponds to a time of 6000 years. Therefore, a stretch of the y-values of the p-y curves for the soft clays by a factor of 1.2 seems satisfactory for analyzing the response of the pile group. The stretching is quite conservative when consideration is given to the presence of the improved soil at the ground surface that, as noted above, was not present during the sustained-loading tests.

After the long-term testing was completed, excavation was made around the pile to a depth of 1.5 meters. Concrete was cast around the pile at the bottom of the excavation to provide partial fixity against rotation. Strain gauges were attached at intervals of 0.3 m along the exposed length and on opposite sides of the pile on the plane that passed through the axis and the point of application of the load. Measurements were taken as loads of 22, 44, 67, 89, and 133 kN were applied. Using the length from the applied load to each gauge location, the bending moments at the position of the strain gauges were accurately known. The bending stiffness $E_p I_p$ could then be obtained from the following equation.

$$EI = \frac{M}{\phi} \tag{6.11}$$

where M = bending moment, and ϕ = curvature at the section of interest.

A linear distribution of strain was assumed between the strain gauges on opposite sides of the pile. The experimental values of $E_p I_p$ as a function of the applied bending moment are shown in Figure 6.35. The considerable amount of scatter in the data is not surprising; however, the results generally confirm the results from analysis (see Chapter 4), particularly for the moments of relatively low magnitude. The experimental values of $E_p I_p$ are

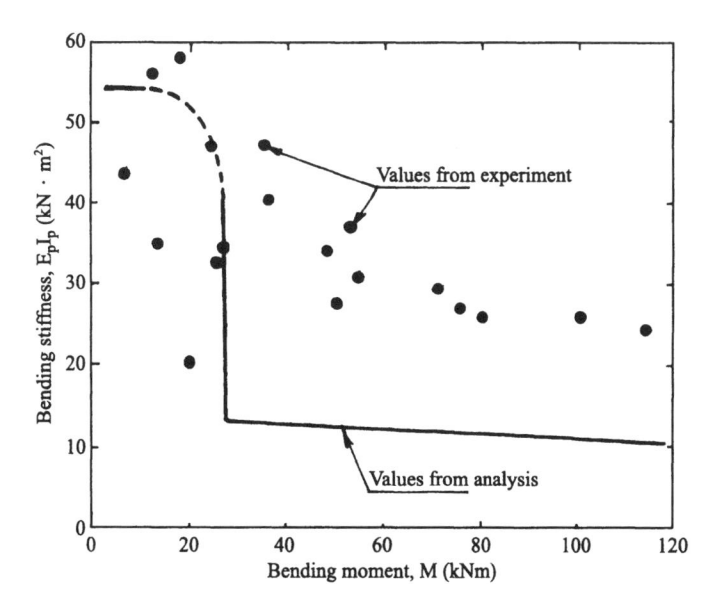

Figure 6.35. Experimental and analytical values of bending stiffness of a CFA pile, Pyramid Building.

244 Piles under lateral loading

considerable higher than the analytical values for the relatively large values of bending moment, probably because a fully cracked section is assumed in the analytical computation of E_pI_p. An average value of E_pI_p of 38,700 kN-m was selected for analyses.

In order to have a value of ultimate bending moment at which a plastic hinge would develop in the pile, the geometry of the cross section of the pile was employed, along with the strength of the steel and concrete, and an interaction diagram was computed with a computer code. The results are shown in Figure 6.36 where M_{ult} is shown as a function of the applied axial load. The values of M_{ult} range from about 150 kN-m to over 250 kN-m, depending on the magnitude of the applied load.

After developing the information shown above, Computer Program GROUP can now be used to analyze the response of the pile-supported mat or raft. Using the loading shown in Figure 6.30 the results in Table 6.24 were obtained. Reference to Figure 6.32 shows that the piles in Sub-Groups 1 and 2 have an apparent factor of safety of more than two. Referring to Figure 6.36 show that the heaviest-loaded piles also have an apparent factor of safety of more than two against the formation of a plastic hinge.

Table 6.24. Results of analysis of the pile group at the pyramid building under combined loading.

Pile group.	Axial load kN	Lateral load kN	Maximum bending moment kN-m
1	1040	63.4	58.2
2	1050	63.3	58.1
3	788	66.3	60.7
4	828	66.2	60.8
5	466	68.8	62.2
6	529	68.8	62.4
7	592	68.7	62.7
8	619	68.7	62.8
9	646	68.7	62.9

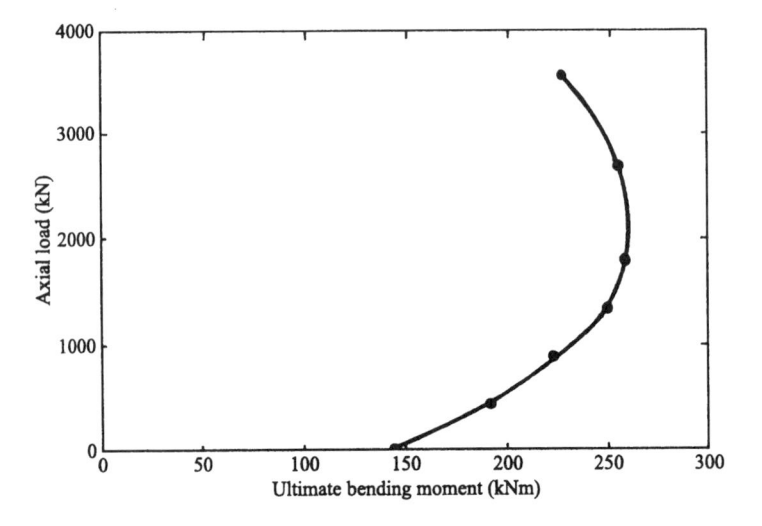

Figure 6.36. Interaction diagram for a CFA pile, Pyramid Building.

The loading of the pile-supported raft were factored upward by 1.8 and then by 2.0 and, while failure was not indicated, the axial load of 2120 kN on the pile in Sub-Group 2 showed that the piles were nearing failure under axial load. Furthermore, the maximum bending moment of 178 kN-m showed that a substantial portion of the bending moment was developed that would cause a plastic hinge to develop. Thus, a global factor of safety of two can be confirmed for the group.

6.5.3.4 Discussion of results
The computed value of axial settlement from a computer code for the working loads was 1.9 mm and the computed value of rotation of the raft was 5.5×10^{-5} radians. Measurements were made at the site and the corresponding values were found to be 2.9 mm and 4.2×10^{-5} radians. The measurements are considered to confirm the computations within expectations.

The analysis of the foundations of the Pyramid Building illustrates the usefulness and desirability of performing field tests of full-scale, single piles. Methods of analysis of single piles under static loading, both axial and lateral, served to confirm the validity of the results of the experiments. The results of sustained, lateral loading of a single pile, while not suitable for the development of a theory, did allow modification of p-y curves for short-term loading to simulate long-term effects. The experimental difficulty for the determination of the value of bending stiffness, $E_p I_p$, as a function of applied bending moment was demonstrated. Nevertheless, the results of the experiment, even with a large amount of scatter, served to confirm the computational technique.

The piles at the Pyramid Building extended to less than one meter into the mat or raft and the free-head condition was used in the analysis of the corner mat. Computation with the assumption of the fixed-head condition could have been shown; however, results of computations, not included here, showed the overall behavior of the corner raft was fairly close to that for the free-head condition. Research is awaited on the prediction of the rotational restraint of a pile head as a function of amount of penetration into a cap or mat, taking into account the various structural considerations of the piles and the mat.

6.5.4 Piles for stabilizing a slope

6.5.4.1 Introduction
Many slopes in distress or with a potential for sliding consist of clays with properties that complicate analyses. An example is a strain-softening clay with a residual strength much less than the peak strength. Therefore, the determination of the relevant soil properties may require a major effort. With the defining soil parameters, various methods can be used for improving the factor of safety of a slope, including drainage, decreasing the driving force and/or increasing the resisting force by grading, using lagging and tie backs, and installing piles. The following paragraphs present guidance if piles are selected as the remedial method.

Fukouka (1977) presented three cases where piles were used to improve the factor of safety of slope. 1. Heavily-reinforced, steel-pipe piles were used at Kanogawa Dam to stabilize a landslide. A series of piles, 458 mm in diameter and 5 m apart, were driven in pairs through prebored holes near the toe of the slide. A plan view of the slide showed a length of about 1100 m and with a generally circular pattern. The installation

246 *Piles under lateral loading*

of the piles and a drainage tunnel apparently stabilized the slide. 2. A slide developed at the Hokuriku Expressway in Fukue Prefecture when a cut was made to a depth of 30 meters. After movement of the slope was noticed, a row of steel *H*-piles was installed but the piles were subsequently damaged because the velocity of the slide suddenly increased because of torrential rains. Drainage was improved and four rows of piles, parallel to the slope, were installed. Analyses showed that the factor of safety was increased from unity to about 1.3. 3. Piles were used to stabilize a slide at the Higashi-tono landslide. The length of the slide along the slope was 130 m, its width was about 40 m, and the sliding surface was found to be about 5 m below the ground surface. A total of 100 steel-pipe piles, 319 mm in diameter, were installed in the slide over a period of three years. The factor of safety was computed to be increased by a factor of 0.18, which was sufficient to stabilize the slide.

Hassiotis & Chameau (1984) and Oakland & Chameau (1986) present brief descriptions of a large number of cases where piles were used to improve the stability of a slope. Detailed descriptions were presented about the use of driven piles and bored piles.

If piles are selected as the remedial measure, bored piles are frequently selected because pile driving can cause an initial decrease in the resistance. Furthermore, the drilling machine can be arranged to allow the drilling to be done a considerable distance from the drilling machine so that the equipment forces can be kept from the slope.

6.5.4.2 *Method of computation (Reese et al. 1992)*
A sketch of a slope with a pile is shown in Figure 6.37a. The potential sliding surface is indicated at a distance of h_p below the ground surface. The pattern of forces on the piles is shown in Figure 6.37b with the summation of the unit forces on the pile, in the horizontal direction, indicated by the force F_s. The forces on the portion of the pile below the potential sliding surface are shown in Figure 6.37c.

Two procedures are possible with respect to the earth pressure against the piles: 1. The piles are relatively weak in bending; and 2. The piles are so strong in bending that the ultimate soil forces can act without causing a failure in bending. For piles that are weak in

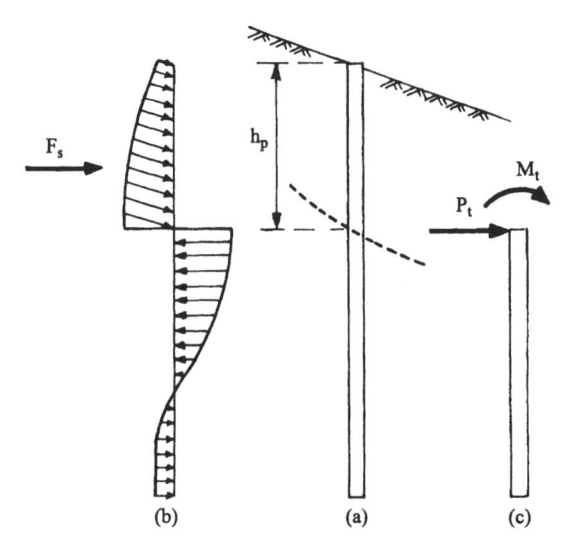

Figure 6.37. Forces from soil against a pile in a moving slope (after Reese et al. 1992).

Analysis of single piles and groups of piles subjected to active and passive loading 247

bending, two procedures are possible with respect to the distribution of earth pressure. With the view that the deformation is maximum at the sliding surface and decreases to zero at the ground surface, a triangular distribution of earth pressure may be assumed. The pressure is zero at the ground surface and increases linearly with depth to a value that will develop the ultimate bending moment in the pile. The pattern of the distribution of forces from the soil for a pile weak in bending is shown in Figure 6.38a. The shear and moment at the top of the portion of the pile below the sliding surface is just sufficient to develop the ultimate moment in the pile with the greatest load.

The second approach for a pile that is weak in bending is to assume that the deformation is relatively constant throughout the sliding mass. The ultimate values of soil resistance, employed for the analysis of piles that are strong in bending, are reduced by a percentage to a value that will develop the ultimate bending moment in the pile with the greatest load. For piles that are strong in bending, the soil is assumed to have moved a sufficient distance that the limiting values of lateral pressure have develop against the portion of the pile above the potential sliding surface. These lateral pressures may be taken at the p_{ult} values from the recommendations for p-y curves for the particular soil in the slope. The use of the full values of p_{ult} may be indicated often because the reinforcing steel in a bored pile can usually be augmented to yield the ultimate bending moment that is desired. The pattern of the distribution of forces from the soil for piles that are strong in bending is shown in Figure 6.38b. The capacity of the pile in bending is assumed to be greater than the maximum bending moment caused by the application of the shear and moment at the top of the portion of the pile below the sliding surface.

The first step in the analysis is to use an appropriate computer code to find the most critical sliding surface with no piles in place. The factor of safety against sliding is found. 2. With the given pile the value of M_{ult} is found. 3. With the relevant properties of the soil, the criteria for formulating p-y curves are consulted to obtain values of p_{ult} that act against the pile. 4. The values of p_{ult}, or from the distribution of reduced soil pressures, are assumed to act parallel to the slope, and values of F_s and h_p are computed. 5. The values of

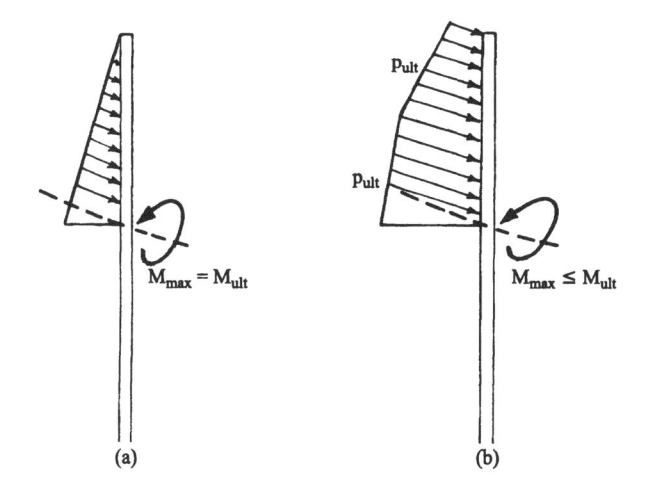

Figure 6.38. Patterns of soil forces against piles in a moving slope, (a) pile weak in bending ; (b) pile strong in bending.

248 *Piles under lateral loading*

M_t and P_t are computed from F_s and h_p^* and a computer code is employed to find the maximum bending moment in the pile. 6. The maximum bending moment is compared with M_{ult}. 7. If necessary, the diameter, bending strength, and spacing of the piles is adjusted with the view that the resistance of the piles against sliding will not move the sliding surface a great deal. Steps (3) through (7) are repeated with the revised data. 8. The computer code for the analysis of the slope is run again, with forces against the pile or piles used as input to resist sliding. 9. The new position of the sliding surface is found, new soil forces are found, and the analysis of the slope is repeated. The procedure is continued until the solution converges to the proper piles and correct sliding surface 10. The piles may be checked and modified again if necessary and the analytical procedure is continued until piles of appropriate size and spacing are found that yield an improved factor of safety of appropriate magnitude.

If the piles of a particular bending strength are already in place, the problem is to find the increased factor of safety when the soil loading is such to just bring the piles to their ultimate bending moment. On other occasions, piles of a specific size and strength are selected for installation in the slope, regardless of their bending strength. In the example that follows, bored piles of a specific diameter are employed and reinforcement is computed for the case of full values of p_{ult} and for reduced values. The latter case proved to be the most efficient.

6.5.4.3 *Example for solution*
A bridge was to be constructed over a river where unstable banks existed. Several slides had occurred at places along the river, especially at times of sudden drawdown. As could be expected, the properties of the soil varied both vertically and horizontally, as shown in Figure 6.39. Values of ε_{50} were unavailable from test data and values for the analyses followed the recommendations shown in Chapter 3.

6.5.4.4 *Step-by-step solution*
Shown in Figure 6.39 is the computed sliding surface for the sudden-drawdown case. The potential slide shows to be deep-seated and the factor of safety was found to be 1.04, a factor that indicated a potential failure. The results of the computations agree with observations of slides that had occurred in the past. In the studies that follow, bored piles were found that would increase the factor of safety of the slope to a reasonable value. The piles are not necessarily the most economical ones. The solution could then be compared to other methods of stabilization, such as changing the geometry of the slope. If the use of the bored piles was shown to be the best method of improving the factor of safety, further studies would be undertaken where the diameter and spacing of the piles are varied to find the best combination. This final step is not shown in the computations that follow.

A concept for the stabilization of the slope is shown in Figure 6.40. Preliminary analyses showed that the piles were so deep seated that the piles must be restrained at their tops. Piles 3 and 4 serve as anchors with grade beams restraining the tops of Piles 1 and 2 against deflection. Only a distance along the river of 6.5 m was considered initially with

---

* The step-by-step procedure is based on the assumption that the tops of the piles are unrestrained and will behave as cantilever beams above the sliding surface. In the example that follows, the sliding surface is so deep that the piles must be restrained against deflection at their tops; therefore, a computer code must be employed in the analysis of the piles from the ground surface downward.

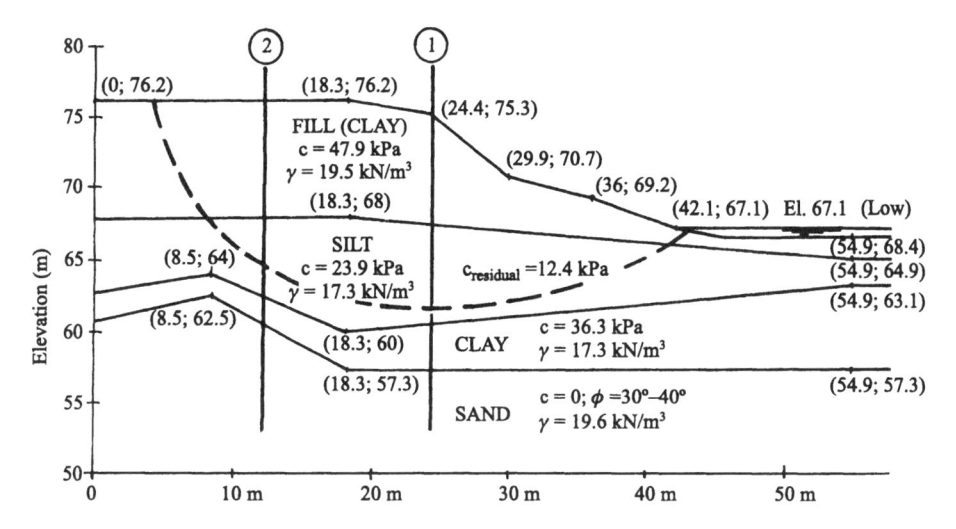

Analysis of single piles and groups of piles subjected to active and passive loading 249

Figure 6.39. Soil properties at a slope near a river, showing computed failure surface with water at low level (after Reese et al. 1992).

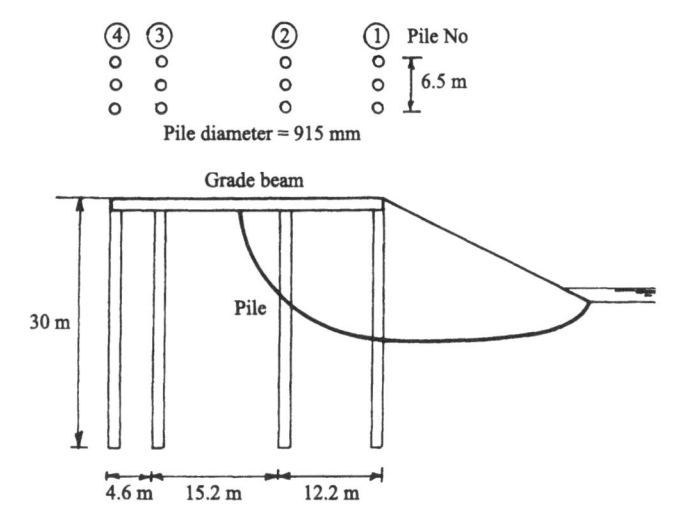

Figure 6.40. Arrangement of piles for stabilizing a slope near a river (after Reese et al. 1992).

the view that the field engineer would decide the total distance along the river that required stabilization. The ratio of spacing to diameter is 2.8; therefore, the full value of the soil forces and resistances are assumed to be acting. The analysis deals with the center row of piles, away from the river as indicated in the elevation view in the figure.

The piles are 915 mm in diameter and penetrate to a depth of 30 m below the ground surface. The river-ward pile was at the top of the slope and about 15 m from the point where the computed sliding surface would exit so the drilling could be done without the drilling equipment adding to the driving forces.

The equations for the computation of p_{ult} for soft clay were employed, assuming that the depths to the potential sliding surface could be taken from Figure 6.39 where no piles were in place. The results for a pile in Row 1 and in Row 2 are shown in Table 6.25. With the piles in place, the potential sliding surface will move upward, necessitating iterations.

The data in Table 6.25 were employed as the distributed loading for Pile 1 and a computer code was run to find the maximum bending moment in the pile. The assumption was made that the deflection of the top of the pile will be small because the grade beam will be supported by three anchor piles. The maximum bending moment was computed to be 7240 kN-m, making use of the gross moment of inertia for the stiffness of the pile. To sustain such a large moment the use of 20 rebars with diameters of 40 mm along with an exterior shell with a thickness of 12 mm was investigated and found to be inadequate. Therefore, attention was turned to the use of a pile with less bending strength and with a reduced value of the driving force.

The assumption was made that the pile would be reinforced with 20 bars of 20 mm in diameter. The strength of the concrete was assumed to be 35,000 kPa and the yield strength of the steel was assumed to be 420,000 kPa. A computer code was used to compute the ultimate bending moment of the section and a value of 843 m-kN was found. A computer code was used to find the soil forces that would just cause a failure in bending of Pile 1. Only 13% of the tabulated values in Table 6.25 for Pile 1 was found to lead to a maximum bending moment of 820 m-kN, a value approximating the failure moment.

The distributed loading to Pile 2 was reduced similarly and a computer program was run to compute the improved factor of safety with the piles in place. The factor of safety was found to be 1.24, a value that may be satisfactory for the particular design. If a somewhat higher factor of safety is desired, the reinforcement could be increased which, in turn, would increase the percentage of the values of p_{ult} that could be developed which, in turn, would increase the factor of safety against sliding. The increased amount of reinforcing steel would be needed along the piles only in the region of the maximum bending moment.

The shear at the top of Pile 1 was 213 kN and at the top of Pile 2 was 203 kN. Thus, the load on the anchor piles was 416 kN. The assumption is made that both the anchor piles will deflect the same and share the load equally. With the lateral load of 208 kN at the top of each of the anchor piles, the pile-head deflection was computed to be 6.3 mm, a small enough value that the use of a zero deflection in computing the response of the piles in the slope did not cause much error. The maximum bending moment was computed to be 450 kN-m, well below the ultimate bending moment of 843 kN-m.

Table 6.25. Values the ultimate resistance against Piles 1 and 2.

Depth below ground surface m	Pile 1 p_{ult}, kN/m	Pile 2
0	131.5	131.5
8.1	394.5	
8.1	196.8	
8.2		394.5
8.2		196.8
12.3		196.8
14.4	196.8	

Analysis of single piles and groups of piles subjected to active and passive loading 251

6.5.4.5 *Discussion of results*
A rational approach to the use of bored piles in stabilizing a slope is presented. For the particular example, more detailed studies can be undertaken to find the optimum size and spacing of the piles. Where slides are shallow, piles can act as cantilevers and can be used with good economy. Construction must obviously be done with care in order to refrain from triggering a slide with construction loads.

6.5.5 *Piles in a settling fill in a sloping valley*

6.5.5.1 *Introduction*
In many areas of the world the construction of facilities must be carried out at sites that are unfavorable. One such site is where a fill must be made in a sloping valley. Observations show that engineered-soil fills will experience time-related settlement in spite of the use of good specifications and excellent control to ensure desirable qualities of the compacted soil. If the valley is level, the settlement of the fill will lead to downdrag which must be considered and dealt with appropriately. If the valley is sloping, the settlement of the fill will result in lateral movements in addition to settlement. The lateral forces on piles supporting a structure can be very large. The example presented here concerns the design of piles to support a shopping mall build over fill in a sloping valley (Reese et al. 1989).

6.5.5.2 *Example for solution*
The construction of a shopping mall in Birmingham, Alabama, required that 2,300,000 m^3 of soil and rock be excavated from a hill and placed in an adjacent valley. Compaction of the select material was carried out to rigorous specifications and a gently sloping site of 28.3 hectares was created. A profile through the site is shown in Figure 6.41.

Near the surface, the fill consisted of sandy clays (CL) to sandy silts (ML) with values of the plasticity index (PI) ranging from 5 to 9. At depth, the fill consisted of clayey sands (SC) to silty sands (SM) with PI's ranging from 0 to 5. The optimum moisture content of the materials in the fill was about 16% and the maximum dry density ranged from 16.5 to 17.8 kN/m^3. The fill was placed in loose lifts of 230 mm and compacted with a tamping roller to 98% to 103% of Standard Proctor density at moisture contents ranging from −1% to −3% of optimum. A number of studies were made of the properties of the fill and the following results were obtained: $c = 86$ kPa; $\phi = 29$ deg.; $\gamma = 18.9$ kN/m^3. The properties were assumed to be constant with depth and in the horizontal

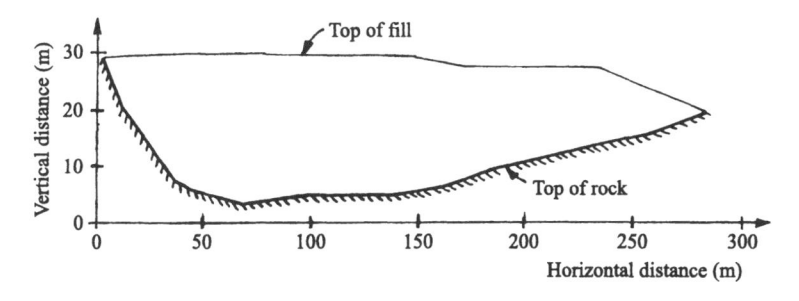

Figure 6.41. Typical profile through a fill in a valley (after Reese et al. 1989).

252 *Piles under lateral loading*

direction. Bedrock existed below the fill and borings at the site showed the rock to have a compressive strength of 39.2 MPa. Except for the displacements of the origins of *p-y* curves, the remainder of the curves followed recommendations in Chapter 3 for the various soil strata.

The shopping mall was to be founded on bored piles that penetrated the fill and 3 m into the rock. Preliminary studies showed the bored piles to have a diameter of between 2 and 3 m, assuming the piles to act as cantilever beams. A more detailed analysis was indicated.

6.5.5.3 *Step-by-step solution*
The technical literature (Boughton 1970, Charles 1975) reveals that a relatively wide range of post-construction settlements in compacted fills, ranging from 0.1% to 1% of the depth of the fill. Data were unavailable on the settlement as a function of the properties of the fill material; therefore, the articles cited above were judged to provide upper-bound and lower-bound values for the settlement to be expected at the construction site.

A considerable amount of unpublished results concerning the settlement of such fills was available in the files of a commercial firm. A review of the data led to the selection of 0.40% of the height of the fill as a reasonable upper-bound settlement of the fill at the site. Thus, the maximum, time-related settlement of the fill, with a height of 25.9 m, was 104 mm and ranged downward for the portions of the fill with lesser heights.

Information on the lateral movement of fills that settle has been presented by Wilson (1973). Data, showing the ratio of horizontal to vertical movements, from the results of inclinometer readings are presented for five dams. The scatter in the results is too large to serve as the basis for further analyses, but the data suggest that the ratio is largest near the edge of the fill and decreased gradually from the edge of the fill toward the center.

Finite-element analyses were performed, assuming a linear response of the fill with the aim of developing an estimate of the lateral movement with respect to depth and with the lateral position along the fill. Two-dimensional, plane-strain computations were made for a vertical cross section through the fill along the axis of the proposed shopping mall. The fill was modeled by two-dimensional, quadrilateral finite elements and the fill-rock contact was assumed to be rigid and perfectly rough with no slippage. Body forces due to the weight of the fill were used in a 'gravity turn-on' analysis. Such analyses are known to indicate unrealistic settlements during construction but are believed to yield realistic results for the post-construction period.

Three sets of properties of the fill were used in the computations: 1. A constant value was used for Young's modulus E and a value of 0.25 was used the Poisson's ratio v, 2. E was assumed to be constant and v was assumed to have a value of 0.45, and 3. E was assumed to be equal to kx. where x is depth, and v was assumed to have a value of 0.25. In each case the value of E or k was adjusted to yield the value of settlement (0.4% of the height of the fill) for the particular depth being analyzed.

Some of the results of the finite-element analyses are presented in Figures 6.42 and 6.43. In Figure 6.42, the horizontal movements along the top of the fill are plotted as a function of the distance x along the fill, where the start of the fill has a value of x of 1.83 meters. As may be seen, the results from the three sets of finite-element computations agree reasonably well. As may be seen, the horizontal movement is zero at about 85 m along the fill, where the thickness of the fill is virtually constant. The horizontal movement of the fill is obviously zero where the thickness of the fill is zero, and has a maximum value of about 25 mm at about 30 m along the fill.

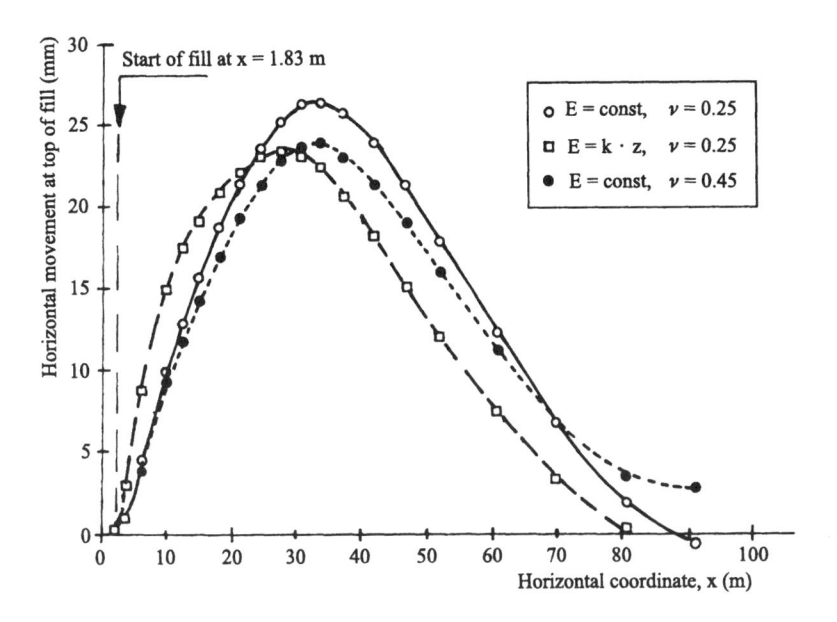

Analysis of single piles and groups of piles subjected to active and passive loading 253

Figure 6.42. Computed horizontal movements at top of fill as a function of distance along till (after Reese et al. 1989).

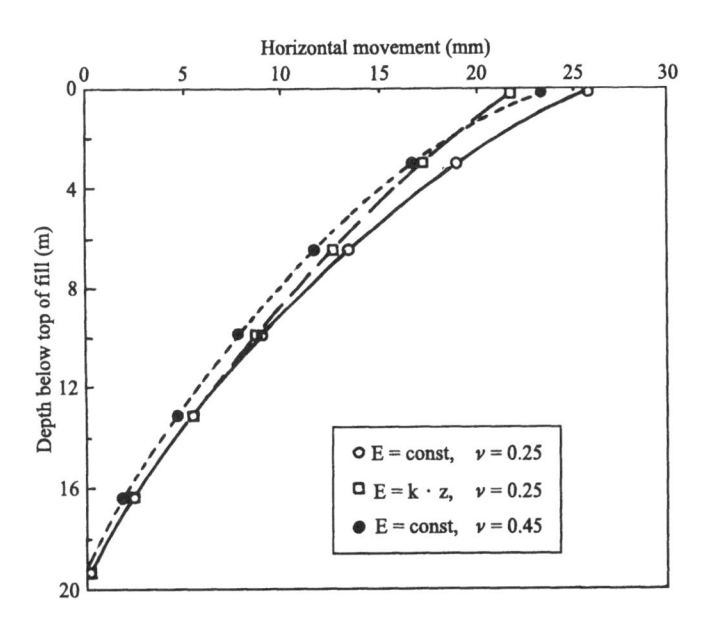

Figure 6.43. Computed horizontal movement of fill as a function of depth at a point 30 m from beginning of fill (after Reese et al. 1989).

254 *Piles under lateral loading*

The horizontal movement of the fill as a function of depth is shown in Figure 6.43 at the point of maximum horizontal movement of the surface. As stipulated the horizontal movement for a depth of 20 m is zero. While the movement is nonlinear with depth and the rate is greater near the surface of the fill, not much error is introduced by assuming the horizontal movement is linear with depth, an assumption made for later computations.

The elevation for a section through the center of the mall, showing the bored piles that were analyzed, is presented in Figure 6.44. Piles 1 and a few others not shown were founded in rock and are connected through a tie beam to Pile 2. An expansion joint through the tie beam makes Piles 3 through 10 independent of the section of the foundation at Pile 2 and beyond. Therefore, separate analyses were made of two sections of the foundations.

The following procedure was followed in each of the analyses: 1. the horizontal movements of the surface of the fill was tabulated at the location of each set of the bored piles (other piles are placed in or out of the section shown in the figure to accommodate the width of the mall); 2. assuming a linear movement with depth and no movement at the top of the rock, the horizontal movements were tabulated point to point with depth for the locations of each of the bored piles; 3. *p-y* curves were formulated for closely-spaced depths at the locations of each of the bored piles *with the origins of the curves displaced a distance y equal to the horizontal displacement at that depth;* and 4. a computer code was used to analyze each pile, using the displaced *p-y* curves and boundary conditions at the heads of the piles consistent with structural considerations. The result of Step 3 is that distributed lateral forces, equal to the *p*-values corresponding to the *y*-values from the finite-element analyses, will be applied to the pile with zero lateral deflection. The lateral displacements from the zero lateral deflection will be found from the results of the *p-y* method.

An example of the large number of *p-y* curves that were developed is presented in Figure 6.45. Using the groundline movement of 25.4 mm, a linear variation of movement with depth, and a depth of fill of 25 m the movement at a depth of 3.05 m would be 22.4 mm. The *p-y* curves for piles with diameters of 0.915 m and 1.22 m were plotted as

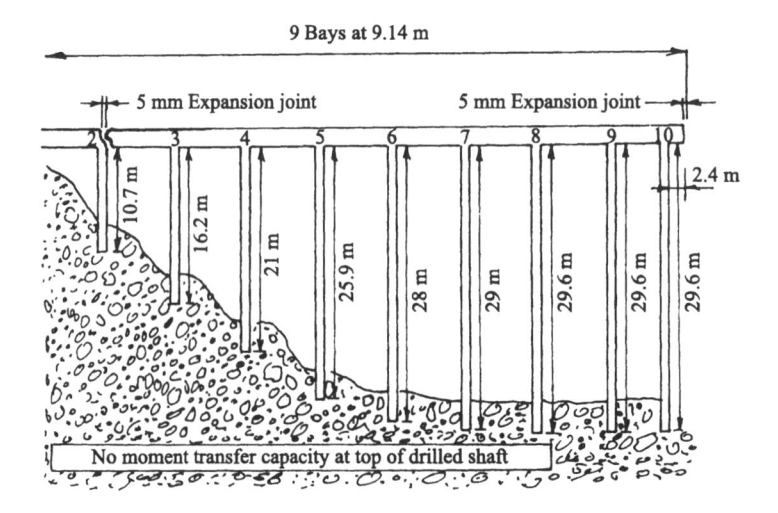

Figure 6.44. Section through bored piles placed in fill to support building (after Reese et al. 1989).

Analysis of single piles and groups of piles subjected to active and passive loading 255

shown, using the selected properties of the fill. Such curves are required as closely spaced depths for each of the piles that must be analyzed.

Some preliminary analyses were done and the diameters of the bored piles were selected, as shown in Table 6.26. The strengths of the concrete and of the steel were used, along with the amount and placement of the steel for each of the two diameters of the piles, and bending stiffnesses $E_p I_p$ were computed for use in the analyses. Average values were selected, and after the responses of the piles were computed, evaluation of the results showed that re-computation with more precise values of $E_p I_p$ was unnecessary.

Also shown in Table 6.26 are movements of the ground surface from the finite-element method. Because of the variation in results due to sets of parameters selected to describe

Table 6.26. Data used in analysis of bored piles.

Bored pile	Horizontal distance from Pile 1 m	Diameter m	Length m	Groundline movement mm
1	0	0.91	3.1	0
2	9.1	0.91	10.7	10.
3	18.3	0.91	16.2	12.7
4	27.4	0.91	21.0	25.4
5	36.6	1.22	25.9	25.4
6	45.7	1.22	28.0	25.4
7	54.9	1.22	29.0	8.9
8	64.0	1.22	29.6	5.1
9	73.2	1.22	29.6	1.3
10	79.9	1.22	29.6	0

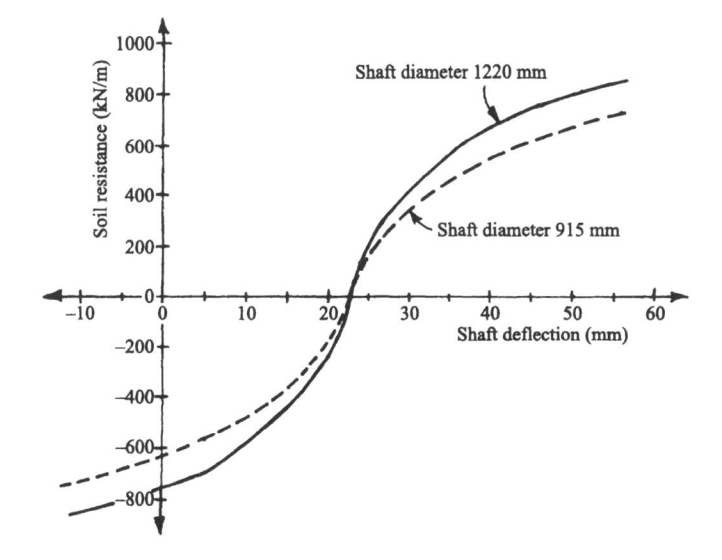

Figure 6.45. Computed p-y curves for a depth of 3.05 m below top of fill and for a computed soil movement of 22.4 mm (after Reese et al. 1989).

256 Piles under lateral loading

the characteristics of the fill, as shown in Figures 6.42 and 6.43, judgment was used in the selection of groundline movements in Table 6.26.

With respect to Pile 2, the assumption is made that the tie beam is held so rigidly by Pile 1 and the others that there is zero pile-head deflection at Pile 2. A computer code was used with the following two boundary conditions; 1. Pile deflection y_t is zero and 2. Pile-head moment M_t is zero. Regarding the second boundary condition, the structural engineer decided to design the tops of the piles to transmit shear but no bending moment; therefore, the joint at the pile heads was designed to allow free rotation. The results for Pile 2 are shown in Figure 6.46. The influence of the rock on the shear and moment is striking; even though the deflection is small. The point where the pile enters the rock may be seen by the abrupt change in the slope of the curve for shear; the deflection of the pile at that point is about one millimeter.

The tie beam that connected Piles 3 through 10 was designed to be stiff under axial load so that the deflection of those piles would be identical. Trials were made to find the common deflection of the piles such that the tie beam was in equilibrium; some round-off was employed and the deflection of all of the pile heads was found to be 11.4 mm. When the ground-line deflection was computed to be more than 11.4 mm, as for Piles 3 through 6, a lateral load must be put at the top of the pile to pull it back, causing a negative shear according to the adopted sign convention. The opposite is true for the Piles 7 through 10, where the ground surface movements was computed to be less than 11.4 mm. The results of the computations are shown in Table 6.27.

The results in Table 6.27 shows that the maximum tensile load in the tie beam occurs between Piles 6 and 7 and has a magnitude of 2219 kN. The solution for each of the piles with the boundary conditions of P_t as show above and M_t equal to zero yielded the values of V_{max} M_{max} shown in the table. The full curves of deflection, moment, shear, and soil reaction can be tabulated or plotted if desirable for design.

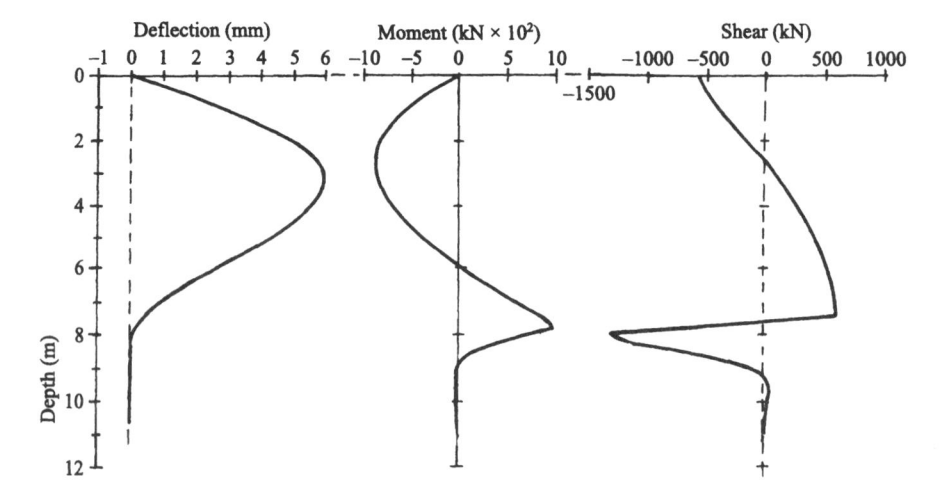

Figure 6.46. Results of computations for response of Bored Pile 2 in a settling fill (after Reese et al. 1989).

Table 6.27. Resulting loads on heads of Piles 3-10.

Bored pile	Pile-head load P_t kN	Maximum shear V_{max} kN	Maximum moment M_{max} m-kN
	−108	560	446
4	−543	940	817
5	−797	948	1806
6	−771	948	1806
7	+275	358	605
8	+504	504	1128
9	+688	688	1584
10	+752	752	1747

6.5.5.4 Discussion of results

A rational solution to an important problem in foundation engineering was possible because of several factors: 1. Valid empirical information was available of the long-term settlement of a fill of the sort that was installed; 2. One-dimensional, finite-element codes were available that allowed the lateral displacement of the fill to be predicted; and 3. The code for computing the response of piles under lateral loading could be modified to reflect the predicted horizontal displacement as a function of depth of the fill next to the piles. The piles that were designed had much smaller diameters than those designed by assuming the piles to behave as beams with rigid supports at their tops and bottoms.

The example illustrates that the results from high quality observations in the field can be coupled with advanced analytical techniques to solve a complex problem in the design of deep foundations.

CHAPTER 7

Case studies

7.1 INTRODUCTION

The procedures for the analysis of single piles under lateral loading presented in the preceding chapters are employed herein to enable comparisons to be made between results from experiments and from computations. Not only will the comparisons provide information on the accuracy of the analytical methods but the techniques of analysis will be demonstrated.

Many tests have been reported in literature on the results of field-testing of full-scale piles; however, in some instances critical information is missing. The following data are either necessary or desirable. Where some data are missing, estimates can sometimes be made to achieve a comparison. Examples of such estimates are given in the cases that follow.

Pile
– Length and penetration into the soil
– Detailed description of each cross section as a function of penetration
    Occurrence and location of steel, concrete, and any other material
    Strength f'_c and modulus of elasticity of the concrete
    Yield strength f_y and modulus of elasticity of the steel
    Similar values for any other materials in the cross sections

Soil
– Classification of soils with Atterberg Limits and with any other necessary soil tests
– Identification of rock and classification by *RQD* and other data
– Position of the water table
– Undrained shear strength of clays and stiffness from ε_{50} (ε_{50} may be estimated if necessary)
– Friction angle for cohesionless soils (or data from penetration tests that can be correlated with the value of ϕ), unit weight, and information on the structure of the grains (e.g., resistance of grains to crushing)
– Compressive strength of rock and data on secondary structure

Loading and Pile Head Restraint
– Arrangement for applying load and point of application of lateral load with respect to the ground surface
– Nature of loading, whether static, cyclic, or sustained
– Sufficient magnitude of loading to achieve a nonlinear response.
– Free-head or partially restrained.

260 *Piles under lateral loading*

Instrumentation
– Methods and details of measuring loading, pile-head deflection, and rotation
– Nature and details of internal instrumentation in pile

Results
– Presented in tabular form or such that data can be tabulated
– Any special observations

The above information will allow the bending stiffness (E_pI_p) of the pile to be computed as a function of applied moment and will also allow the computation of the bending moment at which the pile will develop a plastic hinge (M_{ult}), or will just reach plastic behavior at the extreme fibers (M_y) in the case of metal piles. The *p-y* curves can be developed according to the methods of prediction that were presented earlier.

The cases of more importance are those where the piles were instrumented so that the bending moment can be found along the length of the piles and where both static and cyclic loading were employed. Cyclic loading has proven to be of considerable importance when the soil is cohesive and water is above the ground surface. A limited number of such cases are available. Some of these provided the data on which the recommendations for *p-y* curves are based, but it is important to determine how well the results from the use of the *p-y* criteria agree with the results from the experiments.

The cases are separated into categories: clays, sands, layered soils, *c*-ϕ soils, and weak rock. Unfortunately, only a few cases are available for the last two categories. Where data are available on bending moments along the pile, curves are presented comparing the values of the maximum bending moments from experiment and from computation.

The computations were done with the professional version of the computer program described in Appendix D. The student version of computer Program LPILE can be used to make computations for several of the cases; the professional version of the program is needed for the more complex cases.

7.2 PILES INSTALLED INTO COHESIVE SOILS WITH NO FREE WATER

7.2.1 *Bagnolet*

Kerisel (1965) reported the results of three, short-term, static lateral-load tests of a closed-ended 'bulkhead caisson'. The cross section of the pile is shown in Figure 7.1. Two sheet-pile sections were welded together to form the pile. The three tests were performed on the same pile which was recovered and reinstalled for all tests following the first one. The bending stiffness E_pI_p was given as 25,500 kN-m² and, if E is selected as 200,000 MPa, the value of I is 0.0001275 m⁴.

Insufficient information is available from which to compute the ultimate bending moment. However, if the assumption is made that the steel has a yield strength of 248 MPa, the bending moment at which the extreme fibers of the pile will just reach yielding (M_y) is at 204 kN-m.

The equation for the equivalent diameter of the circular section can be applied but an examination of the shape showed that the selection of the equivalent diameter of 0.43 m was appropriate.

Different boundary conditions and depths of embedment were used in the three tests. In each case, the pile head was free to rotate but the lateral load was applied at different distances above the groundline.

The tests were performed east of Paris in a fairly uniform deposit of medium stiff clay, classified as CH by the Unified Method. The reported properties of the clay are shown in Table 7.1, and were found from unconfined compression and cone tests. In the absence of stress-strain curves, the value of ε_{50} was selected from Table 2.5. The water table was below the tips of the piles, but the degree of saturation was over 90% and it is assumed that the undrained shear strength can be employed in the analyses.

The results from the tests at Bagnolet are shown in Figures 7.2 through 7.4. The analyses were performed with the criteria for predicting p-y curves for stiff clay with no free water. As may be seen, for both the pile-head deflection and the maximum bending moment, the agreement between results from experiment and from computation is good to excellent. The agreement for bending moment is somewhat better but deflection is predicted with good accuracy.

Of interest, however, is that the maximum bending moment from experiment was just over 60% of that which is expected to cause yielding of the extreme fibers of the steel in the pile. In general, where possible, loading should be increased to a maximum bending moment that is just below the yield moment, assuming that the pile may be used at another location. Further, the application of the lateral load at, or near, the ground surface will result in larger values of pile-head deflection for the same value of maximum bending moment.

A review of the figures, from the results at Bagnolet, gives some insights into the comparative behavior of piles under lateral loading. Even though the ground-line moment is smaller for Case 1, the ground-line deflection is larger at the same final value of shear than for the other two cases. The deflected shape of the pile (not shown here) reveals that

Table 7.1. Reported properties of soil at Bagnolet.

Depth	Water content	Undrained shear strength	ε_{50}*	Total unit weight
m	%	kPa		kN/m³
0	–	100	0.005	17.9
3.96	31.5	125	0.005	17.9
4.69	29.0	130	0.005	17.9

*Obtained from Table 3.5

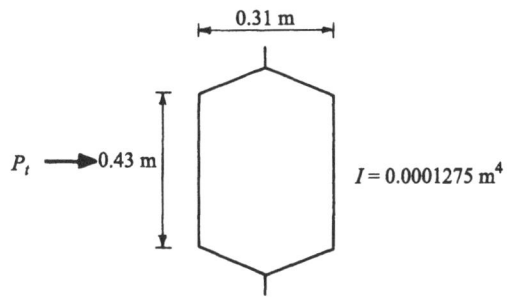

Figure 7.1. Cross section of pile at Bagnolet.

262 *Piles under lateral loading*

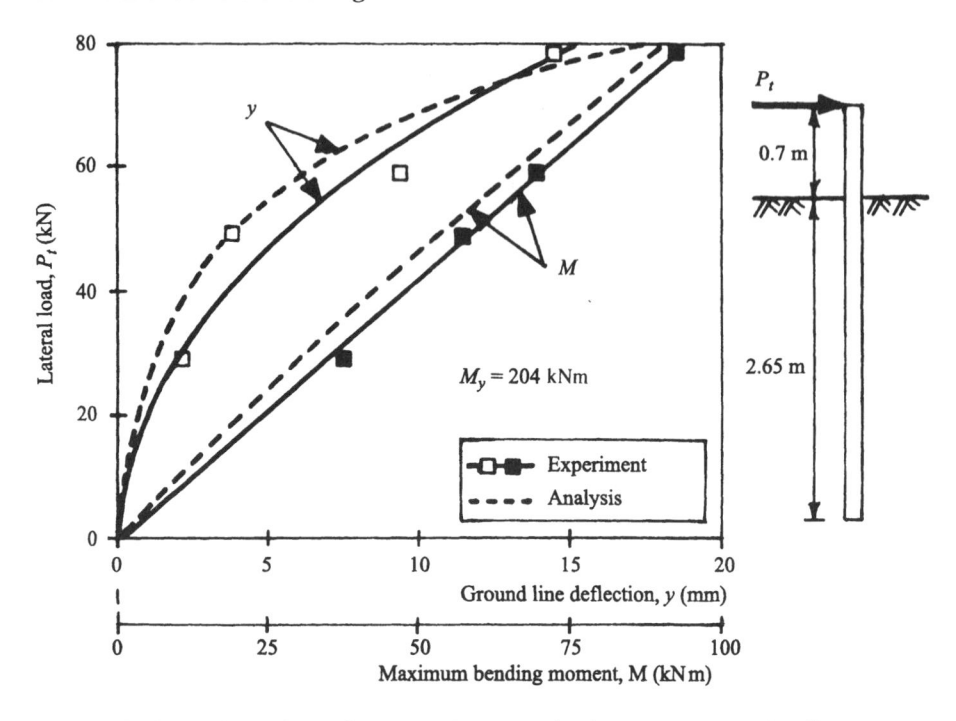

Figure 7.2. Comparison of experimental and computed values of maximum bending moment and deflection, Case 1, Bagnolet.

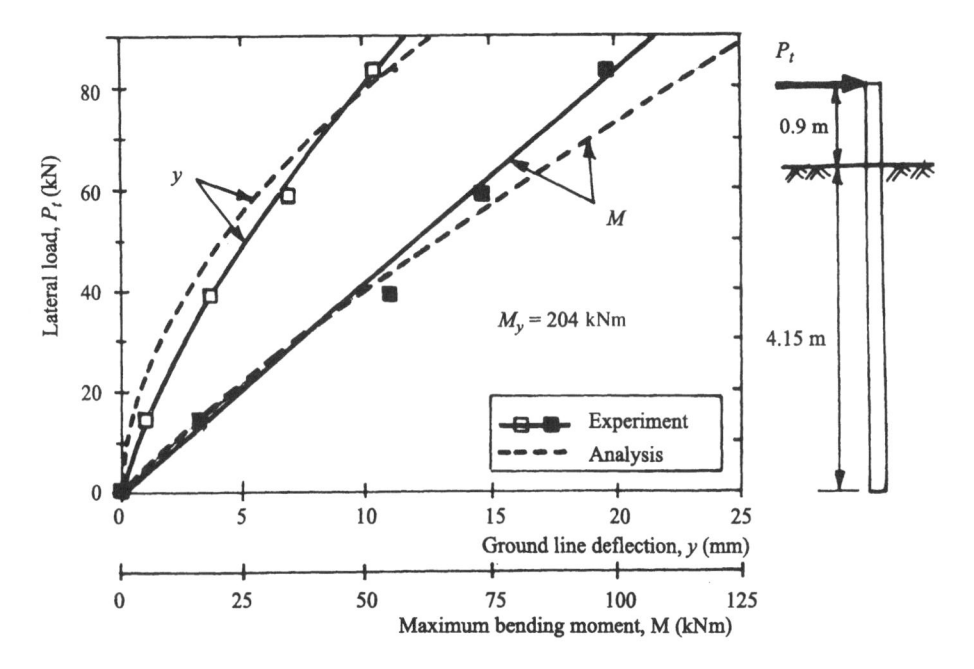

Figure 7.3. Comparison of experimental and computed values of maximum bending moment and deflection, Case 2, Bagnolet.

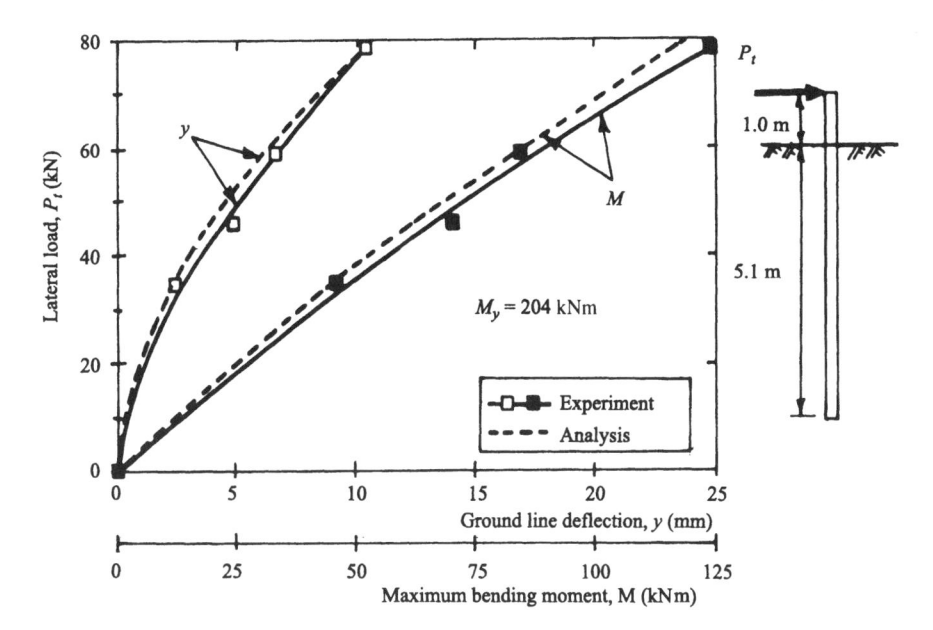

Figure 7.4. Comparison of experimental and computed values of maximum bending moment and deflection, Case 3, Bagnolet.

the small penetration allows the bottom of the pile to deflect and is accompanied by only one point of zero deflection. However, as may be seen, there was not a corresponding increase in the maximum moment for the pile in Case 1, as might be expected.

In comparing the results for the three cases, as the ground-line moment increases, due to the increased moment arm, the maximum bending moment goes up for the same final value of ground-line shear. This results suggests that, if a lateral load is applied to a pile at a great distance above the ground-line, the behavior of the pile will depend to a lesser degree on the soil characteristics.

7.2.2 *Houston*

Reese & Welch (1975) reported the results from a test of a bored pile with a diameter of 0.762 m and a penetration of 12.8 m. An instrumented steel pipe, with a diameter of 0.260 m and a wall thickness of 6.35 mm, formed the core of the pile. A rebar cage, consisting of 20 bars with diameters of 44.5 mm, had a diameter of 0.610 m. The yield strength of the steel was 276 MPa and the compressive strength of the concrete was 24.8 MPa. The value of the bending stiffness $E_p I_p$ was measured during the testing by reading the output from strain gauges on opposite sides on the instrumented pipe and was computed to be 4.0×10^5 kN-m^2. The bending moment at which a plastic hinge would occur was computed to be 2030 kN-m. The test was performed in Houston, Texas, under the sponsorship of the Texas Department of Transportation and the Federal Highway Administration.

The soil was overconsolidated clay, called Beaumont clay locally, and had a well-developed secondary structure. The water table was at a depth of 5.5 m at the time of the

264 *Piles under lateral loading*

field tests. Tube samples with a diameter of 100 mm were taken, observing the necessary precautions to reduce sampling disturbance. The properties of the clay are shown in Table 7.2. The undrained shear strength was measured by unconsolidated-undrained triaxial compression tests with confining pressure equal to the overburden pressure. The values of ε_{50} shown in the table were obtained from a review of the laboratory tests and agree with values shown in Table 3.5 except that the top 0.4 m is shown to be slightly stiffer that presented in Chapter 3. Some samples were subjected to repeated loading and the effect on the stress-deformation relationships was observed.

The lateral loads were applied at 0.076 m above the ground surface and loads were both static and cyclic. The same pile was used without redriving to obtain results for both types of loading. The successive loads were widely separated in magnitude so that the cycling at the previous load was assumed to have no effect on the first cycle at the next load. At each increment of lateral load, readings were taken at one cycle, 5 cycles, 10 cycles, and 20 cycles for the larger loads. The results from the cycling were analyzed and a method of predicting the effect of cyclic loading was developed, as shown in Chapter 3, based on the stress level and the number of cycles. For computation of the response of the pile to static lateral loading, the *p-y* curves were developed based on the criteria for stiff clay with no free water.

Comparisons of the pile-head deflection and maximum bending moment for static loading are shown in Figure 7.5. The comparison for deflection is excellent and the analysis is conservative in the computation of bending moment. The conservatism in the computation of bending moment is also reflected in Figure 7.6, which shows the bending moments as a function of depth for the lateral load of 445 kN for static loading. The depth to the point of maximum bending moment agrees well between experiment and computation; however, the depth to the point of zero bending moment is underpredicted by one or two meters.

Comparisons of the pile-head deflection and maximum bending moment for cyclic loading are shown in Figures 7.7. The results are for 20 cycles. The comparison for deflection is excellent, except for the larger loads, and the analytical method is conservative in the computation of bending moment.

7.2.3 *Brent Cross*

Price & Wardle (1981) reported the results of a test of a steel pipe in London Clay. The diameter of the pile was 0.406 m and its penetration was 16.5 m. In addition to the analysis of the test by the original authors, Gabr et al. (1994) did a further study. The moment

Table 7.2. Reported properties of soil at Houston.

Depth m	Water content %	Undrained shear strength kPa	ε_{50}*	Total unit weight kN/m^3
0	18	76.0	0.005	19.4
0.4	18	76.0	0.005	19.4
1.04	22	105.0	0.005	18.8
6.1	20	105.0	0.005	19.1
12.8	15	163	0.005	19.9

*Values from review of laboratory stress-strain curves

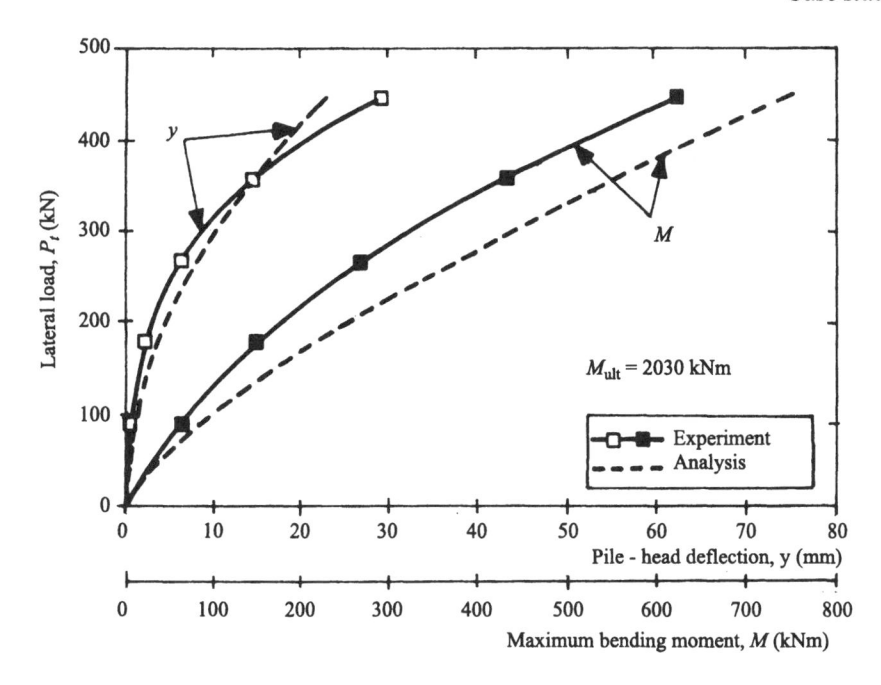

Figure 7.5. Comparison of experimental and computed values of maximum bending moment and pile-head deflection, static loading, Houston.

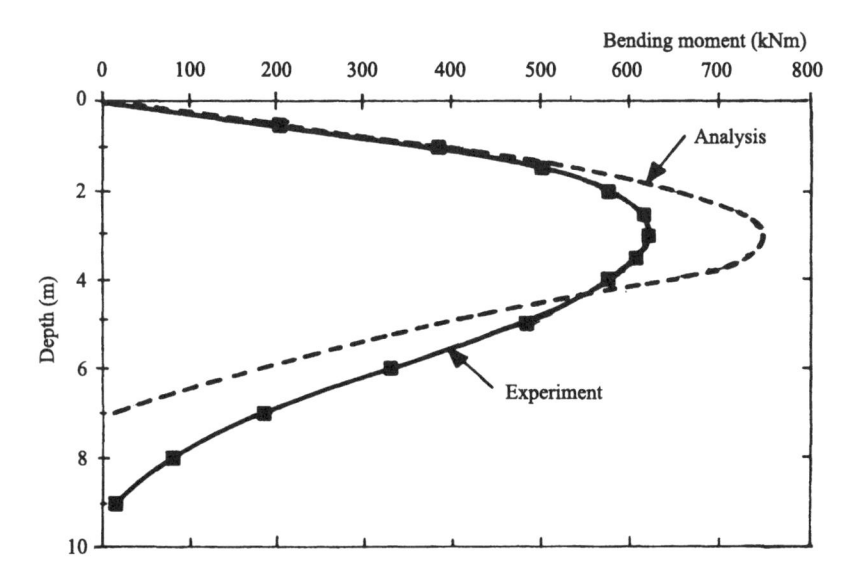

Figure 7.6. Comparison of curves of bending moment versus depth for P_t of 445 kN, static loading, Houston.

266 *Piles under lateral loading*

of inertia, I_p, of the pile was reported as 2.448×10^{-4} m^4; the bending stiffness, $E_p I_p$, used in the analyses that follow was 5.14×10^4 kN-m^2. The bending moment at which the extreme fibers would reach yield was computed to be 301 kN-m, and the ultimate bending moment, at which a plastic hinge would develop, was computed to be 392 kN-m.

The data on the properties of the London Clay at the site was obtained from the testing of specimens taken with thin-walled tubes with a diameter of 98 mm. The water table was presumably some distance below the ground surface. The values, shown in Table 7.3, of undrained shear strength were scaled from a plot presented by the authors. The strength of the clay near the ground surface seems low for over-consolidated clay. Data on the stiffness of the soil were not reported so computations were done with values of ε_{50} were obtained from Table 3.5. The suggested values of ε_{50} are in the ranges of values obtained experimentally by Jardine et al. (1986) for low plasticity clays. The *p-y* curves for the analyses were obtained by using the criteria for stiff clay with no free water.

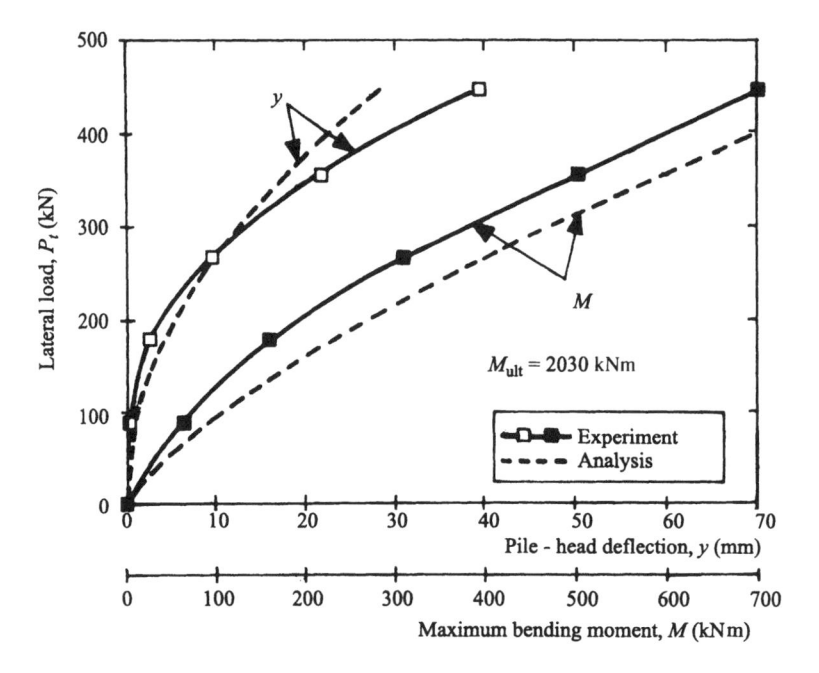

Figure 7.7. Comparison of experimental and computed values of maximum bending moment and pile-head deflection, cyclic loading, Houston.

Table 7.3. Reported properties of soil at Brent Cross.

Depth m	ε_{50}*	Undrained shear strength kPa
0	0.007	44.1
4.6	0.007	85.2
6.2	0.007	80.6
19	0.005	133.3

* From Table 3.5

The lateral load was applied at 1.0 m above the groundline and both static and cyclic loads were applied. The static loads were of a larger magnitude than the cyclic loads and were applied in a re-loading state after cycling had been done. The assumption is made that the cycling with the smaller loads did not affect the subsequent static results. Only the results from the static loading are reported below. The results from the experiment and from computations, with the methods presented herein, are shown in Figure 7.8. The agreement between the experiment and analysis is reasonable with the analytical method yielding results that are somewhat conservative.

The value of maximum bending moment for the largest lateral load of 100 kN was computed to be 198 kN-m, which is significantly below the computed value of bending moment at first yield.

7.2.4 Japan

The Committee on Piles Subjected to Earthquake (1965) reported the results from the testing of a steel-pipe pile with a closed end that was jacked into the soil. The pile was 305 mm in outside diameter with a wall thickness of 3.18 mm, and its penetration was 5.18 meters. The moment of inertia, I_p, was 3.43×10^{-5} m^4, and the bending stiffness, $E_p I_p$, was 6868 kN-m^2. The bending moment M_y at which yielding of the extreme fibers would occur was computed to be 55.9 m-kN, and the ultimate bending moment M_{ult} was computed to be 71.8 kN-m.

The soil at the site was a soft, medium to highly plastic, silty clay with a high sensitivity. The undrained shear strength and stiffness of the soil was obtained from undrained triaxial shear tests. The strains at failure were generally less that 5%, and failure was by brittle fracture. The properties of the soil are shown in Table 7.4. The values of the undrained shear strength are typical of those for normally consolidated clay. Therefore, in the absence of stress-strain curves from the laboratory, values of ε_{50} shown in the table were taken from Table 5.3. The p-y curves for static loading for the

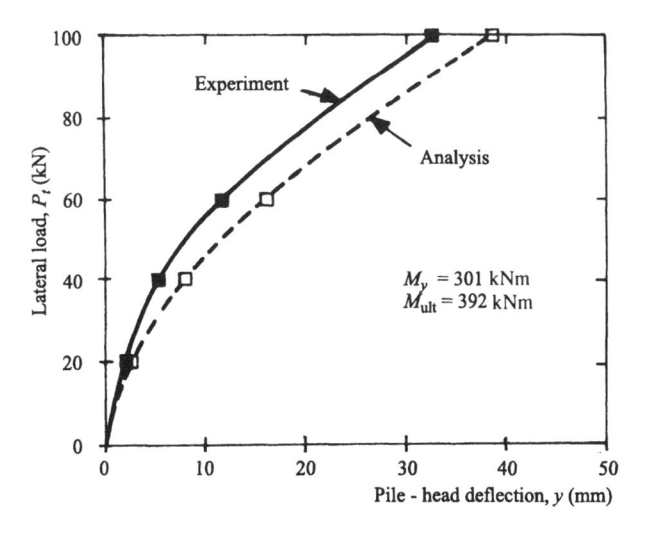

Figure 7.8. Comparison of experimental and computed values of pile-head deflection, Brent Cross.

268 *Piles under lateral loading*

analyses were obtained by using the criteria for soft clay with free water. However, the *p-y* curves for static loading using the criteria for stiff clay with no free water yielded almost identical results.

The loading was applied at 0.201 m above the groundline; the maximum lateral load was a moderate value of 14.24 kN which produced a pile-head deflection of 4.83 mm and a maximum bending moment of 17.34 m-kN. Thus, with respect to a failure due to yielding of the extreme fibers, the factor of safety was 3.2 and the factor of safety against a failure in plastic yielding was 4.1.

The loading was static. A plot of the comparison between experimental and computed deflections is shown in Figure 7.9. The computations were carried to a maximum lateral load of 45 kN, a load that caused a maximum bending moment approximately equal to that to cause the first yield of the extreme fibers of the steel. Good agreement was found between experimental and computed values of pile-head deflection for the range of loads that were applied.

Table 7.4. Reported properties of soil at Japan.

Depth	Undrained shear strength	ε_{50}*	Submerged unit weight
m	kPa		kN/m³
0	27.3	0.02	4.9
5.18	43.1	0.02	4.9

*Obtained from Table 3.3

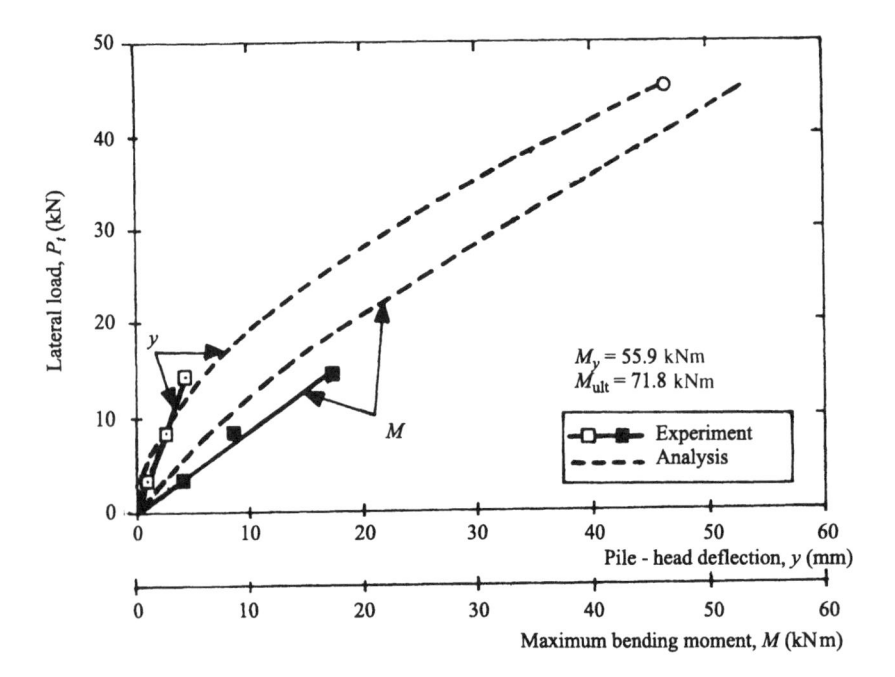

Figure 7.9. Comparison of experimental and computed values of maximum bending moment and pile-head deflection, Japan.

Bending moments were measured at the site but information is unavailable on the techniques that were used. A plot of the comparison between experimental and computed maximum bending moment is also shown in Figure 7.9. Again, agreement was good for the range of loads that were applied.

Even though the length to diameter ratio was relatively small at 17, examination of the results for deflection and bending moment along the pile showed that the pile would have failed in bending rather than by excessive deflection.

7.3 PILES INSTALLED INTO COHESIVE SOILS WITH FREE WATER ABOVE GROUND SURFACE

7.3.1 *Lake Austin*

Matlock (1970) presented results from lateral-load tests employing a steel-pipe pile that was 319 mm in diameter, with a wall thickness of 12.7 mm, and a length of 12.8 m. The bending stiffness was 31280 kN-m^2. The bending moment at which the extreme fibers would first yield was computed to be 231 kN-m, and the bending moment for the formation of a fully plastic hinge was computed to be 304 kN-m.

The pile was driven into clays near Lake Austin, Texas, that were slightly overconsolidated by desiccation, slightly fissured, and classified as CH according to the Unified System. The undrained shear strength was measured with a field vane; was found to be almost constant with depth, and $(c_u)_{vane}$ averaged 38.3 kPa.

A comprehensive investigation of the soil was undertaken and the computations shown herein are based on tests with the field vane. The vane strengths were modified to obtain the undrained shear strength of the clay. The values of the soil properties employed in the following computations are shown in Table 7.5. The value of ε_{50} was found from triaxial tests and averaged 0.012. In view of the almost constant value of c_u with depth, a constant value of ε_{50} appears reasonable. The submerged unit weight was determined at several points below the mudline and the average value was found to be 10.0 kN/m^3. Water was kept above the ground surface during all of the testing.

Table 7.5. Properties of soil at Lake Austin.

Depth	Water content*	Undrained shear strength
m	%	kPa
0	29.0	30.2
1.14	33.5	32.2
1.14	33.5	42.3
3.39	50.1	17.5
3.70	49.6	30.1
4.30	48.3	23.4
5.69	46.1	51.8
7.25	54.5	29.8
9.47	55.5	32.6
15.0	–	32.6

*Average values

270 *Piles under lateral loading*

The pile was tested first under static loading, removed, redriven, and tested under cyclic loading. The load was applied at 0.0635 m above the groundline.

The pile was instrumented internally, at close spacings, with electrical-resistance strain gauges for the measurement of bending moment. Each increment of load was allowed to remain long enough for readings of strain gauges to be taken by an extremely precise device. A rough balance of the external Wheatstone bridge was obtained by use of a precision decade box and the final balance was taken by rotating a drum, 150 mm in diameter, on which a resistance alloy wire had been wound into a spiral groove in the drum. A contact on the resistance alloy wire was read on the calibrated drum when a final balance was achieved. The accuracy of the device was better than one microstrain but, by choice in 1955, some time was required for readings to be taken from the top of the pile to the bottom and back up again. Because of creep of the soil at relatively moderate to high loadings, the pressure in the hydraulic ram that controlled the load was adjusted as necessary to maintain a constant load. The two sets of readings at each point along the pile were interpolated to find the reading at a particular time.

During cyclic loading, readings of the strain gauges were taken at various numbers of cycles of loading. The load was applied in two directions, with the load in the forward direction being more than twice as large as the load in the backward direction. After a significant number of cycles, when successive readings of deflection were the same, an equilibrium condition was assumed.

The p-y curves for the analyses were obtained by using the criteria for soft clay with free water. Comparisons of the pile-head deflection and maximum bending moment for static loading are shown in Figure 7.10. The comparison is satisfactory for both sets of results.

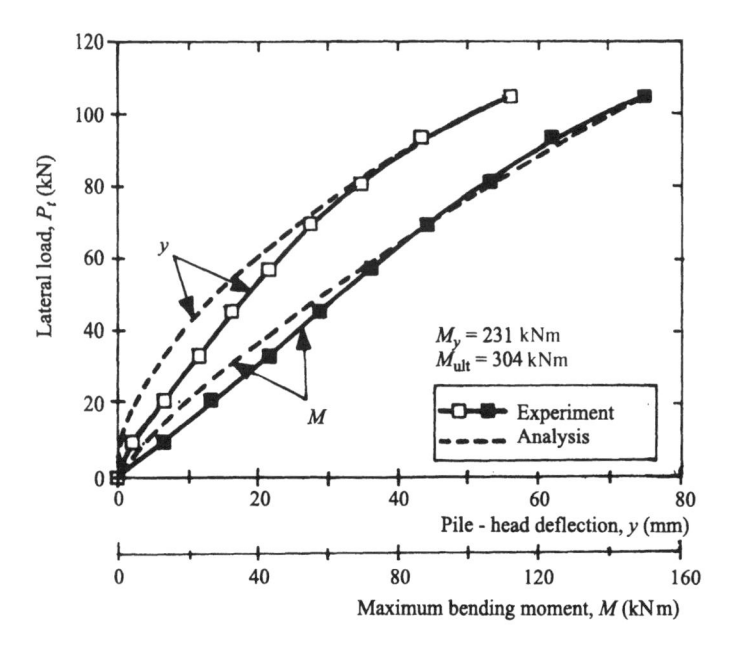

Figure 7.10. Comparison of experimental and computed values of maximum bending moment and pile-head deflection, static loading, Lake Austin.

Case studies 271

A lateral load of 80.9 kN caused a maximum bending moment of approximately one-half of the 231 kN-m that would cause the first yield. Figure 7.11 shows a comparison of the experimental and computed bending moment curves as a function of depth. The agreement in the curves in all respects is excellent.

Comparisons of the pile-head deflection and maximum bending moment for cyclic loading are shown in Figure 7.12. The results from the analytical method are unconserva-

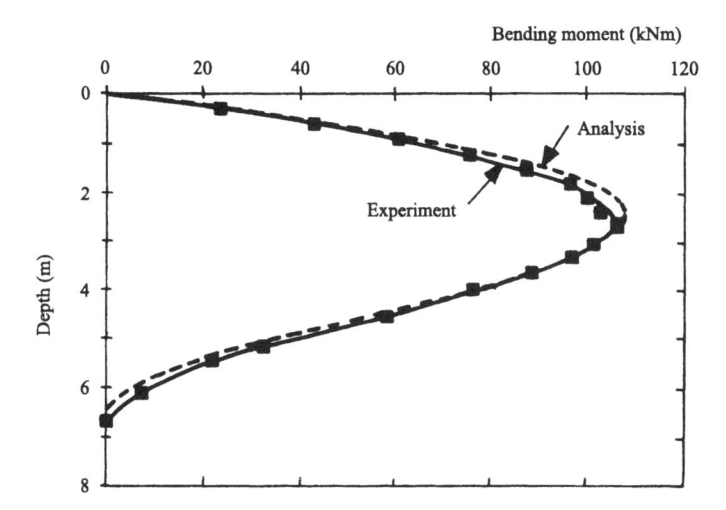

Figure 7.11. Comparison of curves of bending moment versus depth for P_t of 80.9 kN, static loading, Lake Austin.

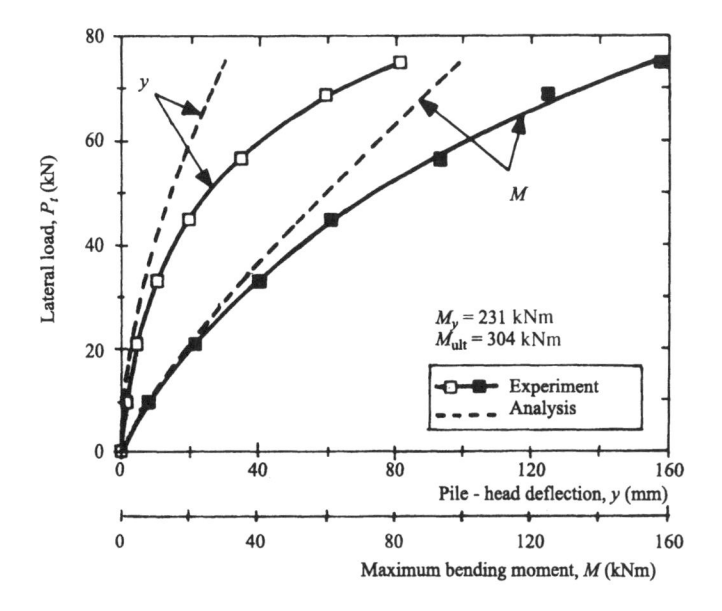

Figure 7.12. Comparison of experimental and computed values of maximum bending moment and pile-head deflection, cyclic loading, Lake Austin.

272 *Piles under lateral loading*

tive for both sets of computations, but are considered to be satisfactory in the lower ranges of load that are most relevant to design.

A historical comment is of interest, Terzaghi commented in 1955 that the use of strain gauges to get the response of the soil to the lateral loading of the pile was not possible; therefore, emphasis was placed on obtaining the necessary revision to get the response of the soil. Terzaghi visited the test site during a trip to the University of Texas in 1956 and appeared to be impressed with the progress of the research.

7.3.2 *Sabine*

The pile tested at Lake Austin was removed and installed at Sabine where the soil was a soft clay. As before, the pile was tested both under static and cyclic loading. Also, testing was conducted with the pile head free to rotate and restrained against rotation.

Meyer (1979) analyzed the results of testing the soil at Sabine and reported that the clay was a slightly overconsolidated marine deposit, had an undrained shear strength of 14.4 kN/m^2, and a submerged unit weight of 5.5 kN/m^3. Computations were made with values of ε_{50} of 0.02, as suggested in Table 3.3. The *p-y* curves for the analyses were obtained by using the criteria for soft clay with free water.

The lateral loads were applied at 0.305 m above the ground line. Comparisons of the pile-head deflection and maximum bending moment for static loading are shown in Figure 7.13. The results from the analytical method for deflection are conservative, and the agreement is satisfactory for maximum bending moment.

Comparisons of the pile-head deflection and maximum bending moment for cyclic loading are shown in Figure 7.14. The comparisons show excellent agreement for both deflection and maximum bending moment.

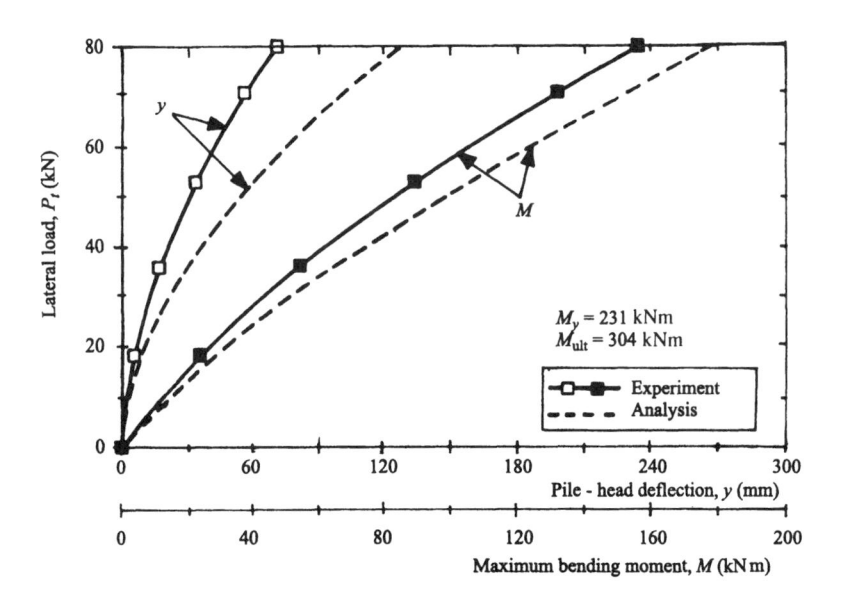

Figure 7.13. Comparison of experimental and computed values of maximum bending moment and pile-head deflection, static loading, Sabine.

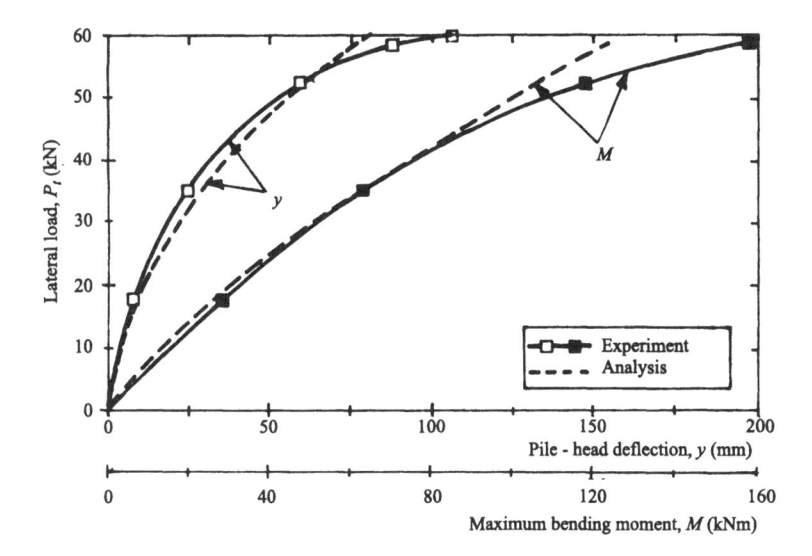

Figure 7.14. Comparison of experimental and computed values of maximum bending moment and pile-head deflection, cyclic loading, Sabine.

7.3.3 Manor

Reese et al. (1975) describe lateral-load tests employing two steel-pipe piles that were 15.2 m long, with a diameter of the upper section of 0.641 m and of the lower of 0.610 m. The piles were driven into stiff clay at a site near Manor, Texas. The piles were calibrated prior to installation and the mechanical properties of each of the piles are shown in Table 7.6. The bending moment, M_y, when yield stress develops at the extreme fibers and the ultimate bending moment, M_{ult}, are shown only for the top sections, where the ultimate bending moment occurs during loading. The experimental p-y curves in Figures 1.5 and 1.6 were derived from data from the tests at Manor.

The clay at the site was strongly overconsolidated and there was a well developed secondary structure. The undrained shear strength of the clay was measured by unconsolidated-undrained triaxial tests with confining pressure equal to the overburden pressure. The properties of the clay are shown in Table 7.7. The site was excavated to a depth of about 1 m and water was kept above the surface of the site for several weeks prior to obtaining data on soil properties.

The values of ε_{50} were found from experiment but scatter was great, probably because of the secondary structure of the clay. Values of ε_{50} were found from Table 3.5; the results in Table 7.7 are generally in agreement with experimental values.

Both of the piles were instrumented with electrical-resistance strain gauges for measurement of bending moment. The gauge readings were taken with an electronic data-acquisition system, and a full set of readings could be taken in about one minute. The point of application of the load for both piles was 0.305 m above the groundline. Pile 1 was tested under static loading with the load being increased in increments; loading ceased when the bending moment was near the yield moment.

Pile 2 was tested under cyclic loading and the loads were cycled under each increment until deflections were stabilized. The number of cycles of loading was in the order of 100 and

Table 7.6 Mechanical properties of piles at Manor.

Pile	Section m	I m^4	EI kN-m^2	M_y kN-m	M_u kN-m
1	Top 7.01	0.002335	493,700	1757	2322
	Bottom 8.23	–	168,400	–	–
2	Top 7.01	0.002335	480,400	1757	2322
	Bottom 8.23	–	174,600	–	–

Table 7.7 Reported properties of soil at Manor.

Depth m	Water content %	Undrained shear strength kPa	ε_{50}*	Total unit weight kN/m^3
0	–	25	0.007	–
0.9	37	70	0.007	18.1
1.52	27	163	0.005	19.4
4.11	22	333	0.004	20.3
6.55	22	333	0.004	20.3
9.14	19	1100	0.004	20.8
20.00	–	1100	0.004	–

*From Table 3.5 and in general agreement with experiment

applied at a rate of about two cycles per minute. Observations during and after testing revealed that the erosion was significant in response to the cyclic loading. A gap was revealed in front of the pile after removing a load, except for the loads of very low magnitude; the gap became filled with water that was ejected during the next cycle of loading. The water was caused to rush upward at a high velocity and carried particles of clay. During the testing, radial cracks developed in the ground surface in front of the pile and a subsequent examination showed that erosion has occurred outward through the cracks as well as in front of the pile.

While the equilibrium condition was reached with about 100 cycles of loading, as noted above, it is probable that the application of hundreds or thousands of cycles would have caused additional deflection. As noted in Chapter 3, the question of the expected number of cycles of loading, particularly for clay soils below free water, needs careful attention in any problem of design. Also, the gapping around a pile is plainly related to the loss of resistance during cyclic loading, and gapping may be related to diameter other than by the first power as implied by the recommendations for p-y curves.

The data were analyzed by use of the criteria for stiff clays below free water. The comparisons of ground line deflection and maximum bending moment for Pile 1, for static loading, are shown in Figure 7.15. The agreement in both instances is excellent, with the analytical method being slightly conservative. The maximum bending moment of 1271 kN-m that was measured was substantially less than the yield moment and the ultimate moment.

The comparisons of groundline deflection and maximum bending moment for Pile 2, for cyclic loading, are shown in Figure 7.16. Excellent agreement was found between results from experiment and from analysis, with computations slightly conservative for both ground-line deflection and maximum bending moment. The maximum bending moment of 1385 kN-m that was measured was less than the yield moment, and the ultimate moment.

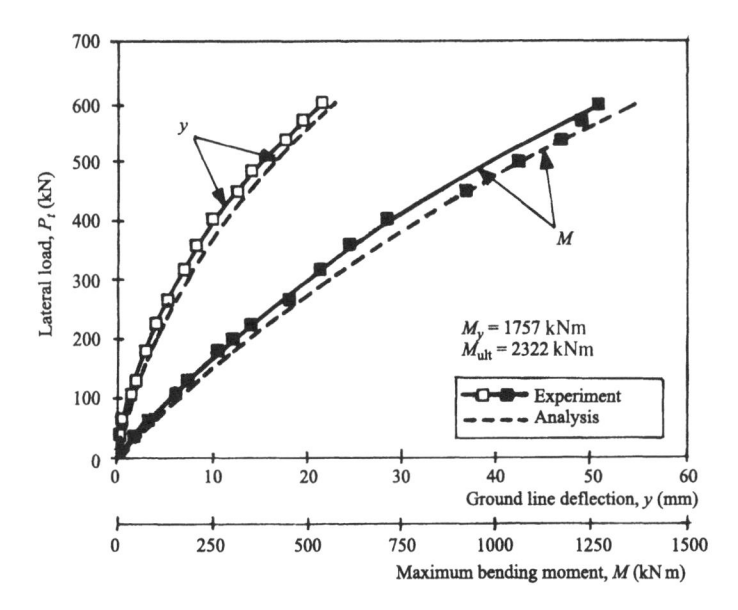

Figure 7.15. Comparison of experimental and computed values of maximum bending moment and deflection, static loading, Manor.

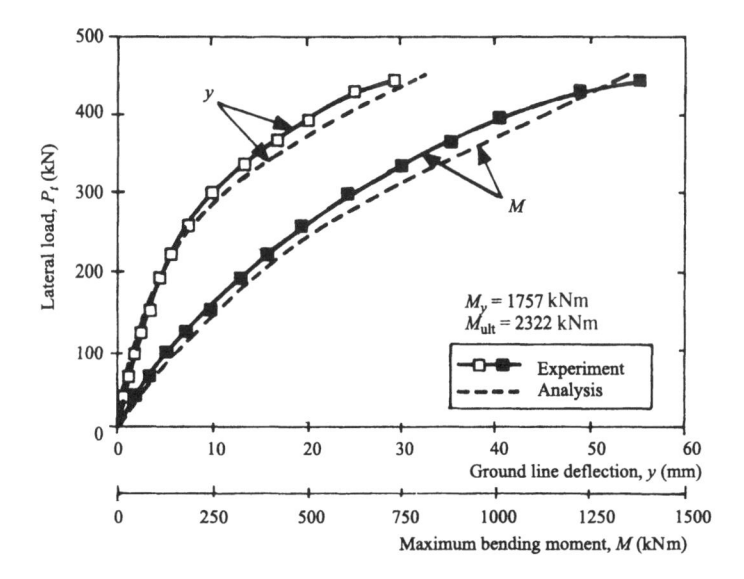

Figure 7.16. Comparison of experimental and computed values of maximum bending moment and deflection, cyclic loading, Manor.

In total, the experimental and computed values for the series of tests at Manor agree well. In some other cases, the analytical method appears to yield a conservative result for piles in overconsolidated clay that is under water. The erosion due to cyclic loading is a critical matter and other results are needed because the soil at Manor may have had

276 *Piles under lateral loading*

some characteristic that made it more erodible than other overconsolidated clays. However, a series of laboratory studies, not reported here, failed to reveal the nature of any such characteristic.

7.4 PILES INSTALLED IN COHESIONLESS SOILS

7.4.1 *Mustang Island*

Cox et al. (1974) describe lateral-load tests employing two steel-pipe piles that were 21 m long. The piles were driven into sand at a site on an island near Corpus Christi, Texas. The piles were identical in design and had diameters of 0.610 m. They were calibrated prior to installation and had the following mechanical properties: $I_p = 8.0845 \times 10^{-4}$ m^4; $E_p I_p =$ 163,000 kN-m^2; $M_y = 640$ kN-m; and $M_{ult} = 828$ kN-m. Both piles were instrumented internally with electrical-resistance strain gauges for the measurement of bending moment. The piles were loaded separately; Pile 1 was subjected to static loading, and Pile 2 to cyclic loading. The load on both piles was applied at 0.305 m above the mudline.

The soil at the site was a uniformly graded, fine sand with a friction angle of 39 degrees. The submerged unit weight was 10.4 kN/m^3, and the relative density averaged about 0.9. The water surface was maintained at 150 mm or so above the mudline throughout the test program. The piles were driven open-ended and the modification of the sand was perhaps less than would have occurred if full-displacement piles had been installed.

The data were analyzed by use of the criteria for cohesionless soils (Reese et al. 1974). The comparisons of groundline deflection and maximum bending moment for Pile 1, for static loading, are shown in Figure 7.17. The agreement in both instances is excellent.

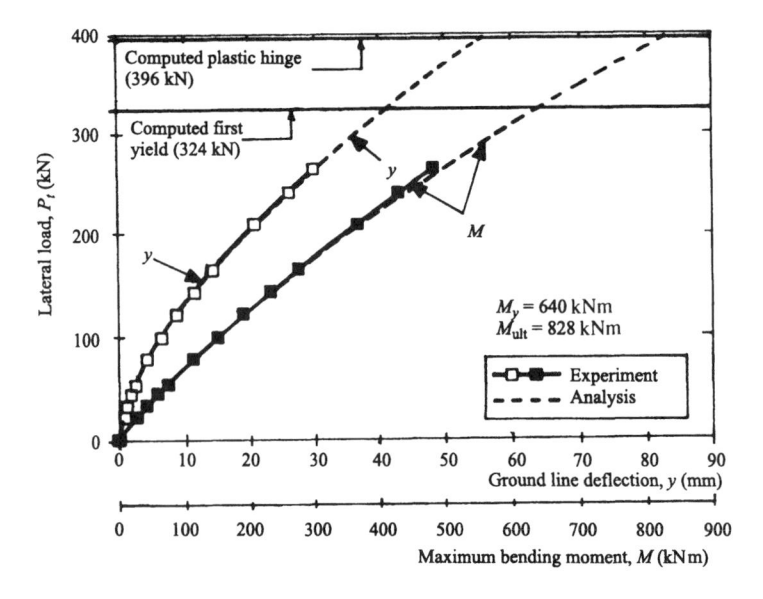

Figure 7.17. Comparison of experimental and computed values of maximum bending moment and deflection, static loading, Mustang Island.

Case studies 277

The engineer is not only interested in the ground-line deflection and maximum bending moment but also on the accuracy of the distribution of the computed bending moment with depth. Such information will allow a possible reduction, below a particular depth, in the wall thickness of a driven pile or in the number of rebars in the reinforced-concrete pile. For the Mustang Island experiment, the comparison is presented in Figure 7.18 for a static, lateral load of 210 kN, which would reflect a factor of safety of about 1.5 with respect to first yield. The curves agree well and show that the special requirement for bending strength no longer exists after a depth of about 5 meters.

The comparisons of groundline deflection and maximum bending moment for Pile 2, for cyclic loading, are shown in Figure 7.19. The agreement in both instances is excellent.

In total, the experimental and computed values for the series of tests at Mustang Island agree extremely well. It is of interest to note, however, the characteristics of the sand and the method of installation of the pile. Other sands and other methods of installation could well produce a different set of results.

7.4.2 *Garston*

Price & Wardle (1987) reported the results of lateral-load tests of a bored pile, identified as TP15, with a length of 12.5 m and a diameter of 1.5 m. The location of the tests was not given and is listed as the location of the Building Research Establishment for convenience. The reinforcement consisted of 36 round bars, 50 mm in diameter, on a 1.3-m-diameter circle. The yield strength of the steel was 425 N/mm². The cube strength of the concrete was 49.75 N/mm². The bending moment (M_{ult}) at which a plastic hinge would occur was computed to be 15,900 kN-m at concrete strain or 0.003, the value of strain explicitly defined as corresponding to the failure of the concrete.

The authors installed highly precise instruments along the length of the pile. The readings allowed the determination of bending moment with considerable accuracy.

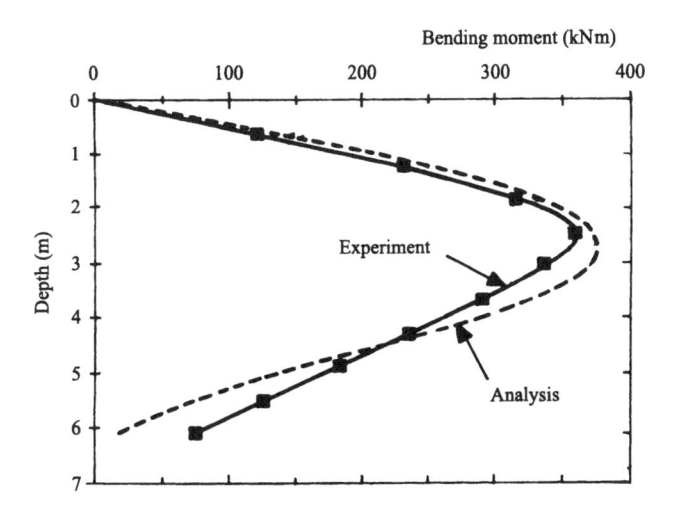

Figure 7.18. Comparison of curves of bending moment versus depth for P_t of 210 kN, static loading, Mustang Island.

278 *Piles under lateral loading*

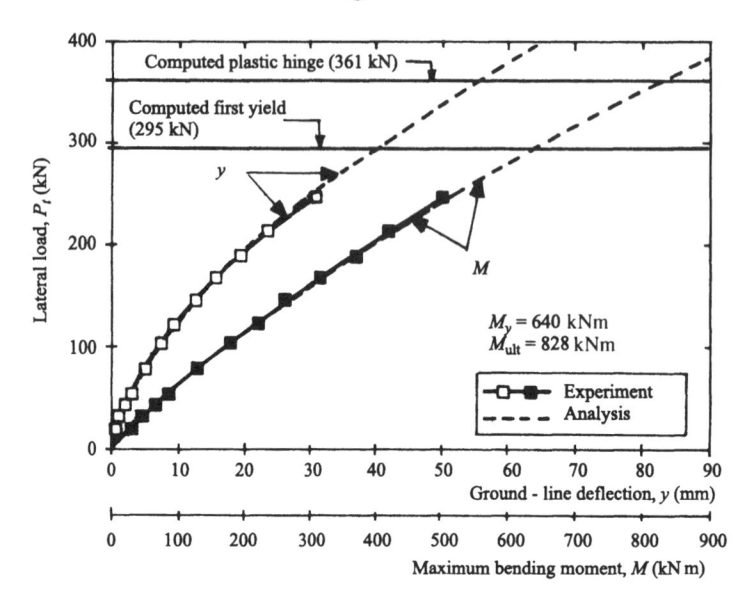

Figure 7.19. Comparison of experimental and computed values of maximum bending moment and deflection, cyclic loading, Mustang Island.

The properties of soil reported by the authors, and the interpretations used for the following analyses, are shown in Table 7.8. The fact that granular soil has been shown to increase in stiffness with an increase in strain could well influence the values that are shown

The lateral load was applied at 0.9 m above the ground line. Each load was held until the rate of movement was less than 0.05 mm in 30 minutes. The load was reduced to zero in stages and held at zero for one hour.

The p-y curves for static loading for the analyses were obtained by using the criteria for sand. The comparisons of pile-head deflection and maximum bending moment are shown in Figure 7.20. The curves for deflection show that the computation is about 20% unconservative for the larger loads and in good agreement for the smaller loads. The reduction of the shear strength with the increase in strain was not implemented in the analyses and could well account for the slight unconservatism. The maximum bending moment from the experiment is about 12% higher than the computed value at the same lateral load. The computer yielded a lateral load of 4520 kN to cause a plastic hinge.

7.4.3 *Arkansas River*

Mansur & Hunter (1970), and Alizadeh & Davisson (1970) reported the results of lateral-load tests for a number of piles in connection with a navigation project. Pile 2 with a penetration of 15 m was selected for analysis. The pile was formed of a steel pipe with a diameter of 0.406 m and a wall thickness of 8.153 mm. As shown in Figure 7.21, four steel angles were added to the pile at equal spacings to carry instruments, giving the pile an effective diameter of 0.48 m, a moment of inertia of 3.494×10^{-4} m^4, and a bending stiffness of 69,900 kN-m^2. Estimating the value of yield strength of the steel at 248,000 kPa, the value of the moment at first yield of the steel (M_y) was computed to be 361 kN-m.

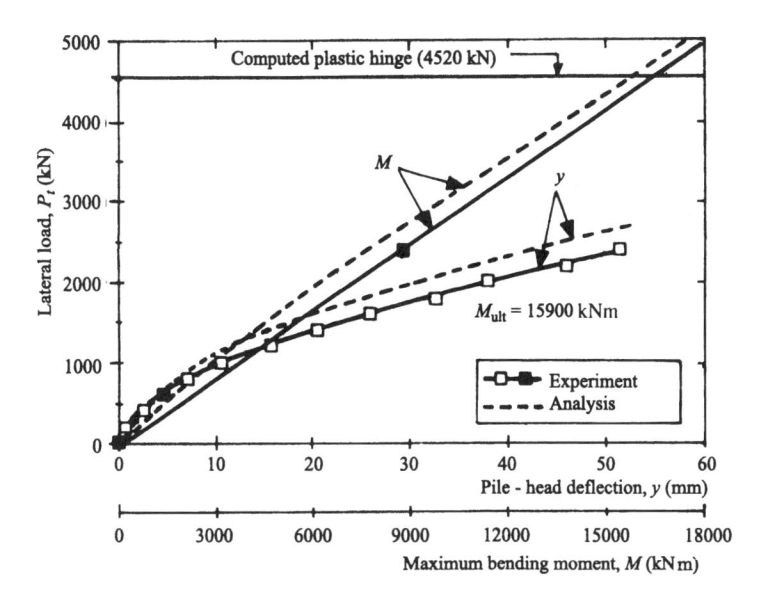

Figure 7.20. Comparison of experimental and computed values of maximum bending moment and pile-head deflection, static loading, Garston.

Table 7.8. Reported properties of soil at Garston.

Depth m	Description	N_{SPT}	Unit weight kN/m^3	Friction angle degrees
0-0.36	Fill	18	–	–
0.36-3.5	Dense sandy gravel	≈65	21.5	43
3.5-6.5	Coarse sand and gravel	30	9.7	37
6.5-9.5	Weakly cemented sandstone	≈61	11.7	43
9.5-	Highly weathered sandstone	≈140		

Several borings were made at the site and there was a considerable variation in the properties across the site. The soil in the top 5.5 m was a poorly graded sand with some gravel and with little or no fines. The underlying soils were fine sands with some organic silt. The water table was at a depth of 0.3 m. The total unit weight above the water table was 20.0 kN/m^3, and below the water table was 10.2 kN/m^3. A study of the soil borings indi-

280 *Piles under lateral loading*

cated that the water table was at a depth of 1.5 m. Data from the site showed that the site had been preconsolidated due to the presence of 6 m of overburden that was removed prior to testing.

Data and analysis of the sand at the site are given in Table 7.9. The first three columns in the table show the depth below ground surface, the computed value of vertical overburden pressure, and the blow counts from the Standard Penetration Test. The contributions of Robertson, shown in Figure 7.22, were consulted and the fourth column was obtained, showing values of q_c in MPa from the static cone test based on correlations with the values of N_{SPT}. Values of q_c/σ'_{v0} were then computed and used to obtain values of ϕ from a chart (Fig. 7.23) proposed by Durgunoglu & Mitchell (1975). The contributions of Van Impe (1986) (see Fig. 7.24) and Jamiolkowski et al. (1993) were used to obtain values shown in the last two columns, estimates of the magnitude of the soil modulus E_s in MPa and a value of n_E, a multiplier of E_s, based on the degree of overconsolidation. These last two columns served to give an insight into the selection of the initial slopes of the *p-y* curves. The *p-y* curves for static loading for the analyses were obtained by using the criteria for sand.

The loading was applied at the groundline and the loading was static. The comparison of the results from the experiment and from the computations are shown in Figure 7.25. Both cases of shear angle gave results that are somewhat conservative in the higher ranges

Table 7.9 Penetration resistance and analysis of soil at test site, Arkansas River

Depth M	σ_{v0} Mpa	N_{SPT}	q_c Mpa	q_c/σ'_{v0}	ϕ deg	E_s MPa	nE
0	0	12	5.0	–	–	15	4
0.61	0.012	12	5.0	417	45	15	4
2.4	0.039	14	5.5	183	42	15	4
4.0	0.056	20	10.0	179	42	22.5	3
4.6	0.062	17	8.0	129	41	19.5	3
5.0	0.071	25	13.0	183	42	27	3
7.0	0.086	28	14.0	163	42	28.5	3
8.5	0.102	18	12.0	118	40	19.5	3
10.0	0.117	27	15.0	128	41	30	2.5
11.6	0.133	29	15.0	113	40	30	2.5
20.0	0.219	29*	15.0	68	36	30	2.5

*(estimated)

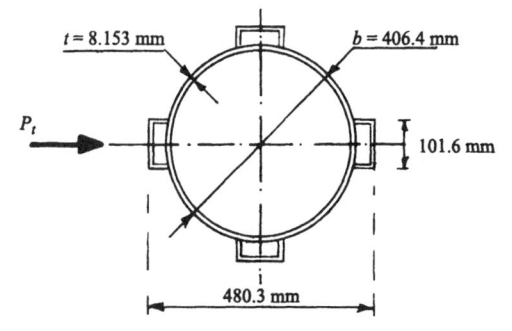

Figure 7.21. Cross section of pile at Arkansas River.

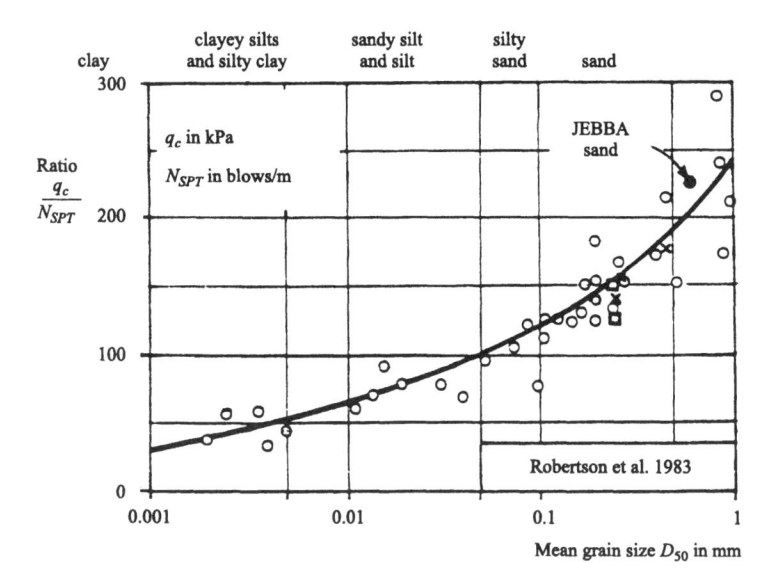

Figure 7.22. Ratio of q_c/N_{SPT} as a function of D_{50} (after Robertson et al. 1983).

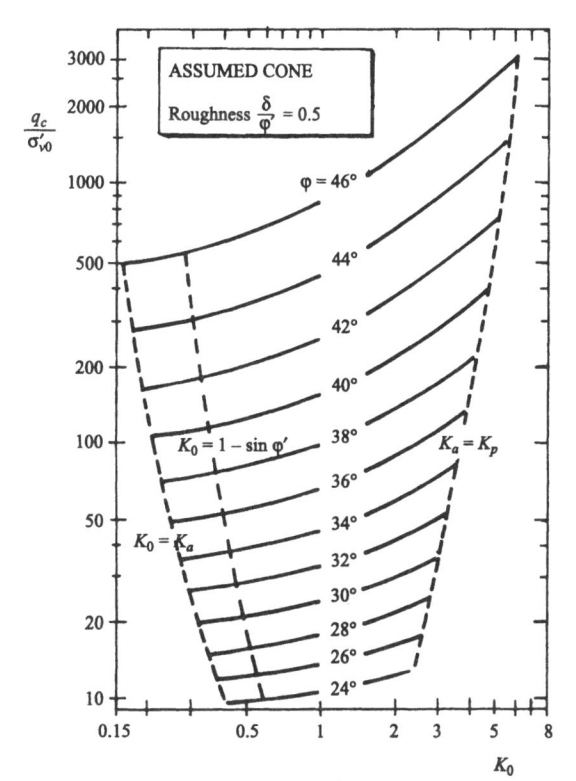

Figure 7.23. Proposed values of friction angle as a function of results from cone tests considering overburden pressure and coefficient of lateral earth pressure (after Durgunoglu & Mitchell 1975).

282 *Piles under lateral loading*

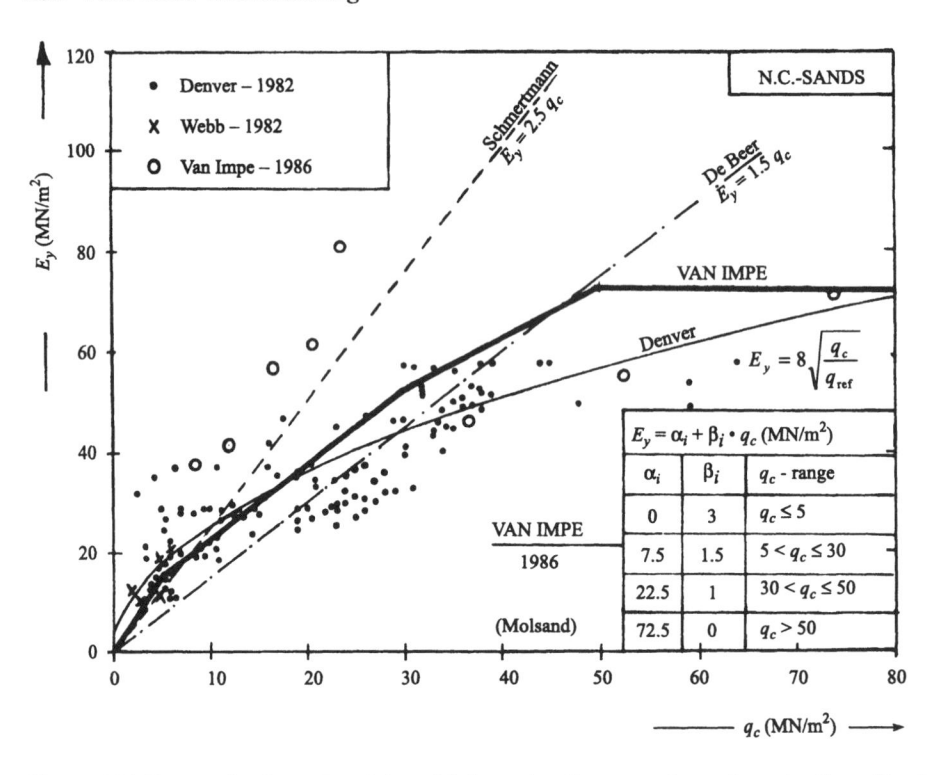

Figure 7.24. Proposed values of modulus of deformation from experimental results (from Van Impe 1985).

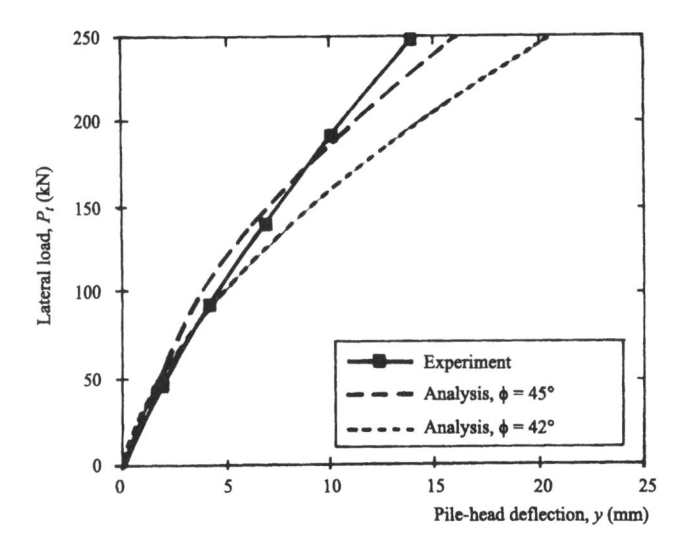

Figure 7.25. Comparison of experimental and computed values of ground line deflection, Arkansas River.

of loading but the computations with the higher value of ϕ gave, in general, very good agreement. The lateral load to cause the first yield of the steel at the extreme fiber was computed to be 324 kN.

7.5 PILES INSTALLED INTO LAYERED SOILS

7.5.1 Talisheek

Gooding et al. (1984) describe experiments for the Louisiana Power and Light Company where a steel-pipe pile was tested at Bogalusa, Louisiana, under conditions to simulate the foundation for a transmission tower. A sketch of the loading arrangement is shown in Figure 7.26. As may be seen, the pile at the groundline was subjected to lateral load, bending moment, and axial load.

The outside diameter of the pipe was 0.9144 m, its wall thickness was 0.009525 m, and its penetration was 4.27 m. The moment of inertia, I, was computed as 0.002772 m^4, its bending stiffness, E_pI_p, was 554,400 kN-m^2, its bending moment at first yield of the steel, M_y, was 1516 kN-m, and its ultimate bending moment, M_{ult}, was 1950 kN-m.

The upper layer of soil was classified as a stiff sandy clay and consisted principally of clay-sized particles but included some granular particles. It was classified as a CL with the Unified System (see Table 7.10). The shear strength was found from the unconfined compression tests of specimens that were either 76 mm or 127 mm in diameter. In the absence of data on ε_{50} a value of 0.007 was selected by reference to Table 3.5.

The second layer was a dense fine sand where the uncorrected values of N from the Standard Penetration Test averaged 71. The shear angle was based on values from the Standard Penetration Test and the overburden pressure, leading to a value of $\phi = 50°$. Employing the procedure illustrated with the tests at Arkansas River to confirm the value of ϕ, the following analysis can be made. Starting with $N_{SPT} = 71$, the following value can be computed for the penetration resistance at 60% of the driving energy: $N_{SPT(60)} \cong 71/0.6 \cong 120$. Using the correlations for dense sand presented earlier, the value of 120 corresponds to a cone value (CPT) of $q_c \approx 48$ MPa (for normally consolidated, clean, very dense sand). Then, the value of $q_c/\sigma_{vo} = 48,000/60=800$ leads to a value of $\phi \approx 50°$ from correlations (see Figure 7.23).

The p-y curves for static loading for the upper layer of clay were obtained by using the criteria for stiff clay with no free water and for the layer of sand by using the criteria for sand. The procedure for layered soil was implemented. The penetration of the pile was small and the anticipation was correctly made that the bottom of the pile would undergo a

Table 7.10 Reported properties of soil at Talisheek.

Depth	Water content	Total unit weight	Undrained shear strength	ε_{50}*	Friction angle
m	%	kN/m^3	kN/m^2		degrees
0	17.3	18.7	59.2	0.007	–
1.83	17.3	18.7	59.2	0.007	–
1.83	21.6	20.1	–	–	50
6.0	21.6	20.1	–	–	50

*From Table 3.5

284 Piles under lateral loading

sensible deflection. Therefore, the decision was made to introduce a set of data at the base of the pile, giving resistance in force as a function of the deflection of the tip. The ultimate force was estimated by multiplying the vertical stress at the tip by the tangent of the friction angle and by the area of the base of the pile. That force was computed to be 218 kN. The force-displacement relationship was estimated by use of the data from load-transfer curves for skin friction for axially loaded piles. The relationships were computed and are shown in Table 7.11.

Table 7.11. Computed force-displacement relationship for lateral deflection of the tip of the pile, Talisheek.

Force kN	Displacement m
0.0	0.00
95	0.001
133	0.002
157	0.003
173	0.004
184	0.005
194	0.006
203	0.007
215	0.009
217	0.010
218	0.0127
218	1.000

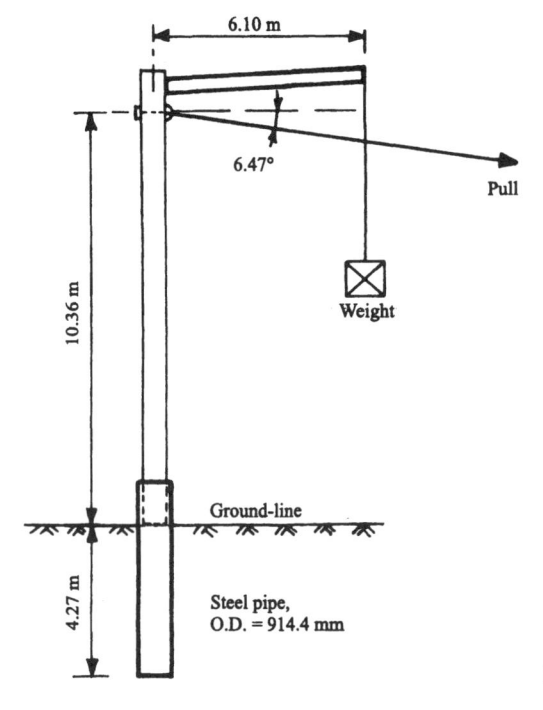

Figure 7.26. Test arrangement at Talisheek.

As noted in Figure 7.26, the loading arrangement could apply simultaneously lateral load, bending moment, and axial load, all in the positive direction. Table 7.12 shows the set of loadings and the observed deflections at the groundline.

The results of the test at Talisheek are interesting for a number of reasons: a combination of loads were employed, the pile was short, the pile failed under load, a resisting force was assumed at the base of the pile in the computations, and the computations indicate excellent agreement with the experiment. A comparison of the results for pile-head deflection from experiment and from computation is shown in Figure 7.27. The deflection is shown as ordinate and the abscissa shows the number of the load. A significant difference in the curves is shown at Load 6 probably because, in the experiment, all loads were reduced to zero after Load 5. The computations for deflection are conservative at the larger loads. The selection of the friction angle from data from the Standard Penetration Test leads to significant approximations. Had a smaller value of friction angle been estimated, the agreement at the larger loads would have been not as good as shown in Figure 7.27.

A further point of interest is that the computations showed that significant values of shear developed at the tip of the pile because of the computed values of deflection. The shear was equal or close to the maximum value of 218 kN for Loads 10 through 16. The shear at the tip of the pile influenced behavior for even smaller loads. The obvious conclusion is that the penetration of the pile was insufficient for the loadings that were applied because a design based on the development of shear at the base of a pile would generally be risky.

Of particular interest is the comparison of the loads at which failure occurred. The pile in the experiment failed by the formation of a plastic hinge at Load 18. The computer showed a failure in bending at Load 16 where the computed value of the maximum bending moment

Table 7.12. Set of loads applied at tests at Talisheek.

Load number	Lateral load P_t kN	Bending moment M_t kN-m	Axial load P_x kN	Deflection y_t mm
1	44.1	456.8	4.5	9.1
2	66.1	685.0	8.9	16.8
3	79.4	822.3	8.9	22.9
4	88.2	913.6	13.3	27.4
5	97.0	1004.9	13.3	30.5
6	0	328.1	107.2	15.2
7	0	641.3	211.9	19.8
8	0	939.5	309.8	25.9
9	13.2	1076.4	314.3	25.9
10	26.4	1213.3	314.3	30.5
11	39.6	1350.2	318.7	35.1
12	52.8	1487.1	318.7	42.7
13	66.1	1624.1	323.1	48.8
14	79.3	1761.0	323.1	56.4
15	88.1	1852.2	323.1	–
16	92.5	1897.9	327.6	–
17	96.9	1943.5	327.6	–
18	103.5	2012.2	327.6	73.2*

*Failure by plastic buckling

286 *Piles under lateral loading*

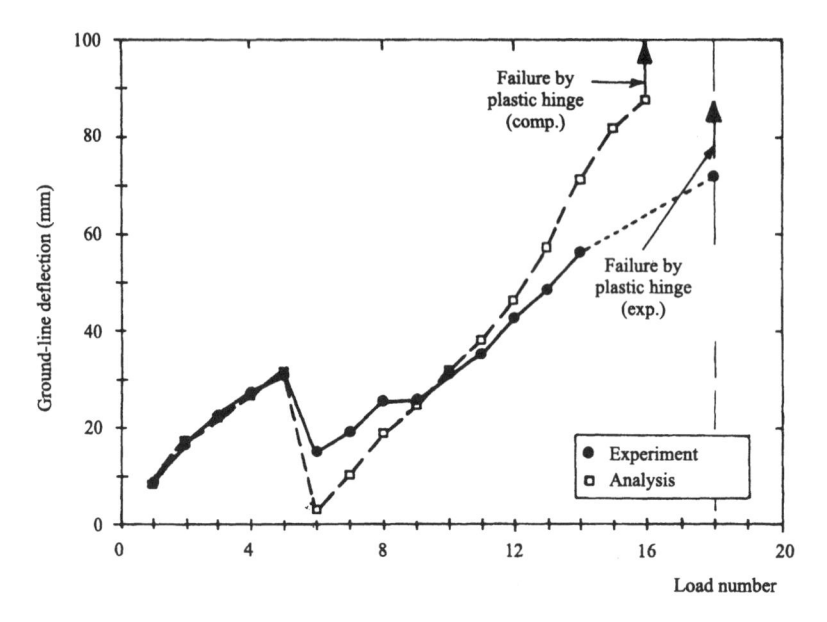

Figure 7.27. Comparison of experimental and computed values of groundline deflection, static loading, Talisheek.

was equal to M_{ult}, 1950 kN-m. For the short pile in this experiment, the agreement between the failure loads from computation and the experiment is remarkable. However, the method of selecting the shear versus deflection at the base of the pile is not validated by experiment and the particular values that were selected may have been fortuitous.

It is of interest to note that a similar experiment was performed at a second site where the pile also failed in the experiment by plastic buckling. However, in that case the agreement was good only if a large resisting force was in existence at the base of the pile. Such a large resisting force was possibly present, due to some obstruction at the tip of the pile, because slope-indicator readings, not shown here, showed that the tip of that pile did not deflect, contrary to the observations presented at Talisheek.

7.5.2 *Alcácer do Sol*

Portugal & Sêco e Pinto (1993) describe the testing of a bored pile at the site of a bridge at Alcácer do Sol. Three piles were tested and the results for Pile 2 are shown here.

The pile was 40 m in length and had a diameter of 1.2 m. It was reinforced with 35 bars with a diameter of 25 mm. The strengths of the concrete and steel were reported to be 33.5 MPa and 400 MPa, respectively. The cover of the rebars was taken as 50 mm. The bending stiffness was computed, and a value of 3.29×10^6 kN-m^2 was selected for use in the analyses. The ultimate bending moment was computed to be 3370 kN-m. The pile was instrumented for the measurement of bending moment along its length.

From the ground surface downward, the soil is described as silty mud, sand, muddy complex, and sandy complex. The properties of the soil were found from SPT, CPT, and vane tests, and the values that were selected for use in the analyses are shown in Table 7.13. The *p-y* curves for the upper layer of clay were obtained by using the criteria for stiff clay with no free water. The subsequent layers used the criteria for sand, for stiff clay with no free water, and for sand. The criteria for layered soils were implemented.

The lateral load was applied at 0.2 m above the ground line. Bending moment was measured along the length of the pile but information is unavailable on the techniques that were used. Ground-line deflection and maximum bending moment were reported for three values of lateral load: 100, 200, and 300 kN.

The position of the water table was not reported but it is assumed that the water table was close to the ground surface. The data were analyzed by use of the criteria for clay with no free water and the criteria for sand. The comparisons of groundline deflection and maximum bending moment for Pile 2 are shown in Figure 7.28. The analytical method over-predicts deflection but the maximum bending moment is computed with appropriate

Table 7.13 Reported properties of soil at Alcácer do Sol.

Depth	Water[+] content	Total unit weight	Undrained shear strength	ε_{50}*	Friction angle
m	%	kN/m³	kN/m²		deg
0	62.5	16.0	20.0	0.020	–
3.50	62.5	16.0	20.0	0.020	–
3.50	28.6	19.0	–	–	30
8.50	28.6	19.0	–	–	30
8.50	62.5	16.0	32.0	0.020	–
23.0	62.5	16.0	32.0	0.020	–
23.0	28.6	19.0	–	–	35
40.0	28.6	19.0	–	–	35

[+]Computed *From Table 3.3

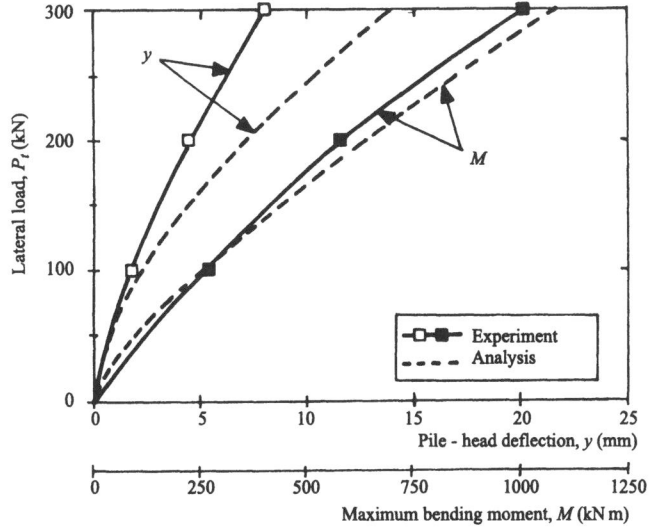

Figure 7.28. Comparison of experimental and computed values of maximum bending moment and pile-head deflection, static loading Alcácer do Sol.

accuracy. The maximum bending moment of 1007 kN-m that was measured was much less than the ultimate moment.

7.5.3 Florida

Davis (1977) described the testing of a steel-pipe pile that had a diameter of 1.42 m and a penetration of 7.92 m. The tube was filled with concrete to a depth of 1.22 m, and a utility pole was embedded so that the lateral loads were applied at 15.54 m above the ground line. Meyer (1979) analyzed the results of the test and reported the bending stiffness to be 5,079,000 kN-m^2 in the top 1.22 m, and 2,525,000 kN-m^2 below. The ultimate bending moment was reported to be 6280 kN-m in the top 1.22 m, and 4410 kN-m in the lower portion.

The soil profile consisted of 3.96 m of sand above saturated clay. The sand had a total unit weight of 19.2 kN/m^3, and a friction angle of 38 degrees. The water table was at a depth of 0.61 m. The undrained shear strength of the clay was 120 kPa, and its submerged unit weight was 9.4 kN/m^3. A value of ε_{50} of 0.005 was selected for the analyses, following values shown in Table 3.5.

The p-y curves for static loading for the upper layer were obtained by using the criteria for sand. Some discussion is desirable about selecting the criteria for the layer of clay. Because the clay is below sand, no loss of resistance would occur because of gapping. Then, for static loading, the options are stiff clay with no free water (no erosion will occur) or stiff clay with free water. The latter criteria were selected. However, in similar designs, the engineer might try both sets of recommendations to gain some insight into possible differences in response. The computed curves for deflection with depth are not presented but the pile deflections in the zone of the clay would undoubtedly be quite small. The criteria for layered soil were implemented.

A comparison of the experimental and computed values of pile-head deflection is shown in Figure 7.29. The curves agree well for the early loads but start to deviate strongly above a lateral load of 160 kN (and a moment at the ground line of 286 kN-m, considering the point of the application of the load). The output from the computer code was examined and it was noted that the bottom of the relatively short pile was deflecting. The data presented in Table 7.14, on lateral resistance of the soil at the base of the pile as a function of lateral deflection, were used as input for another series of computations. As may be seen in the figure, the agreement between the experimental and computed values improved considerably with the use of the base shear. The selection of a stiffer curve for the base shear, which could be justified, could have brought the experimental and computed values into very close agreement.

Overall, the agreement between the experimental and computed curves is excellent. The need to make use of base shear for short piles, as noted in a previous example, is of considerable interest.

7.5.4 Apapa

Coleman (1968), and Coleman & Hancock (1972) describe the testing of Raymond step-tapered piles near Apapa, Nigeria. The results were analyzed by Meyer (1979). Two piles, identical in geometry, were driven and capped with concrete blocks. A hydraulic ram was placed between the caps and the piles were loaded by being pushed apart. The reported properties of the piles are presented in Table 7.15.

Case studies 289

The soil at the site consisted of 1.52 m of dense sand underlain by a thick stratum of soft organic clay. The friction angle of the sand was obtained in the laboratory by triaxial tests of reconstituted specimens and was found to be 41 degrees. The strength of the soft organic clay was obtained from in situ-vane tests and was found to be 23.9 kPa; the value of ε_{50} for the clay was assumed to be 0.02. The water table was at a depth of 0.91 m. The unit weight of the sand above the water table was 18.9 kN/m³ and the submerged unit weight of the clay was 4.7 kN/m³.

Table 7.14. Computed force-displacement relationship for lateral deflection of the tip of the pile, Florida.

Force kN	Displacement m
0.0	0.00
76	0.002
118	0.004
143	0.006
164	0.008
178	0.010
183	0.012
188	0.014
190	0.015
190	1.00

Table 7.15. Mechanical properties of piles at Apapa.

Section m	Diameter, m	EI kN-m²
0-2.44	0.442	22,400
2.44-6.10	0.417	20,100
6.10-15.3	0.391	18,700

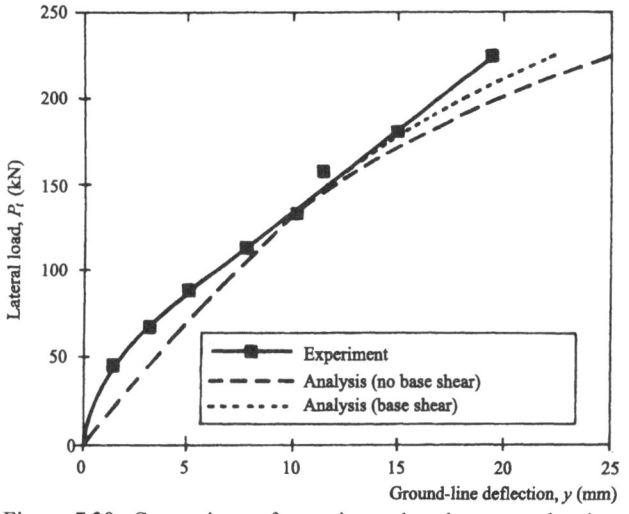

Figure 7.29. Comparison of experimental and computed values of groundline deflection, static loading, Florida.

290 *Piles under lateral loading*

The *p-y* curves for static loading for the upper layer were obtained from the criteria for sand and for the lower layer from the criteria for soft clay. The procedure for layered soil was implemented.

The lateral loads were applied at 0.61 m above the ground line, and the deflection was measured at that point. A comparison of the experimental and computed values of pile-head deflection is shown in Figure 7.30. The computed curve agrees well with the experimental results. The difference between the experimental results for the two piles, which presumably were identical, probably reflects differences in the structural characteristics of the piles, and possibly differences in the effective properties of the soil due to effects of installation.

7.6 PILES INSTALLED IN c-ϕ SOIL

7.6.1 *Kuwait*

A field test for behavior of laterally-loaded, bored piles in cemented sands (c - ϕ soil) was conducted in Kuwait (Ismael 1990). Twelve bored piles that were 0.3 m in diameter were tested. Piles 1 to 4 were 3 m long, while piles 5 to 12 were 5 m long. The study was directed at the behavior of both single piles and piles in a group. Curves showing measured load-versus-deflection at the pile head for 3-m-long single piles and 5-m-long single piles are presented in the paper. Only results from loading the 5-m piles are studied by using the soil criteria for c-ϕ soil.

The 5-m piles were reinforced with a 0.25 m-diameter cage made of six 22 mm bars, and a 36 mm-diameter reinforcing bar was positioned at the center of each pile. The piles were instrumented with electrical-resistance strain gauges. After lateral-load tests were completed, the soil was excavated to a depth of 2 m to expose strain gauges. The pile was

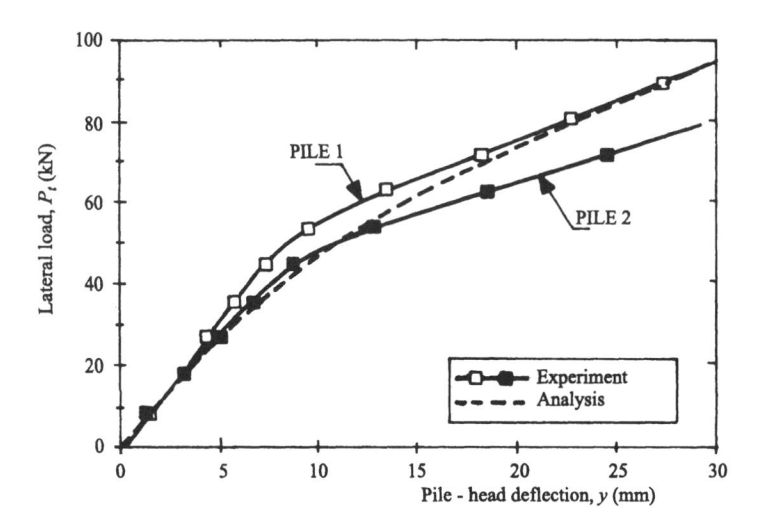

Figure 7.30. Comparison of experimental and computed values of pile-head deflection, static loading, Apapa.

reloaded and the curvature was found from readings of strain. The flexural rigidity was calculated from the initial slope of the moment-curvature curves as 20.2 MN-m^2. The experimental value of E_pI_p is significantly larger than values computed from mechanics for reasons that cannot be identified. Because of the inability to apply mechanics to the analysis of the cross section of the pile, the bending moment to cause a plastic hinge to develop could not be computed. The experimental value was judged to be superior and is used in the analyses shown below.

The subsurface consisted of two layers as shown in Table 7.16. The first layer described as medium dense cemented silty sand, was about 3.5 m in thickness. The values of c and ϕ for this layer were found by drained triaxial compression tests and were 20 kPa and 35°, respectively. The unit weight averaged 17.9 kN/m^3. Using Equation 3.67 and Figures 3.30 and 3.31, k_c was found to be 90,000 kN/m^3 (at the beginning of the curve) and k_ϕ was found to be 80,000 kN/m^3 (at the end of the curve) yielding a value of k_{py} of 170,000 kN/m^3.

The first layer was underlain by medium dense to very dense silty sand with cemented lumps. The values of c and ϕ were zero and 43°, respectively. The unit weight averaged 19.1 kN/m^3. The value of k_{py} for the soil was found from Figure 3.31 as 80,000 kN/m^3. The p-y curves for static loading for the upper layer were obtained from the criteria for c-ϕ soils and for the lower layer from the criteria for sand. The procedure for layered soil was implemented.

A computer code, employing c-ϕ criteria, was used to predict curves of load versus deflection at the pile head for the 5-m pile. Good agreement was found between measured and predicted behavior, as shown in Figure 7.31. Because of the inability to apply mechanics to the analysis of the cross section of the pile, the bending moment to cause a plastic hinge to develop could not be computed.

7.6.2 Los Angeles

A field test for behavior of laterally loaded, bored piles in mostly c-ϕ soil was conducted in Los Angeles in 1986 by Caltrans (California Department of Transportation). The pile was 1.22 m in diameter and had a penetration of 15.85 m. The load was applied at 0.61 m above the groundline. The pile was instrumented with electrical-resistance strain gauges and with Carlson cells for the measurement of bending moment; however, information on the calibration of the instruments was not given.

Table 7.16. Properties of soil at Kuwait.

Depth m	Desc.	Average properties of soil at site								
		N_{SPT}	w %	Unit wt. kN/m^3	L.L.	P.I.	S.L.	Sand %	Silt %	Clay %
0-3	Med. Dense silty sand	21	3.0	17.9	20.4	3.1	14.6	80.0	12.6	7.4
3-5.5	Med dense to very dense silty sand with cemented lumps	75	3.7	19.1	None	N.P	–	82.9	17.1	0

292 *Piles under lateral loading*

The compressive strength of the concrete was 24,800 kPa and the tensile strength of the reinforcing steel was 413,700 kPa. The area of the steel reinforcement was 2% of the area of the pile. A total of 24 rebars were used, each with an area of 0.001065 m². The distance between the outside of the rebar cage and the wall of the pile was 76.2 mm.

The kinds of soil and the properties are given in Table 7.17. The water table was well below the ground surface. The *p-y* curves for static loading for the first layer were obtained by using the criteria for stiff clay with no free water; for the second, third, and fourth layers by using the criteria for c-ϕ soil; and for the fifth layer by using the criteria for stiff clay with no free water. The procedure for layered soil was implemented.

The lateral loads were applied at 0.61 m above the groundline and the loads were applied in increments. The analysis was done by modifying $E_p I_p$ according to the magnitude of the bending moment as presented in Chapter 4. Computations with the analytical technique shown there, as noted earlier, reveal a sudden and dramatic decrease in $E_p I_p$ when the first tension crack appears in the concrete. For the present analyses, the precipitous decrease occurs in the range of the applied loads, leading to the computations of very large deflections. The computations for comparison with experimental results (Figure 7.32) were made with values of $E_p I_p$ that were believed to be appropriate for progressive cracking of the concrete in tension. The ultimate bending moment was computed to be 4400 m-kN and the maximum applied moment was computed to be close to the ultimate.

Table 7.17. Kind of soil and properties of soil at Los Angeles.

Depth m	Soil	Cohesion kPa	Friction angle deg	Unit weight kN/m³
0–1.5	Plastic clay	179		19.2
1.5–7.4	Sand and some clay	4.8	30	19.8
7.4–10.5	Sandy clay	19.2	35	19.8
10.5–13.4	Sandy silt	19.2	21	18.9
13.4–20	Plastic clay	110		18.4

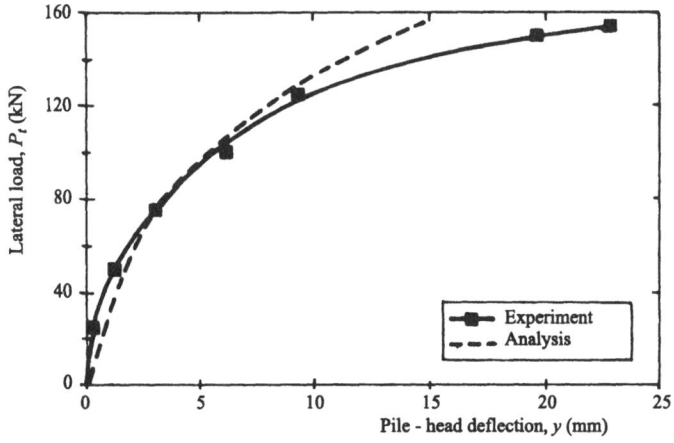

Figure 7.31. Comparison of experimental and computed values of pile-head deflection, static loading, 5-m-long pile, Kuwait.

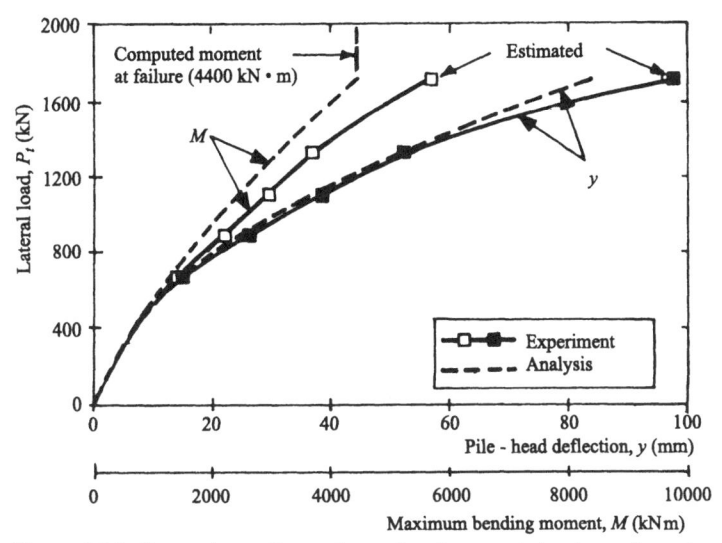

Figure 7.32. Comparison of experimental and computed values of maximum bending moment and pile-head deflection, static loading, Los Angeles.

Comparison between the experimental and computed values of maximum bending moment and pile-head deflection are shown in Figure 7.32. Comparative values for pile-head deflection, show excellent agreement, taking into account the modifications that were made in values of $E_p I_p$. The experimental values of maximum bending moment are higher that the computed values; however, that no field calibration of the instruments used to measure bending moment may be significant.

The magnitude of the lateral load sustained by the pile is significant. The load would have been even larger if the load had been applied at the ground line rather that 0.61 m above the ground. Also, had axial load been applied the pile would have performed more favorably because the axial load would have caused the closing of tensile cracks.

7.7 PILES INSTALLED IN WEAK ROCK

7.7.1 Islamorada

The test was performed under sponsorship of the Florida Department of Transportation and was carried out in the Florida Keys (Reese & Nyman, 1978). The rock was a brittle, vuggy, coral limestone, allowing a steel rod to be driven into the rock to considerable depths, apparently because the limestone would fracture and the debris would fall into the vugs. Cavities in the order of a third of a meter in diameter existed in the limestone in some regions.

Two cores were obtained for compressive tests. The small discontinuities at the outside surface of the specimens were covered with a thin layer of gypsum cement to minimize stress concentrations. The ends of the specimens were cut with a rock saw and lapped flat and parallel. The compressive strengths were found to be 3.34 and 2.60 MPa. The axial deformation was measured during the testing and the average value of the initial reaction modulus of the rock was found to be 7240 MPa.

The rock at the site was also investigated by in-situ-grout-plug tests under the direction of Dr. John Schmertmann (1977). A 140-mm diameter hole was drilled into the limestone, a high-strength-steel bar was placed to the bottom of the hole, and a grout plug was cast over the lower end of the bar. The bar was pulled to failure, and the hardened grout was examined to ensure that failure occurred at the interface of the plug and the limestone. Tests were performed at three locations and the results are shown in Table 7.18. A compressive strength of 3.45 MPa was selected as representative of the rock in the zone near the rock surface where the deflection of the pile was most significant.

The bored pile was 1.22 m in diameter and penetrated 13.3 m into the limestone. A layer of sand over the rock was retained by a steel casing, and the lateral load was applied at 3.51 m above the surface of the rock. A maximum lateral load of 667 kN was applied. The curve of load versus deflection was nonlinear but there was no indication of failure of the rock.

In the absence of details on the strengths of the concrete and steel and on the amount and placement of the rebars, the bending stiffness of the gross section was used. The following values were used in the equations for p-y curves presented in Chapter 3: q_{ur} = 3.45 MPa; E_{ri} = 7,240 MPa; k_{rm} = 0.0005; b = 1.22 m; L = 15.2 m; and $E_p I_p$ = 3.73 × 10^6 kN-m^2.

The comparison of pile-head deflection for results from experiment and from analysis (Reese, 1997) is shown in Figure 7.33. The figure shows excellent agreement between values from experiment and from analysis for lateral loads up to about 350 kN, using unmodified values of the bending stiffness. A sharp change in the load-deflection curve occurs at 350 kN. A possible reason is the decrease in bending stiffness $E_p I_p$ at the larger loads.

The use of values of $E_p I_p$ from mechanics would be desirable; however, analysis shows a sudden decrease in $E_p I_p$ when strain in the concrete reaches the point where the concrete cracks in tension. However, if a pile (or beam) is considered, cracking does not occur at every point along the member but initially at wide spacing. Thus, the net effect is that $E_p I_p$ reduces gradually for the section, as a function of bending moment, and not suddenly as from analysis.

As shown in Figure 7.33, values of $E_p I_p$ were reduced gradually to find deflections that would agree fairly well with values from experiment. The following combination of values of load and bending stiffness were used in the analyses; in units of kN and kN-m^2, respectively: 400, 1.24 × 10^6; 467, 9.33 × 10^5; 534, 7.46 × 10^5; 601, 6.23 × 10^5; and 667, 5.36 × 10^5. The assumption that the decrease in slope of the curve of y_t versus P_t at Islamorada can be explained by reduction in values of $E_p I_p$ is reasonable. Also, a gradual re-

Table 7.18. Results of grout-plug tests by Schmertmann.

Depth range m	Ultimate resistance MPa
0.76-1.52	2.27
	1.31
	1.15
2.44-3.05	1.74
	2.08
	2.54
5.49-6.10	1.31
	1.02

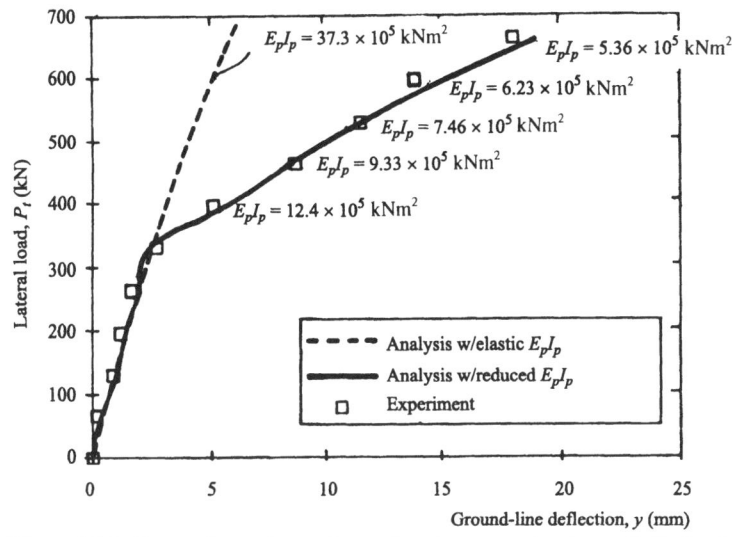

Figure 7.33. Comparison of experimental and computed values of pile-head deflection, static loading, Islamorada (after Reese 1997).

duction in values of $E_p I_p$, as shown, yields values of deflection by computation that agree closely with measured values. However, the Islamorada example gives little guidance to the designer of piles in rock except for the early loads. The example from San Francisco that follows is more instructive.

7.7.2 San Francisco

The California Department of Transportation performed lateral-load tests of two bored piles near San Francisco and the results of the tests, while unpublished, have been provided through the courtesy of Caltrans (Speer, 1992).

As is often typical in the investigation of the engineering properties of rock, the secondary structure led to difficulty in sampling. The sandstone was found to be medium to fine grained (0.10 to 0.5 mm), well sorted and thinly bedded (25 to 75 mm thick). In most of the corings, the sandstone was described as very intensely to moderately fractured with bedding joints, joints, and fracture zones. Cores of insufficient length were available for compression tests. Pressuremeter tests were performed at the site and the results, as might be expected, were scattered. The plotted results of the values obtained for the moduli of the rock are shown in Figure 7.34. The averages that were used for analysis are shown as a function of depth by the dashed lines. The following values were estimated for the compressive strength of the rock: 0 to 3.9 m, 372 kPa; 3.9 to 8.8 m, 1290 kPa; and below 8.8 m, 3210 kPa.

Two piles, 2.25 m in diameter, were tested simultaneously, and the results for Pile B will be analyzed. Pile B exhibited a large increase in deflection for the last load, probably signaling a failure of the pile due to a plastic hinge. High-strength steel bars were passed through tubes, transverse and perpendicular to the axes of the piles, and loads placed by hollow-core rams. The load was measured by load cells, and the piles were instrumented with slope indicators and strain bars. Deflection was measured by transducers, and slope and deflection of the tops of the piles were obtained by readings of the slope indicator.

296 Piles under lateral loading

The load was applied in increments at 1.24 m above the ground line for Pile B, and deflection was measured at 36.5 mm above the groundline.

In addition to the values of initial reaction modulus and compressive strength of the rock, shown above, the following values were used in the analyses: $k_{rm} = 0.00005$; $b = 2.25$ m; $L = 1.7$ m, and $E_p I_p = 32.2 \times 10^6$ kN-m^2 for the beginning loads. The compressive strength of the concrete 34.5 MPa, the tensile strength of the rebars was 496 MPa, there were 28 bars with a diameter of 43 mm, and the cover thickness was 0.18 m. The ultimate bending moment was computed to be 17,740 kN-m.

The importance the use of appropriate values of bending stiffness in analyses has been emphasized. Accounting for the reduction of $E_p I_p$ with increasing magnitude of bending moment is critical in the computation of the response of a pile. Three methods were used to predict the values of $E_p I_p$ as a function of bending moment: the analytical method, the approximate method (or ACI method, which can be used because on axial load was applied to the pile during the testing), and the experimental method. The three plots for $E_p I_p$ versus bending moment are shown in Figure 7.35.

The experimental method made use of the average of the observed deflections, the applied load, and iteration to find the values $E_p I_p$ and the corresponding values of the maximum bending moment that fitted the results. The analytical method and the ACI method are presented in Chapter 4.

All three curves in Figure 7.35 show a sharp decrease in $E_p I_p$ with increase in bending moment, but the analytical method yields a precipitous drop. All the values of $E_p I_p$ begin from 35.15×10^6 kN-m^2 (value from the analytical method was somewhat larger because of the presence of the steel). The values of I from the ACI method were multiplied by a constant value of E of 28.05×10^6 kPa to get values of $E_p I_p$. The analytical method is based on the assumption that all concrete is cracked when the first crack appears; thus, it is surprising that values from the ACI method and the experimental method fall below values from analysis for over half the range of loading. The logical explanation for such a result is not immediately available.

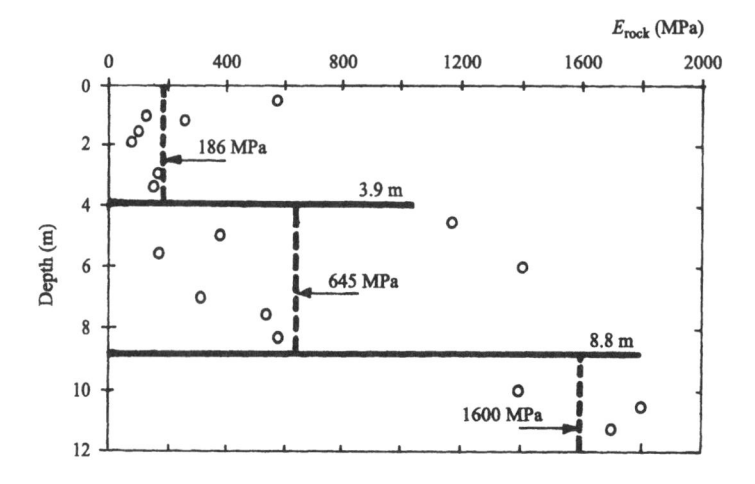

Figure 7.34. Initial modulus of rock from pressuremeter, San Francisco (after Reese 1997).

The curves of deflection as a function of lateral load, using the values of E_pI_p from Figure 7.35, are shown in Figure 7.36. The experimental values yielded precise agreement, as was ensured by the fitting technique. The values from computations using the E_pI_p from the ACI procedure fits the experimental values better than do the values from the analytical method (Reese 1997). However, as will be demonstrated later, if a load factor of 2.0 is selected, applied to the load that causes a plastic hinge, the deflections from experiment and from analysis would range from about 2 mm to 4 mm. Such differences are thought to be unimportant in regard to the deflection at service loads.

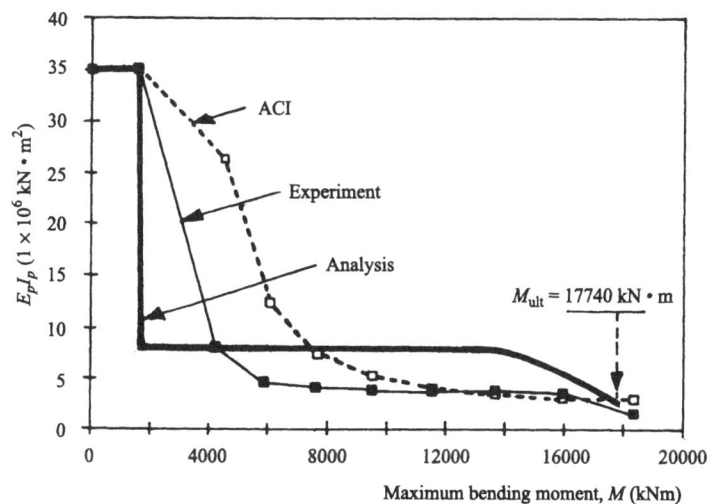

Figure 7.35. Values of bending stiffness as a function of applied moment for three methods (after Reese 1997).

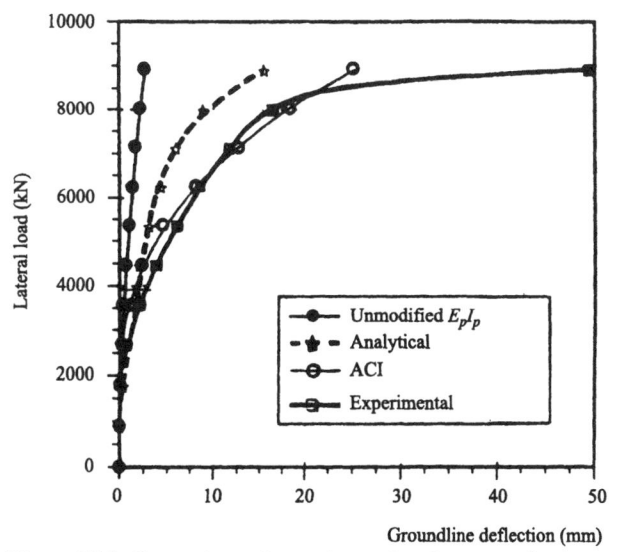

Figure 7.36. Comparison of experimental and computed values of pile-head deflection for different values of bending stiffness, static loading, San Francisco (after Reese 1997).

298 *Piles under lateral loading*

Also plotted in Figure 7.36 is a curve showing the deflections that were computed with no reduction in the value of $E_p I_p$. While deflections at service loads may be computed with little significant error, the use of the unmodified $E_p I_p$ is unacceptable because the computation of the load to cause a plastic hinge would be grossly in error.

The values of $E_p I_p$ from the various procedures were used to compute the maximum bending moment as a function of applied lateral load. The results are shown in Figure 7.37. As may be seen, the experimental values of maximum bending moment were predicted quite well with any of the methods. Assuming the computed value of M_{ult} is correct, as computed from the properties of the cross section, the analytical methods predict the ultimate bending moment with good accuracy.

7.8 ANALYSIS OF RESULTS OF CASE STUDIES

The results of case studies are analyzed here in a manner to provide guidance to the engineer who wishes to design a pile to sustain lateral loading. The steps the engineer takes in making a design are presented to guide an evaluation of the results of the studies.

1. A pile is selected and its cross section is analyzed to obtain bending stiffness as a function of bending moment and axial load.
2. The ultimate bending moment M_{ult} is found during the computations in (1), along with the bending moment M_y that will cause the extreme fibers of a steel shape to just reach yielding.
3. The properties of the soil are evaluated, p-y curves are computed for the kind of loading to be applied, the point of application of the loads is specified, and other boundary conditions are selected for the service loads.
4. The loads are incremented in steps and computer solutions are made to find the loads that cause failure in bending (M_{ult} or M_y), or in rare cases the loads are found that cause excessive deflection.
5. The service loads are checked, employing partial-safety factors or a global factor of safety, and a revised section is selected, if necessary, for a new set of computations.

The above steps were implemented in the analysis of the results of the case studies with the assumption that only the ultimate bending moment will control; that is, deflection is assumed not to control the design. A review of the combined results presented later in this section reveals that the assumption was valid, except for Case 1 at Bagnolet where the pile plainly failed in deflection, and perhaps for the case of cyclic loading at Lake Austin, as discussed later.

If a steel pile was tested, the value of the lateral load to cause M_y to develop was found by computation, defined as P_t at failure. A load factor of 1.8 was assumed. The value of P_t at the service load was found by dividing P_t at failure by the load factor. Values of computed and experimental deflections and computed and experimental bending moments were tabulated at the service load. The global factor of safety was computed by dividing the bending moment from experiment into the computed value of M_y.

The same procedure was followed for bored piles with the value of P_t at failure found as before as the lateral load that yielded the computed value of M_{ult}. The difference is that the value of P_t at the service load was found by dividing the computed value of P_t at failure by a load factor of 2.2. The logic related to the piles of different materials is that the engineer is able to compute the ultimate bending moment of the steel pile with greater accuracy than that of the reinforced-concrete pile.

The first step in computing the factor of safety for each test was to assume that the computed service load would be applied. Next, the experimental value of bending moment was found that corresponded to the service load. Then, the factor of safety was computed by dividing the experimental bending moment at the service load into the computed value of ultimate bending moment. Other values of the load factor could have been selected and entirely different techniques could have been adopted for computing the factor of safety. However, basing the factor of safety on values of bending moment appears reasonable because design of piles under lateral loading is controlled by the allowable bending moment in most instances.

The cases that were studied are shown in Table 7.19 along with information about each of the cases. The results of the analyses are shown in Table 7.20. All of the cases had experimental values of maximum bending moment that were as large as the moment corresponding to the factored load, except for the test in Japan, where extrapolation was employed. The computed factors of safety, using the procedures outlined above, ranged from 1.82 to 3.38, with an average of 2.48. The average factor of safety is larger than the load factors that were used because of the nonlinear increase of moment with load.

7.9 COMMENTS ON CASE STUDIES

With regard to bending moment, the results from the 17 individual tests show that the method of computation results in an adequate factor of safety. Four of the tests involved cyclic loading and the computed factors of safety for those tests averaged 2.76. Four of the tests were with bored piles and the computed factors of safety for those tests averaged 2.81. The remainder of the tests were of steel piles and the computed factors of safety averaged 2.28. The soils ranged from soft clay to weak rock and only one test, Case 3 at Bagnolet, showed a value of less then 2.

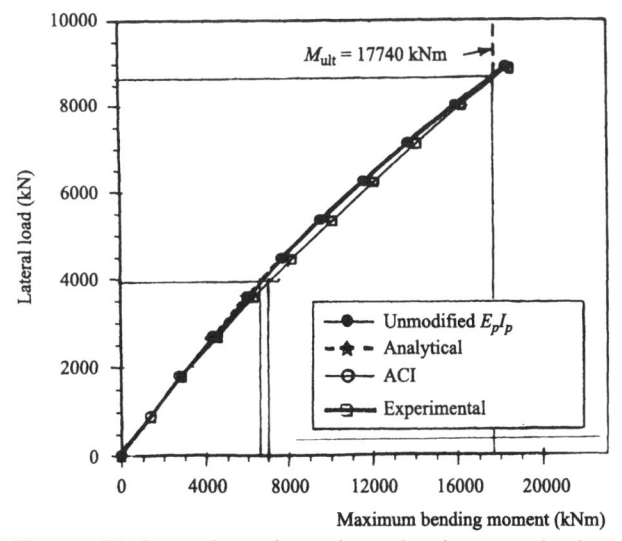

Figure 7.37. Comparison of experimental and computed values of maximum bending moment for different values of bending stiffness, static loading, San Francisco (after Reese 1997).

300 *Piles under lateral loading*

Figure 7.38 presents a comparison of the values of experimental and computed maximum moment at service loads. The agreement is excellent. About as many of the results show larger computed values (on the left of the straight line) as smaller computed values.

With regard to deflection, the comparative values are all so small to be negligible, or are relatively close, except for the test at Lake Austin under cyclic loading. There the ex-

Table 7.19. Cases of piles under lateral loading that were analyzed.

Case	Diameter, m	Kind of Soil	Pile Material	Nature of Loading	Ultimate Moment, Kn-m	Remarks
Bagnolet						
Case 1	0,43	Soft clay	Steel	Static	204	Defl. failure
Case 2						
Case 3						
Houston	0,762	Stiff clay w/o	Rein. Conc.	Static	2030	
				Cyclic	2030	
Japan	0,305	Soft clay	Steel	Static	55,9	Data extrapolated
Lake Austin	0,319	Soft clay	Steel	Static	231	
				Cyclic	231	
Sabine	0,319	Soft clay	Steel	Static	231	
				Cyclic	231	
Manor	0,641	Stiff clay h2o	Steel	Static	1757	
	0,61			Cyclic	1757	
Mustang Island	0,61	Sand	Steel	Static	640	
				Cyclic	640	
Garston	1,5	Layers	Rein. Conc.	Static	15900	
Los Angeles	1,22	Layers	Rein. Conc.	Static	4400	
San Francisco	2,25	Weak rock	Rein. Conc.	Static	17740	

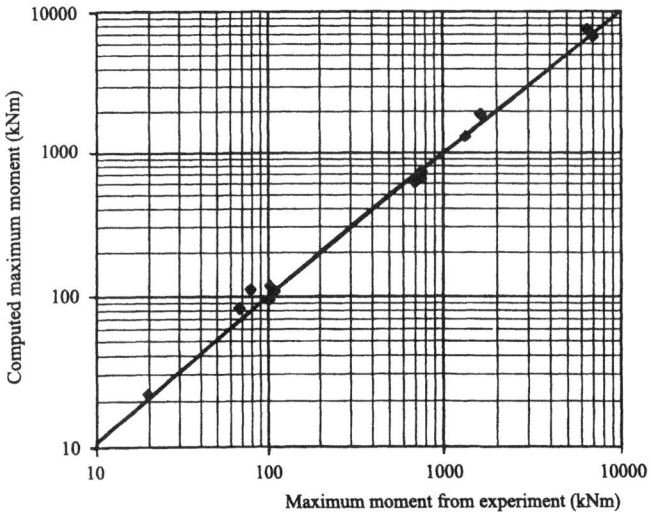

Figure 7.38. Comparison of experimental and computed values of maximum bending moment at service load for various tests.

Table 7.20. Results of analysis of data from tests of piles under lateral loading.

Case	M_{ult} kN-m	P_t fail kN	P_t serv. kN	y_t comp mm	y_t exp mm	M com kN-m	M exp kN-m	Fac.saf.
Bagnolet								
Case 2	204	138	76,7	9,6	9,6	104	95	2,34
Case 3	204	130	72,2	9,4	9,5	105	112	1,82
Houston								
static	2030	950	432	20,2	26	702	600	3,38
cyclic	2030	900	409	26	34	742	642	3,16
Japan								
static	55,9	50	28	22	28**	19,6	21.9**	2,55
LakeAustin								
static	231	145	81	35	35	110	106	2,18
cyclic	231	113	63	22	46	79	110	2,18
Sabine								
static	231	99	55	49	36	103	96	2,4
cyclic	231	72	40	27	41	68	82	2,82
Manor								
static	1757	693	385	11	9,7	760	715	2,46
cyclic	1757	543	302	13,1	10,2	710	610	2,88
Mustang Is.								
static	640	324	180	16	16	305	305	2,1
cyclic	640	295	164	15	15	320	320	2
Garston								
static	15900	4520	2055	33	40	6600	7500	2,12
Los Angeles								
static	4400	1779	809	21	22	1640	1890	2,33
San Francis.								
static	17740	8670	3940	2	3	7030	6640	2,67

**By extrapolation

perimental value was more than twice as large as the computed value and large enough to be of some concern from a practical point of view. A similar result was obtained for the test under cyclic loading at Sabine, where the experimental value was 1.52 times as large as the computed value. On the basis of the results from Lake Austin and Sabine for cyclic loading, the engineer might wish to specify field tests in soft clay under cyclic loading if deflection is a critical parameter. A review of all of the curves showing computed and experimental deflection shows that, in general, the computation yields acceptable results.

Figure 7.39 present a comparison of the values of experimental and computed pile-head deflection at service load. The agreement is fair with about as many showing larger computed values as smaller computed values. The test for cyclic loading at Lake Austin shows the poorest agreement. Except possibly for that test, the differences probably would not lead to experimental difficulties.

The test at Talisheek is the only one where an axial load was applied along with the lateral load. While the data do not yield a computed value of the factor of safety, a review of the results in Figure 7.27 shows that the analytical method was able to predict the results from the experiment with reasonable accuracy. The analytical method appears to account

302 *Piles under lateral loading*

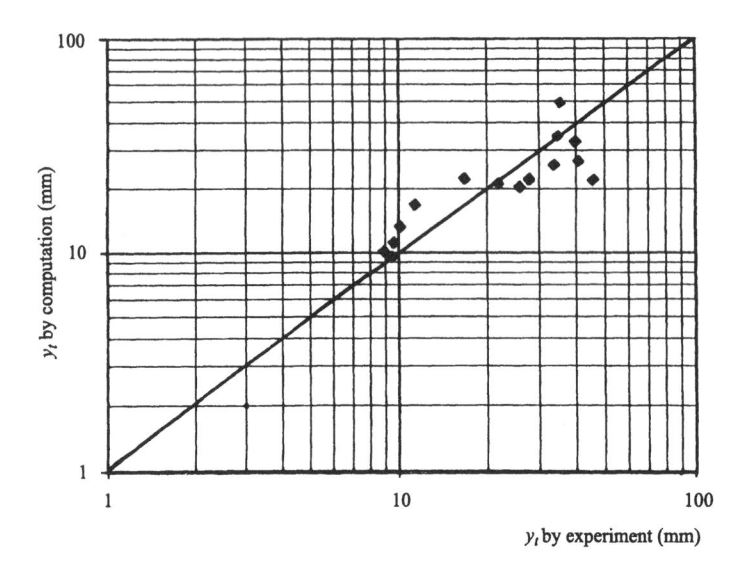

Figure 7.39. Comparison of experimental and computed values of pile-head deflection at service load for various tests.

appropriately for axial loading. Because most axial loads in practice are compressive, the behavior of bored piles will be improved by the axial loading while steel piles will be affected adversely.

Two tests, the ones at Talisheek and at Florida, were of 'short' piles where the tip of the piles deflected in an opposite direction to the pile head. While short piles are to be avoided in practice, if possible, the facility to use a curve at the pile tip showing lateral load as a function of pile deflection is useful.

A review of all of the tests that were analyzed shows that several parameters are critical with respect to the response of a pile to lateral loading. Such a review suggests, in addition to the requirement of providing relevant and good quality soil data, the desirability of having a larger data set of experiments, particularly where bending moment is measured and where the loading is cyclic such as is encountered frequently in practice.

CHAPTER 8

Testing of full-sized piles

8.1 INTRODUCTION

8.1.1 *Scope of presentation*

The presentation here is limited to the testing of single, full-sized piles but acknowledgment is made that numerous investigators have obtained valuable results from testing model piles in the laboratory. For example, Chapter 5 details the contributions of several workers who provided data that allowed numerical values to be assigned to the effects of pile-soil-pile interaction. In regard to pile groups under lateral loading, the work in the laboratory of Franke (1988), Prakash (1962), and Shibata et al. (1989) is worthy of note.

The centrifuge has become a popular tool for the investigation of problems in soil-structure interaction and is in use by many agencies in many countries. The centrifuge has been used to investigate the problem of single piles and groups of piles under lateral loading. Among the investigators who have used the centrifuge to study the response of piles to lateral loading are Kotthaus & Jessberger (1994), Terashi et al. (1989). Bouafia & Garnier (1991), and McVay et al. (1998). The results of centrifuge tests have provided useful information, particularly with respect to groups of piles, and the technique is expected to continue to contribute to the understanding of piles under lateral loading.

8.1.2 *Method of analysis*

The model employed presented herein for design of piles under lateral loading requires an equation solver, a computer code, experimental data on response of soil, and proposals for the response of various kinds of soil to lateral loading. The method has been presented in detail in the preceding chapters. Because of the complexity of the interaction between a pile and the supporting soil, the experimental data must come from load testing of full-sized piles in the field. Valid predictions could not have been made for the soil resistance as a function of the lateral deflection of a pile without well documented data from the testing of instrumented piles in the field, called *research* piles. The testing of *research* piles and *proof* piles is discussed in the following paragraphs.

8.1.3 *Classification of tests*

Research piles are those with comprehensive instrumentation to improve currently available *p-y* curves and to obtain data for *p-y* curves for soil where recommendations are absent or for *p-y* curves to improve existing predictions.

Proof piles are those where lateral-load tests are employed to obtain data for site-specific designs. The assumption is made that the piles to be used to support the proposed structure have been selected, and may be termed *production* piles. The testing of proof piles will normally require only minor instrumentation. The pile should be installed at a representative soil profile in the same manner as is planned for the production piles.

8.1.4 *Features unique to testing of piles under lateral loading*

A feature special to lateral loading is that the replication of conditions at the head of a production pile may not be possible in the field. The production pile may be subjected to a range of lateral loads, axial loads, and bending moments, some of which may be negative in the usual sense. Further, while deflections may be measured with accuracy, failure is more often dependent on the value of the maximum bending moment.

If the production piles are to be subjected to purely axial load, a test pile of the same dimensions and penetration can be installed in a representative soil profile and subjected to some multiple of the design load. If the axial settlement is less that allowed in specifications, the pile is judged to be adequate for the proposed structure, assuming that pile-soil-pile interaction is taken properly into account. Thus, a proof test of an axially loaded pile may provide sufficient information for design and the specific evaluation of the load transferred in side resistance and end bearing need not be made.

Because a test pile subjected to lateral loading cannot practically be loaded with an axial load and a moment (usually negative if the pile head is restrained against rotation), the performance of a proof test as for axial loading is not feasible. Therefore, the lateral-load test must be aimed at gathering data on the response of the soil.

The test pile is not required to be exactly like a production pile. A desirable plan is to use data on soil and on the test pile and to make a prediction of the lateral load versus deflection, using a computer program. The prediction would reveal the lateral load at which a plastic hinge would develop and allow the engineer to select appropriate increments of load. The proof test for lateral loading consists of comparing the experimental values of deflection with predicted values. Differences in the values are assumed to be due to properties of the soil, because the other parameters in the prediction method can be controlled. The properties of the soil can be modified to bring agreement in the measured and computed deflections and those properties can be used to design the production piles. An example of the testing of a proof pile is given later. A detailed example is also presented of a test of a research pile to obtain *p-y* curves.

It is of interest to note that the so-called *continuum* effect, where the response of the soil at an element is influenced by the response at all other elements, is explicitly satisfied during field testing. That is, even though the response of the soil is presented at discrete locations by *p-y* curves, the experimental *p-y* curves correctly reflect the response of the soil as a continuum or a continuous body.

8.2 DESIGNING THE TEST PROGRAM

8.2.1 *Planning for the testing*

The planning for the test of a proof pile may be minimal. For example, the senior writer and his colleagues tested two bored piles that were installed to be production piles but construction was delayed. The test consisted of placing a strut between the two piles, including a load cell and hydraulic ram in series with the strut, and applying the load in increments. Deflection and rotation were measured at each pile head. Predictions of pile-head deflections could be made because soil properties were determined at the location of each of the piles. The Owner required that the piles not be damaged so loading was discontinued when prediction showed the bending moment to be well below the computed value of ultimate moment. The piles had been installed in soft rock and the results reveal two important facts: 1. The difference that can be expected experimentally when piles are installed nearby in the same soil; and 2. The ability of the current recommendations to predict the early portion of the p-y curves for soft rock.

The performance of a test of a research pile, and the testing of some proof piles, requires a major effort and involves the participation of a number of specialists. The following pages present a number of steps that must be completed successfully. The planning is critical because a misfortune at the site can render useless the whole effort. For example, the senior writer was told of a case where the inadvertent application of a very large load by an automatic loading system spoiled an entire test.

The recognition of the factors involved in a test of large bored piles and suggestions for testing were presented by Franke (1973). In many test programs, an important consideration is the satisfaction of the requirements of a standards association. Some such standards are referenced later but certain provisions may not be applicable when the response of the pile is to be analyzed by the p-y method, and especially when a research-oriented program in undertaken.

8.2.2 Selection of test pile and test site

The site selection is simplified if a test is to be performed in connection with the design of a particular structure. However, even in such a case, care should be taken in the selection of the precise location of a test pile. In general, the test location should be where the soil profile reveals the weakest condition. In evaluating a soil profile, the soils from the ground surface to a depth equal to five to ten pile diameters are of principal importance.

The selection of the site where a fully instrumented pile is to be tested for research purposes is usually difficult. The principal aim of such a test is to obtain experimental p-y curves that can be employed in developing predictions of response of soil in a well defined subsurface. Thus, the soil at the site must be relatively homogeneous and representative of a soil type for which predictive equations are needed.

After a site has been selected, attention must be given to the moisture content of the near-surface soils. If cohesive soils exist at the site and are partially saturated, steps probably should be taken to saturate the soils if the soils can become saturated at a later date. If the cohesive soils can be submerged in time and if the piles can be subjected to cyclic lateral loads, the site should be flooded during the testing period.

The position of the water table and the moisture content are also important if the soil at the test site is granular. Partial saturation of the sand will result in an apparent cohesion that will not be present if the sand dries or if it becomes submerged.

If a lateral load test is being performed to confirm the design at a particular site, the diameter, stiffness, and the length of the test pile should be as close as possible to similar properties of the piles proposed for production. Because the purpose of the test is to obtain information on soil response, consideration should be given to increasing the stiffness and

306 *Piles under lateral loading*

moment capacity of the test pile in order to allow the test pile to be deflected as much as is reasonable.

The length of the test pile must be considered with care. As shown in Figure 2.5, the pile-head deflection will be significantly greater if it is in the 'short' pile range. Tests of these short piles can be difficult to interpret because a small difference in pile penetration could cause a large difference in groundline deflection.

The selection of the research pile with full instrumentation involves a considerable amount of preliminary analysis. Factors to be considered are the precision required in the results of the testing, the relevant pile diameter for which the soil response is required, soil parameters before and after pile installation, soil parameters that relate to the energy to install the pile, the kind of instrumentation to be employed for determining bending moment along the length of the pile, the method of installing instrumentation in the pile, the magnitude of the desired groundline deflection, and the nature of the loading.

8.3 SUBSURFACE INVESTIGATION

In Chapter 3 where procedures are given for obtaining p-y curves, recommendations were made for the type of required soil properties. Those recommendations should be consulted when obtaining data on soils for use in analyzing the results of the lateral-load experiments. However, in a number of the experiments, the investigation of the relevant soil parameters was not as detailed as desirable; further, advances in techniques for evaluating soil characteristics are common. The paragraphs that follow provide a summary of appropriate procedures and are not intended to be a definitive presentation.

Essential to a program of testing is a detailed evaluation of the relevant soil parameters along, and beyond, the length of the test pile. Three sets of procedures are available to the engineer for evaluation of soil parameters associated with the behavior of piles under lateral loading. These are: 1. Performance of soil borings and acquisition of tube samples for laboratory testing; 2. Performance of appropriate in situ tests; and 3. Implantation of sensors before pile driving for measurement of change in stresses in the soil. Depending on the specific conditions at the site of the test and the aims of the test, a combination of the procedures is usually indicated. The testing is more comprehensive for the testing of a research pile but a diligent investigation is required if a proof pile is tested.

The subsurface investigation for axially loaded piles usually extends to a depth of 1.2 times the length of the pile. The same rule of thumb can be applied for laterally loaded piles; however, the soils near the ground surface dominate the response of the pile. Other principles that apply to the investigation of the soil when piles are tested under lateral load are: 1. For long and short piles in soil, the properties of the soil of the zone five to ten diameters below the ground surface is of predominant importance; 2. For short piles in soil, characteristics of the soil at the tip of the pile must be determined to allow the formulation of a curve showing lateral load versus deflection at the tip of the pile; 3. For both long and short piles in rock, whatever the depth of the rock, the stiffness of the rock is important; and 4. The correlations that are proposed for obtaining p-y curves from soil properties must be based on procedures that are available to practicing engineers.

The first set of procedures noted above is necessary in allowing a definition of the various strata in the founding zone and in allowing the acquisition of data for classification. Data can also be obtained on the undrained strength and stiffness of cohesive soils.

However, the difficulties in obtaining undisturbed samples are substantial. Hvorslev (1949) did a classic study on procedures for exploration and sampling and presented detailed information on disturbance due to sampling of cohesive soils. Photographs of specimens of cohesive soil showed distortion due to resistance against the walls of the sampler. Such resistance was reduced or eliminated by using a sampler with metal foils (Kjellman et al. 1950). However, even with minimum disturbance due to the friction of sampling, disturbance due to the changes in the state of stress due to sampling and to the removal of the specimen cannot normally be eliminated. With respect to accounting for the effects of sampling disturbance, Ladd & Foott (1974) proposed a laboratory testing procedure for obtaining the undrained strength of soft, saturated clays.

Even though advances have been made in sampling and laboratory testing of cohesive soils, many investigations are still carried out with 'standard' techniques. Inevitably, some subsurface investigations of cohesive soils start from standard techniques and are therefore leading to conservative correlations with respect to the measured p-y behavior. Soil testing going out from advanced and high quality data-collection technology, does require less conservative, carefully choosen, calibrated correlations. Therefore, the engineer must be careful in studying the relevant soil tests that were performed to obtain p-y curves with respect to the methods used to obtain the undrained strength of clays. Techniques employed at sites where designs are to be made must take into account the methods used at the test sites.

Subsurface investigation by the use of in situ tests has received intensive study and is growing in popularity. A detailed discussion of the various tools and their application is given in papers and reports by numerous authors in technical journals and proceedings of conferences and seminars (Stokoe et al. 1988 & 1994). A summary of the various methods and their application for Belgium in the design of axially loaded piles is given by Holeyman et al. (1997).

Some of the methods are mentioned here and discussed briefly. The cone penetrometer test (CPT) was developed in Europe and is used worldwide to gain information on foundation soils, in particular. Because of the extensive use of cone testing in Europe and its increasing use elsewhere in the world, some details are presented here on that testing technique. The CPT is performed with a cylindrical penetrometer with a conical tip, or cone (see Fig. 8.1), and is pushed into the ground at the end of a series of push rods at a constant rate of 20 mm/sec. A load of up to 250 kN can be applied at the top of the push rods with a hydraulic ram reacting against a special truck or against an anchor. Forces on the cone and the friction sleeve are measured during penetration by internal load sensors. Not shown in Figure 8.1 is that an inclinometer can be placed inside the device to measure any tilt at the time of performing a test.

The penetrometer for performing the cone test may be fitted with devices at one or several positions along its length for measurement of pore water pressure and is termed the piezocone. Cone tests with pore pressure measurements are designated CPTU. Guidelines for the cone and other penetrometers were given by the ISSMGE Technical Committee on Penetration Testing (Information 7, Swedish Geotechnical Institute 1989).

The results from cone testing can be interpreted, in principle, to obtain the following information:

    stratification,
    soil type,

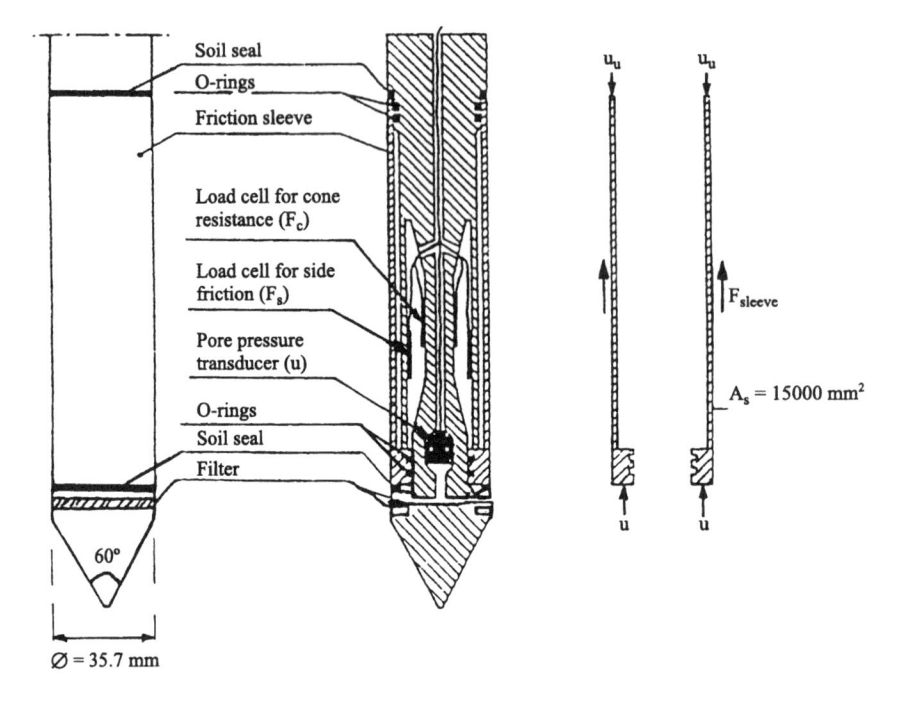

Figure 8.1. Piezocone-CPTU probe.

soil density,
mechanical soil properties
   shear-strength parameters
   deformation and consolidation characteristics

Later in this chapter, where the testing of a research pile is described, details on cone testing are presented where the cone was employed in obtaining changes in properties of sand as the result of pile installation and pile testing.

The pressuremeter test (PMT) is performed by measuring the pressure and deformation as a membrane is expanded against the walls of a borehole. Various methods of foundation design have been based on the results of such tests. The self-boring pressuremeter was designed to eliminate the disturbance due to the relaxation of the soil in the borehole and is used in special circumstances.

The dilatometer test (DMT) is performed by inserting a metal wedge with a membrane built into its side. The membrane is caused to move laterally against the soil while force and deflection are measured. The in situ lateral stress in the soil can be measured and the stiffness of the soil at low strains can be measured.

The Standard Penetration Test (SPT) was developed in the United States and is used there and elsewhere to obtain information on the strength of cohesionless soils. A thick-walled sampler is driven in the soil at the base of a borehole, seated by blows to a depth of 150 mm, and the blows are counted to drive the tool 300 mm, and designated the N-value. Correlations have been made between the N-value and characteristics of granular soil. The disturbed soil in the thick-walled sampling tube can be used to get moisture contents and to classify the soil. The results from the SPT can vary widely with the techniques employed in the testing; thus, the method is used with some caution by many practicing engineers.

A combination of two procedures, boring and laboratory testing and in situ testing, is required for the testing of research piles. The particular tests to be performed will depend on site-specific conditions and in consideration of methods generally available to correlate with p-y curves that are developed. Shear-strength characteristics for cohesive soils can be determined from high quality sampling and laboratory testing. For non-cohesive soils, however, and even for cohesive materials, the relevant small-strain parameters for the soil are found with difficulty from laboratory testing and, thus, modern in situ testing finds an important role. The methods of investigation for the testing of proof piles should be correlated with methods proposed for the design of the production piles.

The installation of sensors to measure the changes in stress in the soil due to pile driving and to lateral loading is highly desirable when research piles are to be tested. Such sensors are readily available and can be installed in the walls of the pile as well as in the surrounding soil. The increase and decay of pore-water pressure in fine-grained soils due to pile driving and due to the subsequent lateral loading provide valuable information related to the response of the soil.

8.4 INSTALLATION OF TEST PILE

For cases where information is required on pile response at a particular site, the installation of the research pile should agree as closely as possible to the procedure proposed for the production piles. It is well known that the response of a pile to a load is affected considerably by the installation procedure; thus, the detailed procedure used for pile installation, including gathering of relevant data on pile driving or on related procedures, is of utmost importance.

With regard to the effects of installation, soil stresses from the installation of piles can be studied by use of the dilatometer. Appendix G presents data from Van Impe (1991) on horizontal soil stresses near piles during installation.

For the case of a test pile in cohesive soil, the placing of the pile can cause excess porewater pressures to occur. As a rule, these porewater pressures should be dissipated before testing begins; therefore, the use of piezometers at the test site may be important. The use of the cone penetrometer with pore-pressure measurement can be considered. The cone can be left in place during some phases of the installation and the loading of the test pile.

The installation of a pile that has been fully instrumented for the measurement of bending moment along its length must consider the possible damage of the instrumentation due to pile driving or other installation effects. Pre-boring or some similar procedure may be useful. However, the installation must be such that it is consistent with methods used in practice. In no case would water jetting be allowed.

It would be desirable to know how the installation procedures had influenced the soil properties at the test site. The use of almost any sampling technique would cause soil disturbance and would be undesirable. The use of non-intrusive methods may be helpful. Van Impe & Peiffer (1997) described the use of the dilatometer in obtaining the effects of pile installation. Testing of the near-surface soils close to the pile wall at the completion of the load test is essential and can be done without undesirable effects.

310 *Piles under lateral loading*

8.5 TESTING TECHNIQUES

Excellent guidance for the procedures for testing a pile under a lateral loading is given by the Standard D 3966, 'Standard Method of Testing Piles Under Lateral Loads,' of the American Society of Testing and Materials (ASTM). Eurocode 7 (1994), also presents information on testing. In respect to standards, Bergfelder & Schmidt (1989) discuss testing to comply with the requirements of the German Standard DIN 4014 (1989). Recent recommendations (Holeyman et al. 1997; and Van Impe & Peiffer 1997) about procedures and principles for horizontal loading show agreement between Eurocode 7 and DIN 4020, 4054, and 4014.

For the standard test as well as for the instrumented test, two principles should guide the testing procedure: 1. The loading (static, repeated with or without load reversal, sustained, or dynamic) should be consistent with that expected for the production piles and 2. The testing arrangement should allow deflection, rotation, bending moment, and shear at the groundline (or at the point of load application) to be measured or able to be computed.

With regard to loading, even though *static* (short-term) *loading* is seldom encountered in practice, the response from that loading is usually desirable so that correlations can be made with soil properties. As noted later, it is frequently desirable to combine static and *repeated loading*. A load can be applied, readings taken, and the same load can be re-applied a number of times with readings taken after specific numbers of cycles. Then, a larger load is applied and the procedure repeated. The assumption is made that the readings for the first application at a larger load are unaffected by the repetitions of a smaller load. While that important assumption may not be strictly true, errors are on the conservative side.

As noted earlier, *sustained loads* will probably have little influence on the behavior of granular materials or on over-consolidated clays if the factor of safety is two or larger (soil stresses are well below ultimate). If a pile is installed in soft, inorganic clay or other compressible soil, sustained loading will obviously influence the soil response. However, loads would probably have to be maintained a long period of time and a special testing program would have to be designed.

The application of a *dynamic load* to a single pile is feasible and desirable if the production piles sustain such loads. The loading equipment and instrumentation for such a testing program would have to be designed to yield results that would be relevant to a particular application and a special study would be required. The design of piles to withstand the effects of an earthquake involves several levels of computation. Soil-response curves must include an inertia effect and the free-field motion of the earth must be estimated. Therefore, p-y curves that are determined from the tests described herein have only a limited application to the earthquake problem. No method is currently available for performing field tests of piles to gain information on soil response that can be used directly in design of piles to sustain seismic loadings.

The testing of battered piles is mentioned in ASTM D 3966. The analysis of a pile group, some of which are batter piles, is discussed in Chapter 5. As shown in that chapter, information is required on the behavior of battered piles under a load that is normal to the axis of the pile. An approximate solution for the difference in response of battered piles and vertical piles is presented in Chapter 3. Figure 3.30 may be used to modify p-y curves as a function of the direction and amount of the batter.

The testing of pile groups, also mentioned in ASTM D 3966, is desirable but is expensive in time, material, and instrumentation. If a large-scale test of a group of piles is pro-

Testing of full-sized piles 311

posed, detailed analysis should precede the design of the test in order that measurements can be made that will provide critical information. Such analysis may reveal the desirability of internal instrumentation to measure bending moment in each of the piles.

It is noted in ASTM D 3966 that the analysis of test results is not covered. The argument can be made, as presented earlier in this chapter, that test results can fail to reveal critical information unless combined with analytical methods. A subsequent section of this chapter suggests procedures that demonstrate the close connection between testing and analysis. A testing program should not be initiated unless preceded and followed by analytical studies.

8.6 LOADING ARRANGEMENTS AND INSTRUMENTATION AT THE PILE HEAD

8.6.1 *Loading arrangements*

A wide variety of arrangements for the test pile and the reaction system are possible. The arrangement to be selected is the one that has the greatest advantage for the particular design. There are some advantages, however, in testing two piles simultaneously as shown in Figure 2 of ASTM D 3966. A reaction system must be supplied and a second pile can supply that need. Furthermore, and more importantly, a comparison of the results of the two tests performed simultaneously will give the designer some idea of the natural variations that can be expected in pile performance. It is important to note, however, that spacing between the two piles should be such that the pile-soil-pile interaction is minimized.

Drawings of the two-pile arrangements are shown in Figures 8.2 and 8.3. In both instances the head is free to rotate and the loads are applied as near the ground surface as convenient. In both instances, free water should be maintained above the ground surface if that situation can exist during the life of the structure.

The details of a system where the piles can be shoved apart or pulled together is shown in Figure 8.2. This two-way loading is important if the production piles can be loaded in this manner. The lateral loading on a pile will be predominantly in one direction, termed

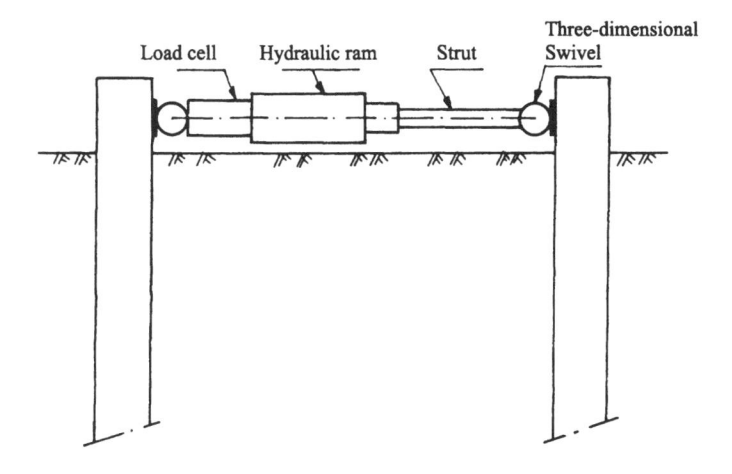

Figure 8.2. Arrangement for testing two piles simultaneously under two-direction lateral loading.

312 *Piles under lateral loading*

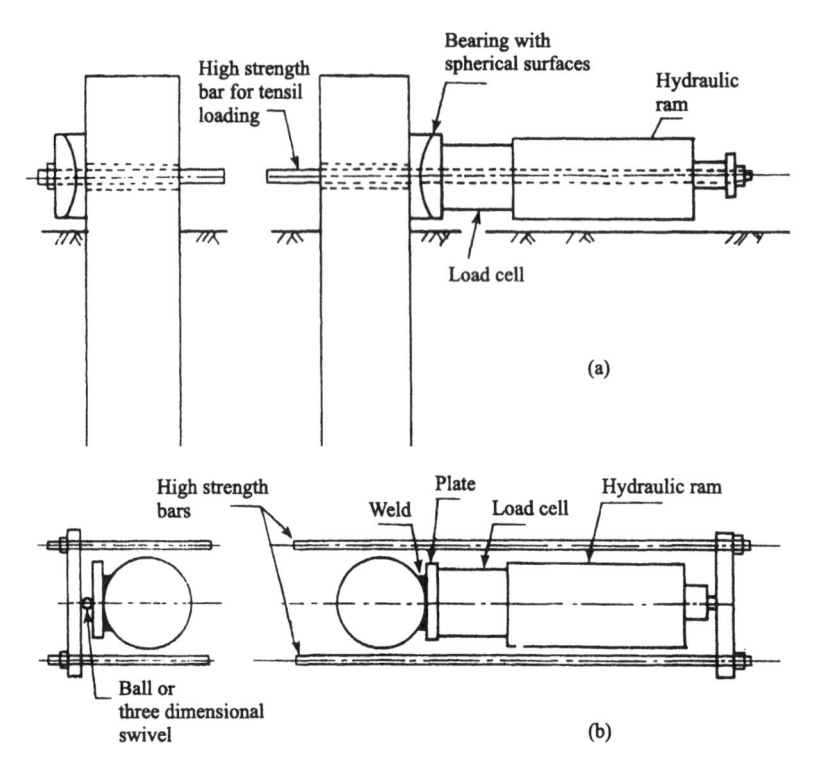

Figure 8.3. Arrangement for testing two piles simultaneously under one-direction lateral loading, (a) elevation view, (b) plan view.

the forward direction here. If the loading is repeated or cyclic, a smaller load in the reverse direction could conceivably cause the soil response to be different than if the load is applied only in the forward direction. As noted earlier, it is important that the shear and moment be known at the groundline; therefore, the loading arrangement should be designed as shown so that shear only is applied at the point of load application.

Figure 8.3 shows the details of a second arrangement for testing two piles simultaneously. In this case, however, the load can be applied only in one direction. A single bar of high-strength steel that passes that passes along the diameter of each of the piles is employed in the arrangement shown in Figure 8.3a. Two high-strength bars are utilized in the arrangement shown in Figure 8.3b. Not shown in the sketches are the means to support the ram and load cell that extend horizontally from the pile. Care must be taken in employing the arrangement shown in either Figures 8.2 or 8.3 to ensure that the loading and measuring systems will be stable under the applied loads.

The photograph in Figure 8.4 is of a test of two large-diameter bored piles that were tested by the arrangement shown in Figure 8.3a (Long & Reese 1984).

The most convenient way to apply the lateral load is to employ hydraulic pressure developed by an air-operated or electricity-operated hydraulic pump. The capacity of a ram is computed by multiplying the piston area by the maximum pressure. Some rams, of course, are double acting and can apply a forward or reverse load on the test pile or piles. The preliminary computations should ensure that the ram capacity and the piston travel are ample.

If the rate of loading is important (and it may be if the test is in clay soils beneath water and erosion at the pile face is important), the maximum rate of flow of the pump is important along with the volume required per inch of stroke of the ram. The seals on the pump and on the ram, along with hydraulic lines and connections, must be checked ahead of time and spare parts should be available.

High pressures in the operating system constitute a safety problem and can cause operating difficulties. On some projects, the use of an automatic controller for the hydraulic system is justified. A backup control must be available to allow the override of the automatic system in case of malfunction. There was one important project where the malfunction of the hydraulic system caused a large monetary loss.

The system shown in Figure 8.2 will require that the load cell and the ram be attached together rigidly and that bearings be placed at the face of each of the piles so that no eccentric loading is applied to the ram or to the load cell. The arrangement shown in Figure 8.2 may require that the points of application of load be adjustable to prevent the application of torsion to one or both of the piles. The loading system shown in Figure 8.3a will ensure that no eccentricity will be applied to the load cell and the hydraulic ram. If the two-bar system shown in Figure 8.3b is employed, care will be necessary to maintain stability in the system.

8.6.2 *Instrumentation*

The investigator has available a wide variety of instrumentation for measurements along the pile and at the pile head. Seco e Pinto & De Sousa Coutinho (1991) describe the use of electrical-resistance strain gauges and a slope indicator for measurements along the

Figure 8.4. Photograph of testing two bored piles using one-direction lateral loading, Sunshire Skyway Bridge, Tampa, Florida.

314 *Piles under lateral loading*

pile. An inclinometer, with a sensitivity of two sexagesimal seconds, was used to measure the rotation of the pile head. While strain gauges on the wall of the pile is the normal instrumentation for measurement of bending moment (Long et al. 1993), innovative techniques have been developed. Matlock et al. (1980) devised strain-measuring devices that could be lowered into a pipe pile and locked into place by the penetration of sharp-pointed bolts.

The sketches in Figures 8.2 and 8.3 show how a load cell may be used for the measurement of applied load (shear at the pile head). Also, a knowledge of the point of application of the load with respect to the groundline will yield the moment at the pile head. The arrangement shown in Figures 8.5 provide a concept for obtaining deflection and rotation of the portion of the pile above the groundline. The device shown in Figure 8.6 has been used successfully for obtaining rotation at the point of application of load. Because

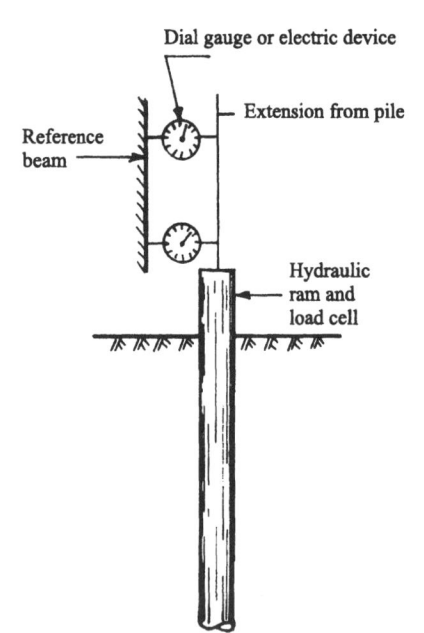

Figure 8.5. Schematic drawing of deflection-measuring system for lateral-load testing of piles.

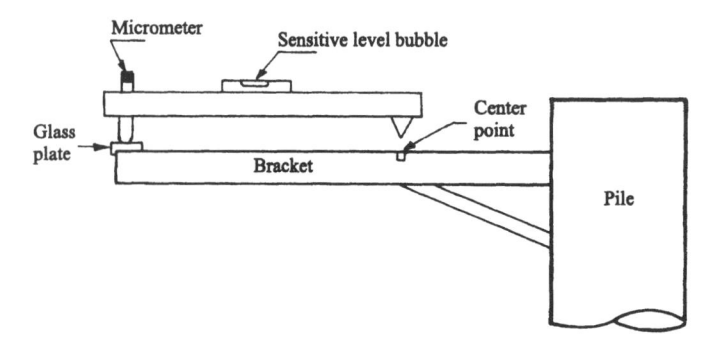

Figure 8.6. Schematic drawing of device for measuring pile-head rotation for lateral load testing of piles.

of the difficulty of applying load exactly at the groundline, analyses can be done more directly by using the data at the point of application of the load and by taking the distance from the load to the groundline into account.

Electronic load cells are available for routine purchase. These cells can be used with a minimum of difficulty and can be tied into a high-speed data-acquisition system if desired.

The motion of the pile head can be measured with dial gauges but a more convenient way is to employ electronic gauges. In either case, gauges with sufficient travel should be obtained or difficulty will be encountered during the test program. Two types of electronic motion transducers are in common use: linear potentiometers, and LVDT's (linear variable differential transformers); in either case the motion transducer should be attached so that there is no binding as the motion rod moves in and out.

Another possible arrangement for the measurement of pile-head deflection and rotation, as discussed in the European work-group recommendations (Bauduin 1997), is shown in Figure 8.7. The recommendations of the work group, with language modified slightly to agree with terminology herein, are given in the following paragraphs.

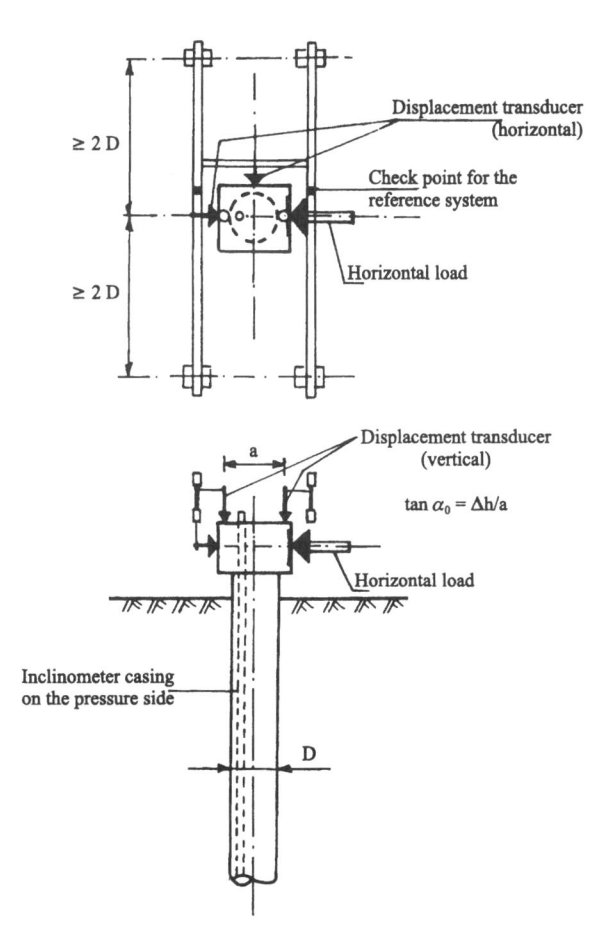

Figure 8.7. Plan and elevation views of the top of a pile showing placement of instrumentation for lateral-load test.

"The supports of the girder, where instruments for the measurement of deflection and rotation of the pile head are attached, must be separated a minimum distance of 1.5 times the pile diameter, and the piles must be a clear distance apart of 4.0 times the pile diameter. In case the pile-head movements are registered by an electronic dial gauge, it is mandatory to control the measurements by mechanical or optical measuring devices. The precision of measurement should be 0.01 mm for electronic dial gauges and 0.10 mm for optical measuring devices. In principle, these measurements should be made by two totally independent methods. At least, the reference points for the measurement of the pile-head movements must be controlled by an independent method before beginning and after completion of the test and while the maximum load is acting. For pile-head rotations, a measurement range of 1.0 degree and an accuracy of 0.05% is adequate.

The deflection curve can be measured by an inclinometer. However, these tests require additional time and extend the test period.

The bending moments can be determined by measuring the strains of the pile shaft both in the compressive zone and in the tensile zone.

For measurement of strains at specific points, electric strain gauges are suitable which can be directly attached to the steel pile or the pile reinforcement, or in the case of cast-in-place concrete piles, they are attached to special measurement casings which are located inside the reinforcement cage. Through pairwise arrangement and proper staggering of the measurement sections, pile deformations and their distribution can be determined sufficiently accurately in pile shafts with constant cross-section and constant modulus of elasticity.

In case of cast-in-place concrete piles, the cross-section as well as the modulus of elasticity vary frequently along the depth within a definite band width. In this case, representative results can be obtained from integral measurement elements, which measure compressive strains for pile sections at about 1.0 m length. For example, strain gauges are arranged inside measurement casings which are attached only at the ends of the measurement sections, and otherwise are not connected to the pile. Longer measurement sections are not advisable considering the variation of the bending moment with depth.

Measurements with micrometers and deformators are relatively expensive because of the necessary preparations and the procedure of measurements which also requires access above the pile head during the test loading. Measurement after completion of construction are not possible for foundation piles with this method.

For the evaluation of effective stresses and bending moments from the pile deformations, it is important that pile cross-section and deformation characteristics of the pile material are known accurately. The uniformity of the concrete quality can be controlled by ultrasonic testing.

In the case of cast-in-place concrete piles, the analysis will be complicated as a result of deviations from the theoretical cross-section due to bulb formation, variability of concrete quality or from cracking of concrete in the tension zone.

Decisions have to be made in each case on the method for which realistic values of flexural stiffness can be obtained. The uncertainties of these analyses require limit considerations, which primarily will affect the results more than the other factors."

Two final comments about the instrumentation are important. The verification of the output of each instrument should be an important step in the testing program. Also, the instruments should be checked for temperature sensitivity. In some cases it may be necessary to perform tests at night or to protect the various instruments from all but minor changes in temperature.

8.7 TESTING FOR DESIGN OF PRODUCTION PILES

8.7.1 *Introduction*

With regard to the design of production piles, three courses of action are dictated, depending on the number of piles to be installed. 1. If the number of piles is small and the soil profile is similar to one of those for which criteria are available, the designer may proceed with no testing and with the selection of an appropriate factor of safety. 2. If a large number of piles are to be installed and particularly is the soil profile is unusual, the designer may wish to test one or more instrumented piles, as described in the a section that follows. 3. In many cases, the designer may elect to test a pile that is essentially uninstrumented. The piles may or may not become a part of the foundation system after testing. The magnitude of the lateral load is relatively modest, compared to the axial load; therefore, the reaction may be accomplished by a simple arrangement. A convenient solution is to install two identical piles and to test them simultaneously.

Special problems with instrumentation are encountered when the piles in an offshore structure are to be instrumented (Kenley & Sharp 1993). Vibrating-wire strain gauges and electrical-resistance strain gauges were installed at a few points near the tops of the piles. The bending moments that were measured served to confirm the methods employed to predict the response of the piles based on published *p-y* curves.

8.7.2 *Interpretation of data*

The interpretation of data from a test of an uninstrumented pile is a straightforward process. The pile or piles will be tested in the free-head condition with the loading applied close to the groundline. No attempt is required to produce a loading to agree with that to be encountered by the production piles. Plots are made for the point of load application of deflection versus applied load and pile-head rotation.

A computer code is then used and computations are made of deflection and rotation for the same loads that were applied experimentally. A convenient procedure is to have a computer in the field and to make the computations simultaneously with the application of the experimental load. If the two sets of data do not agree, the properties of the soil are changed in the computations to reach agreement. Usually, only the shear strength needs to be changed; undrained shear strength for clay and the friction angle for granular soil. The procedure is repeated for each load and average values of shear strength can be selected. The modified values of shear strength can then be used to design the production piles.

8.7.3 *Example Computation*

The test selected for study was performed by Capozzoli (1968) near St. Gabriel, Louisiana. The pile and soil properties are shown in Figure 8.8. The loading was short term. The

318 Piles under lateral loading

soil at the site was a soft to medium, intact, silty clay. The natural moisture content of the soil varied from 35 to 46 percent in the upper 3 m of soil. The undrained shear strength, shown in Figure 8.8, was obtained from triaxial tests. The unit weight of the soil was 17.3 kN/m³ above the water table and 7.5 kN/m³ below the water table.

The results from the field experiment and computed results are shown in Figure 8.9. The experimental results are shown by the closed circles; the results from a computer code with the reported shear strength of 28.7 kPa and with an ε_{50} of 0.01 are shown by the solid line. The soil properties were varied by trial and the best fit to the experimental results was found for an undrained shear strength of 42.5 kPa and an ε_{50} of 0.009. These

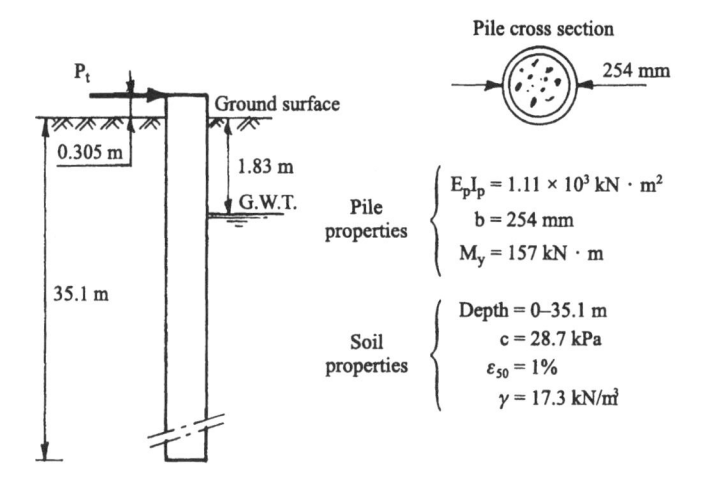

Pile cross section — 254 mm

Pile properties:
$E_p I_p = 1.11 \times 10^3$ kN·m²
$b = 254$ mm
$M_y = 157$ kN·m

Soil properties:
Depth = 0–35.1 m
$c = 28.7$ kPa
$\varepsilon_{50} = 1\%$
$\gamma = 17.3$ kN/m³

0.305 m, 1.83 m, 35.1 m, G.W.T., Ground surface, P_t

Figure 8.8. Information for analysis of results of test at St. Gabriel.

93.4 kN = P_{ult} (computer)

Computer
c = 28.7 kPa
$\varepsilon_{50} = 1\%$

Computer
c = 42.5 kPa
$\varepsilon_{50} = 0.9\%$

● Measured

Lateral load, P_t (kN) vs Groundline deflection, y_t (mm)

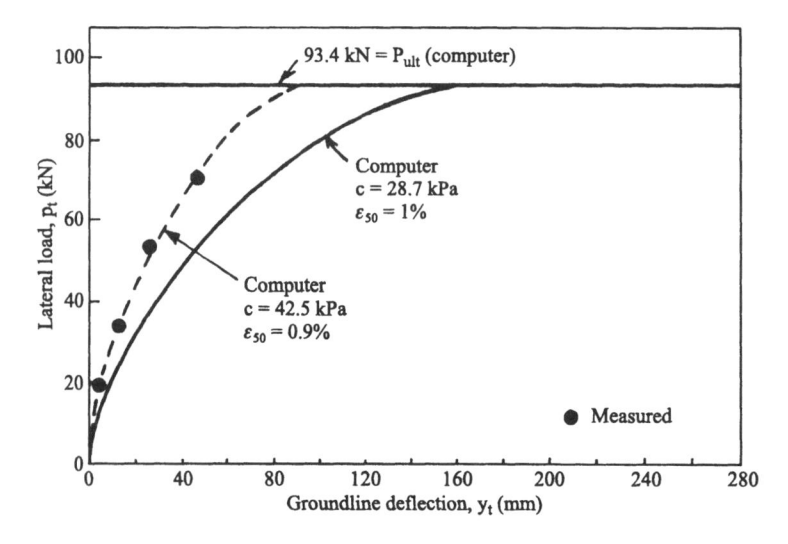

Figure 8.9. Comparison of experimental and computed values of pile-head deflection for St. Gabriel test.

values of the modified soil properties should be used in design computations for the production piles if the production piles are to be identical with the one employed in the load test.

A computer code was employed and an ultimate bending moment for the section that is shown was computed to be 157 kN-m. In making the design computations with the modified soil properties, the computed maximum bending moment should be no greater than the ultimate moment (157 kN-m) divided by an appropriate factor of safety. In computing the maximum bending moment, the rotational restraint at the pile head must be estimated as accurately as possible. If it is assumed that the pile will be unrestrained against rotation and that the load is applied one ft above the groundline, a load of 93.4 kN will cause the ultimate bending moment to develop. The deflection of the pile must be considered because deflection can control some designs rather than the design being controlled by the bending resistance of the section.

Two other factors must also be considered in design. These are: the nature of the loading and spacing of the piles. The experiment employed short-term loading; if the loading on the production piles is to be different, an appropriate adjustment must be made in the p-y curves. Also, if the production piles are to be in a closely spaced group, consideration must be given to pile-soil-pile interaction.

8.8 EXAMPLE OF TESTING A RESEARCH PILE FOR p-y CURVES

8.8.1 Introduction

The performance of tests to obtain experimental p-y curves requires planning and the selection of numerous details. Each such test will be unique, of course, and the establishment of general procedures is not possible. However, an example is presented to illustrate the methods employed in a particular case. Tests were performed, using pipe piles with a diameter of 610 mm, to obtain p-y curves in sand at Mustang Island (Cox et al. 1974). The tests were successful and the details presented in the following paragraphs should be instructive in the planning and performance of similar tests.

The tests were commissioned by the petroleum industry with particular reference to piles supporting offshore structures. An obvious requirement was that the water table was above the ground surface.

Two piles were tested, Pile 1 under static loads and Pile 2 under cyclic loads. The results from the loading of Pile 1 could be correlated with the properties of the soil without any reduction in soil resistance due to the effects of cycling. The loading of Pile 2 simulated that to be expected from wave loadings on an offshore platform. The major load was applied by compression in the system for loading and measurement. Then, for each cycle a minor load of about 25% of the major load was applied by tension in the system for loading and measurement. Loading in the opposite direction to the major load simulated the back pressures from a wave as the crest passed through the platform.

8.8.2 Preparation of test piles

Each test pile consisted of a 11.58-m uninstrumented section, a 9.75 m instrumented section, and a 3.05 m uninstrumented section as shown in Figure 8.10. All pile sections were

320 *Piles under lateral loading*

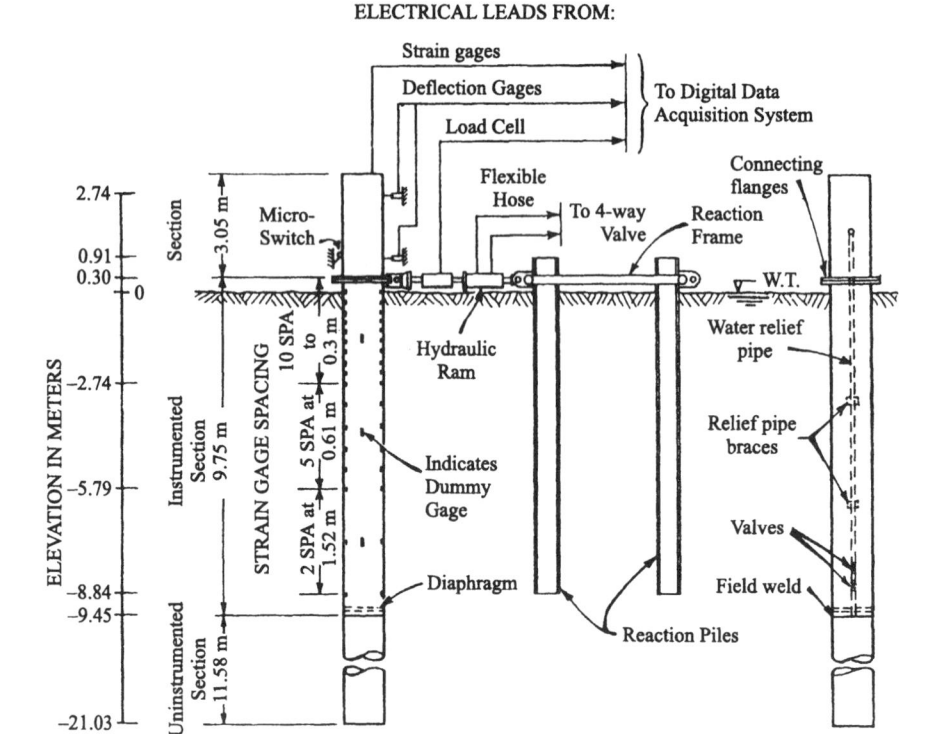

Figure 8.10. Sketch of arrangement for testing piles under lateral loading at Mustang Island.

610 mm in diameter with a 9.53 mm wall thickness. The wall thickness and the lengths of the three different sections were studied by: 1. Estimating the *p-y* curves that were expected, and 2. The performance of numerous computer runs to obtain the expected deflections and bending moments. Such preliminary steps are critical to the success of similar experiments in other soils.

Connecting flanges, 914 × 508 × 38 mm, were welded to the instrumented section and to the 3.05-m section. During driving and testing, the 3.05-m section was bolted to the 9.75-m section at the flange by seven 25-mm diameter bolts.

Twenty-five-mm long pieces of 38 × 38 × 3.2-mm angles were welded on 305-mm centers along the inside of the 9.75-m instrumented section. These angles supported steel straps to which the strain-gauge cables were clamped. Electrical resistance strain gauges were selected for use in determining bending moment along the piles.

Just below the flange on the 9.75-m section, an annular ring was welded inside the pile to which a pressure plate was bolted. A rubber gasket of 3.2-mm thickness was placed between the ring and pressure plate. Strain-gauge cables were brought through 0-ring packing-nuts screwed into the pressure plate.

To install the strain gauges, technicians could slide into the horizontal pipe section while lying face down on a specially made crawler supported by rollers. After installation of strain gauges, a 38-mm thick diaphragm was welded 152 mm above the bottom of the 9.75-m section. The bottom diaphragm and top pressure-plate seal prevented moisture from entering the 9.75-m instrumented section. Excess moisture in the piles could cause damage to the strain gauges.

A 76-mm diameter relief pipe was installed in the 9.75-m section after installation of strain gauges and cables. This relief pipe extended through the diaphragm in the bottom, through the top pressure plate and out the side of the 3.05-m section, as shown in Figure 8.10. The purpose of this pipe was to relieve water pressure created at the bottom diaphragm of the 9.75-m closed section during pile driving. Plans called for removal of the relief pipe after pile driving.

Also shown in Figure 8.10 are parts of the loading and instrumentation systems, which will be discussed later.

The decision was made to load the pile to as high a stress as possible; therefore, tensile tests were performed on specimens cut from the test piles. Analysis of the results of the testing led to the selection of 186 MPa as the highest stress that should be allowed.

8.8.3 Test setup and loading equipment

8.8.3.1 Description of test setup
A drawing of the two test piles and related equipment is shown in Figure 8.10. The load cell and hydraulic ram were placed in series between the reaction frame and the pile being tested. Loads to the free-head piles were applied at the connecting flange between the 3.05- and 9.75-m sections. The connecting flange was located 0.30 m above the mudline. LVDT's were used to measure the pile deflection at two points along the unstrained 3.05-m section. Micro-switches, which would shut off the hydraulic flow to the ram when activated, were placed on either side of the piles to prevent accidental overloading.

8.8.3.2 Hydraulic equipment
The hydraulic equipment was manually operated by controls mounted in a console inside a portable building. Loads were applied to the pile by a 305-mm stroke, double-acting ram that had a capacity of 334 kN in compression and 278 kN in tension. Hydraulic fluid was supplied to the controls by two 5.1-horsepower pumps each having a maximum flow rate of 0.7 m^3 per hour at a maximum pressure of 186 MPa.

The two electrically-driven hydraulic pumps were connected together so that one or both pumps could be used. A cooling system in the pump reservoirs kept the oil from overheating.

The hydraulic system had several unique features which prevented excessive shock loads on the system due to the pile snapping back as the direction of the load was reversed. One of these features was the location of the flow-control valves which controlled the rate of load application. By placing the flow controls so that fluid was metered out of the side of the piston that was not being loaded, stability of load application and damping of shock loads was achieved. Relief of the pressure which was built up by the pile energy was accomplished by two solenoid-operated, two-way valves that opened as the load direction was reversed, thereby allowing bleed-off of pressure in the lines to the ram. These two-way valves had little effect on the system in the field and it appeared that the flow controls and relief valves were the keys to the smooth operation of the system.

322 *Piles under lateral loading*

8.8.4 *Instrumentation*

8.8.4.1 *Spacing of strain gauges*
Simulated field tests using computer programs were used to arrive at a strain-gauge spacing which would yield sufficient strain values for accurate determination of the bending-moment curves along the test pile. The selected gauge spacing is shown in Figure 8.10.

8.8.4.2 *Installation and waterproofing of strain gauges*
Interior surfaces of the piles were sandblasted to remove mill scale and rust, permitting bonding of strain gages to the bare metal. Grinding of the internal surface of the pile at most locations of strain gauges was required because of pits in the metal from the manufacturing process.

A total of 40 strain gauges, 34 active gauges and 6 dummy gauges, were placed in each pile. Dummy gauges were to be used to complete the circuits where active gauges might become inoperative. The metal-foil strain gauges had a length and width of 13 mm. The nominal gauge resistance was 120 ohms ± 0.5 percent and the gage factor was 2.02.

The gauges were bonded to the pile with a two-part epoxy that maintains bond under dynamic strains. Immediately after laying the gage with the epoxy, a weight was placed on the gauge which resulted in a normal pressure of 34 kPa. After one hour of curing at room temperature, heat lamps were used to heat the gauge area to 400 F. The epoxy was cured for two hours at this temperature and the weight was removed. To prevent absorption of moisture, the gauges were then coated with a synthetic-resin waterproofing compound.

After soldering short pigtails to the terminal strip, the gauge installation was checked by measuring its resistance to ground and by testing the bond between the gauge and the pile. A volt-ohmmeter was used to check resistance to ground and continuity. All gages checked 1000 megohms or higher with 500 megohms considered a minimum acceptable value. By wiring the gauge as one arm of a Wheatstone bridge, the gauge bond was checked by rolling with light pressure a rubber eraser across the gauge grid and noting the resulting strain and zero stability on a strain indicator. Gauges that showed high strains or instability due to these pressure tests were not securely bonded and were removed and replaced.

After check-out of the gauges, a second coat of waterproofing compound, a two-part epoxy, was applied. Final waterproofing and mechanical protection was ensured by covering the entire gauge installation with a 3-mm thick neoprene pad bonded to the pile surface with rubber to metal cement.

8.8.4.3 *Installation of strain-gauge cables*
Cables were installed inside the piles by clamping them to brackets placed on 0.3 m intervals, as shown in Figures 8.11 and 8.12. The cable used was an 8 conductor wires covered by shielding and a tough neoprene-rubber exterior. Each cable carried output from four strain gauges, resulting in a total of five cables along opposite diameters of the piles. A total of 330 m of wire was installed in both piles. In order to reach from the pile to the instrument van, each cable extended 12 m beyond the top pressure plate.

With long lead wires, problems are frequently encountered with temperature effects and instability of the strain-gauge signal. To reduce the effects of temperature changes, cables of equal length were installed to diametrically opposed gages. Theoretically, with cables of equal length which are wired into adjacent arms of a Wheatstone bridge, changes in cable resistance due to temperature effects are canceled.

Testing of full-sized piles 323

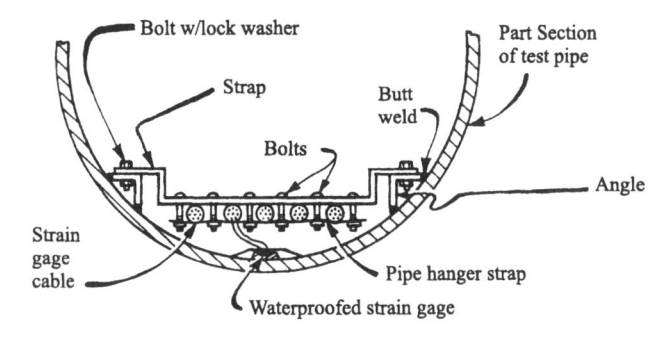

Figure 8.11. Sketch of clamping system for strain-gauge cables.

Figure 8.12. View of inside of pile installation of strain-gauge cables for tests of piles at Mustang Island.

Long lead-wire effects which could cause attenuation of the strain-gauge signals were also considered. Diametrically opposed gauges were wired as adjacent arms of a Wheatstone bridge in a half-bridge arrangement as would be done during field testing. Each gauge was then shunted separately first with a 150,000 ohm and then with a 60,400 ohm resistor. The apparent strains due to these shunts were measured with a strain indicator. After the long lead wires were attached, the same process of shunting the gages was repeated. The difference in apparent strains for the two conditions was the amount of attenuation caused by the long lead wires. The average decrease in apparent strain after adding the cables was

324 *Piles under lateral loading*

about 1 percent of the strain measured without the cables and, therefore, was not considered a problem in the recording of output from the gauges.

8.8.4.4 *Recording equipment for strain gauges*
The selection of instrumentation was given careful consideration. Effects of changes in pile strains due to possible creep of the soil and recording of data during cyclic loading led to the consideration of an instantaneous recording instrument. Such an instrument was rejected, however, because of its low precision and the involved process for data reduction. The decision was made to use a digital system that featured high recording speed, high precision, and convenient record of data.

A 20-channel digital-data-acquisition system was used for recording the output of the strain gages, deflection gages, and the load cell. The equipment was mounted in vertical racks inside the instrument van. The equipment scanned all 20 channels of information and printed the strain data on adding machine tape in units of micro-m per m. The time for the system to balance and print was from 0.4 to 1.5 seconds per channel, depending upon the range change or variation between readings. During lateral load tests, the system was clocked at approximately 17 seconds for scanning and recording all 20 channels of data.

Resolution of the system was to the nearest microstrain and the quoted accuracy was 0.1 percent. However, when the range of the system had to be changed, i.e., for strains greater than 999 micro-m per m, a multiplier had to be used which decreased the resolution.

8.8.4.5 *Circuits*
Strain gauges in the pile were wired into conventional Wheatstone-bridge circuits. Diametrically opposed strain gauges in the piles were wired as adjacent arms of a bridge that was completed with two strain-gauge dummies inside the digital-strain indicator. This bridge arrangement has the advantage that measured strains are twice the actual pile strains and temperature effects are compensating. It should be noted that each gauge was wired separately so that if one gauge became inoperative, one of the six unstrained dummies could be used as a substitute and data could still be obtained from that particular location.

The LVDT's for measuring deflection were also used in a bridge arrangement so that data on voltage output could be obtained with a recording system. The deflection gauges were wired into one arm of an external half-bridge arrangement.

8.8.4.6 *Reference bridge*
Due to the expected long duration of tests, an unstrained reference bridge was used to check possible zero-drift of the balancing and recording equipment. Two strain gauges out of the same lot used in the piles were mounted on a piece of steel which was positioned inside a protective box. A precision resistor and ten-turn potentiometer was wired into the half-bridge circuit with the unstrained gauges. Two separate bridges were made up in case one became defective.

The reference bridge was used in the following manner to check instrument drift. After all data channels had been null-balanced before a test and the balance controls locked in place, the input channels were disconnected. Then the reference bridge was plugged into the switch-and-balance unit and the ten-turn potentiometer was turned until the circuit was null-balanced and the number on the dial of the potentiometer was recorded. By repeating this for each channel the drift in the instrument could be checked by disconnecting any input channel at any time and substituting the reference bridge with the proper potentiometer setting for that channel. If the same setting on the potentiometer did not produce balance of the bridge, then the instrument had drifted or the balance-control knob for the channel had inadvertently been turned. The balance-control knob then could be used to return to the original datum.

Stability and repeatability of the reference bridge were verified by checks made in the laboratory and therefore a drift in the reference bridge during the above procedures was not considered probable.

8.8.4.7 Measurement of load and deflection
Loads were measured in the calibration and field tests by a universal load cell of 445 kN capacity. Accuracy of the load cell was quoted by the manufacturer at 0.25 percent of the full-scale range of 4000 micro-m per m. The accuracy and the manufacturer's calibration constant were checked on a 534-kN testing machine. The load cell had a full four-arm bridge made up of 120 ohm strain gauges.

Deflections during the field tests were measured at two points above the connecting flange on the unstrained 3.05 m section. The gauges used were LVDT's with 150-mm strokes capable of measuring displacements to 2.5×10^{-5} mm. For these tests, however, the resolution was reduced to 0.025 mm. Because the transducers could not be placed next to magnetic materials, they were held in place on a frame by wooden blocks, and their sliding cores were screwed into an aluminum block attached to the 3.05 m section.

8.8.4.8 Instrument van
The hydraulic control system and the digital-recording system were placed inside a van. All necessary operations of the loading and recording systems could be conducted within the van. At the test site, the van was backed to a position near the pile being loaded so that observations could be made through a window.

8.8.5 Calibration of test piles

8.8.5.1 Calibration of test piles by applying known loads
The output of strain gauges bonded to the inside of a pile could be computed by the equations of mechanics but experimental values may not equal the predictions. Errors may be caused by misalignment of the gauges, variations of gauge constants, and/or variations in the thickness and diameter of the pile. Since small errors in strains measured along a test pile can appreciably affect the derived soil-reaction values, calibration of strain gauges for bending moment was necessary.

An additional purpose was served by the calibration loadings. Any gauges could be located and replaced which were grossly in error due to improper installation or for some other reason. Fortunately, no such gauge replacements were necessary in either of the two test piles.

In the calibration test, a hydraulic ram was placed between the piles and used to apply loads simultaneously to both piles. The piles were connected at the ends with steel straps in order to create simple-beam supports. Loads applied to the piles were monitored by a load cell in series with the hydraulic ram and strain-gauge readings were recorded. The load cell, hydraulic ram, and strain-gage-recording equipment were to be used in the field test. All calibration tests, except those specified as field-calibration tests, were run inside a building to shade the piles from sunlight. The piles were carefully oriented so that loads

326 *Piles under lateral loading*

were applied in the plane of the strain gauges. For each load application, bending-moment values were calculated and strain-gauge output was recorded. Gauge constants were determined by dividing the computed values of moment by the strain-gauge readings.

Two locations for load application were required because stress concentrations invalidated constants for gauges close to the loading point. Points of load application were 3.35 m and 4.57 m below the flange of the test piles. At these locations, a maximum load of 133 kN was applied, which corresponded to a maximum stress of about 105 MPa.

A total of six loading series were run on each pile, three with the ram at the 3.35-m location and three with the ram at the 4.57-m location. Typical results from a load series are shown in Figure 8.13a for Pile 1 and in Figure 8.13b for Pile 2. The slight scatter in the results, as shown by the failure of all points to fall on the straight lines, confirms the need

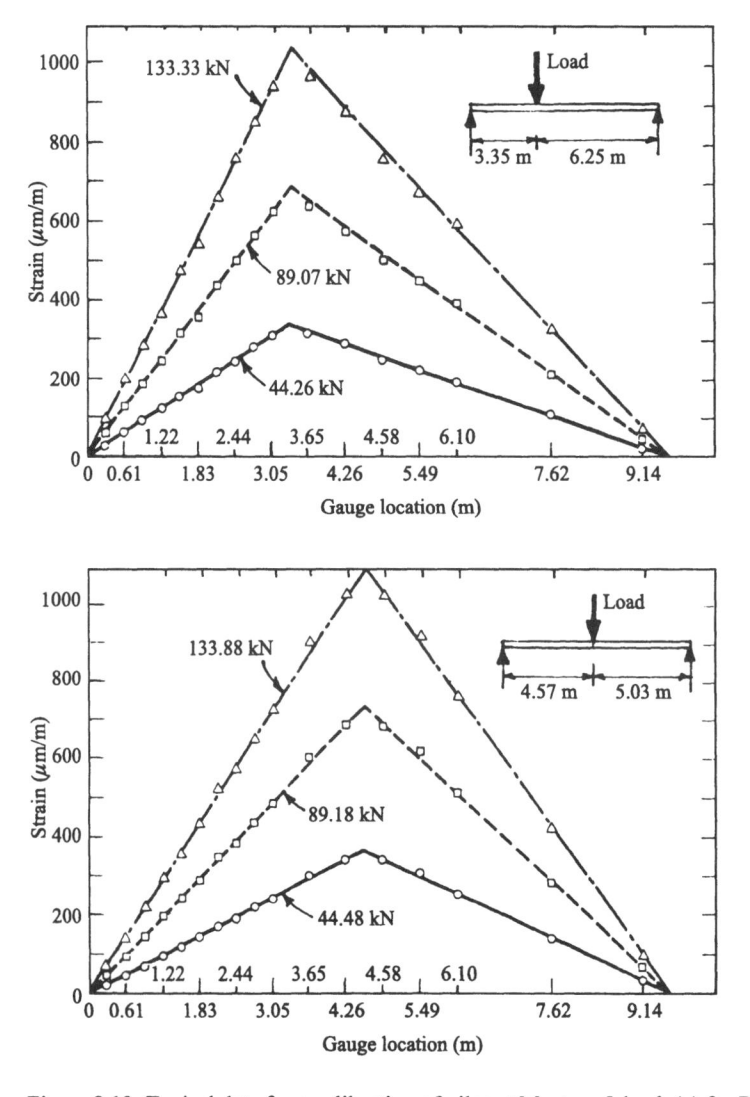

Figure 8.13. Typical data from calibration of piles at Mustang Island, (a) for Pile 1, (b) for Pile 2.

for performing the calibration. That the differences are systematic and not accidental can be seen by the fact that the plotted points maintain their relative positions, either on a line or slightly above or below, for each of the three loadings.

A typical plot of data from which a gauge constant can be computed is shown in Figure 8.14 for the 2.74-m gauge location. Results from two loadings on each pile are shown. A number of such loadings were performed. Figure 8.14 shows a considerable difference between gauge constants (the slope of the line) for the two piles but this is of no consequence since differences in gauge output can result from a number of causes. The slight difference between the slopes of the lines for a particular gauge is of consequence. While the difference is small, repeatable results were expected. The differences for the same pile as shown in Figure 8.14 are considered to result from experimental error rather than being systematic.

All gauge constants except those at 0.305, 0.610 and 0.914 m below the flange may be found from these shop calibration tests

8.8.5.2 *Field calibration of gauges at 0.305, 0.610, and 0.914 m*

Stress concentrations near the flange end of the test piles caused erratic results in gauges nearby during the shop-calibration tests. The nonlinear response was believed to be caused by the support and by discontinuities in the section. Also with the simple-beam support, moments were small causing the output of gages to be small. For these reasons, gages at 1, 2, and 3 ft were calibrated in the field after load tests were performed on the piles.

Field calibration was performed by excavating around each pile to about one meter below the flange. Loads were then applied with the ram, load cell, and all related equipment in the same configuration as during the actual field tests. The bending moment at each gage location was computed and the strain-gage output measured for each load application. A maximum of 156 kN was applied at the flange. The output of gauges was approximately double that recorded in the calibration tests performed in the shop because the bending moments were larger.

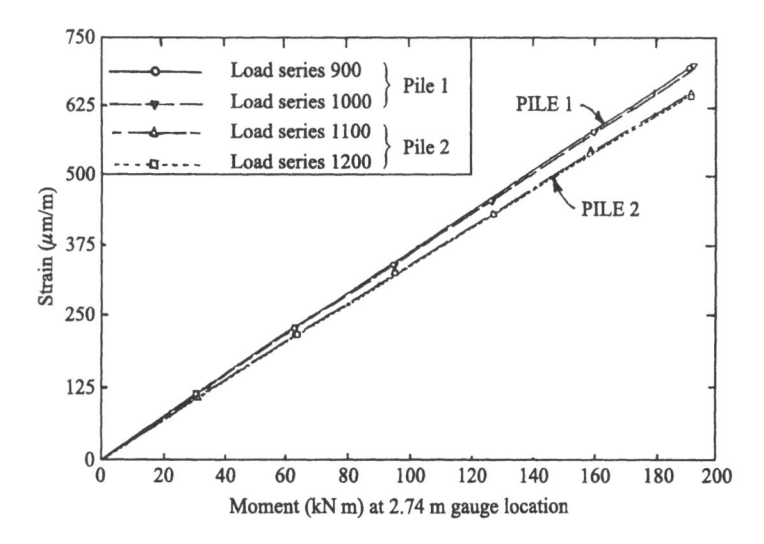

Figure 8.14. Plot to illustrate the determination of the calibration constant for the strain gauges at a specific location in the pile at Mustang Island.

8.8.5.3 *Strain gauge coefficients determined by calibration tests*

Instead of using the graphical method of determining gauge coefficients, as shown in Figure 8.14, the method of least-squares curve fitting was applied to the strain data. The straight-line fit through all the data was forced through zero as was the curves shown in Figure 8.14.

Correction factors for boundary conditions at the top of the pile were applied to the field-calibration data. Strain readings from gauges 0.3 m either side of the ram location were discarded for the shop-calibration series. The calibration coefficients for each gauge for each load series were determined by least-squares fitting, were averaged, and are listed in Table 1.

8.8.5.4 *Determination of flexural rigidity*

An accurate determination of the flexural rigidity (EI) of each pile was necessary for use in the analyses. The EI of the piles was found during the shop-calibration tests assuming that there were no variations in EI along the length of the pile. During loading, deflections were measured by dial gauges (least reading of .0025 mm) placed at several points along the pile. Actual deflection due to bending of the simply supported piles was found by subtracting the translation of the end supports and the distortion of the pile diameter from readings of deflection..

The distortion of the pile diameter was measured from readings of two dial gauges, placed diametrically opposite at the point of load application and in the same plane as the load. For the maximum load of 133 kN the distortion was measured as 0.076 mm. Thus, distortion of the diameter of the piles resulting from soil pressures on the pile during field testing was not believed to be a problem.

As determined from the deflection data, the EI of Piles 1 and 2 was 168,367 kN-m^2 and 174,628 kN-m^2, respectively. These values agree favorably with the EI computed by using the E as determined by the tensile test on coupons from the piles times the computed moment of inertia of the piles.

8.8.6 *Soil borings and laboratory tests*

The shear angle ϕ is the most significant soil property related to the resistance provided by the sand. For any sand, ϕ increases with increasing relative density, and as a pile is driven into a sand the soil becomes denser in the immediate zone surrounding the pile shaft. As the pile is loaded, subsequent changes in ϕ will occur, linked to increases in the level of strain. Determination of ϕ, as well as other properties of sand, is complex because of the difficulty involved in obtaining undisturbed samples. Experiences in testing of soil in Europe would, in such cases, lead to careful in situ testing by the use of methods described earlier, particularly the cone penetrometer test (CPT) and the dilatometer test (DMT). The procedures described below are typical of the traditional approach in the United States, which are being modified in many cases as the European tools become more available and the interpretation of results from in situ tests become better understood.

8.8.6.1 *Soil borings*

Prior to excavation of the test area and installation of the piles, two soil borings were made at the test site. Soil samples were taken from one hole by alternating a 50-mm split-spoon sampler with a 76-mm piston sampler. Between the samples, the hole was drilled and washed from 150 to 230 mm. Samples from the second boring were taken at 1.5-m intervals with a 50-mm split-spoon to 4.6 m. Below 4.6 m, samples were taken at 1.5-m intervals with a wire-line sampler. Additional tests were done with a static-cone penetrometer. Logs of the two borings are shown in Figures 8.15 and 8.16. The depth as shown relates to original ground surface.

The piston sampler used in obtaining samples from the first test boring was considered the best method for obtaining undisturbed samples of a cohesionless material. Samples

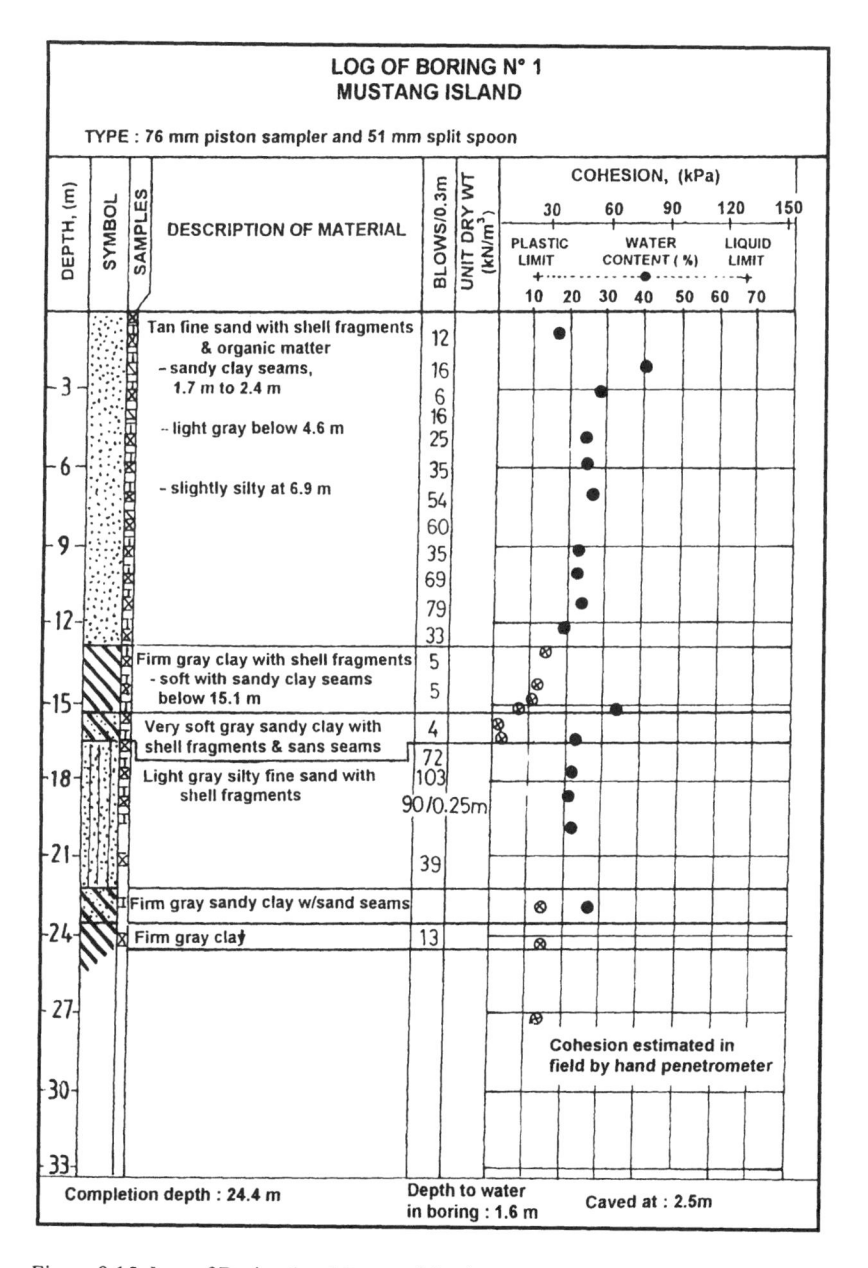

Figure 8.15. Log of Boring 1 at Mustang Island.

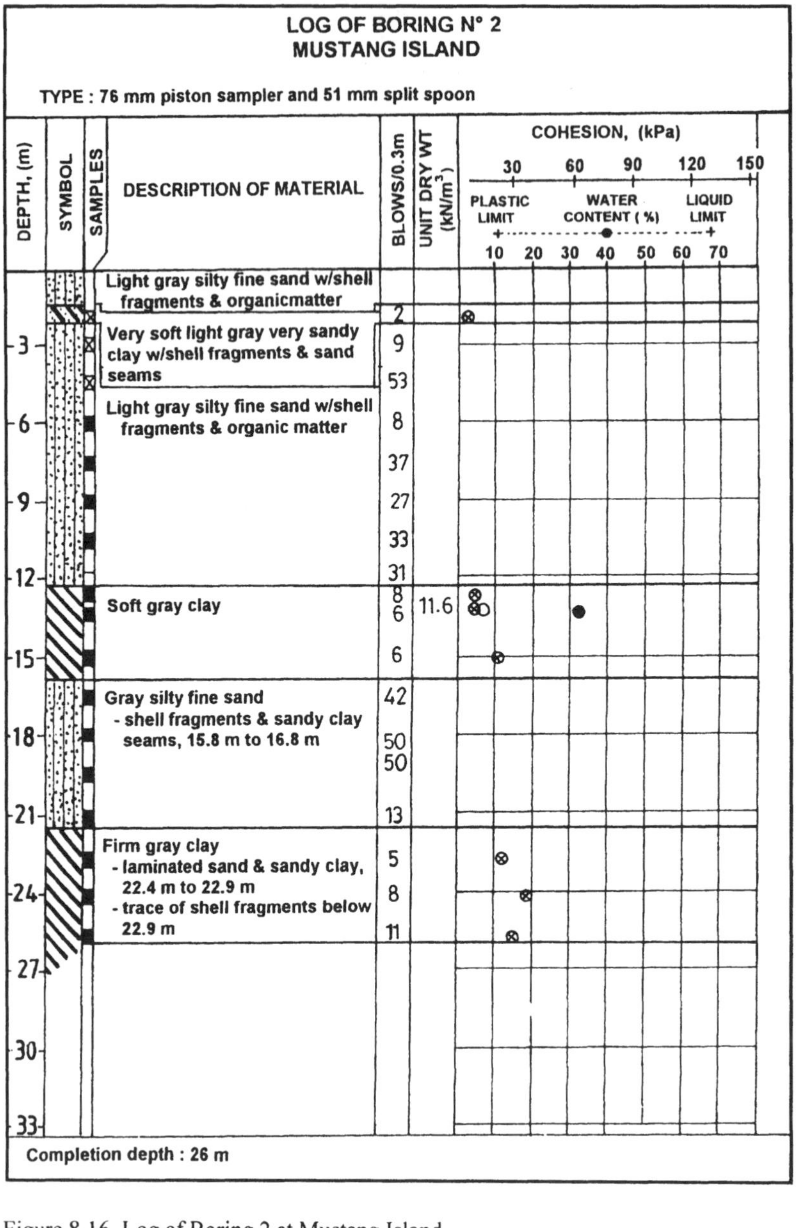

330 *Piles under lateral loading*

DEPTH, (m)	SYMBOL	SAMPLES	DESCRIPTION OF MATERIAL	BLOWS/0.3m	UNIT DRY WT (kN/m³)	COHESION, (kPa)
			LOG OF BORING N° 2 **MUSTANG ISLAND** TYPE: 76 mm piston sampler and 51 mm split spoon			30 60 90 120 150 PLASTIC WATER LIQUID LIMIT CONTENT (%) LIMIT 10 20 30 40 50 60 70
			Light gray silty fine sand w/shell fragments & organic matter	2		
-3-			Very soft light gray very sandy clay w/shell fragments & sand seams	9		
				53		
-6-			Light gray silty fine sand w/shell fragments & organic matter	8		
				37		
-9-				27		
				33		
-12-				31		
			Soft gray clay	8 / 6	11.6	
-15-				6		
			Gray silty fine sand - shell fragments & sandy clay seams, 15.8 m to 16.8 m	42		
-18-				50 50		
-21-				13		
			Firm gray clay - laminated sand & sandy clay, 22.4 m to 22.9 m - trace of shell fragments below 22.9 m	5		
-24-				8		
				11		
-27-						
-30-						
-33-						
			Completion depth: 26 m			

Figure 8.16. Log of Boring 2 at Mustang Island.

were taken by lowering the piston sampler to the bottom of the drilled hole. The inner barrel, which was a 60-mm diameter thin-walled tube, was then advanced very rapidly about 0.3 m into the soil by pressure built up by the mud pump. The sampler was brought back to the surface and the length of the sample was found for weight-volume values. The sample was then extruded into a tin, sealed and sent to the laboratory for determining water content, relative density and grain-size distribution.

Disturbed samples were taken with a split-spoon sampler and wire-line sampler. The split-spoon sampler is a conventional soil sampling tool. The wire-line sampler, was developed for work offshore, and is seldom used outside the oil industry. The main advantage of the wire-line sampler lies in the saving of time in obtaining a sample without bringing the drill string out of the hole. A split-spoon sampler can be lowered inside the drill pipe by a wire line. A drop hammer, sliding on a rod on top of the sampler, is also attached to a wire line. The sampler can be driven by alternately raising and dropping the hammer.

Standard-penetration tests (SPT) were also performed. Results from the SPT are expressed as the number of blows, N_{SPT}, required to advance the sampler 0.3 m into the soil. For samples of cohesive soil, unconfined compressive strengths were measured with a hand penetrometer. The values of compressive strength are shown on logs of the borings.

A series of static-cone tests were performed and will be presented in a later section. The aims of the tests was to investigate the effects on the soil properties of driving the piles and also of performing the testing.

8.8.6.2 *Discussion of soil borings and laboratory tests*

Comparisons of the logs of Borings 1 and 2, Figures 8.15 and 8.16, indicate a slight variation in the soil profile between the two locations. In the top 12 m, the sand strata at Boring 2 had enough silt particles to be classified as a silty fine sand and in Boring 1 the soil was classified as fine sand. The difference in material is shown by a plot of the N_{SPT} – values in Figure 8.17. The values of N_{SPT} – values at Boring 2 are generally lower than those at

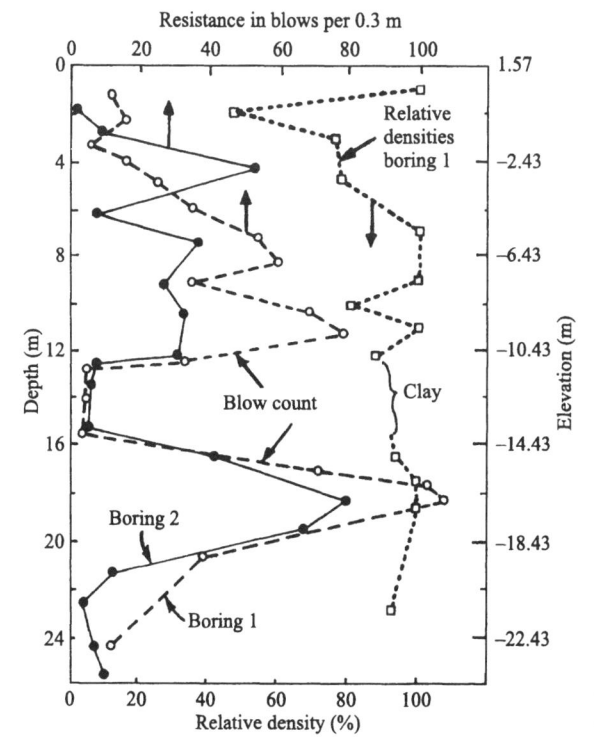

Figure 8.17. Results of Standard Penetration Test and relative densities from piston samples, Mustang Island.

332 Piles under lateral loading

Boring 1. The properties of sands from 0 to 6 m are classified as medium dense, from 6 to 12 m as dense, and from 15 to 21 m as dense.

Laboratory tests were run on samples from Boring 1 that were obtained with the piston sampler. Soil properties determined from these tests included grain-size distribution, natural densities, and minimum and maximum densities. Curves of grain-size distribution in Figure 8.18 indicate uniformly graded sands with the percentage of fines passing the number 200 sieve varying from 0 to 15 percent. Results of density tests are plotted in Figure 8.19.

In general, the sand at the test site varied from clean fine sand to silty fine sand, both having high relative densities. The sand particles by inspection through a microscope were found to be subangular with a large percentage of flaky grains.

8.8.7 Installation of test piles

Prior to the installation of the piles, a pit with a depth of 1.68 m was excavated to reach the water table. The two test piles were driven in the excavated pit and four steel H-piles were driven to a penetration of 9.1 m through the corners of a reaction frame. After the reaction piles were in place, the frame was jacked up and held in place by welded angles. A Delmag-12 diesel hammer was used for driving all the piles.

The 11.6 m open-ended sections of both test piles were driven through a template which held the piles plumb and at a specified distance away from the reaction frame. After the sections were driven to grade, the soil plug was removed by an auger in order to prevent contact of the soil plug against the diaphragm at the bottom of the instrumented sec-

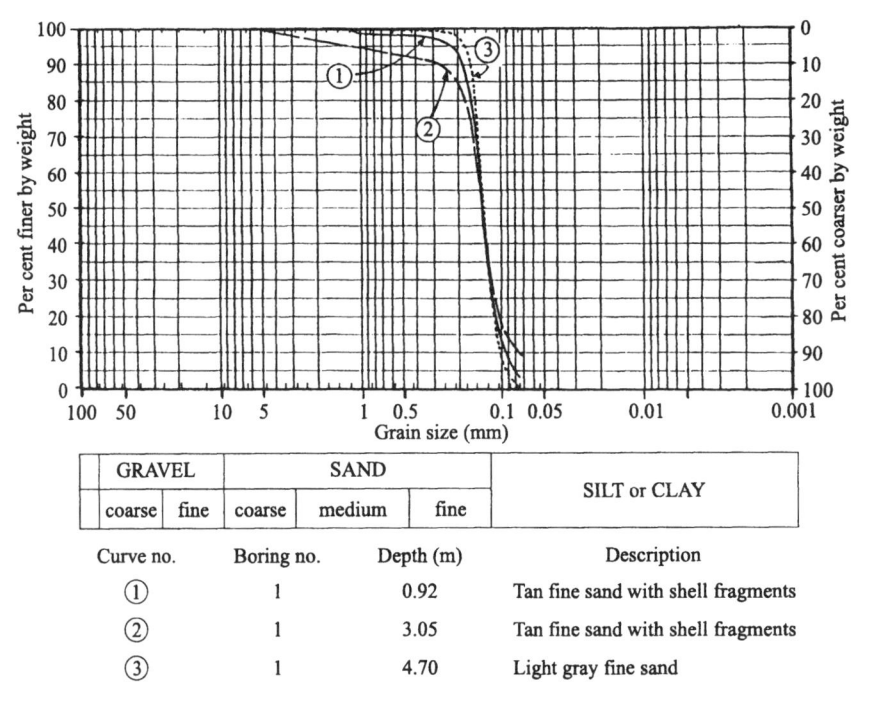

Curve no.	Boring no.	Depth (m)	Description
①	1	0.92	Tan fine sand with shell fragments
②	1	3.05	Tan fine sand with shell fragments
③	1	4.70	Light gray fine sand

Figure 8.18. Grain-size distribution curves for sand from Boring 1 from the depths of 1 to 4.7 m, Mustang Island.

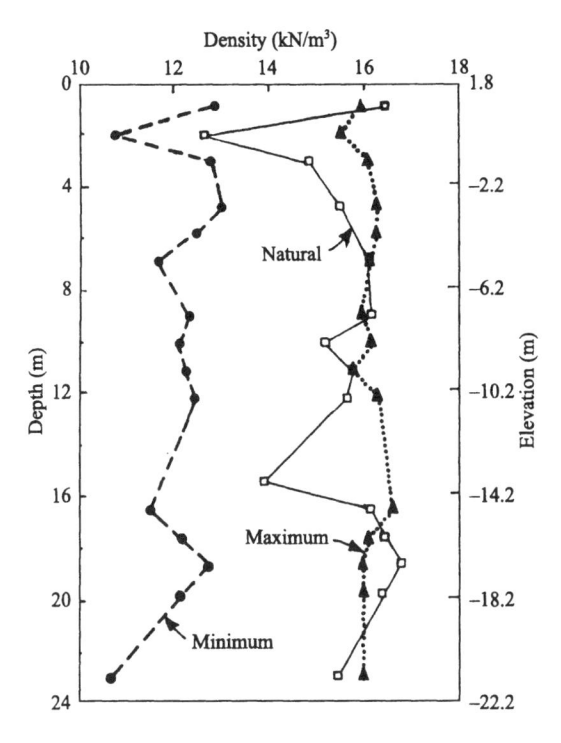

Figure 8.19. Results from tests of density of sand from piston samples, Mustang Island.

tion. The bottom of the 11.6-m section was driven into the 3-m-thick layer of clay shown in the soil borings. Water did not seep into the inside of the 11.6-m section when the soil plug was removed. Had seepage occurred, water relief pipes would have been employed to relieve water pressure during driving.

Pile 2 was driven first and pile driving was stopped after driving 3 m and 6 m in order to check the continuity of the strain gauges. The checking required that a technician be lowered into the top of the 3-m to reach the strain-gauge cables that were coiled on top of a pressure plate. Had there been an excessive number of gages damaged, a re-evaluation of the method of installation would have been initiated. Only a total of 4 gauges in both piles were damaged as a result of driving.

After the test piles were driven, plans called for removal of the 76-mm diameter water-relief pipe. Valves near the lower ends of the relief pipe would be closed, and the threaded connection between the valves would be broken to allow removal of the pipe above the lower valve. With the lower valve closed, water could not enter the instrumented section.

After driving the Pile 2, it was discovered that the lower valve had broken. This freed the 3-in, pipe to jar up and down inside the instrumented section. This hammering action in time broke the weld between the diaphragm and the 3-in. pipe section below the threaded connection. The hole in the diaphragm was subsequently closed with a plumber's 3-in. test plug. No water had entered the instrumented section and no damage occurred to strain gages and cables inside the pile. To prevent a similar failure from occurring in Pile 1, the relief pipe was removed before driving and the hole in the diaphragm was welded shut. Small holes were then cut in the pile just below the diaphragm to allow water and air to exit from the anchor section. As the pile was driven, there was continuous evidence at the ground surface that air was being pushed out of the anchor section.

334 *Piles under lateral loading*

During driving of the last 3 m of Pile 1, excessive rotation of the pile was noted while checks on alignment were being made with a transit. The contractor applied a torque at the flange with winch lines from two trucks. The torque was held on the pile while driving was resumed and the pile was rotated back into correct position.

After installation of both piles, the test sections were purged with a vacuum pump and then pressurized to 100 kPa with nitrogen. Unfortunately, several bolts on the pressure plates in both piles had been stripped during driving and most of the pressure was lost over a 12-hour period. Bags of calcium chloride had been placed in the piles to absorb moisture, and attempts to repair the pressure plates were abandoned.

The setup for the field tests was completed with the assembly of a prefabricated building. This building provided protection of equipment and improved working conditions.

8.8.8 *Test procedures and details of loading*

The plan was to use Pile 1 for static loading and Pile 2 for cyclic loading. The plan was followed but additional loadings were applied to both of the piles, principally to determine the effects of pre-loading. Those additional loadings will not be addressed.

8.8.8.1 *Procedures and static loading of Pile 1*

The plan was to use Pile 1 for static loading and Pile 2 for cyclic loading but a series of static loads were applied to both Pile 1 and Pile 2. Procedures that were employed in the static load tests are presented below.

Prior to beginning a test, calibration constants for the linear displacement transducers were determined by displacing the core of the transducer by a known amount and noting the change in strain measured by the data acquisition system.

The calibration constant for the load cell was furnished by the manufacturer, but because of the importance of the load measurement it was decided to check this constant in the laboratory. The load cell was placed in a calibrated testing machine and the constant was found to agree exactly with the value provided by the manufacturer.

Prior to running a load test, each input channel was balanced with the data-acquisition system. Loads were monitored by direct reading of the output of the load cell. At a given load for which strain measurements were to be made, the load was stabilized, then the output of the load cell was switched into the data-acquisition system for recording along with the deflection and strain gauge readings.

Immediately after recording the readings with the data-acquisition system, a direct reading was made of the output of the load cell. If the measured load had changed due to creep in the pile-soil system during the time required for the readings, the load was again stabilized, and readings taken again. This repetition of readings due to creep was necessary only in a few cases.

After a set of readings was taken, the load was increased to the next load level and the process repeated. It should be noted that the procedure used in these static tests perhaps differed from some static-test procedures in that creep in the pile-soil system had not ceased before readings were taken.

Strains along the pile and deflection, as measured by the top gauge, were plotted during testing. No further increases in loading were made when the measured strain approached the maximum allowable strain for the pile. As loads were decreased, strain data were taken at various load levels in order to measure the bending moment that was locked into the pile.

The testing of Pile 1 began 16 days after the pile had been driven. Data were taken as loads were applied in increments of 11.1 kN to 66.7 kN and then in increments of 5.56 kN from 66.7 kN to a maximum of 266.9 kN. Data were recorded as loads were decreased in increments of 44.8 kN back to 0.

The data that were taken for each increment of load include: lateral load, deflection, and rotation at point of application of load (moment was zero); and bending moment as a function of distance along the pile. The procedure for the analysis of the data is presented below.

8.8.8.2 *Procedures and cyclic loading for Pile 2*

The data-acquisition system had malfunctioned on occasions in performing the static-load tests and the apparatus was repaired prior to performing the static-load tests. Further, plans were made to use a manually operated strain indicator with a switch-and-balance unit in case of necessity.

The first loads applied to Pile 2 in the cyclic-load tests were referred to as 'seasoning loads.' These were small loads (maximum of 13.3 kN) applied equally in two directions to simulate the gradual buildup of wave action prior to a storm. After these loads were applied, loads were increased with the ratio of major to minor load equal to 4.

As for the static tests, the cyclic load was monitored directly and as a function of time. The cycle time was between 16 and 20 seconds and averaged about 17 seconds. The number of cycles applied to the pile was monitored on an electric counter wired into the hydraulic system.

At a particular load level and after a specific number of cycles, strain readings from the load cell, deflection gauges, and strain gauges were recorded for loads in both the major and minor load directions. The decision to proceed to the next load level after a particular number of cycles was based on the data on deflection recorded during testing. If the deflections stabilized, i.e., repeated for successive cycles of load, data were recorded on the fifth cycle of loading. The load was then increased.

If the deflection did not stabilize, the load was increased only if it appeared that the rate of increase in deflection had become constant. After reaching a maximum stress level in the pile, the load was reduced in steps to zero with readings taken at intermediate load levels.

Prior to lateral loading of Pile 2, a pull-out test was performed on the pile in its virgin condition at 52 days after the pile had been driven. Vertical loads up to 1780 kN were applied to the pile causing the pile to move up about 25 mm. The pull-out tests are assumed to have had no significant effect on subsequent lateral loading.

The maximum value of cyclic load applied to Pile 2 was 244.6 kN. The values of the various major loads applied during the cyclic loading of Pile 2, in kN, were: 4.4, 8.9, 13.3, 22.2, 33.4, 44.5, 55.6, 66.7, 77.8, 100.1, 122.3, 144.6, 166.8, 189.0, 211.3, 233.5, 244.6, 211.3, 122.3, 66.7, and then back to 0.

8.8.9 *Penetrometer tests*

As noted earlier, penetrometer tests were performed to study the effects on the sand of pile driving and pile testing.

8.8.9.1 *Penetrometer apparatus*

The cone employed is similar in shape to that shown in Figure 8.1 but no electronic system was included. A test was performed by loading and measuring the deflection of push

336 *Piles under lateral loading*

rods. The push rods were loaded with a scissors jack and the load was measured with a load cell at the top of the push rods. No pore-pressure measurement was done.

8.8.9.2 *Modified penetrometer procedures*
The procedures described below are termed *modified* because the measurement of forces were done externally rather than by sensors located inside the cone, and because the rate of penetration employed did not conform to the suggested standard. A section of the drill string with the cone attached to was driven in the soil. A weight of 34 kg was dropped 0.46 m to drive the drill string and the blow count required to advance the drill stem 0.3 m was recorded. The cone was then advanced 38 mm into the soil at the rate of 50 mm/min. by using the scissors jack with reaction applied by dead weight. After completing soundings in a hole, the drill string and cone were jacked out of the ground with two bumper jacks.

8.8.9.3 *Schedule of penetrometer tests*
A series of penetrometer tests were run at the test site prior to excavation and pile driving. Two tests were run 4-ft away from the bore holes to a depth of approximately 15 ft. These tests were run in an attempt to correlate cone resistance with some of the physical properties determined from the boring samples. After excavation, several more tests were run near the pile locations. These were followed by tests near the pile after the pile was driven and after cyclic loading. A total of twelve penetrometer tests were performed.

8.8.9.4 *Results of penetrometer tests*
Generally speaking, the cone resistance values and resistance to driving, Figure 8.20, increased with depth. Exceptions to this were in strata where there was a probable increase in the amount of fines in the sand. A layer of soft clay was located with the penetrometer at 1.8 to 2.4 m below the original ground surface. Presence of this layer is indicated in soundings P-l, P-2, P-3, and P-4 in Figure 8.20. Because of this undesirable soil near the proposed elevation of the top of the test pile, the soil was removed and backfilled with clean sand. A zone of low penetration resistance was located below 3.7 m, as indicated by the plots in Figure 8.20.

Of particular interest was the amount of increase in cone resistance due to densification of the soil resulting from pile driving. For Pile 1, this increase was determined by comparing results of penetrometer tests P-5 and P-6 done before driving with the results of tests P-9 and P-10 done after driving (see Fig. 8.20). The results of the comparison are shown in Figure 8.21a. This plot indicates that a significant amount of densification due to driving occurred in the upper 5-pile diameters. A similar study was made for Pile 2 by comparing the results of penetrometer tests P-7 and P-8 done before driving with the results of test P-11 done after driving. The results of this comparison are shown in Figure 8.21b, and are nearly the same as those for Pile 1. These studies show a definite improvement of the properties of the sand occurs as a result of pile driving.

A penetrometer test was performed at Pile 2, test P-12 (see Fig. 20), to allow a study to be made of the effect of cyclic loading on soil properties. The results of test P-12 were compared with results of test P-11 taken after pile driving to learn the magnitude of additional densification due to cyclic loading. The results of this study are shown in Figure 8.21b.

It is rather interesting to note that the most significant increases due to lateral loading occur below about 5 pile diameters and are fairly constant from there on down. In Fig-

Testing of full-sized piles 337

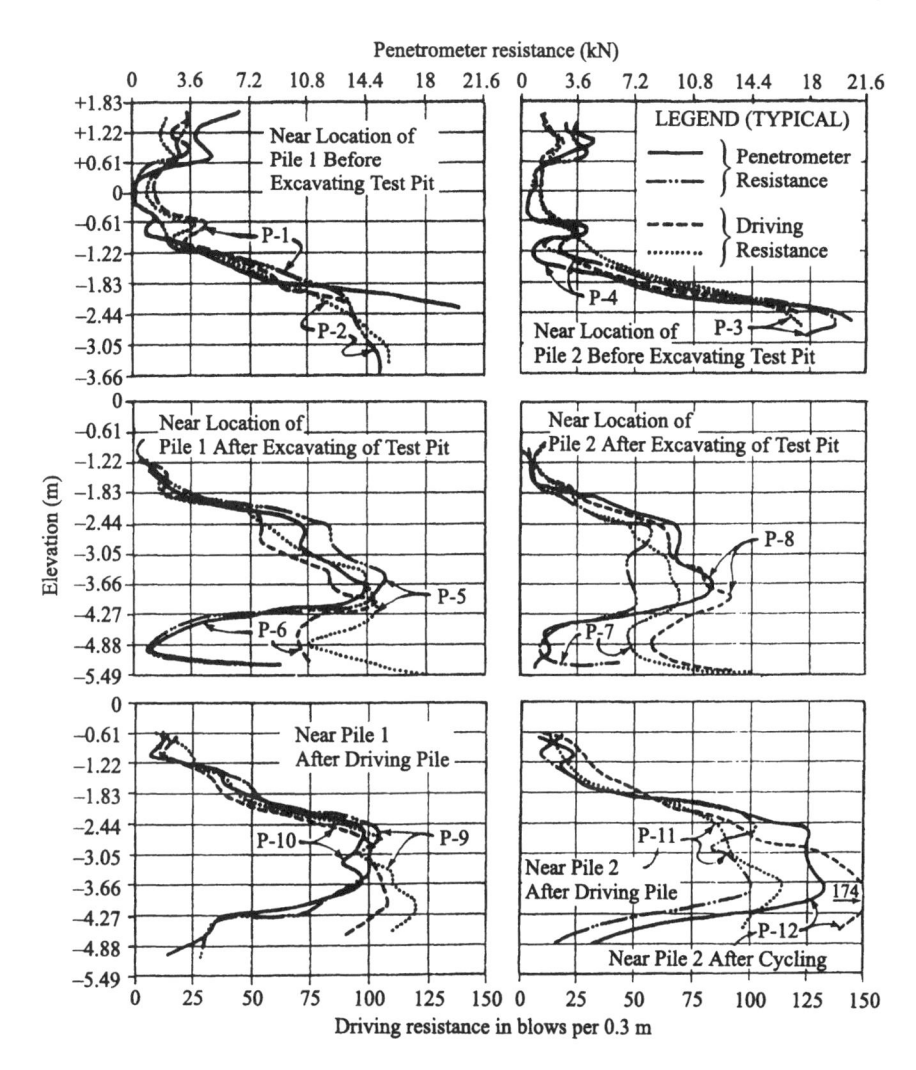

Figure 8.20. Results from penetration tests at various locations and at time related to installation of test piles, Mustang Island.

ure 8.21b the curve for cone resistance after cyclic loading may give support to an observation made during cyclic loading. This observation was that a deep wedge of soil actually followed the pile back and forth during the lateral movement. The wedge appeared to behave like a dense fluid and therefore would not be expected to change significantly in density.

Logs of soil borings near each pile show that the material above a depth of 12.2 m near the vicinity of Pile 2, Figure 8.16, was classified as a slightly silty sand versus a clean sand at Pile 1, Figure 8.15. Further evidence of the variation in properties horizontally is indicated by comparing results of penetrometer tests P-1 and P-2 with the results of P-3 and P-4, as shown in Figure 8.20. The cone resistance values and driving resistance values are much lower near Pile 2 as shown in P-3 and P-4 tests. After installing Pile 2, subse-

338 *Piles under lateral loading*

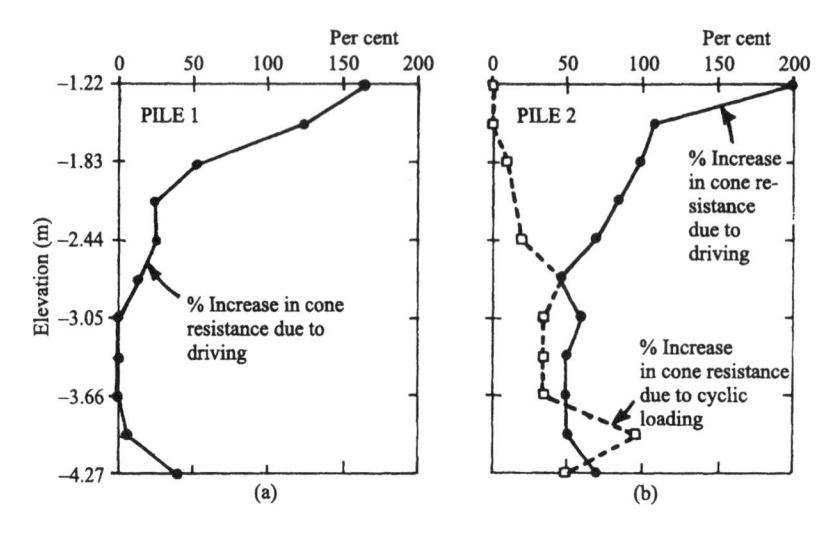

Figure 8.21. Changes in cone resistance due to installation of piles and to performance of cyclic loading, Mustang Island.

quent densification of the soil increased cone-resistance values to slightly more than the values prior to driving at Pile 1. From this comparison, indications are that the soil properties at Pile 2 after driving were very near those of the soil at Pile 1 prior to driving. The difference is not felt to be significant enough to cause a marked difference in behavior of the two piles.

Penetrometer tests were run adjacent to the boring holes in which Standard Penetration Tests had been performed. The cone resistance was compared with N_{SPT} values.

Studies were made to derive a direct correlation factor between measured cone resistance values and the N_{SPT} values. The ratio of q_c, from cone tests, to N_{SPT}, varied erratically from 0.24 to 3.7, indicating no direct correlation. Subsequent studies to relate this variation to depth likewise proved unsatisfactory. While a direct correlation between results of cone penetrometer tests and the SPT was not obtained for the tests, the data which were obtained are sufficient to yield reasonably qualitative information about the soil properties.

8.8.10 *Ground settlement due to pile driving*

Settlement of the ground surface due to driving the two 11.6-m uninstrumented sections was determined from elevations taken with an engineer's level. Data points for which elevations were determined were laid out in a polar coordinate system. Settlement around the two sections is shown in the contour plots in Figure 8.22. Settlement around Pile 2 was observed to be greater than that around Pile 1. As Pile 2 was being driven to grade, the 0.81-m-diameter shell around the hammer penetrated several inches into the soil, causing additional settlement. The contractor's operations interfered with attempts to take settlement data after the 9.75-m-instrumented sections were added to the 11.6-m sections. Visual observations indicated that very little additional settlement occurred during final driving after addition of the 9.75-m section.

8.8.11 Ground settlement due to lateral loading

At Pile 1, data were taken in order to determine the amount of ground settlement around the pile due to lateral loading. Differences between elevations before starting the static test and elevations as the pile was loaded with the maximum static load of 266.9 kN were used to plot the contours in Figure 8.23a. The contour plot in Figure 8.23b was drawn from differences between elevations taken before starting the static test and elevations taken after relaxing the loads back to zero.

Data were taken during the cyclic test on Pile 2 in order to determine the heave and slump in the major and minor load directions. The contours shown in Figure 8.23c were drawn after the maximum number of cycles at 244.6 kN and as the load was held on the pile. The contours plotted in Figure 8.23d were drawn as the minor cyclic load of 61.2 kN was held on the pile. In both Figures 8.23c and 8.23d the ground heave in the direction of the major load resembled a large U. During cyclic loading, a shear plane was observed in the approximate location as the right side of the U shape.

Another method of observation was used to gain insight into the behavior of the ground surface around Pile 2. Pieces of a wooden folding rule were placed vertically in the soil at various distances away from the pile. From these markers, horizontal movement was noted as far out as 760 to 890 mm from the pile surface and vertical movement was noted as far out as 1140 to 1270 mm.

8.8.12 Recalibration of test piles

Changes in the original gauge constants may have occurred after the driving so it was considered worthwhile to check the calibration constants after completing the tests and after the piles had been removed. The piles were sealed to eliminate intrusion of moisture until testing.

The piles were calibrated by the same procedure discussed earlier. Loads were measured by the same load cell used in the original calibration and in the field operations.

Strain readings were made for two series of loadings on each pile. Each series of loads consisted of 44.5, 89.0, and 133.4 kN applied at 3.35 m below the pile flange. The only

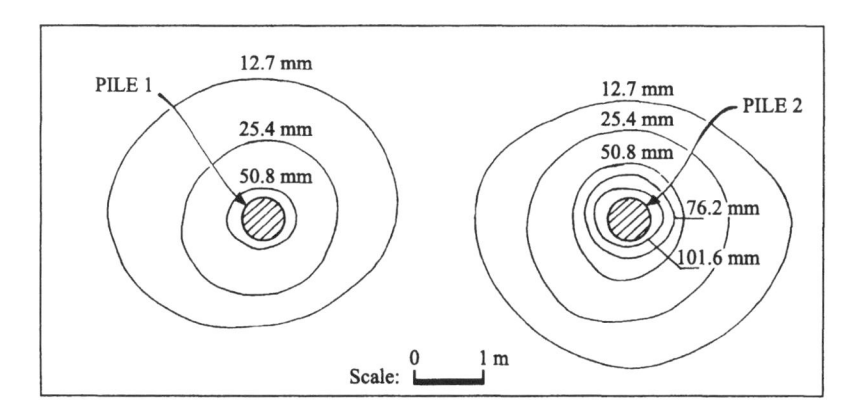

Figure 8.22. Sketches of settlement of ground around Pile 1 and Pile 2 due to driving of first section of 12 m, Mustang Island.

340 *Piles under lateral loading*

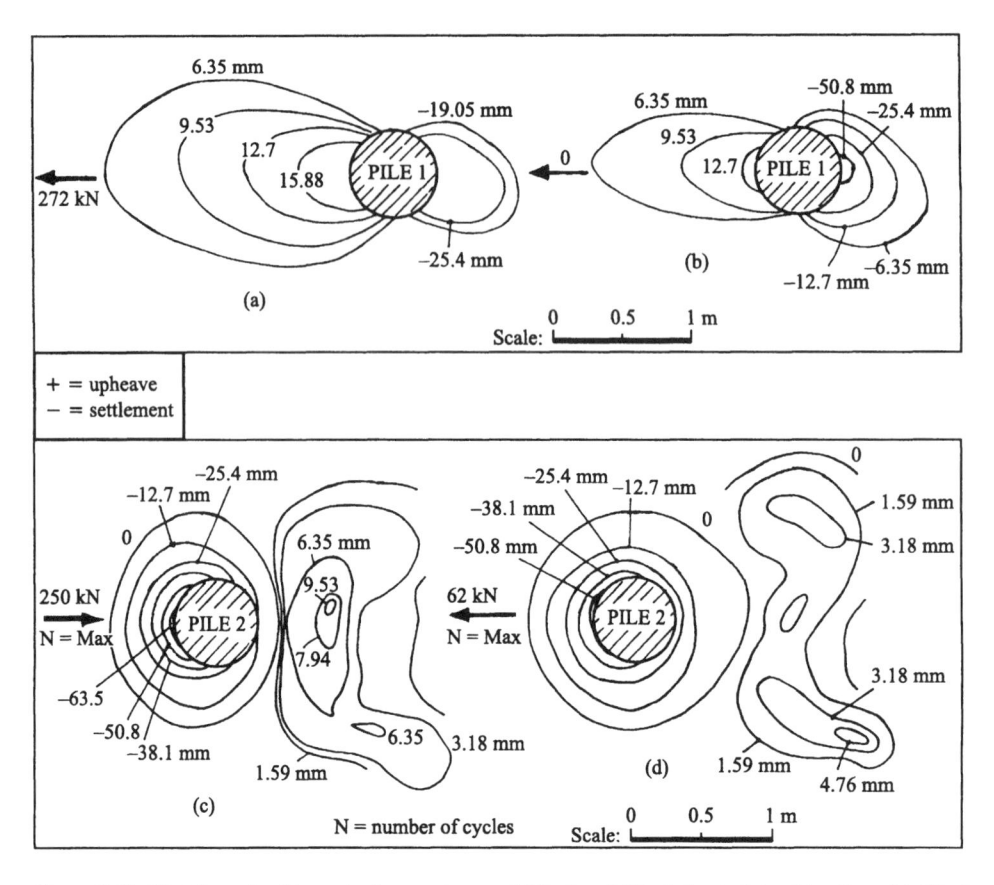

Figure 8.23. Sketches of settlement of ground around Pile 1 and Pile 2 due to lateral loading, (a) Pile 1 during static loading, (b) Pile 1 after static loading, (c) and (d) Pile 2 during static loading, Mustang Island.

gauge calibration coefficients that showed any deviation from the original ones were for gauges located at 1.52, 1.83, and 2.13 m in Pile 1. A slight tendency for these gauges to creep under load was noticed during the recalibration tests. But a study of the field test data indicated that the gauges did not creep during field-loading tests.

8.8.13 *Graphical presentation of curves showing bending moment*

8.8.13.1 *Bending moment curves from static load tests of Pile 1*
Curves of bending moment obtained from the static-load tests of Pile 1 are shown in Figure 8.24 (Reese et al. 1974). The static loads were increased progressively until stresses in the pile reached a maximum.

The moment curves in Figure 8.24 show that the maximum moment increased almost linearly with applied load. The curves also show that the depth to the point of maximum moment increased progressively with applied load until at 266.9 kN the depth was 0.3 m lower than shown by the curve for the 33.4 kN load.

Testing of full-sized piles 341

8.8.13.2 Bending moment curves for cyclic load test of Pile 2

Moment curves for cyclic load tests of Pile 2 are shown in Figure 8.25. The solid curves in Figure 8.25 represent the moment resulting from the first application of load ($N = 0$) at the indicated magnitude of load.

The dashed curves represent moment obtained from data for the maximum number of cycles (N = Max). As shown in Figure 8.25, the depth to the point of maximum moment increased progressively with the load until at the maximum load it was about 0.46 m lower than for the initial loading. A comparison of the depths to the maximum moments in Figure 8.24 with those in Figure 8.25 indicates that the depths to the maximum moment were about 0.3 m lower for the cyclic loads on Pile 2 than for the static loads on Pile 1. This result was expected as the soil resistance deteriorated due to cyclic loading.

The moment curves for the cyclic test shown in Figure 8.25 indicate that the maximum moment increased by a slightly larger amount for the same load increment than for the static test shown in Figure 8.24.

8.8.13.3 Comparison of bending- moment curves for static and cyclic loading

The relation between the moment curves for the static loading of Pile 1 and the cyclic loading of Pile 2 is shown in Figure 8.26. The comparison is presented to show graphically the influence of cyclic loading with respect to static loading.

8.8.14 Interpretation of bending moment curves to obtain p-y curves

8.8.14.1 Introduction

The direct method of analyzing bending-moment curves to obtain *p-y* curves is to integrate and differentiate the curves point-by-point along the length of the pile. The boundary conditions at the head and tip of the pile must be known. Four boundary conditions were

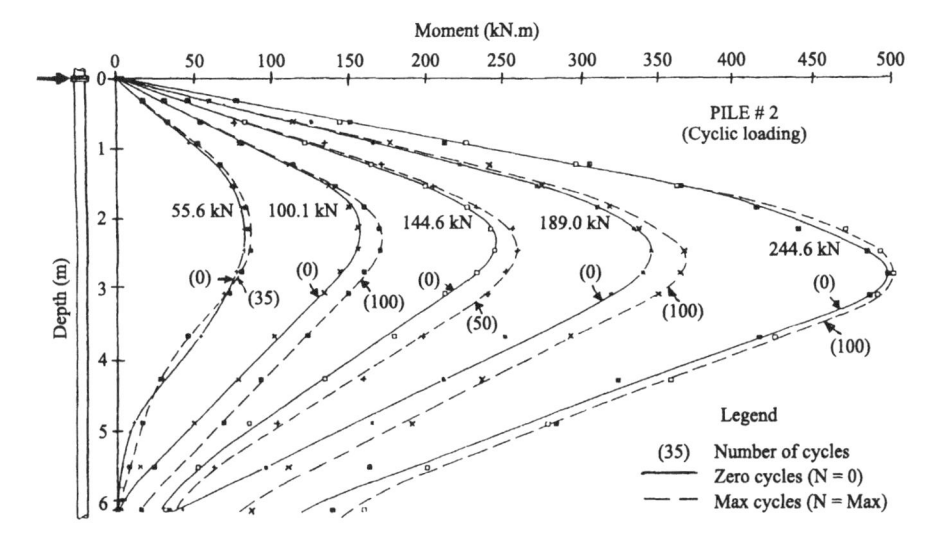

Figure 8.24. Curves showing bending moment as a function of depth from static loading of Pile 1, Mustang Island.

342 *Piles under lateral loading*

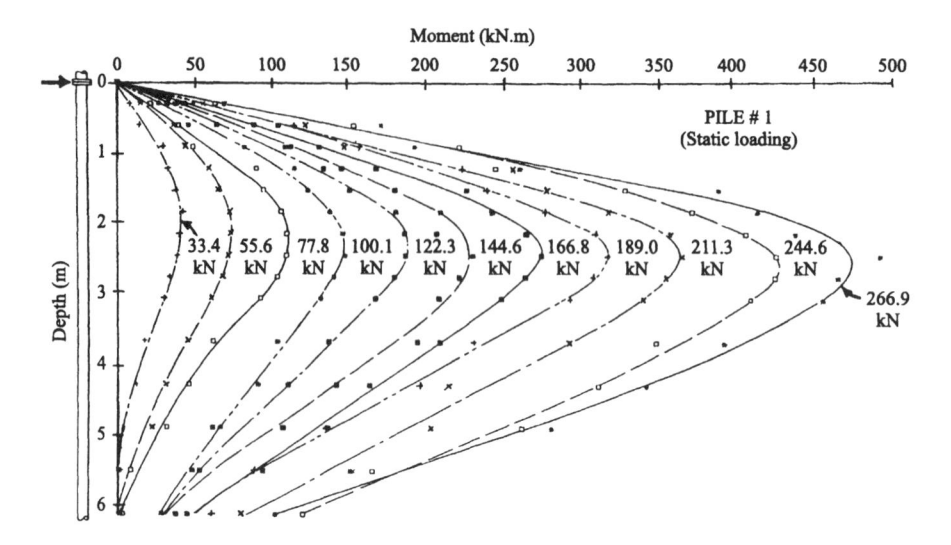

Figure 8.25. Curves showing bending moment as a function of depth from cyclic loading of Pile 2, Mustang Island.

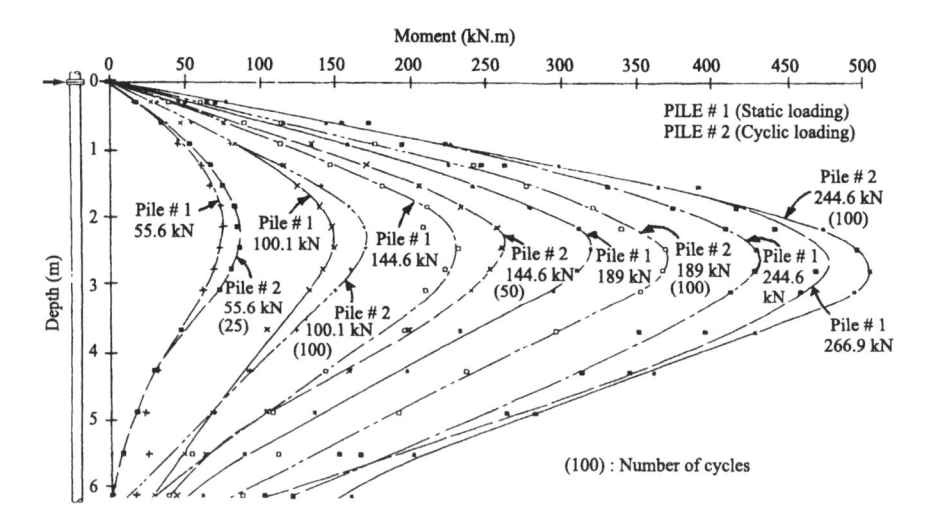

Figure 8.26. Curves showing bending moment as a function of depth from static loading of Pile 1 and from cyclic loading of Pile 2, Mustang Island.

measured at the head of the pile during the testing and the moment and shear at the tip of the pile can be taken as zero. The piles at the present site were long enough that the deflection and rotation at the tip were so small as to cause little analytical difficulty.

The deflection of the pile may be obtained by successive integrations and the soil resistance may be obtained by successive differentiations, as shown by the following equations. The result are sets of y-values and p-values, a set for each of the increments of load. The y and p values may be cross-plotted at discrete points along the length of the pile to obtain the experimental *p-y* curves.

The integrations can normally yield values of y with good accuracy. The problem is with the differentiations. Slight errors in the experimental values of bending moment will cause large and unacceptable errors in the values of p.

$$y = \iint \frac{M}{EI} dx \tag{8.1}$$

$$p = \frac{d^2 M}{dx^2} \tag{8.2}$$

Matlock & Ripperger (1956) solved the problem by constructing a special device by balancing and reading the modified Wheatstone bridge. Electrical-resistance strain gauges were installed at close spacings along a steel-pipe pile, accurate fixed resistances were used for crude balance of the bridge, and the final balance was accomplished with a feeler against a wire on a plastic drum. The drum was about 150 mm in diameter and 600 mm long. A spiral groove had been cut into the plastic and a resistance alloy wire, about 3 mm in diameter, was set into the groove. After selecting a strain level, a crude balance was achieved with the fixed resistances, and the drum was rotated by hand with the feeler against the resistance alloy wire to achieve the final balance. Values of micro-strain had been scribed on the drum. A displacement of about 15 mm along the resistance alloy wire meant one micro-strain.

The system yielded highly accurate readings but a set of readings required about one hour; the gauges were read down the pile and back up. Because of creep in soft clay into which the pile was installed, the readings had to be averaged for a particular specific time. Thus, the lateral load had to be monitored and held very accurately during the readings for each particular lateral load. Creep of the pile was not a problem for light loads but had to be considered for the larger loads.

Differentiations were performed step-by-step along the pile by using the data without adjustment. The resulting p-y curves formed the basis for recommendations for soft clay (Matlock 1970).

Welch & Reese (1972) performed tests on a bored pile that was constructed with a steel pipe down its center. The pipe was instrumented with electrical-resistance strain gauges, spaced closely. Calibration of the near-surface gauges was performed after the test was completed by excavating the soil around the upper portion of the pile, reloading the pile with measured loads, and reading the strain gauges as a function of the known distance from the point of application of load. Data were taken during the testing by an automatic data-recording system and a set of readings could be taken rapidly and without any significant change in the lateral load.

After the data had been corrected by use of the constants from the experimental calibration, a considerable amount of scatter remained. Welch fitted the experimental moment curves with analytical curves defined by polynomials of an appropriate order. The analytical curves could then be differentiated without difficulty. The final step was to use the p-y curves obtained in analyses to compute the bending moments that were observed. Welch found good agreement but not perfect agreement. The predictions for the p-y curves were then tested against other experiments and found to produce close to reasonable agreements between results from analysis and from experiment.

The data from the tests at Mustang Island, as noted earlier, were taken with an automatic data-acquisition system. As shown in Figures 8.24 and 8.25, bending-moment curves can be fitted through the data with good accuracy. However, the data did not have the high preci-

344 *Piles under lateral loading*

sion found by Matlock (1970) and some curve fitting was necessary. Reese & Cox (1968) had developed a method of analysis of uninstrumented piles to yield *p-y* curves where only deflection and rotation of the pile head were available, both as a function of lateral load and applied moment. An example employing the method is presented in Appendix H.

With values of y_t and S_t, one of the following two equations may be employed to find the unknown parameters, k_{py1} and k_{py2} in the case of Equation 8.3 and k_{pt1} and n_1 in the case of Equation 8.4. For the case of soft clay and for cohesionless

$$E_{py} = k_{py1} + k_{py2}x \qquad (8.3)$$

$$E_{py} = k_{py1}x^{n_1} \qquad (8.4)$$

material, as for Mustang Island, Equation 8.4 is plainly most applicable because E_{py} will approach zero as x approaches zero. Accordingly, Equation 8.4 was found to be helpful in fitting the experimental data obtained at Mustang Island, including the values of bending moment with depth and the various boundary conditions.

8.8.14.2 *p-y curves for static and cyclic loading*

The *p-y* curves at various depths for the static test on Pile 1 are shown in Figure 8.27. Curves for the cyclic test on Pile 2, corresponding to the maximum number of cycles are shown in Figure 8.28.

In the *p-y* curves shown in Figures 8.27 and 8.28, both the initial slope and the ultimate resistance increase with depth. A comparison of the *p-y* curves for the static test with the *p-y* curves for the cyclic test shows that the soil resistance decreases as a result of cyclic loading.

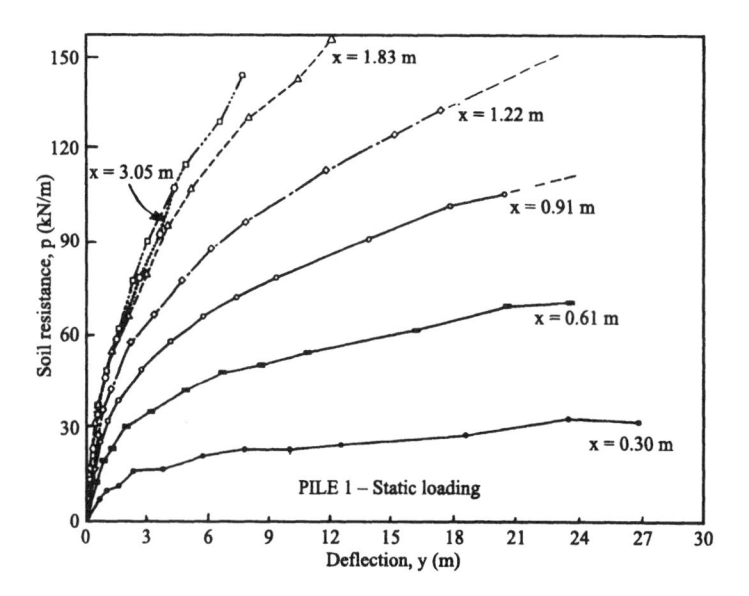

Figure 8.27. Curves showing soil resistance versus deflection (*p-y* curves) from analysis of results of the static loading of Pile 1, Mustang Island.

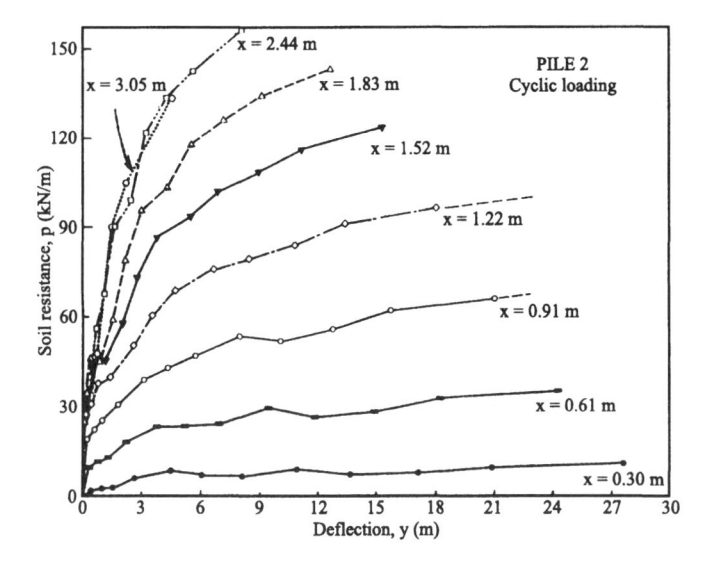

Figure 8.28. Curves showing soil resistance versus deflection (*p-y* curves) from analysis of results of the cyclic loading of Pile 2 at the maximum number of loading cycles, Mustang Island.

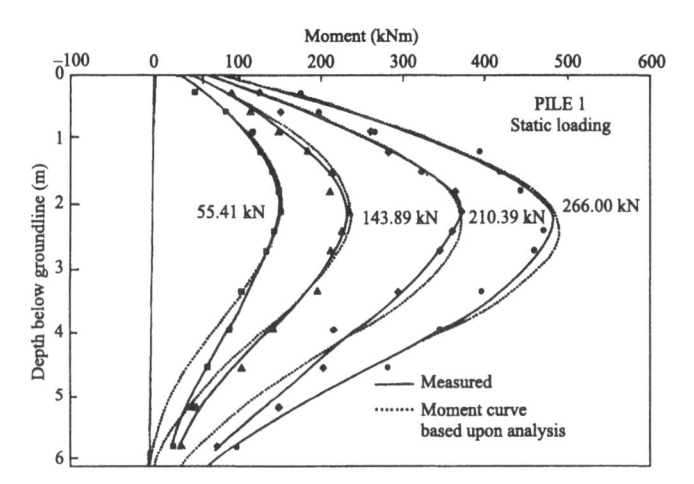

Figure 8.29. Comparison of experimental and computed values of bending moment as a function of depth for various lateral loads, static loading of Pile 1, Mustang Island.

For both sets of data, the greatest depth for which a *p-y* curve could be obtained was 5 diameters. The maximum deflection at that depth was about 4 mm.

8.8.14.3 *Comparison of computed and experimental moment curves for static load test on Pile 1 and cyclic load test on Pile 2*

In order to test the validity of the methods of analysis used to produce the *p-y* curves for Piles 1 and 2, those curves were used as input into a computer code and the bending moments as a function of depth were computed for loads for which experimental curves were available. The results of the comparisons are shown in Figures 8.29 and 8.30. All of the

346 *Piles under lateral loading*

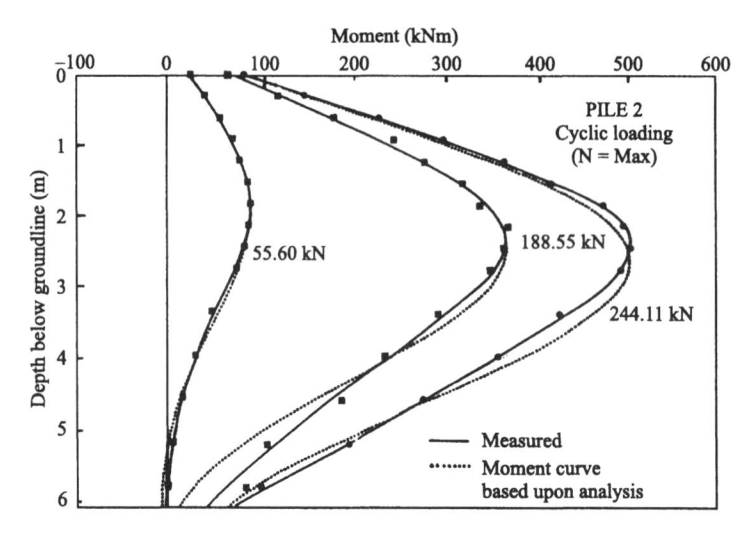

Figure 8.30. Comparison of experimental and computed values of bending moment as a function of depth for various lateral loads, cyclic loading of Pile 2, Mustang Island.

experimental and analytical curves agree closely from the ground surface down to and past the points of maximum bending moment. Most of the curves deviate at low values of bending moment at depths below the groundline of 4 to 6 m. In general, the bending moments for both static and cyclic loading are predicted with good accuracy.

8.9 SUMMARY

Methods of performing tests of piles under lateral load are presented and the use of the methods demonstrated. The results of two test programs are presented with the view of finding the resistance from the soil as a function of the deflection of a pile. The results from each of the test programs were successful, with the greatest success achieved where the instrumentation was most complete and where great care was exercised in the use of the instrumentation.

CHAPTER 9

Implementation of factors of safety

9.1 INTRODUCTION

Paramount in the mind of the engineer is to make a design with safety; secondly, with nearly as much importance, is to have the design perform well without excessive cost. The following sections will direct attention to the many of factors that enter into the selection of pile dimensions. The engineer may be guided by codes and standards of a building authority, but, even so, the engineer is left with considerable latitude. For example, in evaluating the data from a subsurface investigation, a very conservative or a less conservative set of values could be selected.

The *global* approach, where an overall factor is selected, is the usual method for the selection of a safety factor. A more recent method, the *component* approach, is where separate factors are given to various properties required for the design. Both methods will be discussed in some detail. However, a number of topics will first be discussed that are common to both methods.

Modern approaches to the design of pile foundations emphasize deformation as well as ultimate capacity. Deformation of a pile or a group of piles is an important aspect of some of the methods presented, which parallels to some extent the procedures in this book. Examples are present where successive solutions are made by incrementing the loading, nonlinear load transfer mechanisms are employed, and sometimes nonlinear pile material, and the load is found that causes collapse or excessive deformation. Such a procedure is valid, regardless of the method employed to select safety coefficients.

9.2 LIMIT STATES

The design concept using *limit states*, introduced in the early 1960s, is aimed at implementing more rationality regarding safety and is characterized principally by emphasis on allowable deformation. Thus, the concept of limit states broadened the view of the appropriate response of a structure to loading and to the environment. Table 9.1 is presented in consideration of the ways a pile under loading may fail to perform. The limit states in the table are related both to piles under both lateral and axial loading, because in several respects there is a close relationship between the two types of behavior. Chapter 5 demonstrates, for example, that a group of piles under inclined and eccentric loading cannot be designed properly without considering the response under both lateral and axial loading.

Considering the variety of loads to which a pile is subjected and the combinations that dictate design, there are a number of reasons that a single pile of a pile foundation will fail

Table 9.1. Limit states for a pile subjected to lateral and axial loading*.

Ultimate limit states	Most probable conditions
Sudden punching failure under axial loading of individual piles	Pile bearing on thin stratum of hard material
Progressive failure under axial loading of individual piles	Overloading of soil in side resistance and bearing
Failure under lateral loading of Individual piles	Development of a plastic hinge in pile
Structural failure of individual piles	Overstressing due to a combination of loads; Failure in buckling due principally to axial load
Sudden failure of foundation of structure	Extreme loading due to earthquake causing liquefaction or other large deformations; loading on a marine structure from major storm, or from an underwater slide
Serviceability limit states	
Excessive axial deformation	Design of large diameter pile in end bearing Foundation on compressible soils
Excessive lateral deformation	Design with incorrect p-y curves; incorrect assumption about pile-head restraint
Excessive rotation of foundation	Failure to account for effect of inclined and eccentric loading
Excessive vibration	Foundation too flexible for vibratory loads
Heave of foundation	Installation in expansive soils
Deterioration of piles in foundation	Failure to account for aggressive water; poor construction
Loss of esthetic characteristics	Failure to perform maintenance

*Peck et al. 1974, Feld 1968, Szechy 1961, Wright 1977.

to perform properly. Table 9.1 gives limit states under two categories; ultimate limit states and serviceability limit states. The table in not intended to be comprehensive but meant to indicate categories of catastrophic failures in the first instance and failures of adequate performance in the second instance. In a particular design, the engineer either sets down a formal limit state that will control the design or will implicitly have the condition in mind as the design proceeds.

9.3 CONSEQUENCES OF A FAILURE

The previous section presents some of the ways a foundation can fail. It is of interest to consider the consequences of a failure should one occur. A question usually addressed: will a failure cause a minor monetary loss with no possible loss of life, or will the failure be catastrophic with a large monetary loss and loss of life? On a particular design, the answer selected by the engineer plainly will guide the steps in the planning and computations. Sometimes, but not always, guidance for a design is given by codes and standards; even so, the engineer has a considerable measure of personal responsibility.

Some quantitative data, based on an estimation of historical events, is given if Figure 9.1 (slightly modified from Whitman 1984, and based on a private communication with him by G.B. Baecher). An interesting point shown in the figure is that the consequence of failure of a dam and a mobile drill rig is roughly the same but a greater probability of failure is acceptable for the drill rig. Many more people will be affected by the failure of the dam than by the failure of the drill rig; thus, the reaction of the public to the failure of a structure is important.

A comparison of the factors related to the design of a dam and a mobile drill rig shows the following: *subsurface investigation,* far more effort goes into the dam than the drill rig; *loading,* the loads on the dam are more predictable; *design methods,* those for the dam are more straightforward; *length of service,* far longer for the dam; and *design team,* the number of engineers on each team might be roughly the same. The list of factors above is not comprehensive but does serve to illustrate the complexity of selecting a numerical value of the probability of failure of a given installation.

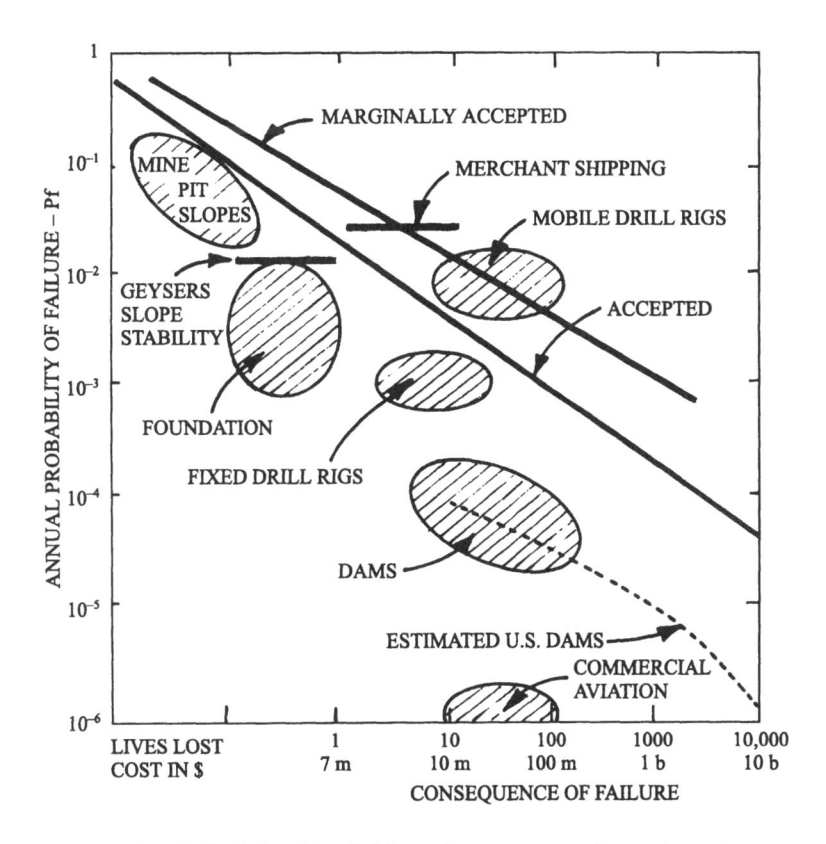

Figure 9.1. Historical relationship of risks and consequences for engineered structures (after Whitman 1984).

350 *Piles under lateral loading*

9.4 PHILOSOPHY CONCERNING SAFTEY COEFFICIENT

With respect to loads characterized by the curve on the left in Figure 9.2, the actual loadings and frequencies for each structure are unique and time-dependent. Thus, if one considers an offshore platform for the production of oil, different curves for loadings could be conceived for the construction period when resistances could be low, for the production period, and for the time after de-commissioning. With respect to the platform, the dead loads presumably could be computed with accuracy; however, the workers operating the platform have sometimes added steel members to facilitate some activity not originally anticipated.

With respect to live loads, operations could result in a wide range of loads on the deck with uncertain frequency. Live loads from a storm are predictable assuming the storm occurs with an assumed frequency of once in 100 years. In the Gulf of Mexico, for example, a considerable amount of data is available on the magnitudes of hurricanes that have occurred. However, the data become sparse in relation to a particular structure at a specific location. The characterization of the loading is complex when considering sea-floor slides, ship impact, marine growth, scour, and unanticipated events.

Characterizing the resistance to the loading on an offshore structure should present less difficulty than outlined above because, in the context of this book, the resistance in provided by deep foundations. However, the example computations presented earlier reveal that many factors enter in to the computation of the response of a pile, either to axial or lateral loading. In general, the factors are the properties of the pile, the properties of the soil, and the theories for behavior of the pile. Some of the weaknesses of the theories for computing pile behavior were discussed and improvements are to be expected. The properties of a pile can be predicted in many cases with accuracy, but piles of cast-in-place reinforced concrete can be expected to vary considerably in character from point to point along the length.

With respect to the properties of the soil, a discussion related to the offshore structure is pertinent. Many sites are investigated with wire-line techniques, where the sampling tool is driven with a drop hammer. Samples of clay are known to exhibit an unknown amount of loss of strength by such sampling. Sands may not be sampled at all, except in a very disturbed state, with the strength of the sand determined from the number of blows required to drive the sampling tool a given distance. Thus, at best, the properties determined from a wire-line investigation lie within a wide range of values. Then the properties

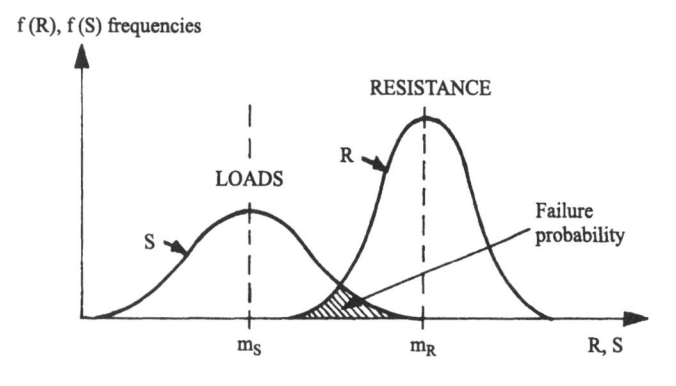

Figure 9.2. Probability frequences of loads and resistance.

of the soil are modified by the installation of a pile and, with clay, such changes are strongly time-dependent. The time-dependent response of the soil is further affected by the nature of the loading, whether short-term, repeated, or dynamic, and by such things as scour and soil deposition. Therefore, a family of curves, such as the one on the right in Figure 9.2 will be required with a multitude of factors taken into account.

With respect to selecting a coefficient of safety, the nonlinear response of the pile foundation to loading is of primary importance. In some approaches in structural engineering, an appropriate method is to select a value of strength of steel that includes an appropriate amount of safety. In foundation engineering, the nonlinear nature of the soil, and frequently the pile, requires that the safety must reside in the magnitude of the load because of the nonlinearity of response.

With respect to philosophy, the formalization of the process of computing the safety of a particular structure with mathematical equations is possible but a preferable approach is to rely on competent engineering. A competent engineer does not build a foundation that will suffer damage in the normal course of events but also will not fail by requiring far more expensive construction than necessary.

9.5 INFLUENCE OF NATURE OF STRUCTURE

The selection of a factor of safety to be implemented will depend on the type and purpose of the foundation (Wright 1977). When failure would possibly result in the loss of life, Meyerhof (1970) has suggested that a probability of failure of less than 10^{-2} percent will be acceptable. The problem, not addressed herein, is to translate a probability of failure, certainly a useful concept, into the selection of a global factor of safety or partial safety factors.

In terms of the nature of structure with respect to the length of time of satisfactory performance, Pugsley (1966) made the proposal that follows. 1. Monumental: life 200-500 years e.g., churches and large bridges, 2. Permanent: life 75-100 years, e.g. large buildings, ordinary rail and road bridges, and 3. Temporary: life 25-50 years, e.g. industrial buildings. The list is useful in that some guidance is given regarding the type of structure being designed. The list is not comprehensive and the engineer will need to expand the list to include a wider variety of structures.

9.6 SPECIAL PROBLEMS IN CHARACTERIZING SOIL

9.6.1 *Introduction*

The shear strength and the stiffness-strain of soil are the principal characteristics needed in the design for lateral loading. If a single method of subsurface investigation is employed, some considerable scatter may occur in the estimation of values of significant parameters. With intelligent and careful evaluation, the scatter will be less pronounced from a wealth of data, from a combination of common and special tests, such as triaxial in the laboratory, and such field tests as cone penetrometer, cone penetrometer with pore-pressure measurement, seismic cone, field-vane, spectral analysis of surface waves, in situ vane, dilatometer, and pressuremeter. The engineer should consider carefully the details of the tests performed before developing recommendations for p-y curves. Later on, in addition to assigning a factor for the quality of the soil-testing methods, another factor should account for the technique for evaluating the soil data itself. Safety coefficients, indeed, are not only related to the soil-testing quality and quantity, but are also linked to the method of design and the specific way of implementing the required, specific soil parameters.

The derivation of a relevant stress-strain level in selecting soil parameters for design from results of in-situ testing and laboratory testing has been the topic of many keynote lectures, reports from technical committees, and remarkable contributions from geotechnical-testing specialists all over the world, (Jamiolkowski et al. 1985, 1991, 1994; Robertson et al. 1983, 1990, 1993; Stokoe et al. 1988 and 1989 and Baldi et al. 1989). Further, with respect to geotechnical parameters, blending (statistically) results of tests from several locations, even on the same site, at an early stage in the design, may mask the crucial and essential variability of a geotechnical parameter and, thus, not allow detection of the most important phenomena of soil-structure interaction (Bauduin 1997).

An additional and even more relevant problem with respect to the selection of soil data is linked to the method of taking into account the influence of pile installation, because installation can affect significantly soil properties close to the pile and pile group (see Chapter 8 and Appendix G). Excess porewater pressures are caused by driving piles into fine-grained and cohesive soils and will dissipate with time, cohesionless soils can be densified due to the driving of piles, non-silica sands can be crushed during a displacement-pile installation. Thus, for driven piles, the initial-stress conditions and stress history change continuously during installation and perhaps for a significant time after driving. For bored piles or continuous-flight-auger (CFA) piles, stresses change due to the excavation and placement of concrete. The effects of installation are more pronounced for piles that are spaced closely together. Many comments are presented in technical literature about the effects of the pile installation on soil parameters and suggestions are given for design (Van Impe 1991, 1994, and 1997 et al.). Consequently the computation of the capacity of piles from soil-test data should include an installation coefficient α_b for the tip capacity and a coefficient ξ_f for the shaft capacity. Both coefficients can be either lower than 1 in case of bored or CFA-type of piles, and either higher or lower than 1 in case of soil-displacement type of piles. A thorough evaluation of those installation parameters for the given pile type at the test site can also be the outcome of dilatometer testing during installation, as described in Chapter 8 (Van Impe & Peiffer 1997).

The geotechnical engineer, from a review of the technical literature, from experience, and from engineered observations at construction sites, should evaluate the expected effects of each type of installation of piles in assigning soil data relevant to a single pile or a pile group.

9.6.2 *Characteristic values of soil parameters*

The commonly applied student-t distributions or standard-deviation estimates are established by assuming a normal distribution of the relevant geotechnical parameter.

However, in most cases the log-normal distribution (i.e. logarithm of the parameter in a normal distribution) is a more reliable procedure:

$$X_i' = \log X_i \tag{9.1}$$

$$\overline{X'} = \frac{1}{n}\Sigma X_i' = \frac{1}{n}\Sigma \log X_i \tag{9.2}$$

$$s = \sqrt{\frac{1}{(n-1)}\Sigma\left(\overline{X'} - X'_i\right)^2} \tag{9.3}$$

As geotechnical parameters are always positive in value, they are in fact as such not normal-distributed values, very seldom are enough tests available to make a final, reliable choice of the relevant distribution. Large values of the variation coefficient therefore require the choice of a log-normal distribution.

For many geotechnical designs, one usually has no more than a few field investigation tests (cone penetrometer, or Standard Penetration, for example) or a boring log with some classification tests. Characteristic values can be obtained using tables (regional experience) in which the field measurement (e.g. cone resistance) or the classification-test results are used as an input to obtain the value of the soil property required. The numbers in such tables are, of course, conservative estimates. Further steps are to compare the value of the soil parameter with an existing geotechnical map and/or to use experience with similar soils in the area.

The choice of characteristic values becomes much more complicated for complex geotechnical problems in which sophisticated codes and models are used. Standard charts in such case provide only a first estimate, and usually leads to a conservative design.

Assessing a characteristic value from the test-derived data for soil parameters should cover the uncertainties related to stochastic variations, taking into account the soil volume that was tested, the nature of the building or structure, the soil-structure-interaction stiffness, the type of sampling, and the overall engineering practice.

9.7 LEVEL OF QUALITY CONTROL

The degree of responsibility accepted by the engineer weights heavily in view of a litigious society. The assurance of quality of the work of each of the contractors and designers plays an important role in accomplishing a successful project. Related to a successful project is the amount and quality of the external inspection of the work.

The first aim must always be to avoid omissions. Liability can never remedy the damage caused during design and implementation by failure to address all of the necessary and relevant elements of the project. Strong professional management by a responsible party, and the participation of experts from all required disciplines (forming interdisciplinary design teams) are seen as two very important factors for success.

One approach frequently used is to tender on the basis of specifications and a description of functions, instead of on a bill of quantities. The so-called *systems approach* encourages the contractor to bid alternative technical and economic solutions.

9.8 TWO GENERAL APPROACHES TO SELECTION OF FACTOR OF SAFETY

The first of the two general methods is termed the *global* approach, The engineer will consider all of the factors at hand, including such things as the quality of the subsurface investigation, the statistical nature of the loading, and the expected competence of the contractor, and an overall factor of safety is selected for individual piles and for the group of piles.

354 *Piles under lateral loading*

The second of the two methods is termed the *component* approach. For a particular design, the components of loads and resistances are identified and an independent factor of safety is selected for each. The independent factors can be combined to yield and overall factor. Two examples are presented for the component approach: the method of *partial safety factors* and the method of *load and resistance factors*. The first of these two methods has been used informally in Europe for many years and discussions are currently underway relative to formal acceptance. The second of the two methods was accepted formally in 1994 by the American Association of State Highway and Transportation Officials (AASHTO) as a standard.

The global approach to selection of the factor of safety and the two component approaches will be discussed in the following sections.

9.9 GLOBAL APPROACH

9.9.1 *Introductory comments*

Engineers have traditionally used a global factor of safety for the design of piles, giving careful consideration to all pertinent parameters influencing behavior. The value in the use of such an overall factor is that the engineer may use judgment in the selection of relevant parameters. For example, the shear strength of the soil may be chosen more liberally or more conservatively, depending on the entire character of the design. Examples of the use of global factors of safety for various geotechnical structures have been discussed by many geotechnical engineers (Feld 1968, Meyerhof 1970, Peck 1965, Pugsley 1966, Szechy 1961, Terzaghi 1962, De Beer 1961, 1965, 1976, 1981 et al., Franke 1973, 1990 and Costanzo-Lancellotta 1997).

Plainly, the engineer aims to prevent a failure of the structure. However, the precise definition of *failure* may be difficult, leading to possible misunderstandings in communicating with the owner and others. Therefore, the need for the structure to perform as expected by the owner over its life needs to be understood by all relevant parties.

With the participation of the owner, risk of 'failure' should be reduced to an acceptable level. The model in Figure 9.2 shows, in elementary form, the distribution of loads and resistances. The probability of 'failure' is governed by the overlapping zone. One has to select the global factor of safety F_c in order to keep the probability of a failure to an acceptable level.

$$F_c = \frac{m_R}{m_S} \qquad (9.4)$$

where m_R = mean value of resistance R, and m_S = the mean value of load S.

Eurocode presents a number of principles related to the design of geotechnical facilities, as shown in Appendix I. Even though the material in Appendix I is marginally related to the design of single piles and pile groups under lateral loading, the ideas there are useful in considering the selection of a factor of safety for a particular design.

Examples can be given of many agencies that use a global factor of safety. The API was selected in the presentation shown below because the use of piles is so important in their offshore operations. Further, petroleum companies sponsored research in the United States that led to some of the methods used in design.

9.9.2 *Recommendations of the American Petroleum Institute (API)*

9.9.2.1 *Design considerations*

The kinds of loading to be employed in design were mentioned in Chapter 6 and a brief discussion was given of means of computing the magnitude of the various kinds. The API (1993) suggests that estimated interval for the recurrence of the storm selected for computation of loadings should be several times the expected life of a structure. Such experience as available in the Gulf of Mexico, for example, suggests that the storm expected to occur once in 100 years (the 100-year storm) be used for design. Thus, after a particular location for a structure is selected, the engineer is faced with estimating the maximum wave height for the 100-year storm, the likely directions of the storm, and other factors to allow the computation of the magnitude of the vertical, horizontal, and overturning forces, as a function of time.

The storm loadings allow the engineer to formulate a collection of loadings for the life of the structure: fabrication, transportation, installation, normal operation, special operation, and removal. Considering the suite of loadings, various piles in the foundation may have critical loads at one time or another. In some instances, a risk analysis for the structure is recommended.

The API recommends an appropriate soil investigation, including the necessary borings and laboratory testing. For both axial and lateral loading on a pile, the development of data showing load versus deflection is required.

9.9.2.2 *Design of piles under axial loading*

The design of piles under axial loading follows the usual procedures of computing load transfer in skin friction and end bearing from pile dimensions and soil properties. Limiting values of the load-transfer coefficients are required, probably because of maximum values developed in full-scale experiments of axially loaded piles. With respect to the penetration of piles to develop the required capacity in compression and tension, the factors of safety in the following table are given.

Load condition	Factor of safety
Design environmental conditions with appropriate drilling loads	1.5
Operating environmental conditions during drilling operations	2.0
Design environmental conditions with appropriate producing loads	1.5
Operating environmental conditions during producing operations	2.0
Design environmental conditions with minimum loads (for pullout)	1.5

The above presentation of the design of axially loading piles according to the recommended practice of API is by no means comprehensive but does serve as an example of

356 *Piles under lateral loading*

the use of global factors of safety. Presumably, the factors are applied to augment the axial loads.

9.9.2.3 *Design of piles under lateral loading*

The use of *p-y* curves is recommended in solving for the capacity of piles under lateral loading and details on several sets of curves are presented in the API manual. With regard to stresses in the piles, API recommends the use of the equations from the American Institute of Steel Construction. The equations include a factor of safety, based on the yield strength of the particular steel being used and based on the particular combination of stresses due to axial and lateral loading.

The writers have called attention to nonlinear nature of the response of soil; thus, relying on a safe level of stress in some cases can lead to a quite low factor of safety. In many of the examples presented herein, loads are factored to find the failure condition, usually a plastic hinge, which leads to a better idea of the global factor of safety for the particular design. Experience shows that many designers are adopting this latter approach, in addition to achieving compliance with allowable stresses.

9.10 METHOD OF PARTIAL SAFTEY FACTORS (PSF)

9.10.1 *Introduction*

A discussion of implementation of safety factors implies a preliminary agreement on the definition of failure. The suggestion of Franke (1991) can be introduced by considering, for example, the design of axially loaded single piles. A comprehensive discussion is presented in Appendix J, in which failure is related to a collapse load and to excessive settlement. While partial safety factors are not used by Franke in Appendix J, the emphasis of pile settlement is consistent with the general aim when partial factors are used.

9.10.2 *Suggested values for partial factors for design of laterally loaded piles*

Table 9.2 presents a list of suggested coefficients for partial safety factors for the design of piles under lateral load and under axial load. Earthquake loadings are omitted because of their special nature. For design of foundations in seismic zones, the engineer will follow the codes and specifications with regard to designing to resist the effects of earthquakes.

In terms of the kind of structure that is being designed, the assumption is made that the list applies to permanent or temporary structures only. If a monumental structure is being designed, the presumption is made that the engineer will undertake special studies regarding loading and the selection of materials, and that construction will be done under extra precautions. The columns of poor control, normal control, and good control are assumed to reflect the quality of the soil investigation and the quality of construction.

With regard to partial safety factors, the resistance (R^*) for design is given by Equation 9.2.

$$R^* = r_m(\gamma_m \gamma_f \gamma_p) \tag{9.5}$$

where r_m = the mean resistance or strength; γ_m = partial safety factor to reduce the material to a safe value; γ_f = partial safety factor to account for fabrication and construction; and γ_p = partial safety factor to account for inadequacy in theory or model for design.

Table 9.2 Suggested partial coefficients for analysis of piles. (*NOT FOR DESIGN*)

Partial load factors for	Designation	POOR CONTROL	NORMAL CONTROL	GOOD CONTROL
Dead weight, water pressure, water loads	γ_1	1.0	1.0	1.0
Bulk goods in silos, fluctuating water pressure	γ_1	1.3	1.3	1.15
Braking or equipment forces	γ_1	1.3	1.3	1.15
Wind loads, wave loads	γ_1	1.5	1.5	1.25
Analysis of specific loads	γ_2 (value based on uncertainty of occurrence and for unforeseen change in design assumptions)			
Partial safety factors for design for lateral loading				
p-y curves for soft clay, stiff clay above water, collapse	$\gamma_m\gamma_p$	1.7	1.65	1.6
p-y curves for soft clay, stiff clay above water, deflection	$\gamma_m\gamma_p$	1.85	1.8	1.7
p-y curves for stiff clay, subject to erosion, collapse	$\gamma_m\gamma_p$	2.2	2.0	1.8
p-y curves for stiff clay, subject to erosion, deflection	$\gamma_m\gamma_p$	1.8	1.7	1.6
p-y curves for sand, collapse	$\gamma_m\gamma_p$	1.7	1.65	1.6
p-y curves for sand, deflection	$\gamma_m\gamma_p$	1.85	1.85	1.7
M_{ult} for steel piles	γ_m	1.5	1.4	1.3
M_{ult} for reinforced concrete piles	γ_m	1.9	1.85	1.8
EI for steel piles	γ_m	1.5	1.4	1.3
EI for reinforced concrete piles	γ_m	2.1	2.0	1.9
Partial safety factors for design for axial loading				
for axial loading unit side resistance from load test	$\gamma_m\gamma_p$	1.7	1.6	1.5
Unit side resistance, no load test	$\gamma_m\gamma_p$	1.85	1.7	1.63
End bearing	$\gamma_m\gamma_p$	1.77	1.62	1.54
AE for steel piles	$\gamma_m\gamma_p$	1.5	1.4	1.3
AE for concrete piles	$\gamma_m\gamma_p$	2.0	1.9	1.8
Collapse load for steel piles	$\gamma_m\gamma_p$	1.5	1.4	1.3
Collapse load for concrete piles	$\gamma_m\gamma_p$	2.0	1.9	1.8

With regard to partial load factors, the design load (S^*) is given by Equation 9.6.

$$S^* = s_m \gamma_1 \gamma_2 \tag{9.6}$$

where s_m = mean value of load; γ_1 = partial load factors to estimate the safe level of the loading; γ_2 = implementing modifications during construction that cause an increase in the loads; effects due to temperature and creep; and other similar reasons.

Employing the partial safety factors shown in Equations 9.5 and 9.6, a global factor of safety may be computed as:

$$F_c = (\gamma_m \gamma_f \gamma_p)(\gamma_1 \gamma_2) \tag{9.7}$$

358 *Piles under lateral loading*

9.10.3 *Example computations*

To illustrate the use of Equation 9.7, the design of a pile to support an overhead sign may be considered, where the lateral load is due to wind forces. The axial load is assumed to be negligible. The assumption is made that the factor γ_1 has been studied and assigned a value of 1.5 to account for the increase in wind loading above the values recommended in codes and that the factor γ_2 has been selected as 1.0. The assumptions are further made that a steel pile will be erected in stiff clay above the water table, that the soil investigation was meager, and that construction will be inspected by a technician; thus, poor control is assumed (relating to the selection of γ_f). Because deflection of the sign would not control, only the collapse of the structure is to be considered. With regard to collapse, the engineer made certain that the pile would penetrate a sufficient distance that long-pile behavior would occur. In addition to a value for γ_1, the following values have been selected from Table 9.2: $\gamma_1 = 1.5$; $\gamma_m \gamma_p = 1.7$. Therefore, the value of F_c = that would be selected for the design of the overhead sign would be $(1.5)(1.0)(1.5)(1.7) = 3.83$.

The problem of the overhead sign is now presented with another set of assumptions. Wind loading has been studied in the area and the value of γ_1 can be selected as 1.25; the value of γ_2 will remain at 1.0. An excellent soil investigation is assumed, a concrete slab will be placed at the ground surface (not given a *p-y* curve) to protect against environmental changes, and the construction will be closely inspected by an engineer. Thus, γ_f can be selected as 1.0 and the value of $\gamma_m \gamma_p$ can be reasonably selected as 1.5. Therefore, the value of the overall safety factor can be computed as $(1.25)(1.0)(1.0)(1.5) = 1.88$.

The two computations presented above are not intended to reflect the solution to any particular problem but merely indicate briefly the kinds of analyses necessary to select the size of pile to perform a certain function. The PSF method is a useful tool but, at present, still requires special care at implementation.

9.11 METHOD OF LOAD AND RESISTANCE FACTORS (LRFD)

9.11.1 *Introduction*

As with the method of partial safety factors, the LRFD specifications of AASHTO present methods of modifying the component loads and the component resistances. The basic equation is shown below.

$$\sum \eta_i \gamma_i Q_i \leq \phi R_n = R_r \qquad (9.8)$$

where η = factors to account for ductility, redundancy and operational importance; γ_i = load factor; Q_i = force effect, stress or resultant; ϕ = resistance factor; R_n = nominal (ultimate) resistance; and R_r = factored resistance.

As may be seen, several features of the method are similar to the method of partial safety factors. The engineer, in getting a solution to Equation 9.8, must estimate the loads and load combinations that may be imposed on the structure, and estimate the ultimate resistance available to resist the loading.

9.11.2 Loads addressed by the LRFD specifications

A large number of types of loads are considered in the LRFD specifications, including, dead load of structural components and nonstructural attachments; dead load of wearing surface and utilities; horizontal load from earth pressure; load from earth surcharge; vertical load from earth fill; load from collision of a floating vessel; load from collision of vehicles; load from an earthquake; ice load; vertical load from dynamics of vehicles; load from the centrifugal force of vehicle traveling on a curve; load from the braking of vehicles; live load from vehicles; live load from surcharge; live load from pedestrians; load from water pressure in fill; load from currents in stream; loads due to changes in temperature of structure; wind load on structure; and wind load on vehicles. Each of the types of loads is discussed (NHI 1998) and some guidance is given in making the computation of magnitude of the load.

A number of basic load combinations (called limit states) are identified for use in design. The combinations are shown below.

Strength I, the basic load combination related to the normal vehicular use of a bridge without wind.

Strength II, the load combination relating to the use of the bridge by owner-specified special design vehicles and/or evaluation-permit vehicles, without wind.

Strength III, the load combination relating to the bridge exposed to wind velocity exceeding 90 km/hr without live loads.

Strength IV, the load combination relating to very high dead load to live load force-effect ratios, exceeding about 7.0 (e.g., for spans greater than 75 m).

Strength V, the load combination relating to normal vehicular use of the bridge with wind velocity of 90 km/hr.

Extreme Event I, the load combination including earthquake.

Extreme Event II, the load combination relating to ice load or collision by vessels and vehicles.

Service I, the load combination relating to the normal operational use of the bridge with a wind velocity of 90 km/hr.

Service II, the load combination intended to control the yielding of steel in structures and the slip of slip-critical connections due to vehicular live load.

Service III, the load combination relating only to tension in prestressed-concrete structures with the objective of control of cracking.

Fatigue, the combination of loads relating to fatigue and fracture from repetitive gravitational live load from vehicles and the dynamic responses under a single design truck.

Construction, the combination of loads relating to the live load from construction equipment during the installation/erection of structures.

While all of the load combinations noted above are considered in the design of foundations, the ones that usually control are the Strength I and the Service I. The two conditions are related to the computation of 1. Ultimate capacity and 2. Deformation as would be done if the engineer used specifications based on allowable stress design.

9.11.3 Resistances addressed by the LRFD specifications

The principal emphasis of the LRFD specifications in regard to resistance resides in the determination of values of geotechnical parameters. The process for planning and execut-

360 *Piles under lateral loading*

ing a program of surface investigation is described (NHI 1998); the sources of variability in estimating the properties of soil and rock are described; and the statistical parameters are identified that can lead to the selection of a resistance factor. The various items are discussed that pertain to the selection of the magnitude of the parameter ϕ.

An example is presented below for the design of a steel pile under axial load that illustrates the procedure in selecting the parameters used in design.

9.11.4 *Design of piles by use of LRFD specifications*

With regard to lateral load, the specifications note that the piles must be designed to avoid structural failure and without excessive deflection. The method used in allowable stress design can also be used in design using load and resistance factors. The usual procedures, principally those used in the previous chapters of this book, are noted.

With regard to axial loading, an example that follows will demonstrate the application of the method presented in the LRFD specifications (NHI 1998). The case was a bridge pier supported by steel piles. The dead load of structural components and non-structural attachments (DC) was 4600 kN; the dead load of wearing surface and utilities (DW) was 3900 kN, and the live load from vehicles (LL) was 3450 kN. Referring to the specifications, the value of γ was selected as 1.0 and the various load factors were selected as follows: $\gamma_{DC} = 1.25$; $\gamma_{DW} = 1.50$; and: $\gamma_{LL} = 1.75$. Therefore, the factored load is $1.00[(1.25)(4600) + (1.50)(3900) + (1.75)(3450)] = 17,638$ kN.

The axial capacity of a single pile was estimated from results from the Standard Penetration Test. The length of the pile was 11 m. The computed value of load in end bearing was 1.460 kN and in side resistance was 445 kN. The tabulated value of the factor for end bearing ϕ_{qp} was 0.45 and the factor for side resistance ϕ_{qs} was also 0.45. Thus, the factored axial resistance of the single pile Q_R was $0.45 (1460 + 445) = 857$ kN. Discounting the loss of resistance due to close spacing, the number of piles required for the foundation of the bridge pier was $17,638/857 = 20.6$; use 21. It is of interest to note that the global factor of safety for the example was $(21)(1460 + 445)/(4600 + 3900 + 3450) = 4.95$. If the capacity of a pile is based on the results of load tests, the global factor of safety would have been much smaller.

Only a portion of the procedure for design of the piles for supporting a bridge pier is presented; for example, the check of the settlement of the piles under service load is omitted. However, the example does show the factoring of both load and resistance.

9.12 CONCLUDING COMMENT

An important point must be made about the design of pile foundations, where the response of the soil and the pile material are both nonlinear with load. Fundamentally, *the best estimate of the nonlinear response of the soil and of the pile material must be employed in analyses and the factor of safety must reside only in the loading.* That is, the best estimate of load-distribution curves must be used to find the load on a foundation, perhaps inclined and eccentric, that will cause collapse or will cause deflection that is intolerable. If combined loading is to be sustained by a single pile or by a group of piles, the engineer must exercise judgment in using a factor to increase each of the independent loads.

While a consideration of the concepts involved in the selection of factors from the PSF method or from the LRFD method is useful and desirable, the best estimates of response of single piles and groups of piles can be made by use of p-y, t-z, and q-w curves and from factors giving the interaction between closely spaced piles. Therefore, emphasis in research is recommended in finding the nonlinear curves along full-scale piles in the field in a variety of soils and rocks that account for the influence of installation and the influence of the nature of loading.

CHAPTER 10

Suggestions for design

10.1 INTRODUCTION

As presented in earlier chapters, a single pile or a group of piles under lateral load responds nonlinearly to applied load and previous chapters have demonstrated the importance of incrementing loading to find the load that will cause collapse or cause excessive defection. Further, a number of case studies were presented to demonstrate that the models employed for pile and soil give answers that agree with experiment closely or within reasonable limits.

Two areas of design need further discussion: 1. The broad range of factors that must be considered; and 2. Ensuring the validity of results from the computer code. These two topics will be addressed in the following paragraphs.

10.2 RANGE OF FACTORS TO BE CONSIDERED IN DESIGN

The brief presentation in Chapter 9 presented a number of limit states that must be considered. In connection with an approach to design beyond merely computational, Professor Ralph B. Peck, former President of the International Society of Soil Mechanics and Foundation Engineering, made an important contribution in the opening lecture at a meeting aimed principally at computational methods (1967). Professor Peck made four points that will be presented with a brief discussion.

1. The assumed loading may be erroneous; 2. The soil conditions may differ from those on which the design is based; 3. The theory upon which the calculations are based may be inaccurate or inadequate; and 4. Construction defects may invalidate the design.

Professor Peck gave examples of each of the points that will be reviewed here. A warehouse was being designed for holding petroleum products. When the cans were stacked with a mechanical loader, the floor load was 2.22 kPa which was selected for design. However, when the necessary aisles were taken into account the load was reduced to 1.78 kPa, as significant reduction.

Professor Peck noted that 'by and large throughout the world, soil conditions are erratic rather than homogeneous. Yet, the implications of the heterogeneity of soil deposits are still not properly appreciated. Few soil deposits are uniform enough to warrant an elaborate investigation of their properties.' Further, the mere process of sampling causes changes in properties. At best, the engineer is likely to be confronted with a considerable range of scattering from the average values of strength and compressibility. The response of the engineer to the erratic value of soil properties must reflect the nature of the design.

Professor Peck noted that in many cases the quality of the theory is better that our ability to predict the properties of some soils. However, he mentioned some aspects of theory (1967) that may not have been addressed fully or to some degree: 1. Behavior of soils under cyclic loading of rather high frequency, 2. Behavior of soils under random loads as from earthquakes, 3. Response of soils from extremely large, rapidly applied loads as from blasts; and 4. Prediction of extremely small motions of foundations under cyclic loads.

The problem of the mistakes in the construction of foundations, Professor Peck noted, 'can void the best efforts of the most able designers, even if their knowledge of loads, soil conditions, and theory is virtually perfect.' In regard to the emphasis of Professor Peck on the quality of construction, the senior author has had a close connection with the bored-pile industry in the United States. Several failures have occurred in recent years. With one possible exception, the several failures that have occurred have been due to construction deficiencies. The possible exception is that the excavation for the bored pile allowed water to penetrate to a stratum of expansive clay at the base of the pile. High quality construction probably could have prevented that failure.

10.3 VALIDATION OF RESULTS FROM COMPUTATIONS FOR SINGLE PILE

10.3.1 *Introduction*

At the outset, the engineer must accept that the solution of the problems of single piles under lateral loading and pile groups of piles under inclined and eccentric loading are complex. Even though computer codes are available that yield output with relative ease, most designs will require many trials. Sufficient time must be made available in the design office, in spite of the speed of the computer, to 1. Refer to case studies (Chapter 7) for solutions of problems similar to the one at hand; 2. Make additional computer solutions with varied parameters, for example, with upper-bound and lower-bound values of shear strength; 3. Check computer output to see that boundary conditions are satisfied; 4. Run enough solutions that a 'feel' is developed for the validity of output for a particular problem; 5. Make a hand solution for the problem using nondimensional curves (Chapter 2); 6. Verify the accuracy of the computer output by use of mechanics and other means; and 7. Establish a program of peer review. The last two points are discussed in the following paragraphs.

10.3.2 *Solution of example problems*

Most computer codes, as a matter of standard practice, include example problems with input and output. The examples should be coded and solutions obtained to compare with the output. Then, the engineer will have some valid output for study. Some input parameters can be changed. For example, the influence of varying the bending stiffness on bending moment can easily be investigated. Thus, information will be gained on the importance of determining some parameters with precision.

10.3.3 *Check of echo print of input data*

Most computer codes will include echo print of the input. A good idea is to examine the listing of the input on the computer screen or to print the input for careful study. Experience has shown that entering incorrect data is a frequent error; the coding of the program to allow for echo printing should prevent such errors.

10.3.4 Investigation of length of word employed in internal computations

The assumption is made that the computer being used is capable of double-precision arithmetic, yielding about 10 or 12 significant figures or more. The user will wish to establish that the machine being used has a sufficient length of word before making computations because the difference-equation method requires that a relatively large number of significant figures be employed in order to avoid serious errors.

10.3.5 Selection of tolerance and length of increment

The tolerance is a number that usually is part of the input to be used in making a particular run. For example, values of deflection for successive iterations are retained in memory and the differences at corresponding depths are computed. All of the differences must be less than the tolerance to conclude a particular run with the computer. Most codes will include a default value.

The user has control over the tolerance and the default value needs to be investigated. A large value of tolerance will lead to inaccurate computations; a very small value will cause a significant increase in the number of iterations and could prevent convergence. The engineer can easily investigate the influence of the magnitude of the tolerance. For most problems, accuracy appears adequate with a value of tolerance that leads to 12 to 15 iterations.

The user must select the length of the subdivisions into which the pile is divided. The total length of the pile is the embedded length plus the portion of the pile above ground, if any. The first step in the selection of the length of increment is examine preliminary output and shorten the length of the pile to limit the points of zero deflection to two or three. The behavior of the upper portion of the pile is usually unaffected as the length is reduced as indicated.

A possible exception to shortening the pile to facilitate the computations may occur if the lower portion of the pile is embedded in rock or very strong soil. In such a case, small deflection could generate large values of soil resistance which in turn could influence the behavior of the upper portion of the pile.

With the length of the pile adjusted so that there is no exceptionally long portion at the bottom where the pile is oscillating about the axis with extremely small deflections and soil resistances, the engineer may wish to make a few runs with the pile subdivided into various number of increments and examine the output, say the pile-head deflection. Every solution is unique so rules for the number of sub-divisions are complex, but the value of y_t becomes virtually constant for many problems with the pile subdivided into 50 increments or more. Errors may be introduced if the number of increments is 40 or less.

10.3.6 Check of soil resistance

With the computer output for a particular problem at hand, a check of the correct soil resistance p for the computed deflection is suggested and can be readily done. Values of deflection y and soil resistance p are usually tabulated on the computer output. A value of deflection should be selected where a p-y curve has been input (or printed).

366 *Piles under lateral loading*

A useful exercise is to make a hand computation for the ultimate resistance p_u which is tabulated as a part of the output for a *p-y* curve. The engineer merely needs to refer to the procedures given in Chapter 3.

10.3.7 Check of mechanics

The values of soil resistance that are listed in a computer output may be plotted as a function of depth along the pile. The squares under the curves for negative values and positive values can be counted and multiplied by the value in (kN/m)m for each square. A check that the forces in the horizontal direction can be made by employing the value of pile-head shear.

The next step can be to make a check of the position of the point of the maximum moment. The point of zero shear can be found by finding the area under the shear curve that equates to the applied horizontal load at the top of the pile. The depth found can be compared with the computer output as a further check of the mechanics.

A rough check of the maximum bending moment can be found by estimating the centroid of the area of the *p*-curve above the point of zero shear. Then moments can by computed. Thus, a rough computation should reveal the correctness of the value of maximum bending moment in the computer output.

The next step in verifying the mechanics is to make an approximate solution for the deflection. The assumption can be made that the slope is zero midway between the first two points of zero deflection below the top of the pile. The deflection at the top of the pile can be computed by taking moments of the M/EI diagram about the top and down to the point of zero slope. The moment diagram can be based on the concentrated loads and points of load application found in the plotting of the curve of soil resistance versus depth.

10.3.8 Use of nondimensional curves

Another type of verification can be made by using the *p-y* curves as tabulated by the computer or as found by hand computations. Then nondimensional curves can be employed to solve the problem. The use of the curves is illustrated in Chapter 2 and are limited in several respects. First, a straight line passing through the origin must pass through the points for E_{py} versus depth. Then, no axial load is allowed and the bending stiffness EI must be constant. Even with the limitations noted, the nondimensional method will yield, with careful work, a result that is surprisingly close to that from a computer code.

10.4 VALIDATION OF RESULTS FROM COMPUTATIONS FOR PILE GROUP

The results from the analysis of a group of piles by computer can readily be checked by tabulating the axial load, shear, and bending moment on each pile. The directions of the loads can then be reversed and placed on the pile cap. The equilibrium of the cap can be checked with the three equations of statics using the magnitudes and positions of the applied loads.

If desirable, the response of a single pile can be checked by making use of the procedures outlined in the previous section.

10.5 ADDITIONAL STEPS IN DESIGN

10.5.1 *Risk management*

The design and construction of a significant structure involves many important steps, and the total project may be complex. Neither the engineer, the owner, nor the contractor wishes to encounter unforeseen difficulties, especially those that lead to legal conflicts. Thus, all of the parties in a project have an interest in managing the risks that are involved, with the engineer taking a leading role.

The engineer should confer with the owner to eliminate unnecessary constraints or unusual conditions that would add undesirable risks to the job. For example, with most contractual arrangements, the owner will provide the subsurface investigation. In some instance, a complex investigation is required, which must be done with the concurrence of the owner.

The engineer is required to provide the contractor with a set of plans and a volume of specifications that are unambiguous and without error. A series of pre-construction meetings with the contractor are useful in eliminating uncertainties.

Many engineering firms establish risk-management teams to ensure, as far as possible, that the design of a project is without error and that the construction can proceed without questions or delays.

10.5.2 *Peer review*

The analysis of a single pile or a group of piles under lateral load can be done readily with computer codes that are available. However, the problems are complex and the engineer has to make numerous decisions about loading, soil properties, pile characteristics, and analytical procedures. Furthermore, a significant number of computer runs are necessary as loading is augmented to find collapse or some limiting deflection. A number of trials may be necessary in selecting the best kind and size of pile for the application.

The responsible engineer could very well develop a peer review process so that each critical step in the analysis is checked by another knowledgeable engineer. Cost, of course, is added to the design but owners need to be aware that the design of piles under lateral load cannot be treated casually. The senior author recently was asked to comment on a legal case where, for an extremely small sum, an engineer undertook the responsibility to provide information on the response of an existing pile group under lateral load. The job was treated lightly, incomplete information was provided, with the result that the engineer's firm was required to make a large effort to defend themselves.

10.5.3 *Technical contributions*

Technical articles are being published regularly on piles under lateral loading. If an engineering office is regularly designing piles under lateral load, such articles must be reviewed. Of particular value are articles that include the testing of piles. The results of the tests may be compared with results from in-house computations as a means of validating the models proposed herein.

In the course of designs of piles under lateral load for a large project, the opportunity may arise to recommend and participate in the performance of a field test. Such a test may well be economically feasible. The careful planning, ensuring good construction, acquiring data of high quality, and performance of detailed analyses can be of benefit to the owner. Such information, with the permission of the owner, can be a significant contribution to the technical literature.

10.5.4 *The design team*

The design of piles under lateral loading will usually involve the contributions of a number of specialists. While the design under lateral loading may be a small part of the overall design, free and willing cooperation of those involved can lead to an optimum solution. In some instances, geotechnical engineers have been asked to provide p-y curves for a particular design and are eliminated from any other activity. The complexity of most deposits of soil argues for the inclusion of the geotechnical engineer throughout the design process. A like argument can be made for the inclusion of the structural engineer, representatives of the owner, representatives of the contractor, and for the participation of a number of others. Such cooperation will lead usually to the best decisions in the design of piles or pile groups under lateral loading.

APPENDIX A

Broms method for analysis of single piles under lateral loading

The method was presented in three papers published in 1964 and 1965 (Broms 1964a, 1964b, 1965). As shown in the following paragraphs, a pile can be designed to sustain a lateral load by solving some simple equations or by referring to charts and graphs.

A.1 PILES IN COHESIVE SOIL

A.1.1 *Ultimate lateral load for piles in cohesive soil*

Broms adopted a distribution of soil resistance, as shown in Figure A.1, that allows the ultimate lateral load to be computed by equations of static equilibrium. The elimination of soil resistance for the top 1.5 diameters of the pile is a result of lower resistance in that zone because a wedge of soil can move up and out when the pile is deflected. The selection of nine times the undrained shear strength times the pile diameter as the ultimate soil resistance, regardless of depth, is based on calculations with movement of soil from the front toward the back of the pile.

A.1.1.1 *Short, free-head piles in cohesive soil*

For short piles that are unrestrained against rotation, the patterns that were selected for behavior are shown in Figure A.2. The following equation results from the integration of the upper part of the shear diagram to the point of zero shear (the point of maximum moment)

$$M_{max}^{pos} = P_t(e + 1.5b + f) - \frac{9c_u b f^2}{2} \tag{A.1}$$

But the point where shear is zero is

$$f = \frac{P_t}{9c_u b} \tag{A.2}$$

Therefore,

$$M_{max}^{pos} = P_t(e + 1.5b + 0.5f) \tag{A.3}$$

Integration of the lower portion of the shear diagram yields

370 *Piles under lateral loading*

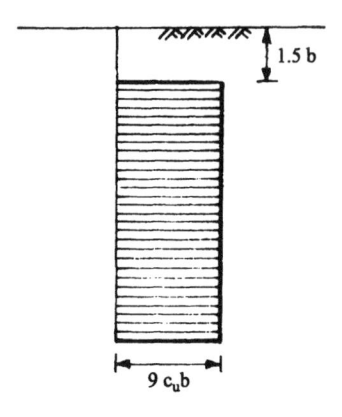

Figure A.1. Assumed distribution of soil resistance for cohesive soil.

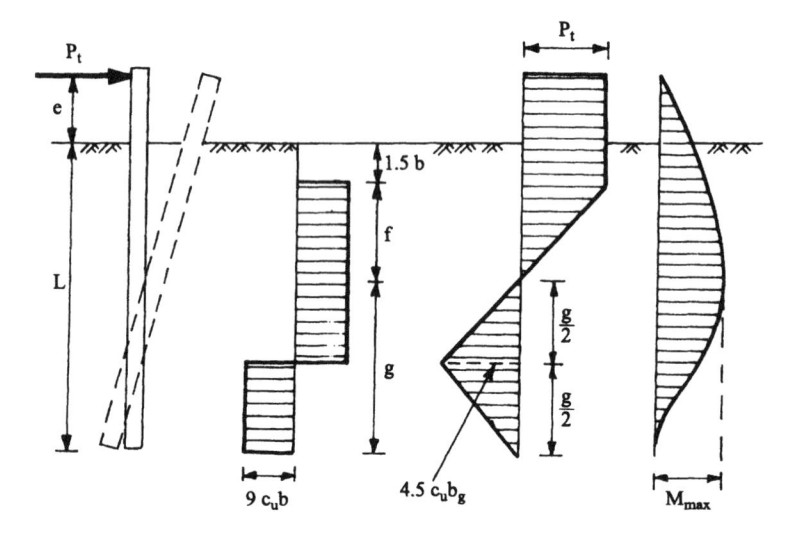

Figure A.2. Diagrams of deflection, soil resistance, shear, and moment for short pile in cohesive soil, unrestrained against rotation.

$$M_{max}^{pos} = 2.25 c_u b g^2 \tag{A.4}$$

It may be seen that

$$L = (1.5b + f + g) \tag{A.5}$$

Equations A.2 through A.5 may be solved for the load P_{tult} that will produce a soil failure. After obtaining a value of P_{tult}, the maximum moment can be computed and compared with the moment capacity of the pile. An appropriate factor of safety should be employed.

As an example of the use of the equations, assume the following:

b = 305 mm (assume 305-mm O.D. steel pipe by 19 mm wall),

$I_p = 1.75 \times 10^{-4}$ m^4, $e = 0.61$ m, $L = 2.44$ m, and

$c_u = 47.9$ kPa.

Equations A.2 through A.5 are solved simultaneously and the following quadratic equation is obtained.

$$P_t^2 + 1,083 P_t - 67,900 = 0$$

$$P_t = 59.4 \text{ kN}$$

Substituting into Equation A.3 yields the maximum moment

$$M_{max} = 59.4 \left[0.61 + 1.5(0.305) + \frac{(0.5)(59.4)}{(9)(47.9)(0.305)} \right]$$

$$= 77 \text{ kN-m}.$$

Assuming no axial load, the maximum stress is

$$f_b = \frac{(77)(0.1525)}{1.75 \times 10^{-4}} = 67,000 \text{ kPa}$$

The computed maximum stress is tolerable for a steel pipe, especially when a factor of safety is applied to P_{ult}. The computations, then, show that the short pile would fail due to a soil failure.

Broms presented a convenient set of curves for solving the problem of the short-pile (see Fig. A.3). Entering the curves with L/b of 8 and e/b of 2, one obtains a value of P_{ult} of 60 kN, which agrees with the results computed above.

A.1.1.2 Long, free-head piles in cohesive soil

As the pile in cohesive soil with the unrestrained head becomes longer, failure will occur with the formation of a plastic hinge at a depth of $1.50b + f$. Equation A.3 can then be used directly to solve for the ultimate lateral load that can be applied. The shape of the pile under load will be different than that shown in Figure A.2 but the equations of mechanics for the upper portion of the pile remain unchanged.

A plastic hinge will develop when the yield stress of the steel is attained over the entire cross-section. For the pile that is used in the example, the yield moment is 430 m-kN if the yield strength of the steel is selected as 276 MPa.

Substituting into Equation A.3

$$430 = P_{\text{ult}} \left(0.61 + 0.4575 + \frac{P_{\text{ult}}}{2.53} \right)$$

$$P_{\text{ult}} = 224 \text{ kN}$$

Broms presented a set of curves for solving the problem of the long pile (see Fig. A.4). Entering the curves with a value of $M_y/c_u b^3$ of 316.4, one obtains a value of P_{ult} of about 220 kN.

372 *Piles under lateral loading*

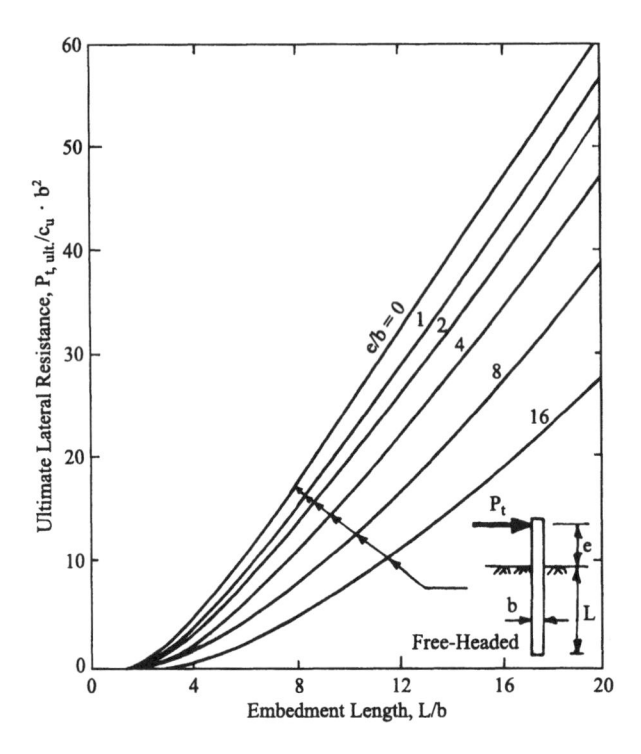

Figure A.3. Curves for design of short piles under lateral load in cohesive soil.

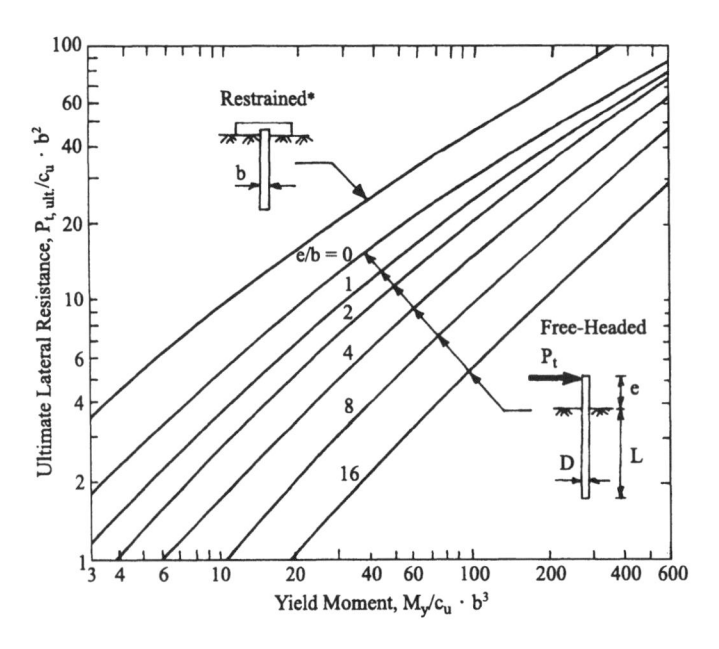

Figure A.4. Curves for design of long piles under lateral load in cohesive soil.

A.1.1.3 Influence of pile length, free-head piles in cohesive soil

Consideration may need to be given to the pile length at which the pile ceases to be a short pile. The value of the yield moment may be computed from the pile geometry and material properties and used with Equations A.2 through A.5 to solve for a critical length. Longer piles will fail by yielding. Or a particular solution may start with use of the short-pile equations; if the resulting moment is larger than the yield moment, the long-pile equations may be used.

For the example problem, the length at which the short-pile equations cease to be valid may be found by substituting a value of P_{rult} of 224 kN into Equation A.2 and solving for f and substituting a value of M_{max} of 430 m-kN into Equation A.4 and solving for g. Equation A.5 can then be solved for L. The value of L was found to be 5.8m. Thus, for the example problem the value of P_{rult} increases from zero to 224 kN as the length of the pile increases from 0.46m to 5.8m, and above a length of 5.8m the value of P_{rult} remains constant at 224 kN.

A.1.1.4 Short, fixed-head piles in cohesive soil

For a pile that is fixed against rotation at its top, the mode of failure depends on the length of the pile. For a short pile, failure consists of a horizontal movement of the pile through the soil with the full soil resistance developing over the length of the pile except for the top one and one-half pile diameters, where it is expressly eliminated. A simple equation can be written for this mode of failure, based on force equilibrium.

$$P_{rult} = 9c_u b(L - 1.5b) \tag{A.6}$$

A.1.1.5 Intermediate length, fixed-head piles in cohesive soil

As the pile becomes longer, an intermediate length is reached such that a plastic hinge develops at the top of the pile. Rotation at the top of the pile will occur and a point of zero deflection will exist somewhere along the length of the pile. Figure A.5 presents the diagrams of mechanics for the case of the restrained pile of intermediate length.

The equation for moment equilibrium for the point where the shear is zero (where the positive moment is maximum) is:

$$M_{pos}^{max} = P_t(1.5b + f) - f(9c_u b)\left(\frac{f}{2}\right) - M_y$$

Substituting a value of f,

$$M_{pos}^{max} = P_t(1.5b + 0.5f) - M_y \tag{A.7}$$

Employing the shear diagram for the lower portion of the pile,

$$M_{pos}^{max} = 2.25 c_u b g^2 \tag{A.8}$$

The other equations that are needed to solve for P_{rult} are:

$$L = 1.5b + f + g \tag{A.9}$$

and

374 *Piles under lateral loading*

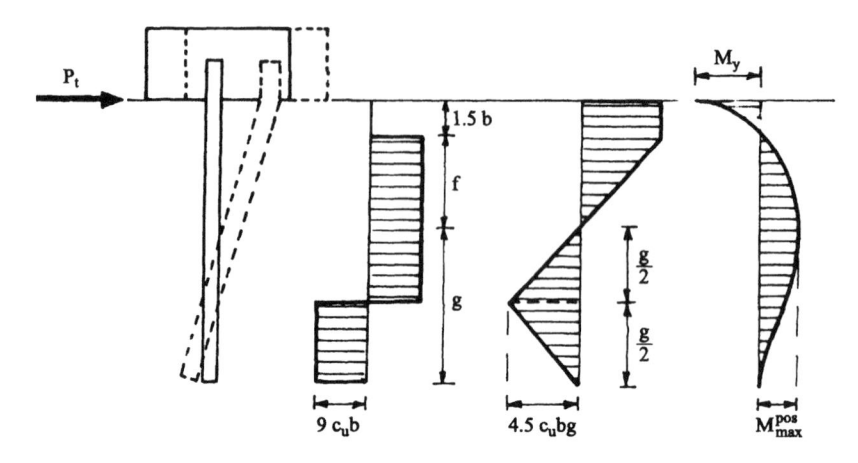

Figure A.5. Diagrams of deflection, soil resistance, shear, and moment for intermediate-length pile in cohesive soil, fixed against rotation.

$$f = \frac{P_t}{9c_u b} \quad \text{(A.10)}$$

Equations A.7 through A.10 can be solved for the behavior of the restrained pile of intermediate length.

A.1.1.6 Long, fixed-head piles in cohesive soil

The mechanics for a long pile that is restrained at its top is similar to that shown in Figure A.5 except that a plastic hinge develops at the point of the maximum positive moment. Thus, the $M^{pos,max}$ in Equation A.7 becomes M_y and the following equation results

$$P_t = \frac{2M_y}{(1.5b + 0.5f)} \quad \text{(A.11)}$$

Equations A.10 and A.11 can be solved to obtain P_{tult} for the long pile.

A.1.1.7 Influence of pile length, fixed-head piles in cohesive soil

The example problem will be solved for the pile lengths where the pile goes from one mode of behavior to another. Starting with the short pile, an equation can be written for moment equilibrium for the case where the yield moment has developed at the top of the pile and where the moment at its bottom is zero. Referring to Figure A.5, but with the soil resistance only on the right-hand side of the pile, taking moments about the bottom of the pile yields the following equation.

$$P_t L - 9c_u b \left(\frac{L - 1.5b}{2} \right) - M_y = 0$$

Summing forces in the horizontal direction yield the next equation.

$$P_{tult} = 9c_u b(L - 1.5b) = 0 \quad \text{(same as Eq. A.6)}$$

The simultaneous solution of the two equations yields the desired expression.

$$P_{ult} = \frac{M_y}{(0.5L + 0.75b)} \tag{A.12}$$

Equations A.6 and A.12 can be solved simultaneously for P_{ult} and for L, as follows

from Equation A.6, $P_{ult} = (9)(47.9)(0.305)(L - 0.4575)$,

from Equation A.12, $P_{ult} = 430/(0.5L + 0.229)$,

then $L = 2.6$ m and $P_{ult} = 281$ kN.

For the determination of the length where the behavior changes from that of the pile of intermediate length to that of a long pile, Equations A.7 through A.10 can be used with M_{max} set equal to M_y, as follows:

from Equation A.7, $P_{ult} = \dfrac{(2)(430)}{(1.5)(0.305) + 0.5f}$

from Equation A.8, $g = \left[\dfrac{439}{(2.25)(47.9)(0.305)}\right]^{0.5} = 3.62$ m

from Equation A.9, $L = (1.5)(0.305) + f + g$

from Equation A.10, $f = \dfrac{P_{ult}}{9}(47.9)(0.305)$

then $L = 7.27$ m and $P_{ult} = 419$ kN.

In summary, for the example problem the value of P_{ult} increases from zero to 281 kN as the length of the pile increases from 0.46 m to 2.6 m, increases from 281 kN to 419 kN as the length increases from 2.6 m to 7.3 m, and above a length of 7.3 m the value of P_{ult} remains constant at 419 kN.

In his presentation, Broms showed a curve in Figure A.3 for the short pile that was restrained against rotation at its top. That curve is omitted here because the computation can be made so readily with Equation A.6. Broms' curve for the long pile that is fixed against rotation at its top is retained in Figure A.4 but a note is added to ensure proper use of the curve. For the example problem, a value of 415 kN was obtained for P_{ult}, which agrees well with the computed value. No curves are presented for the pile of intermediate length.

A.1.2 Deflection of piles in cohesive soil

Broms suggested that for cohesive soils the assumption of a coefficient of subgrade reaction that is constant with depth can be used with good results for predicting the lateral deflection at the groundline. He further suggests that the coefficient of subgrade reaction α should be taken as the average over a depth of $0.8\beta L$, where

376 Piles under lateral loading

$$\beta = \left(\frac{\alpha}{4E_p I_p}\right)^{0.25} \quad (A.11)$$

where α soil reaction modulus and; $E_p I_p$ = pile stiffness.

Broms presented equations and curves for computing the deflection at the groundline. His presentation follows the procedures presented elsewhere in this text.

With regard to values of the reaction modulus, Broms used work of himself and Vesic (1961a, 1961b) for selection of values, depending on the unconfined compressive strength of the soil. The work of Terzaghi (1955) and other with respect to the reaction modulus have been discussed fully in the text.

Broms suggested that the use of a constant for the reaction modulus is valid only for a load of one-half to one-third of the ultimate lateral capacity of a pile.

A.1.3 Effects of nature of loading on piles in cohesive soil

The values of reaction modulus presented by Terzaghi are apparently for short-term loading. Terzaghi did not discuss dynamic loading or the effects of repeated loading. Also, because Terzaghi's coefficients were for overconsolidated clays only, the effects of sustained loading would probably be minimal. Because the nature of the loading is so important in regard to pile response, some of Broms' remarks are presented here.

Broms suggested that the increase in the deflection of a pile under lateral loading due to consolidation can be assumed to be the same as would take place with time for spread footings and rafts founded on the ground surface or at some distance below the ground surface. Broms suggested that test data for footings on stiff clay indicate that the coefficient of subgrade reaction to be used for long-time lateral deflections should be taken as 1/2 to 1/4 of the initial reaction modulus. The value of the coefficient of subgrade reaction for normally consolidated clay should be 1/4 to 1/6 of the initial value.

Broms suggested that repetitive loads cause a gradual decrease in the shear strength of the soil located in the immediate vicinity of a pile. He stated that unpublished data indicate that repetitive loading can decrease the ultimate lateral resistance of the soil to about one-half its initial value.

A.2 PILES IN COHESIONLESS SOILS

A.2.1 Ultimate lateral load for piles in cohesionless soil

As for the case of cohesive soil, two failure modes were considered; a soil failure and a failure of the pile by the formation of a plastic hinge. With regard to a soil failure in cohesionless soil, Broms assumed that the ultimate lateral resistance is equal to three times the Rankine passive pressure. Thus, at a depth z below the ground surface the soil resistance per unit of length P_z can be obtained from the following equations.

$$P_z = 3b\gamma z K_p \quad (A.12)$$

$$K_p = \tan^2\left(45 + \frac{\phi}{2}\right) \quad (A.13)$$

where γ = unit weight of soil; K_p = Rankine coefficient of passive pressure; and ϕ = friction angle of soil.

A.2.1.1 *Short, free-head piles in cohesionless soil*
For short piles that are unrestrained against rotation, a soil failure will occur. The curve showing soil reaction as a function of depth is shaped approximately as shown in Figure A.6. The use of M_a as an applied moment at the top of the pile follows the procedure adopted by Broms. If both P_t and M_a are acting, the result would be merely to increase the magnitude of *e*. It is unlikely in practice that M_t alone would be applied.

The patterns that were selected for behavior are shown in Figure A.7. Failure takes place when the pile rotates such that the ultimate soil resistance develops from the ground surface to the center of rotation. The high values of soil resistance that develop at the toe of the pile are replaced by a concentrated load as shown in Figure A.7.

The following equation results after taking moments about the bottom of the pile.

$$P_t(e+L) + M_t = (3\gamma bLK_p)\left(\frac{L}{2}\right)\left(\frac{L}{3}\right) \tag{A.14}$$

Solving for P_t when M_t is equal to zero,

$$P_t = \frac{\gamma bL^3 K_p}{2(e+L)} \tag{A.15}$$

And solving for M_t when P_t is equal to zero,

$$M_t = 0.5\gamma bL^3 K_p \tag{A.16}$$

Equations A.14 through A.16 can be solved for the load or moment, or a combination of the two, that will cause a soil failure. The maximum moment will then be found, at the depth f below the ground surface, and compared with the moment capacity of the pile. An appropriate factor of safety should be used. The distance f can be computed by solving for the point where the shear is equal to zero.

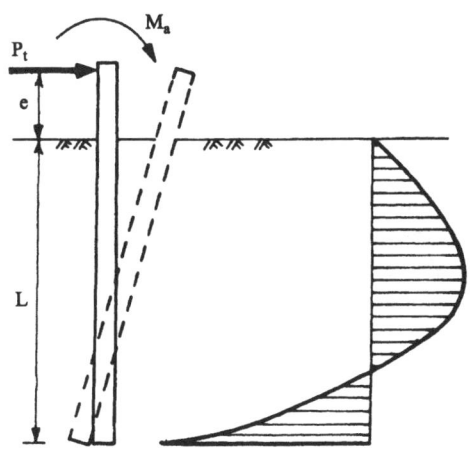

Figure A.6. Assumed distribution of soil resistance for cohesionless soil for a short pile, unrestrained against rotation.

378 *Piles under lateral loading*

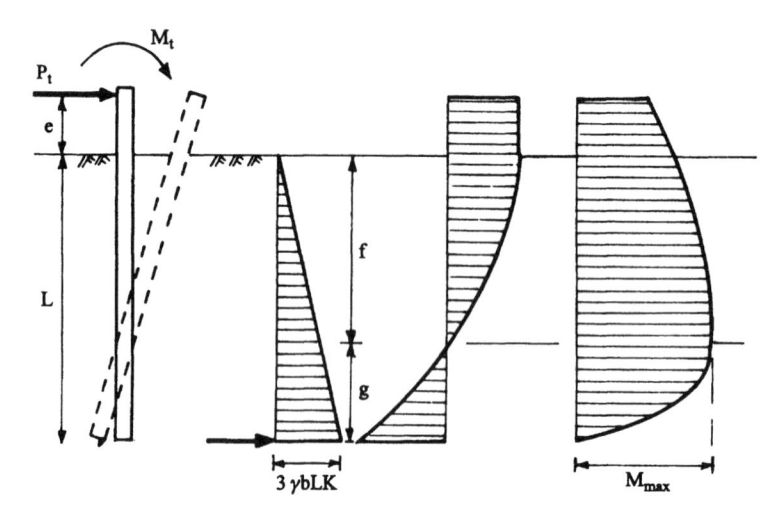

Figure A.7. Diagrams of deflection, soil resistance, shear, and moment for short pile in cohesionless soil, unrestrained against rotation.

$$P_t - (3\gamma b f K_p)\left(\frac{f}{2}\right) = 0 \tag{A.17}$$

Solving Equation A.17 for an expression for f

$$f = 0.816\left(\frac{P_t}{\gamma b K_p}\right)^{0.5} \tag{A.18}$$

The maximum positive bending moment can then be computed by referring to Figure A.7.

$$M_{max}^{pos} = P_t(e+f) - \frac{K_p \gamma b f^3}{2} + M_t$$

Or, by substituting expression for Equation A.17 into the above equation, the following expression is obtained for the maximum moment.

$$M_{max}^{pos} = P_t(e+f) - \frac{P_t f}{3} + M_t \tag{A.19}$$

As an example of the use of the equations, the pile used previously is considered. The friction angle of the sand is assumed to be 34 degrees and the unit weight is assumed to be 8.64 kN/m³ (the water table is assumed to be above the ground surface). Assume M_t is equal to zero. Equations A.13 and A.15 yield the following:

$$K_p = \tan^2\left(45 + \frac{\phi}{2}\right) = 3.54,$$

$$P_{ult} = \frac{(8.64)(0.305)(2.44)^3(3.54)}{(0.61+2.44)} = 22.2 \text{ kN}$$

The distance f can be computed by solving Equation A.18.

$$f = 0.816\left(\frac{22.2}{(8.64)(0.305)(3.54)}\right)^{0.5} = 1.259 \text{ m}$$

The maximum positive bending moment can be found using Equation A.19.

$$M_{max} = (22.2)(0.61+1.259) - \frac{(22.2)(1.259)}{3} = 32.2 \text{ m-kN}$$

Assuming no axial load, the maximum bending stress is

$$f_b = \frac{(32.2)(0.1525)}{1.75 \times 10^{-4}} = 28{,}000 \text{ kPa}$$

The computed maximum stress is undoubtedly tolerable, especially when a factor of safety is used to reduce P_{ult}. Broms presented curves for the solution of the case where a short, unrestrained pile undergoes a soil failure; however, Equations A.14 and A.17 are so elementary that such curves are unnecessary.

A.2.1.2 Long, free-head piles in cohesionless soil

As the pile in cohesionless soil with the unrestrained head becomes longer, failure will occur with the formation of a plastic hinge in the pile at the depth f below the ground surface. It is assumed that the ultimate soil resistance develops from the ground surface to the point of the plastic hinge. Also, the shear is zero at the point of maximum moment. The value of f can be obtained from Equation A.18 shown above. The maximum positive moment can then be computed and Equation A.19 is obtained as before. Assuming that M_t is equal to zero, an expression can be developed for P_{ult} as follows:

$$P_{ult} = \frac{M_y}{e + 0.544\left[\dfrac{P_{ult}}{(\gamma b K_p)}\right]^{0.5}} \qquad (A.20)$$

For the example problem, Equation A.20 can be solved, as follows:

$$P_{ult} = \frac{430}{0.61 + 0.544\left[\dfrac{P_{ult}}{(8.64)(0.305)(3.54)}\right]^{0.5}} = 153 \text{ kN}$$

Broms presented a set of curves for solving the problem of the long pile in cohesionless soils (see Fig. A.8). Entering the curves with a value of $M_y/b^4 \gamma Kp$ of 1926, one obtains a value of P_{ult} of about 160 kN. The logarithmic scales are somewhat difficult to read and it may be desirable to make a solution using Equation A.20. Equations A.19 and A.20 must be used in any case if a moment is applied at the top of the pile.

380 *Piles under lateral loading*

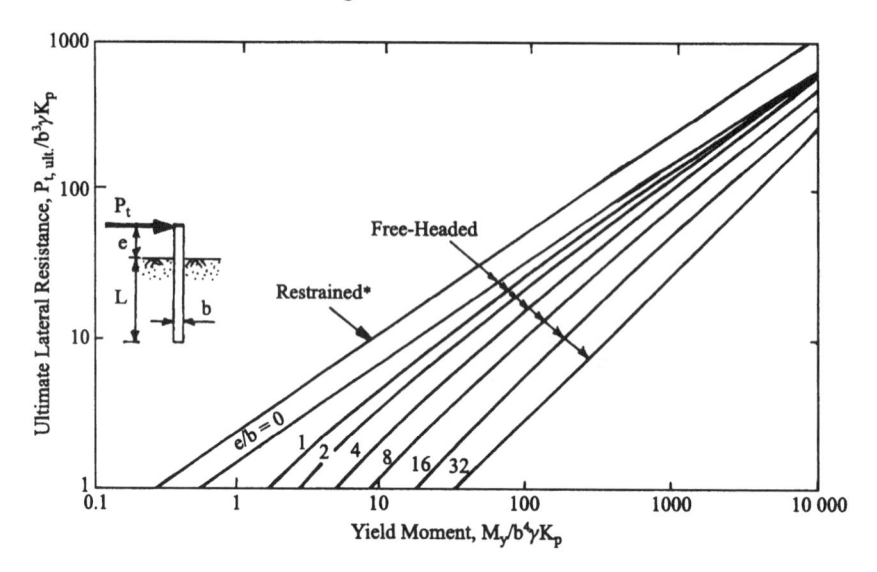

Figure A.8. Curves for design of long piles under lateral load in cohesionless soil.

A.2.1.3 *Influence of pile length, free-head piles in cohesionless soil*

There may be a need to solve for the pile length where there is a change in behavior from the short-pile case to the long-pile case. As for the case of the pile in cohesive soils, the yield moment may be used with Equations A.14 through A.16 to solve for the critical length of the pile. Alternatively, the short-pile equations would then be compared with the yield moment. If the yield moment is less, the long-pile equations must be used.

For the example problem, the value of P_{tult} of 153 kN is substituted into Equation A.16 and a value of L of 6.0 m is computed. Thus, for the pile that is unrestrained against rotation the value of P_{tult} increases from zero when L is zero to a value of 153 kN when L is 6.0 m. For larger values of L, the value of P_{tult} remains constant at 153 kN.

A.2.1.4 *Short, fixed-head piles in cohesionless soil*

For a pile that is fixed against rotation at its top, as for cohesive soils, the mode of failure for a pile in cohesionless soil depends on the length of the pile. For a short pile, the mode of failure will be a horizontal movement of the pile through the soil, with the ultimate soil resistance developing over the full length of the pile. The equation for static equilibrium in the horizontal direction leads to a simple expression.

$$P_{tult} = 1.5\gamma L^2 bK_p \tag{A.21}$$

A.2.1.5 *Intermediate length, fixed-head piles in cohesionless soil*

As the pile becomes longer, an intermediate length is reached such that a plastic hinge develops at the top of the pile. Rotation at the top of the pile will occur, and a point of zero deflection will exist somewhere along the length of the pile. The assumed soil resistance will be the same as shown in Figure A.7. Taking moments about the toe of the pile leads to the following equation for the ultimate load.

$$P_{tult} = \frac{M_y}{L} + 0.5\gamma bL^2 K_p \tag{A.22}$$

Equation A.22 can be solved to obtain P_{tult} for the pile of intermediate length.

A.2.1.6 Long, fixed-head piles in cohesionless soil

As the length of the pile increases more, the mode of behavior will be that of a long pile. A plastic hinge will form at the top of the pile where there is a negative bending moment and at some depth f where there is a positive bending moment. The shear at depth f is zero and the ultimate soil resistance is as shown in Figure A.7. The value of f may be determined from Equation A.18 but that equation is re-numbered and presented here for convenience.

$$f = 0.816 \left(\frac{P_t}{\gamma bK_p} \right)^{0.5} \tag{A.23}$$

Taking moments at point f leads to the following equation for the ultimate lateral load on a long pile that is fixed against rotation at its top.

$$P_{tult} = \frac{M_y^+ + M_y^-}{e + 0.544 \left(\dfrac{P_{tult}}{\gamma bK_p} \right)^{0.5}} \tag{A.24}$$

Equations A.23 and A.24 can be solved to obtain P_{tult} for the long pile.

A.2.1.7 Influence of pile length, fixed-head piles in cohesionless soil

The example problem will be solved for the pile lengths where the pile goes from one mode of behavior to another. An equation can be written for the case where the yield moment has developed at the top of the short pile. The equation is:

$$P_{tult} = \frac{M_y}{L} + 0.5\gamma bL^2 K_p \tag{A.25}$$

Equations A.22 and A.25 are, of course, identical but the repetition is for clarity. Equations A.21 and A.25 can be solved for P_{tult} and for L, as follows:

from Equation A.21, $P_{tult} = 14.0L^2$

from Equation A.25, $P_{tult} = \dfrac{430}{L} + 15.3L^2$

then $L = 3.59$ m and $P_{tult} = 180$ kN.

For the determination of the length where the behavior changes from that of a pile of intermediate length to that of a long pile, the value of P_{tult} from Equation A.22 may be set equal to that in Equation A.24. It is assumed that the pile has the same yield moment over its entire length in this example.

from Equation A.22, $P_{\text{rult}} = \dfrac{430}{L} + 4.664L^2$

from Equation A.24, $P_{\text{rult}} = \dfrac{2(430)}{0.61 + 0.544\left[\dfrac{P_{\text{rult}}}{(8.64)(0.305)(3.54)}\right]^{0.5}}$

then $L = 6.25$ m, $P_{\text{rult}} = 251$ kN

In summary, for the example problem the value of P_{rult} increases from zero to 180 kN as the length of the pile increases from zero to 3.59 m, 180 kN to 251 kN as the length increased from 3.59 m to 6.25 m, and above 6.25 m the value of P_{rult} remains constant at 251 kN.

In his presentation, Broms showed curves for short piles that were restrained against rotation at their top. Those curves are omitted because the equations for those cases are so easy to solve. Broms' curve for the long pile that is fixed against rotation at its top is retained in Figure A.8 but a note is added to ensure proper use of the curve. For the example problem, a value of 300 kN was obtained for P_{rult}, which agrees poorly with the computed value. The difficulty probably lies in the inability to read the logarithmic scales accurately. No curves are presented for the pile of intermediate length with fixed head.

A.2.2 Deflection of piles in cohesionless soil

Broms noted that Terzaghi (1955) has shown that the reaction modulus for a cohesionless soil can be assumed to increase approximately linearly with depth. As noted earlier, and using the formulations of this work, Terzaghi recommends the following equation for the soil modulus.

$$E_{py} = k_{py}z \tag{A.26}$$

Broms suggested that Terzaghi's values can be used only for computing deflections up to the working load and that the measured deflections are usually larger than the computed ones except for piles that are placed with the aid of jetting.

Broms presented equations and curves for use in computing the lateral deflection of a pile; however, the methods presented herein are considered to be appropriate.

A.2.3 Effects of nature of loading on piles in cohesionless soil

Broms noted that piles installed in cohesionless soil will experience the majority of the lateral deflection under the initial application of the load. There will be only a small amount of creep under sustained loads.

Repetitive loading and vibration, on the other hand, can cause significant additional deflection, especially if the relative density of the cohesionless soil is low. Broms noted that work of Prakash (1962) shows that the lateral deflection of a pile group in sand increased to twice the initial deflection after 40 cycles of load. The increase in deflection corresponds to a decrease in the soil modulus to one-third its initial value.

For piles subjected to repeated loading, Broms recommended for cohesionless soils of low relative density that the reaction modulus be decreased to 1/4 its initial value and that the value of the reaction modulus be decreased to 1/2 its initial value for soils of high relative density. He suggested that these recommendations be used with caution because of the scarcity of experimental data.

APPENDIX B

Nondimensional coefficients for elastic piles with finite length, no axial load, constant E_p/I_p constant E_{py}

$\beta L = 10$

βx	A_1	B_1	C_1	D_1
0	1.0000	1.0000	1.0000	0
.1	.9907	.8100	.9003	.0903
.2	.9651	.6398	.8024	.1627
.3	.9267	.4888	.7077	.2189
.4	.8784	.3564	.6174	.2610
.5	.8231	.2415	.5323	.2908
.6	.7628	.1431	.4530	.3099
.7	.6997	.0599	.3798	.3199
.8	.6354	−.0093	.3131	.3223
.9	.5712	−.0657	.2527	.3185
1.0	.5083	−.1108	.1988	.3096
1.2	.3899	−.1716	.1091	.2807
1.4	.2849	−.2011	.0419	.2430
1.6	.1959	−.2077	−.0059	.2018
1.8	.1234	−.1985	−.0376	.1610
2.0	.0667	−.1794	−.0563	.1231
2.2	.0244	−.1548	−.0652	.0896
2.4	−.0056	−.1282	−.0669	.0613
2.6	−.0254	−.1019	−.0636	.0383
2.8	−.0369	−.0777	−.0573	.0204
3.0	−.0423	−.0563	−.0493	.0070
3.5	−.0389	−.0177	−.0283	−.0106
4.0	−.0258	.0019	−.0120	−.0139
4.5	−.0132	.0085	−.0023	−.0109
5.0	−.0046	.0084	.0019	−.0065
10.0	−.0001	.0000	.0000	.0000

$\beta L = 4.0$

βx	A_1	B_1	C_1	D_1
0	1.0015	1.0000	1.0008	0
.1	.9922	.8098	.9009	.0903
.2	.9666	.6395	.8029	.1626
.3	.9282	.4885	.7080	.2189
.4	.8800	.3560	.6176	.2609
.5	.8247	.2411	.5323	.2907
.6	.7645	.1427	.4528	.3097
.7	.7014	.0596	.3795	.3197
.8	.6371	−.0095	.3126	.3221
.9	.5730	−.0659	.2521	.3182
1.0	.5101	−.1108	.1979	.3093
1.2	.3918	−.1712	.1079	.2805
1.4	.2869	−.2001	.0403	.2429
1.6	.1979	−.2060	−.0079	.2020
1.8	.1252	−.1960	−.0399	.1616
2.0	.0682	−.1758	−.0590	.1243
2.2	.0251	−.1501	−.0681	.0916
2.4	−.0059	−.1223	−.0699	.0644
2.6	−.0271	−.0949	−.0664	.0427
2.8	−.0408	−.0696	−.0595	.0263
3.0	−.0488	−.0475	−.0505	.0147
3.5	−.0552	−.0101	−.0239	.0014
4.0	−.0555	.0000	.0038	−.0000

$\beta L = 3.5$

βx	A_1	B_1	C_1	D_1
0	1.0009	1.0000	1.0011	0
.1	.9916	.8098	.9013	.0903
.2	.9660	.6394	.8033	.1626
.3	.9276	.4882	.7085	.2188
.4	.8794	.3557	.6181	.2609
.5	.8240	.2406	.5329	.2906
.6	.7639	.1421	.4535	.3096
.7	.7008	.0588	.3802	.3195
.8	.6366	−.0104	.3133	.3218
.9	.5725	−.0669	.2529	.3178
1.0	.5098	−.1120	.1988	.3088
1.2	.3917	−.1727	.1088	.2797
1.4	.2872	−.2020	.0412	.2418
1.6	.1987	−.2083	−.0071	.2004
1.8	.1268	−.1985	−.0394	.1596
2.0	.0706	−.1785	−.0589	.1217
2.2	.0287	−.1527	−.0686	.0886
2.4	−.0009	−.1246	−.0712	.0608
2.6	−.0207	−.0964	−.0689	.0387
2.8	−.0327	−.0699	−.0634	.0221
3.0	−.0390	−.0459	−.0562	.0106
3.5	−.0424	.0000	−.0354	−.0000

$\beta L = 3.0$

βx	A_1	B_1	C_1	D_1
0	1.0004	1.0000	1.0066	0
.1	.9911	.8087	.9068	.0903
.2	.9655	.6371	.8089	.1624
.3	.9272	.4849	.7142	.2183
.4	.8791	.3512	.6238	.2600
.5	.8240	.2350	.5386	.2892
.6	.7642	.1354	.4591	.3076
.7	.7016	.0510	.3858	.3167
.8	.6380	−.0194	.3189	.3182
.9	.5747	−.0770	.2582	.3133
1.0	.5130	−.1231	.2039	.3032
1.2	.3976	−.1857	.1130	.2717
1.4	.2969	−.2164	.0438	.2311
1.6	.2132	−.2232	−.0069	.1868
1.8	.1474	−.2129	−.0427	.1429
2.0	.0985	−.1906	−.0670	.1024
2.2	.0648	−.1604	−.0831	.0672
2.4	.0438	−.1249	−.0937	.0386
2.6	.0329	−.0858	−.1013	.0175
2.8	.0288	−.0441	−.1073	.0044
3.0	.0282	−.0000	−.1130	−.0000

βL = 2.8

βx	A_1	B_1	C_1	D_1
0	1.0033	1.0000	1.0138	0
.1	.9740	.8073	.9138	.0902
.2	.9685	.6344	.8156	.1621
.3	.9303	.4808	.7205	.2177
.4	.8824	.3459	.6298	.2589
.5	.8275	.2285	.5443	.2875
.6	.7681	.1278	.4645	.3052
.7	.7061	.0423	.3908	.3135
.8	.6432	−.0290	.3233	.3141
.9	.5809	−.0874	.2621	.3082
1.0	.5203	−.1342	.2071	.2970
1.2	.4078	−.1978	.1145	.2632
1.4	.3110	−.2286	.0429	.2201
1.6	.2322	−.2344	−.0111	.1734
1.8	.1721	−.2215	−.0513	.1276
2.0	.1296	−.1948	−.0812	.0857
2.2	.1027	−.1575	−.1042	.0504
2.4	.0882	−.1120	−.1231	.0233
2.6	.0827	−.0593	−.1401	.0060
2.8	.0819	−.0000	−.1565	.0000

$\beta L = 2.6$

βx	A_1	B_1	C_1	D_1
0	1.0119	1.0000	1.0270	0
.1	1.0026	.8047	.9262	.0901
.2	.9771	.6294	.8271	.1616
.3	.9391	.4737	.7312	.2166
.4	.8915	.3367	.6396	.2570
.5	.8371	.2175	.5531	.2845
.6	.7784	.1151	.4723	.3010
.7	.7173	.0282	.3975	.3081
.8	.6557	−.0444	.??88	.3071
.9	.5949	−.1038	.?663	.2996
1.0	.5362	−.1513	.2098	.2868
1.2	.4285	−.2152	.1136	.2495
1.4	.3378	−.2448	.0372	.2030
1.6	.2666	−.2472	−.0229	.1534
1.8	.2149	−.2281	−.0707	.1055
2.0	.1814	−.1917	−.1101	.0633
2.2	.1631	−.1407	−.1443	.0298
2.4	.1560	−.0766	−.1760	.0079
2.6	.1549	−.0000	−.2071	.0000

$\beta L = 2.4$

βx	A_1	B_1	C_1	D_1
0	1.0310	1.0000	1.0493	0
.1	1.0217	.8004	.9465	.0899
.2	.9963	.6213	.8455	.1608
.3	.9585	.4620	.7477	.2148
.4	.9114	.3219	.6541	.2538
.5	.8579	.2001	.5656	.2798
.6	.8003	.0953	.4827	.2944
.7	.7407	.0066	.4056	.2994
.8	.6811	−.0673	.3345	.2962
.9	.6227	−.1276	.2693	.2864
1.0	.5669	−.1754	.2099	.2711
1.2	.4664	−.2381	.1068	.2291
1.4	.3848	−.2633	.0220	.1784
1.6	.3240	−.2576	−.0485	.1258
1.8	.2836	−.2259	−.1089	.0770
2.0	.2612	−.1713	−.1631	.0370
2.2	.2523	−.0958	−.2143	.0099
2.4	.2510	−.0000	−.2646	−.0000

βL = 2.2

βx	A_1	B_1	C_1	D_1
0	1.0680	1.0000	1.0845	0
.1	1.0588	.7938	.9780	.0895
.2	1.0335	.6087	.8733	.1595
.3	.9961	.4442	.7717	.2119
.4	.9498	.2997	.6743	.2490
.5	.8974	.1742	.5819	.2725
.6	.8416	.0666	.4950	.2844
.7	.7843	−.0242	.4137	.2864
.8	.7276	−.0993	.3381	.2801
.9	.6728	−.1598	.2681	.2670
1.0	.6211	−.2069	.2034	.2486
1.2	.5309	−.2647	.0885	.2007
1.4	.4615	−.2798	−.0103	.1456
1.6	.4143	−.2580	−.0976	.0912
1.8	.3876	−.2028	−.1774	.0446
2.0	.3768	−.1165	−.2536	.0121
2.2	.3751	−.0000	−.3288	.0000

$\beta L = 2.0$

βx	A_1	B_1	C_1	D_1
0	1.1341	1.0000	1.1376	0
.1	1.1249	.7838	1.0245	.0890
.2	1.0999	.5901	.9131	.1575
.3	1.0631	.4183	.8049	.2078
.4	1.0179	.2679	.7008	.2419
.5	.9673	.1377	.6015	.2620
.6	.9139	.0269	.5074	.2701
.7	.8599	−.0656	.4187	.2680
.8	.8072	−.1410	.3354	.2575
.9	.7574	−.2001	.2572	.2403
1.0	.7115	−.2442	.1838	.2180
1.2	.6348	−.2903	.0495	.1637
1.4	.5811	−.2855	−.0717	.1053
1.6	.5498	−.2341	−.1844	.0526
1.8	.5370	−.1386	−.2928	.0146
2.0	.5351	.0000	−.3999	.0000

APPENDIX C

Difference equations for step-tapered beams on foundations having variable stiffness[1]

The differential equation to be solved is

$$E_p I_p \frac{d^4 y}{dx^4} = E_{py} y \qquad (C.1)$$

where $E_p I_p$ = flexural rigidity of beam, y = deflection, x = distance along the beam, inches; E_{py} = soil modulus (can be a function of both x and y).

(Note: for ease in presentation, henceforth in this document $E_p I_p$ will be shown as EI and E_{py} will be shown as E_s.)

The first step in the problem is to formulate the necessary difference equations. In difference form, the expressions for slope, moment, shear, and soil reaction at any point m along a beam are

$$S_m = \frac{-y_{m+1} + y_{m-1}}{2(L/t)} \qquad (C.1a)$$

where S_m = slope of beam at point m; L = length of beam; t = number of subdivisions into which beam is divided; y_m = deflection of beam at point m.

$$M_m = \frac{1}{b_2}(y_{m+1} - 2y_m + y_{m-1}) \qquad (C.2)$$

where M_m = moment in beam at point m

$$b_2 = \frac{1}{EI}\left[\frac{L}{t}\right]^2 \qquad (C.3)$$

$$V_m = \frac{1}{b_4}(-y_{m+2} + 2y_{m+1} - 2y_{m-1} + y_{m-2}) \qquad (C.4)$$

where V_m = shear in beam at point m

---

[1] 'Step-tapered beams on foundation having variable stiffness,' by L.C. Reese & A.S. Ginzbarg, unpublished report 1959

396 Piles under lateral loading

$$b_4 = \frac{2}{EI}\left[\frac{L}{t}\right]^3 \tag{C.5}$$

$$p_m = \frac{1}{b_6}(y_{m+2} - 4y_{m+1} + 6y_m - 4y_{m-1} + y_{m-2}) \tag{C.6}$$

where p_m = soil reaction against beam at point m

$$b_6 = \frac{1}{EI}\left[\frac{L}{t}\right]^4 \tag{C.7}$$

For the beam on the elastic foundation, the soil reaction can be defined as

$$p_m = -E_{sm}y_m, \tag{C.8}$$

where E_{sm} = soil modulus. It should be noted that the soil modulus can be any arbitrary function of x and y.

Substituting equation Equation C.8 into Equation C.6 gives

$$-A_m y_m = y_{m+2} - 4y_{m+1} + 6y_m - 4y_{m-1} + y_{m-2}, \tag{C.9}$$

where

$$A_m = \frac{E_{sm}}{EI}\left[\frac{L}{t}\right]^4. \tag{C.10}$$

With regard to the solution of the problem of the laterally loaded pile, the shear and the moment at the pile tip (say, point 0) are both equal to zero;

$$-y_2 + 2y_1 - 2y_{-1} + y_{-2} = 0 \tag{C.11}$$

and

$$y_1 - 2y_0 + y_{-1} = 0 \tag{C.12}$$

In Equations C.11 and C.12, y_{-2} and y_{-1} are two imaginary points beyond the pile tip which are used temporarily.

Using Equations C.9, C.11, and C.12, we can eliminate the y_{-2} and y_{-1} terms. Successive points up the pile are considered, and the following expression is derived:

$$y_m = -B_{2m}y_{m+2} + B_{2m+1}y_{m+1} \tag{C.13}$$

where

$$B_{2m} = \frac{1}{6 + A_m - B_{2m-4} - B_{2m-1}(4 - B_{2m-3})} \tag{C.14}$$

$$B_{2m+1} = B_{2m}(4 - B_{2m-1}) \tag{C.15}$$

Equations C.14 and C.15 hold for all B values, with three exceptions:

$$B_0 = \frac{2}{A_0 + 2} \tag{C.16}$$

$$B_1 = 2B_0 \tag{C.17}$$

$$B_2 = \frac{1}{5 + A_1 - 2B_1} \tag{C.18}$$

The boundary conditions at the top of the pile depend on the loading. For the case in which a lateral load is applied with no moment, they are

$$y_{t+1} - 2y_t + y_{t-1} = 0 \tag{C.19}$$

and

$$-y_{t+2} + 2y_{t+1} - 2y_{t-1} y_{t-2} = b_1 \tag{C.20}$$

where

$$b_1 = \frac{2P_t}{EI}\left[\frac{L}{t}\right]^3 \tag{C.21}$$

and P_t is the lateral load at the top of the pile.

Using Equations C.13, C.19, and C.20, we can derive expressions for deflection at the top of the pile:

$$y_{t+2} = C_1 B_{2t+1} y_{t+1} - C_1 y_t \tag{C.22}$$

$$y_{t+1} = \frac{(2 - B_{2t-1})y_t}{1 - B_{2t-2}} \tag{C.23}$$

$$y_t = \frac{(1 - B_{2t-2})b_1}{C_3(1 - B_{2t-2}) - C_2(2 - B_{2t-1})} \tag{C.24}$$

where

$$C_1 = \frac{1}{B_{2t}} \tag{C.25}$$

$$C_2 = C_1 B_{2t+1} - 2 - B_{2t-2}(2 - B_{2t-3}) \tag{C.26}$$

and

$$C_3 = C_1 - B_{2t-4} - B_{2t-1}(2 - B_{2t-3}) \tag{C.27}$$

A sketch of a pile with a change in flexural rigidity is shown in Figure C.1. For convenience, the deflection of the pile is designated as y below the point where there is a change

398 Piles under lateral loading

in the flexural rigidity and as z above the point of change. Notice the four imaginary points on the y and z curves.

The derivation of the coefficients necessary for calculating deflections by the difference-equation method makes use of the four compatibility equations which state that the deflection, slope, moment, and shear at point k on the y curve are equal to these respective quantities at point k on the z curve. This derivation is shown.

The derivation of the four equations of continuity is as follows: The deflection of both curves must be equal at the point where $m = k$; thus,

$$y_k = z_k \tag{C.28}$$

The slope of both curves must be equal at the point where $m = k$; thus, from Equation C.1,

$$\frac{-y_{k+1} + y_{k-1}}{2(L/t)} = \frac{-z_{k+1} + z_{k-1}}{2(L/t)} \tag{C.29}$$

and

$$-y_{k+1} + y_{k-1} = -z_{k+1} + z_{k-1} \tag{C.30}$$

The moment in the pile with respect to the y curve must equal the moment in the pile with respect to the z curve at the point where $m = k$; thus, from Equation C.2,

$$\frac{1}{b_{2_y}}(y_{k+1} - 2y_k + y_{k-1}) = \frac{1}{b_{2_z}}(z_{k+1} - 2z_k + z_{k-1}) \tag{C.31}$$

and

$$f_1(y_{k+1} - 2y_k + y_{k-1}) = f_2(z_{k+1} - 2z_k + z_{k-1}) \tag{C.32}$$

The shear in the pile with respect to the y curve must equal the shear in the pile with respect to the z curve at the point where $m = k$; thus, from Equation C.4

$$\frac{1}{b_{4_y}}(-y_{k+2} + 2y_{k+1} - 2y_{k-1} + y_{k-2})$$
$$= \frac{1}{b_{4_z}}(-z_{k+2} + 2z_{k+1} - 2z_{k-1} + z_{k-2}) \tag{C.33}$$

and

$$f_1(-y_{k+2} + 2y_{k+1} - 2y_{k-1} + y_{k-2}) = f_2(-z_{k+2} + 2z_{k+1} - 2z_{k-1} + z_{k-2}) \tag{C.34}$$

In addition to the basic expressions and the equations of continuity, it is necessary to use Equation C.9, the basic expression for soil reaction per unit length of the pile.

With the basic expressions and the equations of continuity, it is now possible to eliminate the unwanted terms in the equations.

The deflection at the point $k–2$ can be written in the form of Equation C.13:

$$y_{k-2} = -B_{2k-4}\, y_k + B_{2k-3}\, y_{k-1} \tag{C.35}$$

Since there are no imaginary points in this expression, it can be used in the calculations without change.

The deflection at the point $k-1$ can also be written in the form of Equation C.13:

$$y_{k-1} = -B_{2k-2} y_{k+1} + B_{2k-1} y_k \tag{C.36}$$

Since y_{k+1} is an imaginary point, it is necessary to make substitutions to change Equation C.36 into the following:

$$y_{k-1} = -D_{2k-2} z_{k+1} + D_{2k-1} y_k \tag{C.37}$$

The substitution can be made by solving Equations C.30 and C.32 for y_{k+1} and eliminating z_{k-1}. From C.30,

$$z_{k-1} = z_{k+1} - y_{k+1} + y_{k-1} \tag{C.38}$$

Substituting C.38 into C.32 gives

$$f_1(y_{k+1} - 2y_k + y_{k-1}) = f_2(z_{k+1} - 2z_k + z_{k+1} - y_{k+1} + y_{k-1}) \tag{C.39}$$

From C.39,

$$y_{k+1} = \frac{y_{k-1}(1-\alpha) + 2y_k(\alpha - 1) + 2z_{k+1}}{1+\alpha}, \tag{C.40}$$

where

$$\alpha = \frac{f_1}{f_2} \tag{C.41}$$

Substituting C.40 into C.36 gives

$$y_{k-1} = -B_{2k-2}\left[\frac{y_{k-1}(1-\alpha) + 2y_k(\alpha-1) + 2z_{k+1}}{1+\alpha}\right] + B_{2k-1} y_k \tag{C.42}$$

Collecting terms,

$$\left[1 + B_{2k-2}\left[\frac{1-\alpha}{1+\alpha}\right]\right] y_{k-1} = -\left[\frac{2B_{2k-2}}{1+\alpha}\right] z_{k+1} + \left[B_{2k-1} - 2B_{2k-2}\left[\frac{\alpha-1}{\alpha+1}\right]\right] y_k \tag{C.43}$$

Equation C.43 has the same form as Equation C.37, which is the desired expression. From Equation C.43, the values of the coefficients in Equation C.37 are as follows:

$$D_{2k-2} = \frac{2B_{2k-2}}{1 + B_{2k-2} + \alpha(1 - B_{2k-2})} \tag{C.44}$$

$$D_{2k-1} = \frac{B_{2k-1}(\alpha+1) - B_{2k-2}(2\alpha - 2)}{1 + B_{2k-2} + \alpha(1 - B_{2k-2})} \tag{C.45}$$

400 *Piles under lateral loading*

It is now necessary to find an expression for $y_k = z_k$ in terms of z_{k+1} and z_{k+2}. This can be done most conveniently by employing Equation C.9, the basic expression for soil reaction per unit length of the pile at point k.

$$-A_k' z_k = z_{k+2} - 4z_{k+1} + 6z_k - 4z_{k-1} + z_{k-2} \tag{C.46}$$

The notation change from A to A' is made to clarify the fact that the flexural rigidity of the z portion of the pile should be used in calculating a value for A'.

Solving Equation C.46 for z_{k-2} gives

$$z_{k-2} = -z_{k+2} + 4z_{k+1} + z_k(-6 - A_k') + 4z_{k-1} \tag{C.47}$$

Substituting the value of z_{k-2} from Equation C.47 into Equation C.34 gives

$$f_1(-y_{k+2} + 2y_{k+1} - 2y_{k-1} + y_{k-2}) = f_2[-z_{k+2} + 2z_{k+1} - 2z_{k-1}$$
$$- z_{k+2} + 4z_{k+1} + z_k(-6 - A_k') + 4z_{k-1}] \tag{C.48}$$

Collecting terms,

$$y_{k+2} - 2y_{k+1} + 2y_{k-1} - y_{k-2} - 2\beta z_{k+2} + 6\beta z_{k+1} + z_k(-6\beta - A_k'\beta) + 2\beta z_{k-1} = 0 \tag{C.49}$$

where

$$\beta = \frac{f_2}{f_1} \tag{C.50}$$

The eight terms in Equation C.49 must now be reduced to three by substitution to eliminate the undesired terms. The first step is to eliminate the z_{k-1} term by using Equation C.30.

$$y_{k+2} - 2y_{k+1} + 2y_{k-1} - y_{k-2} - 2\beta z_{k+2} + 6\beta z_{k+1} + z_k(-6\beta - A_k'\beta)$$
$$+ 2\beta z_{k+1} - 2\beta y_{k+1} + 2\beta y_{k-1} = 0 \tag{C.51}$$

Collecting terms,

$$y_{k+2} + y_{k+1}(-2 - 2\beta) + y_{k-1}(2 + 2\beta) - y_{k-2}$$
$$-2\beta z_{k+2} + 8\beta z_{k+1} + z_k(-6\beta - A_k'\beta) = 0 \tag{C.52}$$

The y_{k-2} term can be eliminated by substituting Equation C.35 into Equation C.52.

$$y_{k+2} + y_{k+1}(-2 - 2\beta) + y_{k-1}(2 + 2\beta) + B_{2k-4} y_k$$
$$- B_{2k-3} y_{k-1} - 2\beta z_{k+2} + 8\beta z_{k+1} + z_k(-6\beta - A_k'\beta) = 0 \tag{C.53}$$

Recognizing that $y_k = z_k$ and collecting terms, we have

$$y_{k+2} + y_{k+1}(-2 - 2\beta) + y_{k-1}(2 + 2\beta - B_{2k-3})$$
$$- 2\beta z_{k+2} + 8\beta z_{k+1} + z_k(-6\beta - A_k'\beta + B_{2k-4}) = 0 \tag{C.54}$$

Difference equations for step-tapered beams on foundations 401

In order to make other substitutions into Equation C.54, it is necessary at this point to make derivations for two of the y terms by using Equation C.13, the basic expression for deflection. This expression at the point k-1 is as follows:

$$y_{k-1} = -B_{2k-2}\, y_{k+1} + B_{2k-1}\, y_k \tag{C.55}$$

Solving for y_{k+1} gives

$$y_{k+1} = -\frac{1}{B_{2k-2}} y_{k-1} + \frac{B_{2k-1}}{B_{2k-2}} y_k \tag{C.56}$$

or

$$y_{k+1} = -E_1 y_{k-1} + E_2 y_k \tag{C.57}$$

where

$$E_1 = \frac{1}{B_{2k-2}} \tag{C.58}$$

$$E_2 = E_1\, B_{2k-1} \tag{C.59}$$

The expression for deflection at the point k is as follows:

$$y_k = -B_{2k} y_{k+2} + B_{2k+1} y_{k+1} \tag{C.60}$$

Solving for y_{k+2} gives

$$y_{k+2} = -\frac{1}{B_{2k}} y_k + \frac{B_{2k+1}}{B_{2k}} y_{k+1} \tag{C.61}$$

Substituting C.57 into C.61 gives

$$y_{k+2} = -\frac{1}{B_{2k}} y_k + \frac{B_{2k+1}}{B_{2k}} E_2 y_k - \frac{B_{2k+1}}{B_{2k}} E_1 y_{k-1} \tag{C.62}$$

Collecting terms,

$$y_{k+2} = \frac{1}{B_{2k}}(B_{2k+1} E_2 - 1) y_k - \frac{B_{2k+1}}{B_{2k}} E_1 y_{k-1} \tag{C.63}$$

or

$$y_{k+2} = E_3 y_k - E_4 y_{k-1} \tag{C.64}$$

where

$$E_3 = \frac{1}{B_{2k}}(B_{2k+1} + E_2 - 1) \tag{C.65}$$

402 Piles under lateral loading

$$E_4 = \frac{B_{2k+1}}{B_{2k}} E_1 \tag{C.66}$$

Now Equations C.57 and C.64 can be substituted into Equation C.54.

$$E_3 y_k - E_4 y_{k-1} - E_1(-2 - 2\beta) y_{k-1} + E_2$$

$$(-2 - 2\beta) y_k + y_{k-1} (2 + 2\beta - B_{2k-3})$$

$$-2\beta z_{k+2} + 8\beta z_{k+1} + z_k (-6\beta - A'_k \beta + B_{2k-4}) = 0 \tag{C.67}$$

or

$$E_5 y_{k-1} - 2\beta z_{k+2} + 8\beta z_{k+1} + (-6\beta - A'_k \beta + B_{2k-4} + E_3 - 2E_2 - zE_2\beta) z_k = 0 \tag{C.68}$$

where

$$E_5 = -E_4 + 2E_1 + 2E_1\beta + 2 + 2\beta - B_{2k-3} \tag{C.69}$$

Finally, Equation C.37 can be substituted into Equation C.68.

$$-E_5 D_{2k-2} z_{k+1} + E_5 D_{2k-1} y_k - 2\beta z_{k+2} + 8\beta z_{k+1} + (-6\beta - A'_k \beta + B_{2k-4} + E_3$$

$$- 2E_2 - 2E_2\beta) z_k = 0 \tag{C.70}$$

Collecting terms,

$$z_k(E_5 D_{2k-1} - 6\beta - A'_k \beta + B_{2k-4} + E_3 - 2E_2 - 2E_2\beta) = -(-2\beta) z_{k+2}$$

$$+ (E_5 D_{2k-2} - 8\beta) z_{k+1} \tag{C.71}$$

or

$$z_k = -D_{2k} z_{k+2} + D_{2k+1} z_{k+1} \tag{C.72}$$

where

$$D_{2k} = \frac{-2\beta}{E_5 D_{2k-1} - 6\beta - A'_k \beta + B_{2k-4} + E_3 - 2E_2 - 2E_2\beta} \tag{C.73}$$

$$D_{2k+1} = \frac{E_5 D_{2k-2} - 8\beta}{E_5 D_{2k-1} - 6\beta - A'_k \beta + B_{2k-4} + E_3 - 2E_2 - 2E_2\beta} \tag{C.74}$$

As before, Equation C.9, the basic expression for soil reaction, can be used at point $k + 1$:

$$-A'_{k+1} z_{k+1} = z_{k+3} - 4z_{k+2} + 6z_{k+1} - 4z_k + z_{k-1} \tag{C.75}$$

Collecting terms,

$$z_{k+3} - 4z_{k+2} + (6 + A'_{k+1}) z_{k+1} - 4z_k + z_{k-1} = 0 \tag{C.76}$$

The successive eliminations can begin by using Equation C.30 to eliminate z_{k-1}.

$$z_{k+3} - 4z_{k+2} + (6 + A'_{k+1})z_{k+1} - 4z_k + z_{k+1} - y_{k+1} + y_{k-1} = 0 \tag{C.77}$$

Collecting terms,

$$z_{k+3} - 4z_{k+2} + (7 + A'_{k+1})z_{k+1} - 4z_k - y_{k+1} + y_{k-1} = 0 \tag{C.78}$$

The expression for y_{k+1} from Equation C.57 can be substituted into Equation C.78.

$$z_{k+3} - 4z_{k+2} + (7 + A'_{k+1})z_{k+1} - 4z_k + E_1 y_{k-1} - E_2 y_k + y_{k-1} = 0 \tag{C.79}$$

Collecting terms,

$$z_{k+3} - 4z_{k+2} + (7 + A'_{k+1})z_{k+1} - (4 + E_2)z_k + (E_1 + 1)y_{k-1} = 0 \tag{C.80}$$

The expression for y_{k-1} from Equation C.37 can be substituted into Equation C.80

$$z_{k+3} - 4z_{k+2} + (7 + A'_{k+1})z_{k+1} - (4 + E_2)z_k$$
$$- D_{2k-2}(E_1 + 1)z_{k+1} + D_{2k-1}(E_1 + 1)y_k = 0 \tag{C.81}$$

Collecting terms,

$$z_{k+3} - 4z_{k+2} + (7 + A'_{k+1} - D_{2k-2} E_1 - D_{2k-2})z_{k+1} + E_6 z_k = 0 \tag{C.82}$$

where

$$E_6 = -4 - E_2 + D_{2k-1} E_1 + D_{2k-1} \tag{C.83}$$

The expression for z_k from Equation C.72 can be substituted into Equation C.82.

$$z_{k+3} - 4z_{k+2} + (7 + A'_{k+1} - D_{2k-2} E_1 - D_{2k-2})z_{k+1} - E_6 D_{2k} z_{k+2} + E_6 D_{2k+1} z_{k+1} = 0 \tag{C.84}$$

Collecting terms,

$$z_{k+3} - (4 + D_{2k} E_6)z_{k+2} + (7 + A'_{k+1} - D_{2k-2} E_1 - D_{2k-2} + D_{2k+1} E_6)z_{k+1} = 0 \tag{C.85}$$

The desired expression can now be written

$$z_{k+1} = -D_{2k+2} z_{k+3} + D_{2k+3} z_{k+2} \tag{C.86}$$

where

$$D_{2k+2} = \frac{1}{7 + A'_{k+1} - D_{2k-2} E_1 - D_{2k-2} + D_{2k+1} E_6} \tag{C.87}$$

$$D_{2k+3} = \frac{4 + D_{2k} E_6}{7 + A'_{k+1} - D_{2k-2} E_1 - D_{2k-2} + D_{2k+1} E_6} \tag{C.88}$$

Again, Equation C.9, the basic expression for soil reaction, can be used at point $k + 2$:

404 *Piles under lateral loading*

$$-A'_{k+2} z_{k+2} = z_{k+4} - 4z_{k+3} + 6z_{k+2} - 4z_{k+1} + z_k \tag{C.89}$$

Collecting terms,

$$z_{k+4} - 4z_{k+3} + (6 + A'_{k+2})z_{k+2} - 4z_{k+1} + z_k = 0 \tag{C.90}$$

Equation C.72 can be used to eliminate z_k in Equation C.90.

$$z_{k+4} - 4z_{k+3} + (6 + A'_{k+2})z_{k+2} - 4z_{k-1} - D_{2k}z_{k+2} + D_{2k+1}z_{k+1} = 0 \tag{C.91}$$

Collecting terms,

$$z_{k+4} - 4z_{k+3} + (6 + A'_{k+2} - D_{2k})z_{k+2} + (D_{2k+1} - 4)z_{k+1} = 0 \tag{C.92}$$

Equation C.86 can be used to eliminate z_{k+1} in Equation C.92.

$$z_{k+4} - 4z_{k+3} + (6 + A'_{k+2} - D_{2k})z_{k+2}$$
$$+ (D_{2k+1} - 4)(-D_{2k+2})z_{k+3} + (D_{2k+1} - 4)(D_{2k+3})z_{k+2} = 0 \tag{C.93}$$

Collecting terms,

$$z_{k+4} + (-D_{2k+1} D_{2k+2} + 4D_{2k+2} - 4)z_{k+3}$$
$$+ (6 + A'_{k+2} - D_{2k} + D_{2k+1} D_{2k+3} - 4D_{2k+3})z_{k+2} = 0 \tag{C.94}$$

The desired expression can now be written

$$z_{k+2} = -D_{2k+4} z_{k+4} + D_{2k+5} z_{k+3} \tag{C.95}$$

where

$$D_{2k+4} = \frac{1}{6 + A'_{k+2} - D_{2k} - D_{2k+3}(4 - D_{2k+1})} \tag{C.96}$$

$$D_{2k+5} = D_{2k+4} [4 - D_{2k+2}(4 - D_{2k+1})] \tag{C.97}$$

It is evident that from z_{k+2} forward, the expressions for the coefficients fall into a pattern:

$$z_s = D_{2s}z_{s+2} + D_{2s+1} z_{s+1} \tag{C.98}$$

where

$$D_{2s} = \frac{1}{6 + A'_s - D_{2s-4} - D_{2s-1}(4 - D_{2s-3})} \tag{C.99}$$

$$D_{2s+1} = D_{2s} [4 - D_{2s-2} (4 - D_{2s-3})] \tag{C.100}$$

It should be noted that these expressions for the coefficients are not the same as the basic expressions for the coefficients in the section of the pile below the point of change in section.
The expressions for the C coefficients at the top of the pile are unchanged.

APPENDIX D

Student version of Computer Program LPILE and Student version of Computer Program GROUP

The programs are available to the user on a compact disc (book CD) located in a pocket at the back cover of the book.

D.1 INSTALLATION PROCEDURES FOR BOTH PROGRAMS

D.1.1 *Initial Steps*

The file *setup.exe* is a Windows-based program that will assist the user in installing all the program modules into an user-selected directory with the proper settings for the Windows environment. This student version has been tested to be compatible with the following versions of Windows: 95, 95B (OSR-2), 98, NT 3.51, NT 4.0 and NT 5.0 (Win 2000). Therefore, the computer should be set up to operate under Windows if that has not already been done. The user is assumed to have the basic knowledge for running applications under Windows.

The setup program should run automatically when the CD-ROM is inserted; however, if the program does not run, the following steps should be performed.
1. Start Windows.
2. Insert CD in tray.
3. Click on the Start button and select Run.
4. Type *d:setup.exe* on the Command line where *d:* represents the CD drive which contains the book CD. Press Enter or simply click OK to execute the command and start the installation program.

D.1.2 *Familiarization*

The student is presumed to have studied the textbook and gained a general understanding of the technology employed in the solution of the single pile and the group under lateral loading. Two procedures will be helpful prior to solving an independent solution: (1) read the Help files that accompany LPILE and GROUP, and (2) run the example solutions that accompany each of the codes.

Prior to beginning a solution, all of the relevant input data should be selected and written down. The input can then be entered in a convenient manner and the solution can be obtained rapidly. With input at hand, entering data and obtaining a solution on the computer should take less than ten minutes.

The student should be aware of the importance of (1) using double-precision arithmetic in the solution of the differential equation, (2) the tolerance selected for ensuring convergence to an acceptable solution, and (3) the number of increments into which the length of a pile is sub-divided. Double precision is achieved by the coding and should be no problem except in unusual circumstances. The student can use the default values of tolerance for the initial solutions, but varying the value of the tolerance with successive runs to compare the results of solutions is essential. Further, for various trials, the number of increments into which the pile is divided can be changed and successive results can again be compared to gain an understanding of the importance of increment length.

D.1.3 *Getting started*

The programs are started by double clicking the left mouse in the LPILE or GROUP icon. A new window will appear on the screen, with seven top menu choices: File, Data, Options, Computation, Graphics, Window, and Help. Data for an analytical problem is normally entered by consecutive selection of each item in the Data menu, starting from the top. Certain modifications to the Default method of analysis may be made by selecting items in the Options menu.

D.2 USE OF COMPUTER PROGRAM LPILE

D.2.1 *Example solutions*

Prior to coding the data for a personal problem, the student is urged to run one or more of the three examples that are included in the LPILEP3 sub-directory of the book CD. Each numbered example can be loaded by selecting Open under the File menu and browsing for the desired file in the book CD. A brief description of each example follows.

Example 1. The pile is a 14HP89 (U.S.) structural shape, with a penetration of 15.2 m, loading is against the strong axis, pile is assumed to be supporting a retaining wall; thus, the loading is static. The bending moment at the first yield of the steel was found to be 657 m-kN, considering an axial load of 222 kN. The pile is assumed to be fixed against rotation at the groundline and is subjected to a series of horizontal loads. The soil is a sand with a friction angle of 35°.

The tabulated values of the output may be viewed by clicking on Computations/Edit Output Text However, one of the best features of the program is that the results can be presented graphically. The student may view the following curves: pile-head deflection as a function of lateral load; maximum bending moment as a function of lateral load; family of curves showing deflection versus depth; and a family of curves showing bending moment versus depth.

Example 2. The pile is a leg of an offshore jacket. Below the mudline for 4 m the pile consists of a steel pipe with a diameter of 762 mm inside the jacket-leg extension which is a pipe with a diameter of 838 mm. Below 4 m, the pile is 762 mm in diameter.

The student may examine the input and formulate the balance of the input to the program. Then, the output may be examined to find the loading that will cause a plastic hinge.

Example 3. The pile is a steel pipe with an outside diameter of 373 mm and a wall thickness of 22 mm. The length of the pile is 15.2 m with 2.54 m extending above the ground surface. The soil is a sand with a friction angle of 35°. The lateral load at the pile head is 44.5 kN. The aim of the problem is to find the axial load that will cause buckling of the pile by increasing the axial load in increments and observing the pile-head deflection.

After finding the buckling load, the student may use the Euler formula to find the length of a pile fixed at its lower end and free at the top that will sustain the same load. The demonstration should reveal the necessity of employing nonlinear soil restraint, rather than pure mechanics, in obtaining the buckling load of a soil-supported column.

D.3 USE OF COMPUTER PROGRAM GROUP

D.3.1 *Example solutions*

Again, the student is urged to review one or both of the examples prior to coding the data for a personal problem. Either of the two examples can be loaded in the software by following the same procedure used for the examples in LPILE.

As described in the text, the stiffness of each pile under axial and lateral load is necessary in order to obtain a solution for the group. The stiffness under lateral load, with respect to both deflection and rotation at the pile head, and found by internal coding. However, for a new problem the student is required to enter a curve showing the axial stiffness (axial load versus pile-head movement) of the pile selected.

Example 1. A group of four bored piles is analyzed under an vertical load of 227.6 kips, a horizontal load of 45 kips, and an overturning moment of 2.7×10^6 in-kips (clockwise). The piles are 72 inches in diameter, with a penetration of 660 inches. The graphical feature of the program can be utilized to view plan and elevation views of the group. No reduction in soil resistance due to pile-soil-pile interaction is assumed.

As a minimum in analyzing of the results of the solution, the student should obtain from the output the loading at each pile head and check to see that static equilibrium has been achieved. Also, if desired, the student can take the pile-head loading from the GROUP solution and use LPILE as a check of the detailed behavior of the pile. Input and results may be viewed using International units by selecting the desired system under the Options menu.

Example 2. A group of six bored piles is analyzed under an axial load of 1000 kips, a lateral load of 200 kips, and an overturning moment of 1.0×10^7 in-kips (counter-clockwise). The piles are 30 inches in diameter and with a penetration of 600 inches. Again, the graphical feature of the program can be utilized to see plan and elevation views of the group. Reduction in lateral resistance due to pile-soil-pile interaction is implemented by internal coding in the GROUP program.

The student is urged to check equilibrium of the group as before. The behavior of the individual piles can be examined by reviewing the tabulated output or the graphical output.

APPENDIX E

Nondimensional curves for piles under lateral loading for case where $E_{py} = k_{py} x$

Curves are presented for three cases of loading, where $x = ZT$ and $Z_{max} = L/T$.

Lateral load at top of pile		Moment at top of pile	
Deflection:	$y = (P_t T^3/E_p I_p) A_y$		$y = (M_t T^2/E_p I_p) B_y$
Slope:	$S = (P_t T^2/E_p I_p) A_s$		$S = (M_t T/E_p I_p) B_s$
Moment:	$M = (P_t T) A_m$		$M = (M_t) B_m$
Shear:	$v = (P_t) A_v$		$v = (M_t/T) B_v$
Soil reaction:	$p = (P_t/T) A_p$		$p = (P_t/T^2) B_p$

Pile head fixed against rotation

$y = (P_t T^3/E_p I_p) F_y$

$M = (P_t T) F_m$

$p = (P_t /T) F_p$

409

410 *Piles under lateral loading*

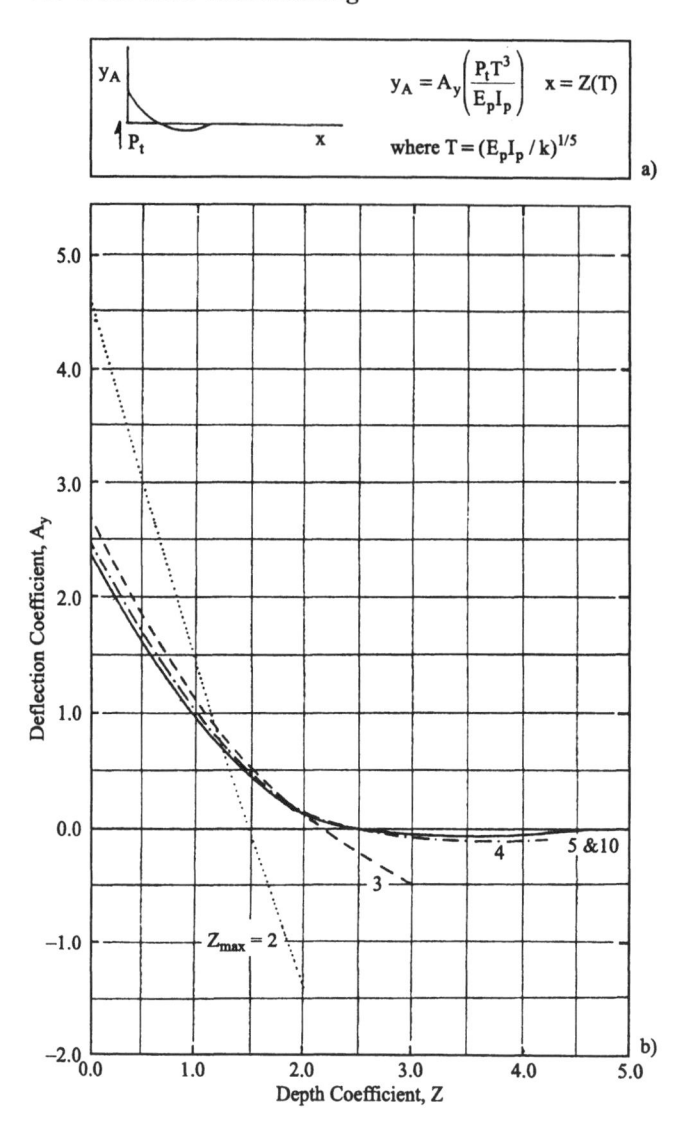

$$y_A = A_y \left(\frac{P_t T^3}{E_p I_p}\right) \quad x = Z(T)$$

where $T = (E_p I_p / k)^{1/5}$

Figure E.1. Pile deflection produced by lateral load at mud-line (Reese & Matlock 1956).

Nondimensional curves for piles under lateral loading 411

$$S_A = A_S\left(\frac{P_t T^2}{E_p I_p}\right) \quad x = Z(T)$$

where $T = (E_p I_p / k)^{1/5}$

a)

b)

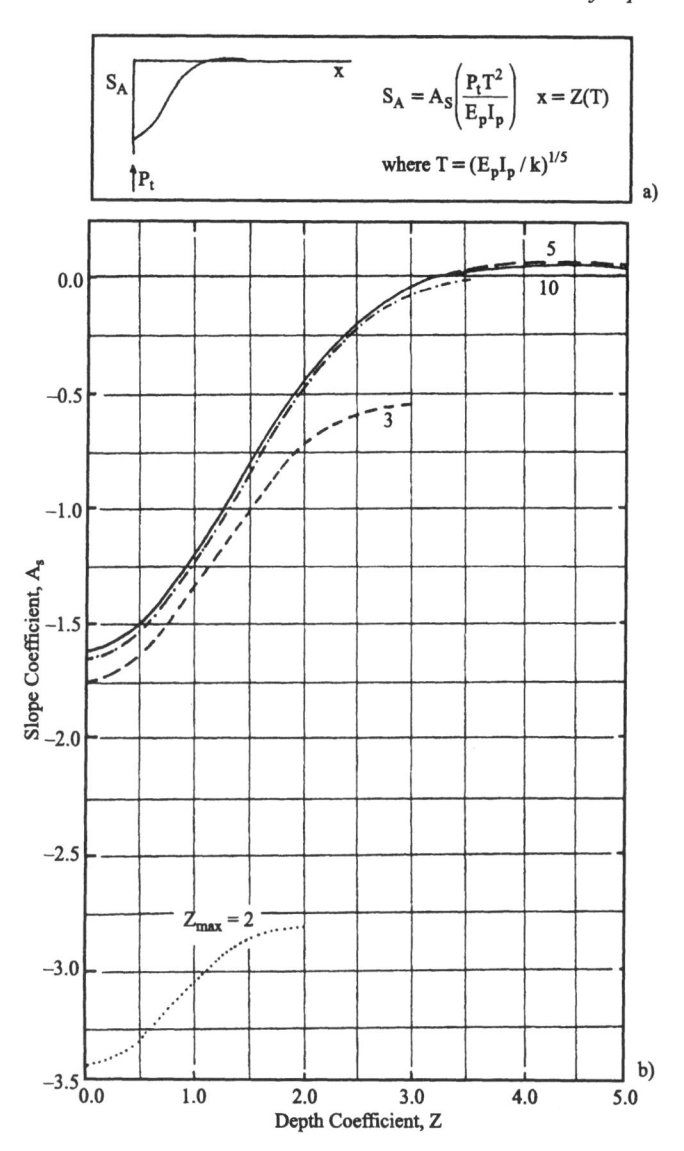

Figure E.2. Slope of pile caused by lateral load at mud-line (Reese & Matlock 1956).

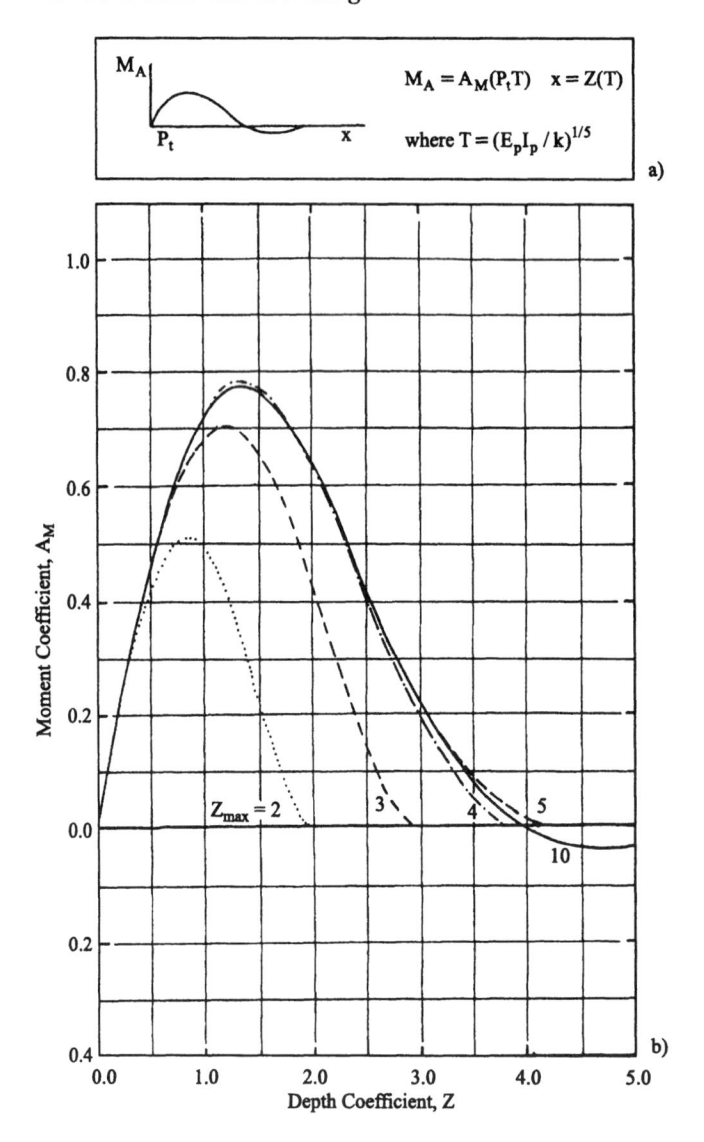

Figure E.3. Bending moment produced by lateral load at mud-line (Reese & Matlock 1956).

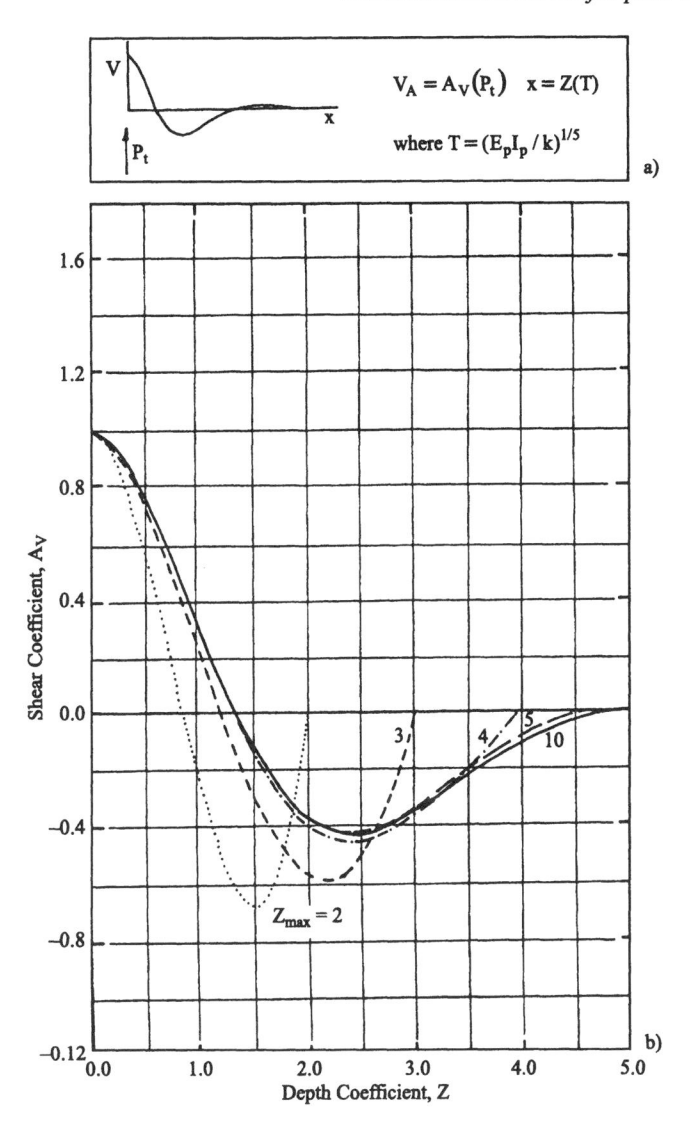

Figure E.4. Shear produced by lateral load at mud-line (Reese & Matlock 1956).

414 *Piles under lateral loading*

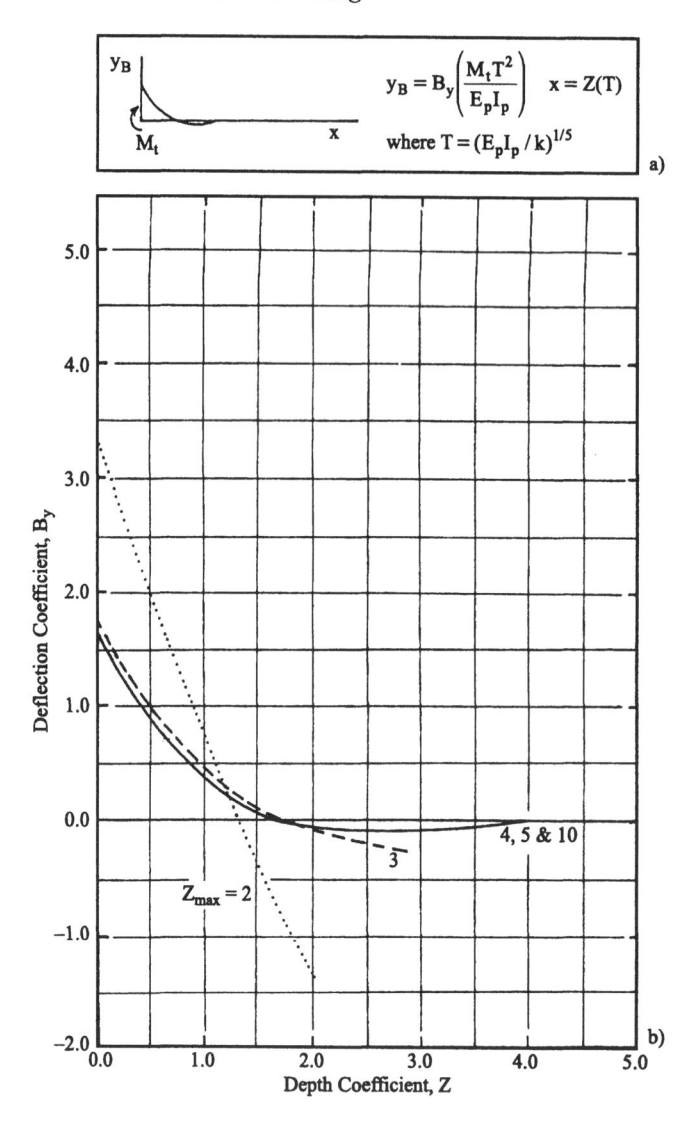

$$y_B = B_y \left(\frac{M_t T^2}{E_p I_p} \right) \quad x = Z(T)$$

where $T = (E_p I_p / k)^{1/5}$

Figure E.5. Pile deflection produced by moment applied at mud-line (Reese & Matlock 1956).

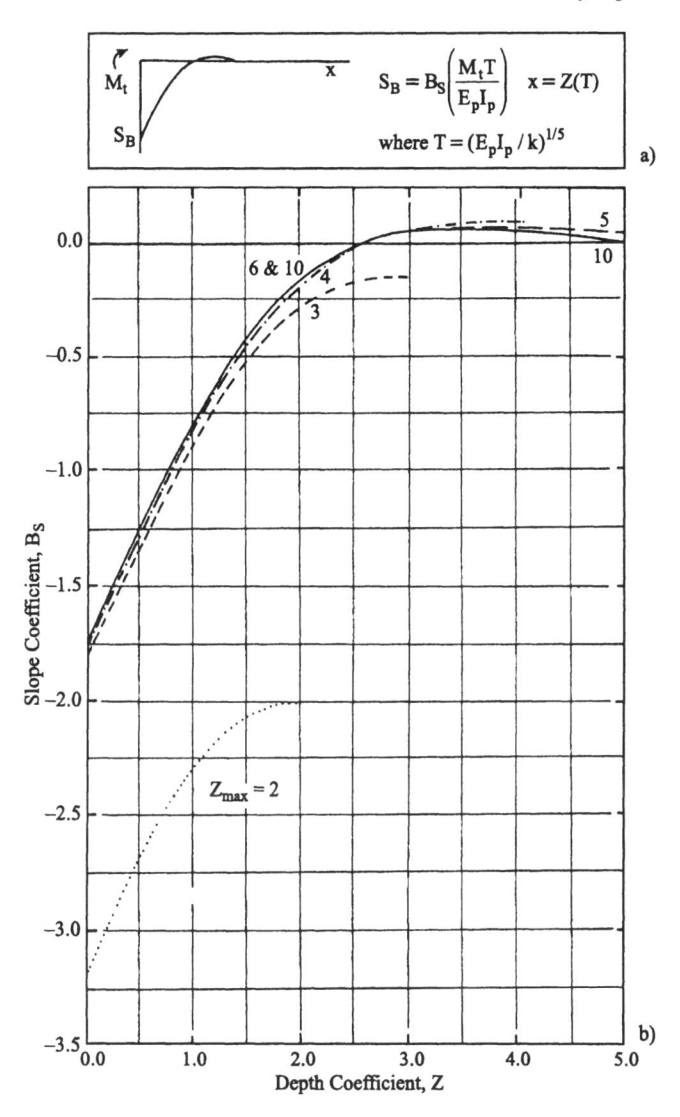

Figure E.6. Slope of pile caused by moment applied at mud-line (Reese & Matlock 1956).

416 *Piles under lateral loading*

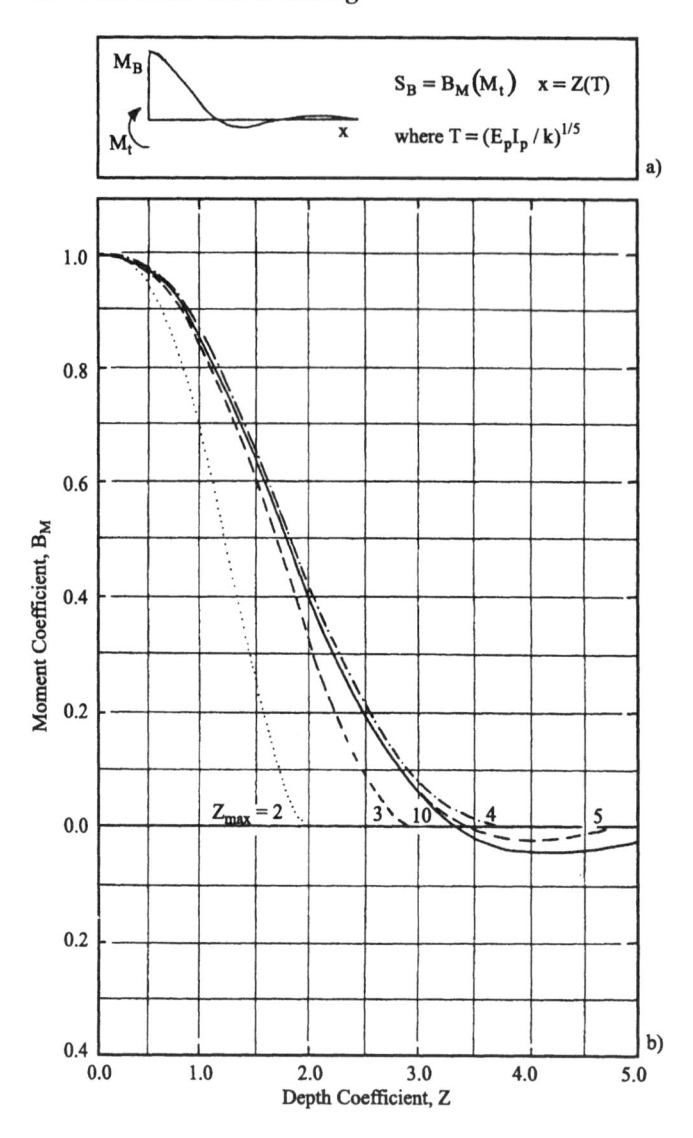

$S_B = B_M(M_t)$ $x = Z(T)$

where $T = (E_p I_p / k)^{1/5}$

Figure E.7. Bending moment produced by moment applied at mud-line (Reese & Matlock 1956).

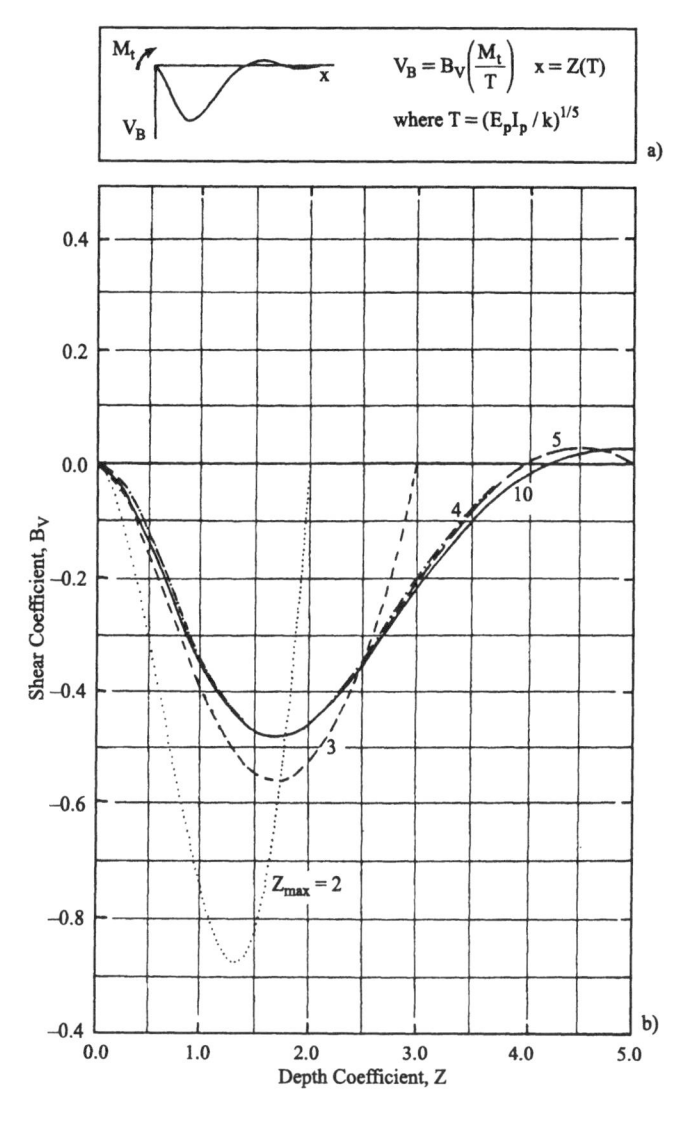

Figure E.8. Shear produced by moment applied at mud-line (Reese & Matlock 1956).

418 *Piles under lateral loading*

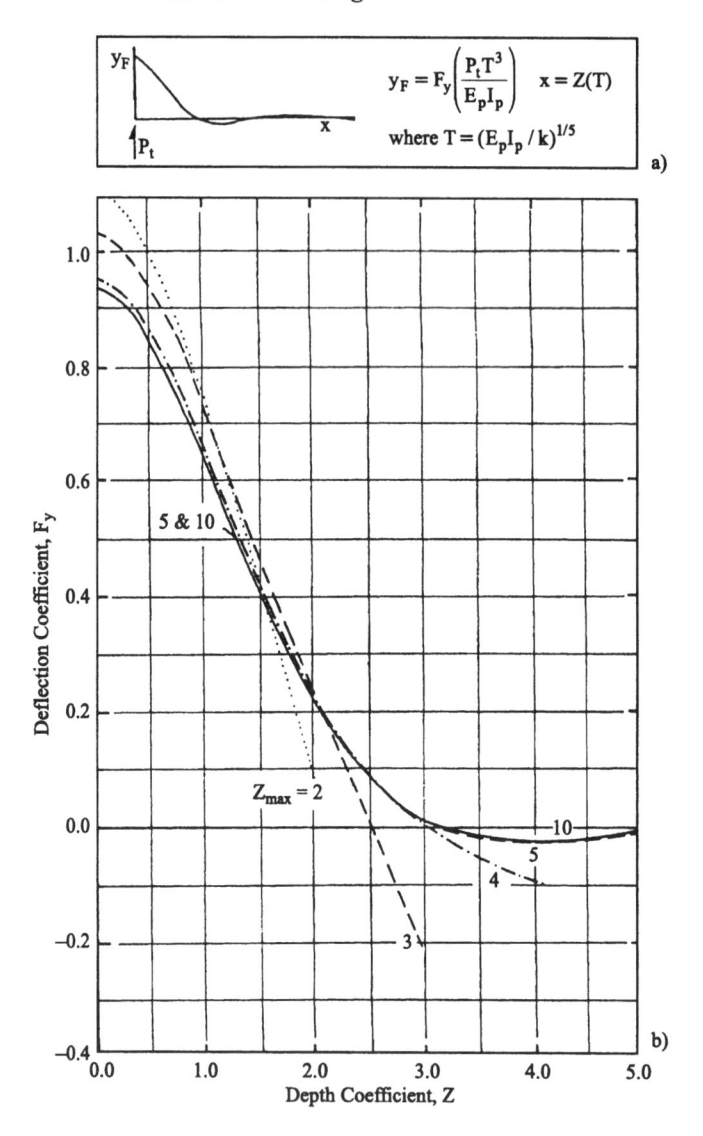

$$y_F = F_y \left(\frac{P_t T^3}{E_p I_p} \right) \quad x = Z(T)$$

where $T = (E_p I_p / k)^{1/5}$

Figure E.9. Deflection of pile fixed against rotation at mud-line (Reese & Matlock 1956).

APPENDIX F

Tables of values of efficiency measured in tests of groups of piles under lateral loading

Table F.1. Results from in-line tests, Cox. et al (1984).

Penetration, diameters	Center-to-center spacing, diameters	Number of piles	Individual pile efficiencies*	Group efficiency
4	1.5	3	0.46,0.56,0.73,	0.58
4	1.5	3	0.50,0.56,0.77	0.61
8	1.5	3	0.52,0.53,0.72	0.59
8	1.5	3	0.52,0.43,0.82	0.60
8	1.5	5	0.54,0.50,0.38,0.53,0.76	0.54
8	1.5	5	0.60,0.43,0.41,0.47,0.78	0.54
4	2.0	3	0.65,0.62,1.03	0.77
4	2.0	3	0.54,0.64,1.01	0.73
8	2.0	3	0.65,0.56,0.92	0.71
8	2.0	3	0.65,0.60,0.84	0.70
8	2.0	5	0.66,0.53,0.53,0.54,0.82	0.62
8	2.0	5	0.63,0.44,0.57,0.52,0.78	0.59
8	3.0	3	0.77,0.77,0.97	0.84
8	3.0	3	0.75,0.73,0.93	0.80
8	3.0	5	0.75,0.75,0.77,0.79,0.98	0.81
8	3.0	5	0.72,0.73,0.77,0.75,0.95	0.78
8	4.0	3	0.83,0.87,0.97	0.89
8	4.0	3	0.85,0.86,0.96	0.89
8	6.0	3	0.92,0.92,1.01	0.95
8	6.0	3	0.92,0.92,1.03	0.95

*Leading pile on right

Table F.2. Results from side-by-side tests, Cox et al (1984).

Penetration, diameters	Center-to-center spacing, diameters	Number of piles	Individual pile efficiency	Group efficiency
4	1.5	3	0.75,0.70.0.83	0.76
4	1.5	3	0.78,0.73,0.77	0.76
4	1.5	3	0.76,0.78,0.78	0.77
8	1.5	3	0.82,0.84,0.85	0.84
8	1.5	3	0.83,0.83,0.83	0.83
8	1.5	5	0.81,0.76,0.69,0.77,0.76	0.76
8	1.5	5	0.83,0.83,0.76,0.82,0.86	0.82
4	2.0	3	0.87,0.80,0.89	0.85
4	2.0	3	0.86,0.87,0.95	0.89
6	2.0	3	0.85,0.80,0.84	0.83
6	2.0	3	0.84,0.84,0.86	.85
8	2.0	3	0.88,0.87,0.85	0.86
8	2.0	3	0.88,0.88,0.87	0.88
8	2.0	5	0.84,0.84,0.80,0.87,0.90	0.85
8	2.0	5	0.87,0.85,0.86,0.87,0.86	0.86
8	3.0	5	0.99,0.95,0.93,0.98,0.98	0.97
8	3.0	5	0.99,0.93,0.95,0.98,0.96	0.96
8	4.0	3	0.98,0.97,0.96	0.97
8	4.0	3	0.99,0.98,0.99	0.99

Table F.3 Efficiencies of piles in groups, Franke (1984).

Center-to-center spacing, diameters	Efficiency of trailing piles	Efficiency of side-by-side piles
1		0,72
1		0.90
1		0.93
1		0.48
1		0.76
2	0.58	0.90
2		0.92
2		0.94
3	0.67	1.00
4	0.80	1.00
5	0.93	1.00
6		1.00
7		1.00
8		1.00

Table F.4. Results from in-line tests, Schmidt (1981,1985).

Penetration, diameters	Center-to-center spacing, diameters	Number of piles	Individual pile efficiencies*	Group efficiency
23.3	2.42	2	0.73,0.94,	0.84
13.3	1.33	2	0.73,0.98	0.85
7.1	2.0	2	0.69,0.83	0.76
7.1	3.0	2	0.83,0.86	0.85

*Leading pile on right

Table F.5. Efficiencies of piles in groups, Shibata et al (1989).

Center-to-center spacing, diameters	Efficiency of leading piles	Efficiency of trailing piles	Efficiency of side-by side piles
2	1.00	0.87	0.64
2.5	1.00	0.80	0.75
5	1.00	0.92	1.00

Table F.6. Efficiencies of piles in side-by-side groups, Wang & Reese (1986).

Center-to-center spacing, diameters	Efficiency in soft clay	Efficiency in dense sand	Efficiency in loose sand
1	0.53	0.54	0.68
1.25	0.60	0.90	0.74
1.5	0.76	0.91	0.75
2	0.79	0.88	1.04
3	0.66	1.19	0.89
4	0.97	1.0	

APPENDIX G

Horizontal stresses in soil near shaft during installation of a pile (Van Impe 1991)

Data were obtained during the installation of two types of piles, the Atlas and the PCS-auger pile.

Figure G.1 shows a typical test-result for the Atlas pile. The normalized earth-pressure coefficient from the dilatometer test (DMT), K_d, was measured before and after the installation of the pile. Mean values of K_d were obtained for two Atlas piles, with the following results where $(K_d\,mean)_{before} / (K_d\,mean)_{after}$ was 1.01 for Atlas pile 36/50 and 1.15 for Atlas pile 51/65. Values were measured for a tube pile 35/65 and $(K_d\,mean)_{before} / (K_d\,mean)_{after}$ was found to be 1.22.

Although these values seems realistic, one has to take care in implementing such ratios. Of course there is an influence of non-homogeneity of the soil. Also a better theoretical understanding of the absolute values of the horizontal stress index is necessary in order to make more reliable interpretations and predictions.

The diagram in Figure G.2, for Atlas Pile 15, 36/50, shows total horizontal stress and pore water pressure as a function of time. Depths are shown to placement of dilatometers and the cone with pore-pressure measurement. Similar information is shown by the diagram in Figure G.3, for Atlas Pile 9, 36/50. Both diagrams shown the effects of the passing of the auger and the placement of concrete. Of interest in Figure G.2 in the clear indication of the passing of the auger.

The gradients of decreasing and increasing values depend on a number of factors, including the stress state at time of initial installation, the depth to the recording instrument, and the overall energy employed in placing the pile. However, the influence of construction operations on stress in the soil is clearly indicated.

The test site for the PCS-auger pile was at Doel. The procedure employed in Europe for installing the PCS pile is shown in Figure G.4. An inner stem diameter of 150 mm is generally used. The rotary table is capable of applying a large torque of 100 kN-m, enabling the installation of the pile without excavating too much soil. Penetration into resistant layers and through hard lenses is possible.

During casting the concrete, an overpressure of 2 to 4 bars is applied on the fresh concrete, while the auger is retained slowly. This procedure doesn't cause vibrations. After casting the concrete, the reinforcement is placed into the pile by using a vibrator. Eventual difficulties can be avoided using a greater inner stem diameter. So some reinforcement can be placed inside before casting the concrete. The degree of soil displacement can be deduced from the overconsumption of concrete (*occ*), defined as:

$$occ = \frac{V_b - V_p}{V_p} \cdot 100 (\%) \qquad (G.1)$$

424 *Piles under lateral loading*

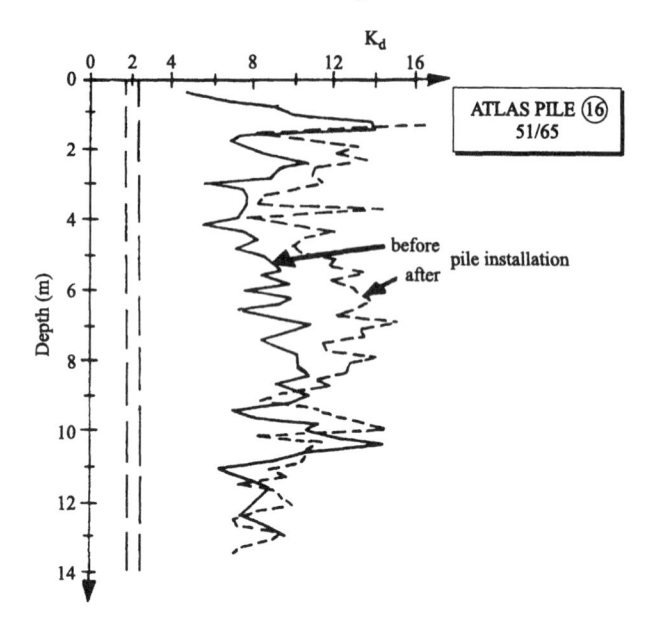

Figure G.1. K_d-profile before and after pile execution.

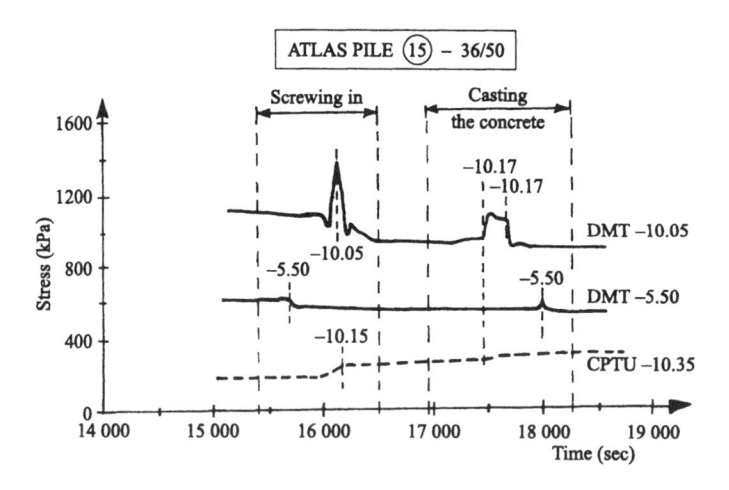

Figure G.2. DMT-measurement during installation of pile 15.

in which V_p = theoretical volume of borehole, and V_b = volume of concrete used.

The parameter, *occ*, gives a mean value and obscures local effects where lenses or layers exist that cause a larger volume of excavation and/or soil displacement. The overconsumption at the test site at Doel was 95%. The high value is due to a soft upper layer (0 to –8.00 m) of dredged material and due to a very soft natural layer (–8.70 to –10.30 m).

Soil tests were performed with the electrical-cone-penetration device prior to the placement of the pile. Dilatometer tests with DMT-1A and DMT-1C) were performed at

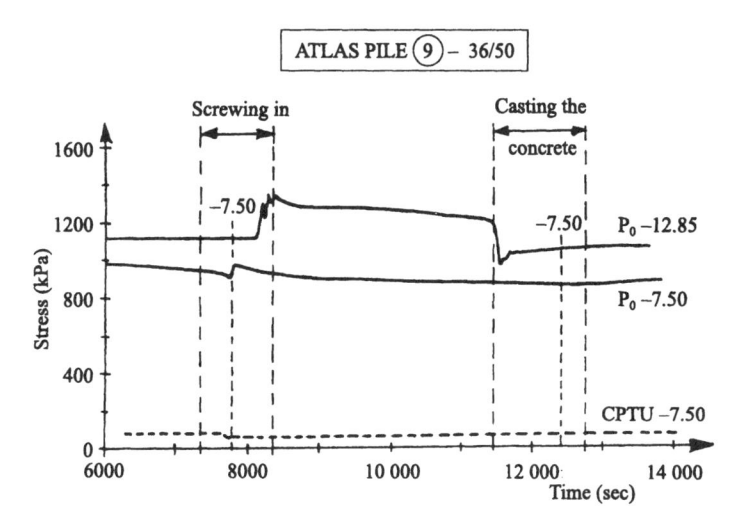

Horizontal stresses in soil near shaft during installation of a pile 425

Figure G.3. DMT-measurement during installation of pile 9.

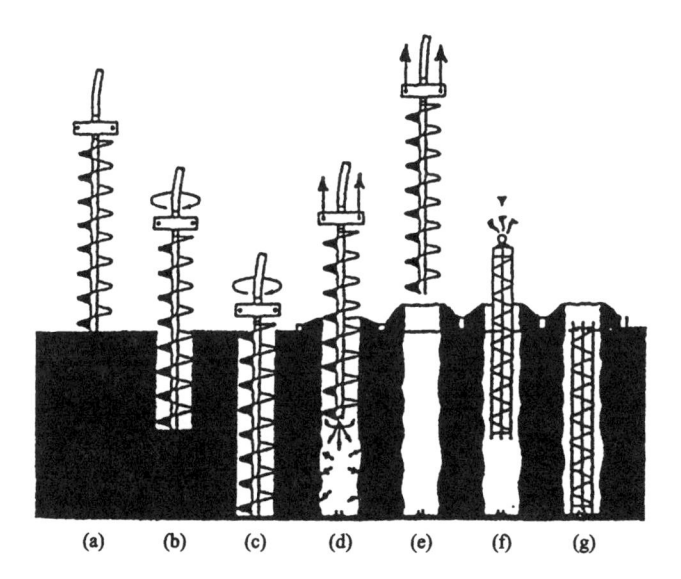

Figure G.4. Phases of execution of PCS pile.

1.5 times the pile diameter from the axis of the pile prior to installation, and the test with DMT-1C was performed in the same location after installation.

Also prior to installation, a dilatometer test (DMT-1B) was placed with the device set at a specific depth. The A-reading of the DMT curve was considered to be equivalent to a time-deformation curve obtained by an odometer. The data were analyzed by the Casagrande log$t$ method of curve fitting and the time was determined at 100% consolidation. Such information allowed the decision to be made to allow pile installation when the A-readings became less that 5 kPa/hour. The assumption was made the re-consolidation and

426 *Piles under lateral loading*

relaxation, due to the installation of the dilatometer, had only a negligible effect on the measurements made during the installation of the pile.

The blade of DMT-1B was oriented tangentially rather than radially with the advantage that installation of the dilatometer caused a smaller disturbance in the initial stress field around the pile. Further, with the tangential orientation, total mean stresses could be measured along arches around the excavation and the pile after placement. However, interpretation of the results from the testing is problematic because no clear relationships exist between the changes in principal stress.

For the tests at Doel, the water table was at 3.20 m below the ground surface, the diameter of the pile was 400 mm, the length of the pile was 19.5 m, Dilatometer 1B was installed at a depth of 8.50 m, and the distance from the blade of the dilatometer to the pile was 600 mm. The results of the tests are shown in Figures G.5a, b, and c.

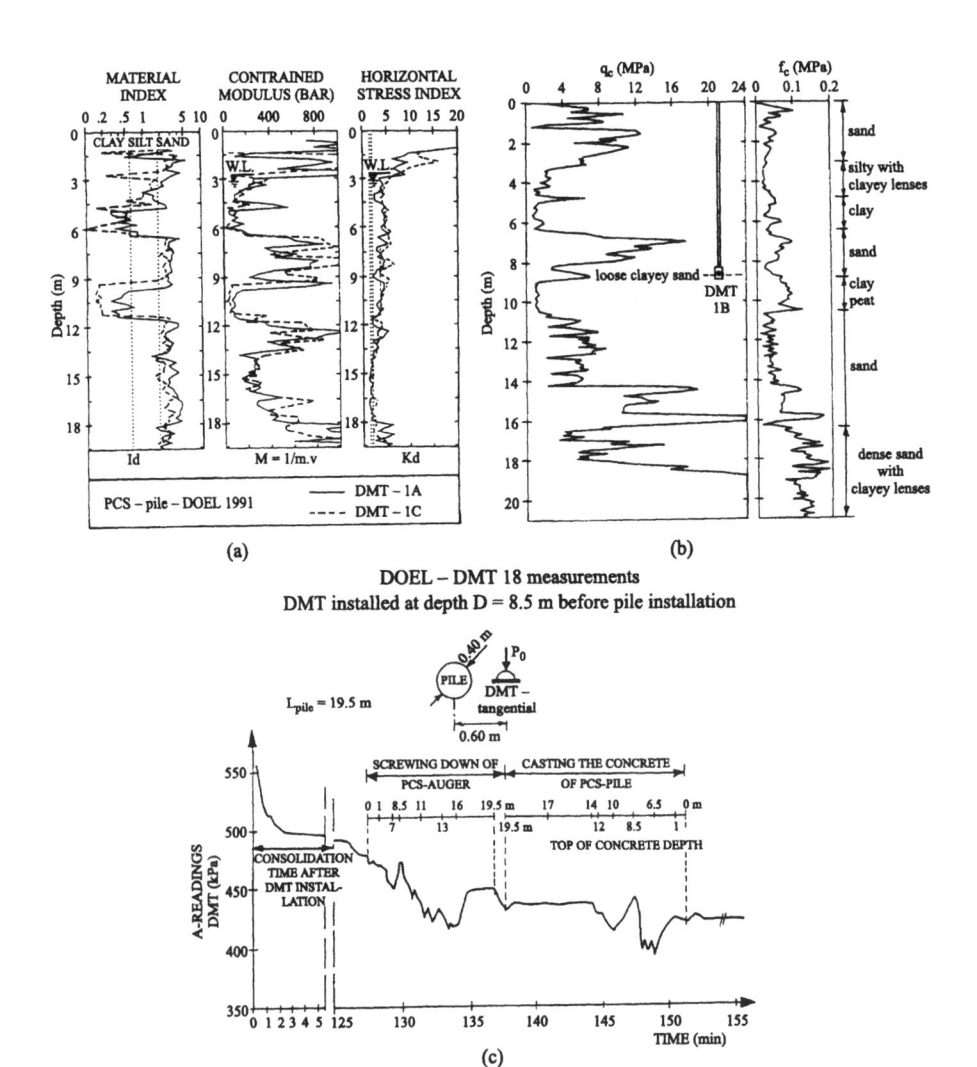

Figure G.5. Results of field tests.

The soil profile is dominated by sand but there are two distinct layers of clay. Some degree of overconsolidation was expected in the stratum of sand from 6-5 to 8 m, as well as the stratum below 12 m. The A-readings from the dilatometer were higher than expected in these strata. However difficulties can arise in implementing the A-readings in quantitative analyses because the readings can be affected significantly by lack of verticality of the blade and by other errors in placement.

Figure G.5c shows that DMT-1B recorded a gradual decrease in the A-readings until a depth of 15 m and after about 2 hours and 15 minutes after starting the excavation. A different result would have been expected had the excavation proceeded at a faster rate as is normal.

The peak in A-readings at 8.5 m of depth is created by the suddenly induced excess pore water pressure at the time of passing by of the auger.

The permeability of the sand layer however was large enough in order to neutralize immediately this effect. At a depth of 15 m, an electrical breakdown resulted in a long stop of activity and resetting of all equipment. This explains the not definable output from the DMT-instrument.

Because the DNT-blade was positioned tangentially, the effect of soil arching around the screwed hole would result in the mean total soil stress becoming constant at some level, which was the starting level at casting the concrete.

This 'arching' total-stress level does not reflect any stress change during the casting phase until the concrete reaches the level of 14 m of depth. Apparently, due to the increased total radial stresses in the bore hole as it is filled up with fresh concrete, the arching tangential total stress is gradually decreasing. The peak value at *8.5* m is only the compensating high excess pore water pressure inducting when the fresh concrete passes the DMT-level. Due to the relatively high permeability, such 'seepage force' influencing the DMT total pressure reading appears quickly.

From a depth of 6.5 m on, the overweight of the column of fresh concrete above the level of the DMT compensates better and better the arching decrease. The values reach a constant at approximately half of the starting A-readings at screwing in.

It seems reasonable to put forward that in a very near surrounding of the auger pile the screwing in causes losses in locked up peak stresses in case of overconsolidated sandy material.

The influence of pile installation on the horizontal stress can be evaluated by comparing results from the DMT before and after pile execution. In Figure G.5a a the difference between results from DMT before (continuous line) and after (dashed line) may be seen. Referring to the horizontal stress index, an increase of total horizontal stress of about 1.30. The increase of DMT horizontal total tangential stress in the top layer and the decrease near the pile tip are remarkable.

The results from the two cases presented here are for particular piles in specific soil profiles and cannot be valid for other types of piles in other soils. However, the results do make certain that the state of stress, and probably the soil properties, around an installed pile are changed by the installation.

APPENDIX H

Use of data from testing uninstrumented piles under lateral loading to obtain soil response

H.1 PROCEDURE

The parameters k_{py1} and k_{py2} or k_{py1} and n_1 may be found by using nondimensional parameters if the values y_t and S_t were measured during the performance of the test. For the example presented below, the soil profile showed that the shearing resistance of the soil at the ground surface was zero or close to zero. Therefore, the second of the two sets of parameters was used in the computations.

If the values of y_t and S_t are known for the ith loading of the pile, the equations presented in Chapter 3 for dimensional analysis may be used to write Equations H.1 through H.3.

$$y_{ti} = \left[\frac{P_{ti}T_i^3}{E_p I_p}\right] A_{yt} + \left[\frac{M_{ti}T_i^2}{E_p I_p}\right] B_{yt} \tag{H.1}$$

$$S_{ti} = \left[\frac{P_{ti}T_i^2}{E_p I_p}\right] A_{st} + \left[\frac{M_{ti}T_i}{E_p I_p}\right] B_{st} \tag{H.2}$$

$$T_i^{n_i+4} = \frac{E_p I_p}{k_i} \tag{H.3}$$

The values of P_{ti} and M_{ti} were measured and $E_p I_p$ is computed from the properties of the section of the pile. The value of $E_p I_p$ must be constant or constant at least over the portion of the pile computed to subjected to bending moment. The values of A_{yt}, B_{yt}, A_{st}, and B_{st} may be obtained by a computer code, such as COM622, as functions of the parameter n. Thus, information is at hand for finding the values of or k_{py1i} and n_{1i} that yields the best fit to the measured values of P_{ti} and M_{ti}.

An appropriate method of solving the above equations for the ith loading is the select a value of n which yields values of T which then allows Equation H.1 to be solved, using the known value of P_{ti}. The procedure may be repeated for the solution of Equation H.2, using the known value of M_{ti}. The values of T and n may then be plotted for the ith loading, as shown in Figure H.1. Frequently, as shown in the figure, a strong intersection is not found between to plots from Equations H.1 and H.2, but the intersection yields values of T_i and n_i which may then be used to solve for values of y and p as a function of depth.

430 *Piles under lateral loading*

The process is repeated for all values of lateral loading, and cross plotting will yield p-y curves at various discrete depths.

H.2 EXAMPLE FOR SOLUTION

Two piles were tested under lateral load in Japan and the results were reported by the Committee on Pile Foundations (1965). The dimensions of the piles and the point of application of the load are shown in Figure H.2.

The piles were designated Pile 7 and Pile 8. The lateral loads applied to Pile 7 were 7.29, 11.20, 15.10, 18.66, and 21.97 kN. The lateral loads applied to Pile 8 were 9.30, 14.35, 18.29, 22.45, and 24.03 kN.

The characteristics of the soil at the test site are given in Figure H.3. A layer of top soil and another of sandy silt with gravel were removed from the site prior to installing the piles. The upper layer of soil around the pile was classified as sandy silt with shells, with the percentage of sand decreasing with depth. The percentage of clay in the soil generally increased with depth from 35% at the top of the stratum to 60% at the tip of the pile. The plasticity index of the clay was about 60. A reasonable classification of the soil at the test site is a silty clay. The soil at the top of the stratum, where the percentage of sand was the

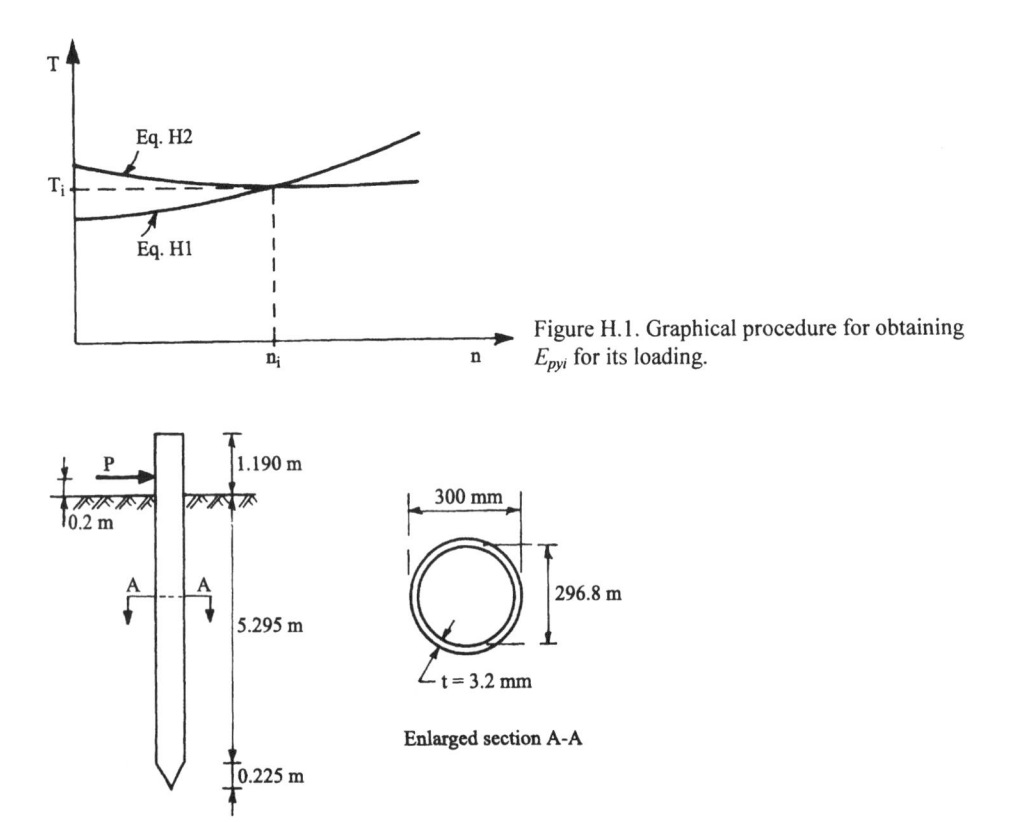

Figure H.1. Graphical procedure for obtaining E_{pyi} for its loading.

Figure H.2. Dimensions and point of loading of test pile.

highest, could be tested in unconfined compression and, as shown in Figure H.3, the average shear strength over the depth of the pile was about 26 kPa.

The piles were loaded cyclically; however, the results from the first cycle, shown in Table H.1, are assumed to represent the results for static loading. The data reported for deflection and rotation at the ground surface for Pile 7 and Pile 8 are shown in the table.

Table H.1. Date on measurement of deflection and slope at ground line, Japanese experiment.

	Lateral load kN	Deflection, mm	Rotation Radians $\times 10^{-3}$
Pile 7	7.29	0.7	1.0
	11.20	1.7	1.8
	15.10	2.6	2.6
	18.66	2.9	3.6
	21.97	4.4	4.6
Pile 8	9.10	0.7	1.3
	14.35	1.4	2.3
	18.29	2.2	3.4
	22.45	3.1	4.5
	24.05	3.5	5.0

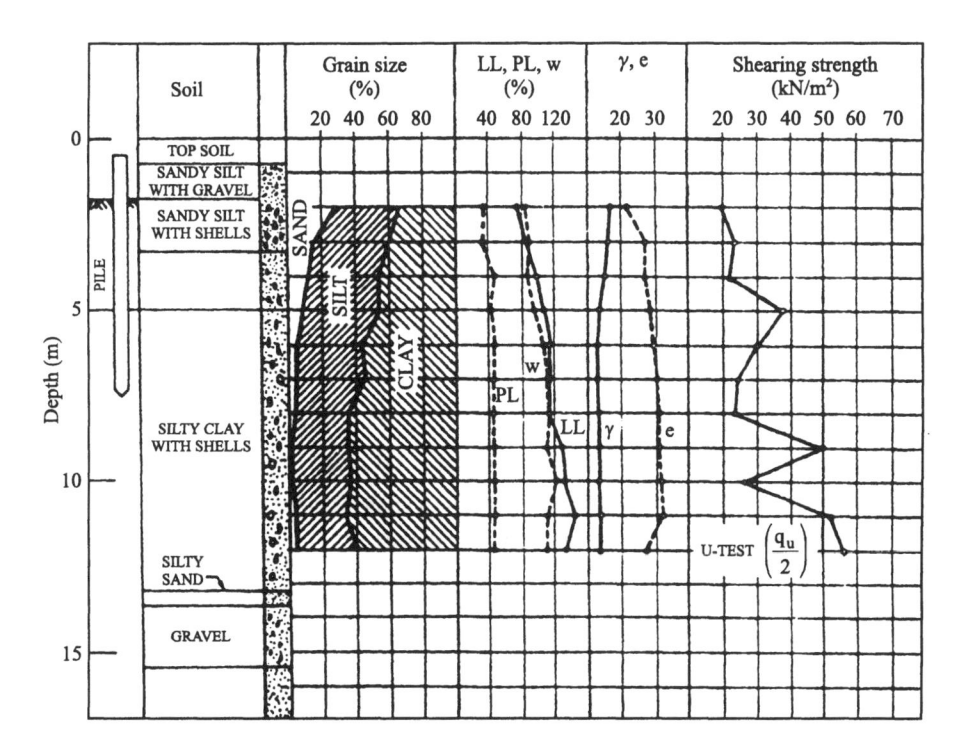

Figure H.3. Soil properties at site where piles were tested.

432 *Piles under lateral loading*

H.3 RESULTS OF ANALYSES

Using data from Table H.1 and the other data as noted in the procedure presented above, the solutions were made for values of k_{pyli} and n_{1i}. The solutions for each set of loadings were made and results were plotted as shown in Figure H.1. The results that were obtained are shown in Figure H.4 for both of the piles. The values obtained for k_{py1} and n_1 were used as input to the computer code and pile-head deflection and rotation were computed for each of the loads. Good agreement between the experimental and computed values of y_t and S_t would suggest that the derived p-y curves are valid. However, as shown in the curves in Figure H.4, good agreement was obtained for the results for Pile 7 but agreement for Pile 8 was not so good.

A review of the results for Pile 8 shows that the computed values were larger that experimental values for deflection but the reverse was true for pile-head rotation. The failure to obtain agreement between the experimental and computed values for Pile 8 show that

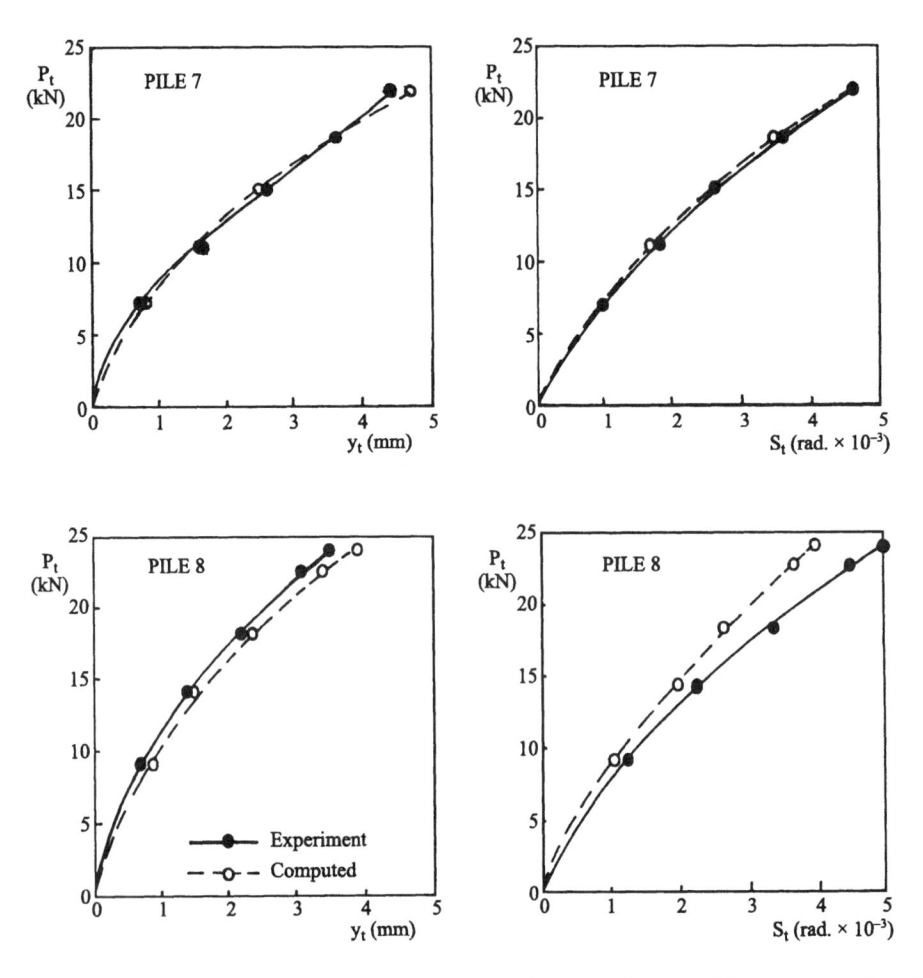

Figure H.4. Sketches showing comparisons of results from experiment and from computations for y_t and S_t for pile 7 and pile 8.

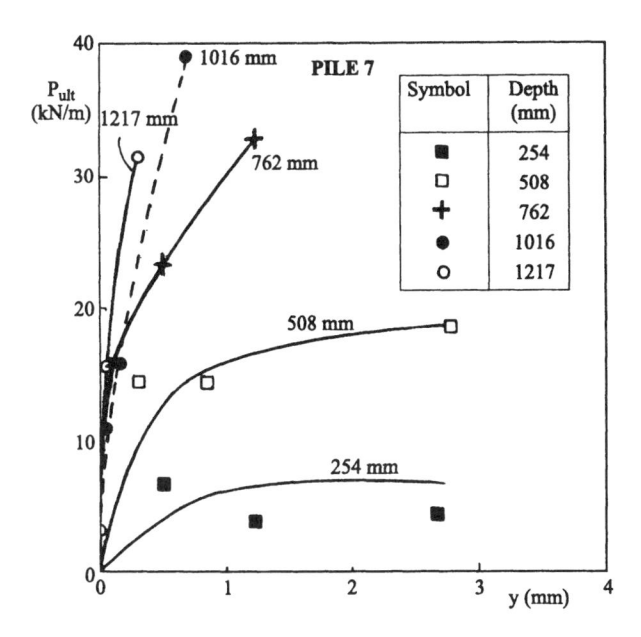

Figure H.5. Plot of p-y curves derived from test of pile 7.

1. The form of the equation selected for E_{py} was not valid or 2. The experimental data were incorrect. In view of the excellent agreement obtained for Pile 7, the lack of agreement for Pile 8 is puzzling. Nevertheless, p-y curves were computed from the results of only Pile 7.

The curves are shown in Figure H.5. The cross-plotting could have been done for any number of depth but five were selected to reflect the change in the curves with depth. The curves exhibit the shape to be expected, and the results may be thought as remarkable in view of the meager amount of data available from the tests of the piles. However, there is some scatter in the plotted data points. The scatter is believed to be due to slight experimental errors and the failure to derive exact values E_{py}. Re-analysis might have improved the results somewhat.

In order to test the validity of the curves in Figure H.5, the criteria for p-y curves for soft clay were employed in the computer code and values of p_{ult} were computed for depths of 254, 508, and 762 mm. The results were 21.6, 25.8, and 30.2 kPa. The first two values are higher than the corresponding values in Fig. H.5 and the third value undoubtedly be lower if deflection had been great enough to develop the value of p_{ult}. Better agreement between the experimental and computed values of p_{ult} had sufficient information been available to implement the criteria for c-ϕ soil. Nevertheless, the overall results from the method of analysis gives some credence to the nondimensional method of computing for p-y curves as presented in Chapter 3.

APPENDIX I – (EUROCODE 7-1994)

Eurocode principles related to geotechnical design

General statements on the design according to Eurocode can be summarized as:

1. In order to establish minimum requirements for the extent and quality of geotechnical investigations, checks on calculations and on construction control, the complexity of each geotechnical design shall be identified together with the risks to property and life. In particular, a distinction shall be made between:

   - Light and simple structures and small earthworks for which it is possible to ensure that the fundamental requirements will be satisfied on the basis of experience and qualitative geotechnical investigations, with negligible risk for property and life;
   - Other geotechnical structures.

2. For projects of low geotechnical complexity and risk, such as defined above, simplified design procedures are acceptable.

3. The following factors shall be taken into consideration when determining the geotechnical design requirements:

   - Nature and size of the structure and its elements, including any special requirements;
   - Conditions with regard to its surroundings (neighboring structures, traffic, utilities, vegetation, hazardous chemicals, etc.),
   - Ground conditions,
   - Groundwater situation,
   - Regional seismicity, and
   - Influence of the environment (hydrology, surface water, subsidence, seasonal changes of moisture).

4. To establish requirements for geotechnical design, three geotechnical categories, 1, 2 and 3 may be introduced.
   A preliminary classification of a structure according to a geotechnical category should normally be performed prior to the geotechnical investigations. The category should be checked and eventually changed at each stage of the design and construction process.
   The various aspects of design of a project may require the use of different geotechnical categories. It is not necessary to treat the whole of the project according to the highest of these categories.
   The procedures of higher categories may be used to justify more economic designs, or where the designer considers them to be appropriate.

a. Geotechnical Category 1
This category only includes conventional types of structures and foundations, with no abnormal risks or unusual or exceptionally difficult ground or loading conditions.

b. Geotechnical Category 2
Structures in Geotechnical Category 2 require quantitative geotechnical data and analysis to ensure that the fundamental requirements will be satisfied, but routine procedures for field and laboratory testing and for design and construction may be used.

The following are examples of structures or parts of structures complying with Geotechnical category 2 :

Conventional types of:
- Spread foundations;
- Raft foundations;
- Pile-supported foundations;
- Walls and other structures retaining or supporting soil or water;
- Excavations;
- Bridge piers and abutments;
- Embankments and earthworks;
- Ground anchors and other tie-back systems;
- Tunnels in hard, non-fractured rock and not subjected to special water tightness or other requirements.

c. Geotechnical Category 3:
This category includes structures, or parts of structures, which do not fall within the limits of Geotechnical Categories 1 and 2. Geotechnical Category 3 includes very large or unusual structures, structures involving abnormal risks, or unusual or exceptionally difficult ground or loading conditions and structures in highly seismic areas.

5. For each geotechnical design, verification is required that no relevant limit state is exceeded.
6. The requirement may be achieved by:
   - Use of calculations;
   - Adoption of prescriptive measures;
   - Use of experimental models and load tests;
   - Use of an observational method.

These four approaches may be used in combination. In practice, experience will often show which type of limit state will govern the design, and the avoidance of other limit states may be verified by a check by observations in the field.

7. The interaction between the structure and the supporting soil shall be considered.
8. Compatibility of strains in the materials involved at a limit state should be considered, especially for materials which are brittle or which have strain-softening properties. Examples include reinforced concrete, dense granular soils, and cemented soils, which exhibit low residual strength. Detailed analysis, allowing for the relative stiffness of the structure and the ground, may be needed in cases where a combined failure of structural members and the ground could occur. Examples include raft foundations, laterally loaded piles, and flexible retaining walls.
9. Buildings shall be protected against the penetration of groundwater or the transmission of vapor or gases to the inner surfaces.

Eurocode principles related to geotechnical design 437

10. When possible, the results from the design shall be checked against comparable experience.
11. In geotechnical design, detailed specifications for design shall include, as appropriate:

- The general suitability of the ground on which the structure is located;
- The disposition and classification of the various zones of soil, rock and elements of construction which are involved in the calculation model;
- Dipping bedding planes;
- Mine workings, caves or other underground structures;
- In the case of structures resting on or near rock, the following shall be included;

    - Interbedded hard and soft strata;
    - Faults, joints and fissures;
    - Solution cavities, such as swallow holes or fissures filled with soft material, and continuing solution processes;
    - The various loads, their combinations, and load cases;
    - The nature of the environment within which the design is set, including the following;

        - Effects of scour, erosion and excavation, leading to changes in the geometry of the ground surface;
        - Effects of chemical corrosion;
        - Effects of weathering;
        - Effects of freezing;
        - Variations in groundwater levels, including the effects of dewatering, possible flooding, failure of drainage systems, etc.;
        - Other effects of time and environment on the strength and other properties of materials; e.g. the effect of holes created by animal activities;
        - Earthquakes;
        - Subsidence due to mining or other causes;
        - The tolerance of the structure to deformations;
        - The effect of the new structure on existing structures or services.

When all of the factors are taken into account and weighed, the engineer with responsibility for design will make a decision about the global factor of safety. A second engineer with the same capability would undoubtedly select a value that differs somewhat from that of the first engineer.

The approach in Europe is to develop and publish numerical values for partial safety factors for each element in the equations for design. A similar approach in the United States is termed load-and-resistance-factor design. While the assigning of individual factors to components of loads and resistances is conceptually attractive and should lead to uniformity in the selection of the global factor of safety, two fundamental difficulties must be overcome: 1. Rarely are two designs the same and the accumulation of statistical data on performance, particularly if a failure has occurred, is not possible; and 2. Nature did not consult engineers when she laid down the soils at a site and defining the properties in precise terms is impossible. The second point is particularly true when attempting to predict the influence of the installation of piles on soil properties.

Reducing the value of the shear characteristics of a layer of soil by a 'partial safety' factor will not change the real soil characteristics and consequently the response of the soil cannot be in agreement with the reality of the kinematics of the failure pattern 'nature' will necessarily imply. The lack of reality in modeling is the greatest conceptual error of the partial factor approach for the case of soils. Nevertheless, because of the today's trend, suggestions are made in the following paragraphs for the use of characteristic soil parameters and partial safety factors in the design of piles under lateral loading. However, as in all other geotechnical proposals for realistic design, the outcome from the use of a combination of partial safety factors should (from back calculation) result in a global safety factor that is consistent with experience.

APPENDIX J – (FRANKE 1991)

Discussion of factor of safety related to piles under axial load

The following presentation shows the complex interrelationships, with respect to failure, if load-settlement curves are available for the base of the pile and the shaft of the pile.

a. The failure load of the single pile may be defined as the occurrence of a zone of failing soil around the pile, deforming in such way so as to induce a well agreed settlement of the pile tip, $s_{(A)}$, with a corresponding drop in the load at the pile head. or
b. The failure load of the single pile may be defined as the occurrence (structurally defined) of excessive differential movements of the building structure.

The assumption is made that the failure load $Q_u^{(A)}$ was obtained from results of previous single-pile, axial-load tests where the pile-tip settlement $s_{(A)}$ reached failure, (for example at 5% or 10 % of the diameter of the tip of the pile). In such case, the design load would become :

$$Q_d^{(A)} \leq \frac{Q_u^{(A)}}{F^{(A)}} \tag{J.1}$$

with $F^{(A)}$ the deterministic total safety factor for capacity under axial load.

If the unit resistance of the pile base and the pile shaft at failure, q_{bf} and q_{sf}, respectively, and the corresponding overall resisting forces of base and shaft, Q_{bf} and Q_{sf}, respectively, were known, the following equation would result:

$$Q_d^{(A)} \leq \frac{q_{bf}}{F_b^{(A)}} A_b + \frac{q_{sf}}{F_s^{(A)}} A_s \tag{J.2}$$

$$= \frac{Q_{bf}}{F_b^{(A)}} + \frac{Q_{sf}}{F_s^{(A)}} \tag{J.2a}$$

where A_b = area of pile tip, and A_s = are of shaft of the pile.

Therefore, $F_b^{(A)}$ and $F_s^{(A)}$ are partial safety factors linked to the capacity of the pile tip and the pile shaft.

$$Q_d^{(A)} \leq \frac{Q_u^{(A)}}{F^{(A)}} = \frac{Q_{bf}}{F_b^{(A)}} + \frac{Q_{sf}}{F_s^{(A)}} \tag{J.3}$$

And

440 *Piles under lateral loading*

$$Q_u^{(A)} = Q_{bf} + Q_{sf} \tag{J.4}$$

From Equation J.3, an interrelationship can be written for $F^{(A)}$:

$$F^{(A)} = \frac{F_b^{(A)} F_s^{(A)} Q_u^{(A)}}{F_s^{(A)} Q_{bf} + F_b^{(A)} Q_{sf}} \tag{J.5}$$

From Equation J.5, at given values of partial safety factors, the value of $F^{(A)}$ is not constant but always depends on the ratio of Q_{bf} to Q_{sf}. Or, from the other way around, the more common application of the selection of a constant value of $F^{(A)}$ would require values of $F_b^{(A)}$ and $F_s^{(A)}$ to vary with the ratio of Q_{sf}/Q_{bf}. A knowledge of the ratio would be required in each actual case, which is not commonly available from the loadings of piles.

In Case b where the load at failure is related to excessive movement, the differential settlements which are excessive have to be linked previously to a total excessive settlement $s^{(B)}$ of the building structure. Therefore,

$$Q_u^{(B)} = Q \text{ at } s^{(B)} \tag{J.6}$$

and is derived from the settlement (test loading) at $s^{(B)}$.

The design load in such case becomes:

$$Q_d^{(B)} \le \frac{Q_u^{(B)}}{F^{(B)}} \tag{J.7}$$

If the unit resistance at the base of the pile and the unit resistance of the shaft of the pile, $q_b(s)$ and $q_s(s)$, respectively, are known, the derivation of $Q_d^{(B)}$ is more complicated than for Case A. Two sub-cases are to be distinguished (see Fig. J.1).

$$s^{(B)} < s_s^*, \text{ and} \tag{J.8a}$$

$$s_s^* \le s^{(B)} < s^{(A)} = 0.1 D_b \tag{J.8b}$$

where s_s^* is the settlement at which the ultimate shaft resistance Q_s will be fully mobilized; therefore

$$Q_s = Q_{sf} \tag{J.9a}$$

$$s_s^* = 0.01....0.02 \cdot D_s \tag{J.9b}$$

$$Q_{sf} = Q(s_s^*) = q_{sf} \cdot A_s \tag{J.9c}$$

And the following equations result at each value of $s^{(B)}$:

$$Q_d^{(B)} \le \frac{q_b^{(s^B)}}{F_b^{(B)}} A_b + \frac{q_s^{(s^B)}}{F_s^{(B)}} A_s \tag{J.10a}$$

$$Q_d^{(B)} = \frac{Q_u^{(B)}}{F^{(B)}} = \frac{Q_b^{(s^B)}}{F_b^{(B)}} + \frac{Q_s^{(s^B)}}{F_s^{(B)}} \tag{J.10b}$$

The following expression can now be written:

$$Q_u^{(B)} = Q_b^{(s^B)} + Q_s^{(s^B)} \tag{J.11}$$

From Equation J10.b another relationship between $F^{(B)}$ and $F_s^{(B)}$, similar to Equation J.5, can be written:

$$F^{(B)} = \frac{F_s^{(B)} F_b^{(B)} (1 + Q_{sf}/Q_{bf})}{F_b^{(B)} Q_{sf}/Q_{bf} + F_s^{(B)}} \tag{J.12}$$

Again, the overall $F^{(B)}$ factor and/or the partial factors are interrelated strongly by means of the ratio of Q_{sf} to Q_{bf}.

The more relevant approach, however, as stipulated in Section 9.1.2, replaces the deterministic safety factor F by partial safety factors γ. These partial safety factors (the so-called basis variables), are comparing loadings S^* to resistances R^*. The safety requirements are fulfilled, as long as the corresponding design values R_d^* and S_d^* do have the following relationship:

$$R_d^* \geq S_d^* \tag{J.13}$$

as derived from

$$R_d^* = \frac{R^*}{\gamma} \geq S^* \gamma_F \geq S_d^* \tag{J.14}$$

As suggested in Sections 9.1.2, there is little meaning in comparing results obtained from implementing partial safety factors, suggested by Eurocode 7, with those derived from conventional procedures. The reason is that the lumped factor used in the conventional design has a much broader meaning, accounting for many engineering aspects of the problem.

References

Alizadeh, M. & Davisson, M.T. 1970. Lateral load test piles, Arkansas River project. *Journal of the Soil Mechanics and Foundations Division*, ASCE 96(SM5): 1583-1604.

Allen, J. 1985. *p-y* curves in layered soils. Dissertation, University of Texas, Austin.

American Association of State Highway and Transportation Officials (AASHTO) 1992. *Standard specifications for highway bridges*, Washington, D.C., 15th edn, 686 pp.

American Association of State Highway and Transportation Officials (AASHTO) 1994. *LRFD Bridge Design Specifications*, Washington, D.C. 1st edn.

American Concrete Institute 1989. *Building Code Requirements for Reinforced Concrete*, ACI 318-89: 97.

American National Standards Institute (ANSI) 1982. Safety Requirements for Concrete and Masonry Work for Construction and Demolition, ANSI 18(A10): 9-82.

American Petroleum Institute (API) 1987. Recommended practice for planning, designing and constructing fixed offshore platforms. *API Recommended Practice* 2A(RP-2A), 17th edn.

American Petroleum Institute (API) 1993. Recommended practice for planning, designing and constructing fixed offshore platforms–working stress design. *API Recommended Practice* 2A(RP-2A WSD), 20th edn, 193 pp.

American Society of Civil Engineers (ASCE) 1990. Minimum design loads for buildings and other structures. *ASCE Standard* (formerly ANSI A58.1): 7-88.

American Society for Testing and Materials (ASTM) 1992a. Standard method of testing, piles under axial compressive load. *Annual Book of ASTM Standards* 1143.

American Society for Testing and Materials (ASTM) 1992b. Standard method of testing piles under lateral loads. *Annual Book of ASTM Standards*, 3966.

Aschenbrenner, R. 1967. Three dimensional analysis of pile foundations. *Journal of the Structural Division*, ASCE 93 (ST1, 5097): 201-219.

Asplund, S.O. 1956. Generalized elastic theory for pile groups. *International Association for Bridge and Structural Engineering* 16: 1-22.

Atkins Engineering Services 1990. Fluid loading of fixed offshore structures. *OTH* 90 322.

Austin American-Statesman 1996. 'Horn muted order,' Captain says 19 December 1996: A12.

Awoshika, K. 1971. Analysis of foundation with widely spaced batter piles. Dissertation, University of Texas, Austin.

Awoshika, K. & Reese, L.C.1971a. Analysis of foundations with widely spaced batter piles. *Research Report* 117-3F, 315 pp., *Center for Highway Research*. University of Texas, Austin.

Awoshika, K. & Reese, L.C. 1971b. Analysis of foundations with widely spaced batter piles. *Proceedings of the International Symposium on the Engineering Properties of Sea-Floor Soils and their Geophysical Identification, Seattle, Washington, 25 July 1971*. University of Washington.

Baecher, G.B. 1985. Geotechnical error analysis. M.I.T. Summer Course, 1.608.

Baguelin, F., Jezequel, J.F. & Shields, D.H. 1978. *The pressuremeter and foundation engineering*. Clausthal: Trans Tech Publications.

Baldi, G., R. Belloti, V.N. Ghionna, M. Jamiolkowski, S. Marchetti & E. Pasqualine 1986. Flat dilatometer tests in calibration chambers. *Proceedings of In Situ '86 conference*, Blacksburg. VA, pp. 431-446.

Baldi, G., R. Belloti, V.N. Ghionna, M. Jamiolkowski & Lo Presti, D.C.F. 1989. Modulus of sands from CPT's and DMT's. *Proceedings of the XII International Conference on Soil Mechancis and Foundation Engineering, Rio de Janeiro, The Netherlands*. Rotterdam: Balkema.

Banerjee, P.K. & Davies, T.G. 1979. Analysis of offshore pile groups. *Proceedings of the Numerical Methods in Offshore Piling, The Institution of Civil Engineering, London, England*: 101-108.

Bang, S. & Taylor, R.J. 1994. Static mooring line configuration analysis tool. *Proceedings of the 1994 Marine Technology Society Conference, Washington, D. C.*

Barltrop, N.D.P., Mitchell, G.M. & Attkins, J.B. 1990. Fluid loading on fixed offshore structures. *Offshore Technology Report, OTH* 90 322, *HMSO*.

Bauduin, 1997. Characteristic values: some background – Discussion paper to W.G. 1 of SC7.

Bergfelder, J. & Schmidt, H.G. 1989. Zur Planung und Auswertung von horizontalen Pfahlprobebelastungen. *Geotechnik* 12(2): 57-61

Bhushan, K., Haley, S.C. & P.T. Fong 1979. Lateral load tests on drilled piers in stiff clay. *Journal of the Geotechnical Engineering Division*, ASCE 105, (GT8): 969-985.

Bhushan, K., Lee, L.J. & Grime, D.B. 1981. Lateral load test on drilled piers in sand. *Preprint of the ASCE Annual Meeting, St. Louis, Missouri, 26-30 October 1981.*

Bieniawski, Z.T. 1984. *Rock mechanics design in mining and tunneling.* Rotterdam: Balkema.

Bogard, D. & Matlock, H. 1983. Procedures for analysis of laterally loaded pile groups in soft clay. *Proceedings of the Specialty Conference on Geotechnical Engineering in Offshore Practice*, ASCE: 499-535.

Bolton, M.D. 1994. Design methods. *Proceedings of the Wroth Memorial Symposium, London, England*: 49-71.

Bouafia, A.A. & Garnier, J. 1991. *Experimental study of p-y curves for piles in sand, centrifuge 91*: 261-268. In Ko & McLean (eds), Rotterdam: Balkema.

Boughton, N.O. 1970. Elastic analysis for behavior of rockfill. *Journal of the Soil Mechanics and Foundations Division,* ASCE 96(SM5): 1715-1733.

Bowles, J.E. 1988. *Foundation analysis and design.* 4th edn. New York: McGraw-Hill.

Bowman, E.R. 1958. Investigation of the lateral resistance to movement of a plate in cohesionless soil. Thesis, University of Texas, Austin.

Bransby, M.F. 1996. Difference between load-transfer relationships for laterally loaded pile groups: active p-y or passive p-y. *Journal of Geotechnical Engineering Division,* ASCE 122(12): 1015-1018.

Briaud, J.-L., Smith, T.D. & Meyer, B.J. 1982. Design of laterally loaded piles using pressuremeter test results. *Symposium on the Pressuremeter and Marine Applications, Paris, France.*

Broms, B.B. 1964a. Lateral resistance of piles in cohesive soils. *Journal of the Soil Mechanics and Foundations Division*, ASCE 90(SM2): 27-63.

Broms, B.B. 1964b. Lateral resistance of piles in cohesionless soils. *Journal of the Soil Mechanics and Foundations Division,* ASCE 90(SM3): 123-156.

Broms, B.B. 1965. Design of laterally loaded piles. *Journal of the Soil Mechanics and Foundations Division,* ASCE 91(SM3): 77-99.

Brown, D.A., Morrison, C. & Reese, L.C. 1988. Lateral load behavior of a pile group in sand. *Journal of the Geotechnical Engineering Division,* ASCE 114(11): 1261-1276.

Brown, D.A. & Reese, L.C. 1985. Behavior of a large-scale pile group subjected to cyclic lateral loading. *Report to Minerals Management Service*, US Department of Interior, Reston, Virginia, Department of Research, Federal Highway Administration, Washington, D.C. and US Army Corps. of Engineers, Waterways Experiment Station, Vicksburg, Mississippi and *Geotechnical Engineering Report* GR85-12, Geotechnical Engineering Center, Bureau of Engineering Research, Austin, Texas.

Brown, D.A., Reese, L.C. & O'Neill, M.W. 1987. Cyclic lateral loading of a large-scale pile group. *Journal of the Geotechnical Engineering Division*, ASCE 113(11): 1326-1343.

Brown, D.A. & Shie, C.F. 1991. Modification of *p-y* curves to account for group effects on laterally loaded piles. *Proceedings of the Geotechnical Engineering Congress 1991*, ASCE 1 (Geotechnical Special Publication 27): 479-490.

Brown, D.A. & Shie, C.F. 1992. Some numerical experiments with a three dimensional finite element model of a laterally loaded pile. *Comp. and Geotechnics* 12: 149-162.

Brown, D.A., Shie, C.F. & Kumar, M. 1989. *p-y* curves for laterally loaded piles derived from three-dimensional finite element model. *Proceedings of the III International Symposium, Numerical Models in Geomechanics (NUMOG III), Niagra Falls, Canada*. New York: Elsevier Applied Sciences: 683-690.

Canadian Geotechnical Society 1978. *Canadian foundation engineering manual, Part III, Deep Foundations*, 108 pp. Montreal: Canadian Geotechnical Society.

Capozolli, L. 1968. Pile test program at St. Gabriel, Louisiana, Louis J. Cappozolli & Associates, (unpublished).

Carter, J.P. & Kulhawy, F.H. 1988. Analysis and design of drilled shaft foundations socketed into rock. Report EL-5918, Geotechnical Engineering Group, Cornell University, Ithaca, New York.

Carter, J.P. & Kulhawy, F.H. 1992. Analysis of laterally loaded shafts in rock. *Journal of the Geotechnical Engineering Division*, ASCE 118(6): 839-855.

CERC 1984. *US Army coastal engineering research center, shore protection manual*.

Charles, J.A. 1975. Strains developed in two rockfill dams during construction. *Geotechnique*, 25(2).

Chow, Y.K. 1986. Analysis of vertically loaded pile groups. *International Journal for Numerical and Analytical Methods in Geomechanics*, 10: 59-72.

Chow, Y.K. 1987. Iterative analysis of pile-soil-pile interaction. *Geotechnique*, 37(3): 321-333.

Coleman, R.B. 1968. Apapa road Ijora causeway reconstruction. *Report on Horizontal Load Tests on Piles, Federal Ministry of Works and Housing, Ijora, Lagos, Nigeria*.

Coleman, R.B. & Hancock, T.G. 1972. The behavior of laterally loaded piles. *Proceedings of the V European Conference on Soil Mechanics and Foundation Engineering, Madrid, Spain*, 1: 339-345.

Comité Euro-International du Béton 1978. Model Code for Concrete Structures. *Bulletin d'Information*, 124/125E, 348 pp., *Federation Internationale de la Preconstrainte, Paris, France*.

Committee on Piles Subjected to Earthquake 1965. Lateral Bearing Capacity and Dynamic Behavior of Pile Foundation. Architectural Institute of Japan (in Japanese): 1-69.

Costanzo, D. & Lancelotta, R. 1997. Eurocode 7, basic requirements for application, Revista Italiana di Geotecnica: 30-37.

Cox, W.R., Dixon, D.A. & Murphy, B.S. 1984. Lateral load tests of 25.4 mm diameter piles in very soft clay in side-by-side and in-line groups. *Proceedings of the Laterally Loaded Deep Foundations: Analysis and Performance*, ASTM, SPT 835.

Cox, W.R., Reese, L.C. & B.R. Grubbs 1974. Field testing of laterally loaded piles in sand. *Proceedings of the Offshore Technology Conference, Houston, Texas*, paper 2079.

Coyle, H.M. & Reese, L.C. 1966. Load transfer for axially loaded piles in clay. *Journal of the Soil Mechanics and Foundations Division*, ASCE 92(SM2, 4702): 1-26.

Coyle, H.M. & Sulaiman, I.H. 1967. Skin friction for steel piles in sand. *Journal of the Soil Mechanics and Foundations Division*, ASCE 93(SM6, 5590): 261-278.

D'Appolonia, E. & Romualdi, J.P. 1963. Load transfer in end bearing steel h-piles. *Journal of the Soil Mechanics and Foundations Division*, ASCE 89(SM2, 3450): 1-25.

Davis, L.H. 1977. Tubular steel foundation. *Test Report* RD-1517 (unpublished). Florida Power and Light Company, Miami, Florida.

De Beer, E.E. 1961. Définition des coefficients de sécurité au glissement de talus à partir des sollutions proviquant la rupture:*Comptes rendus du cinquième congrès International de Mécanique des Sols et des Travaux de Fondations*, Paris, Volume II.

De Beer, E.E. 1965. Influence of the mean normal stress on the shearing strength of sand. *Proceedings of the VI International Conference on Soil Mechanics and Foundation Engineering, Montreal, Canada*, 1.

De Beer, E.E. 1976. Einige Betrachtungen über die Grenztragfahig- keit des bodems unter flachgründungen. Mitteilungen, 32, *Deutsche forschungsgesellschaft für bodenmechanik (Degebo), Berlin* (in German).

De Beer, E.E. 1977. Piles subjected to lateral loads. *State-of-the-Art Report, Proceedings of the Specialty Session 10-IX International Conference on Soil Mechanics and Foundation Engineering, Tokyo, Japan*: 1-14.

De Beer, E.E. 1988. Different behavior of bored and driven piles. *Proceedings of the I International Geotechnical Seminar on Deep Foundations on Bored and Auger Piles, Ghent, Belgium, 7-10 June 1988*: 47-82.

De Beer, E.E. & Lousberg, E. 1961. Définition des coefficients de sécurité au glissement de talus à partir des solicitations provoquant la rupture. *Proceedings of the IX International Conference on Soil Mechanics and Foundation Engineering, Paris, France*, II (in French).

De Beer, E.E., Lousberg, E., De Jonghe, A., Wallays, M. & Carpentier, R. 1981. Partial safety factors in pile bearing capacity. *Proceedings of the X International Conference on Soil Mechanics and Foundation Engineering, Stockholm, Sweden*, 1.

Deere, D.V. 1968. *Geological considerations, rock mechanics in engineering practice.* In K.G. Stagg & O.C. Zienkiewicz (eds). New York: Wiley: 1-20.

Desai, C.S. & Appel, G.C. 1976. 3-d analysis of laterally loaded structures. *Proceedings of the II International Conference on Numerical Methods in Geomechanics, Blacksburg, Virginia.*

de Sousa Pinto, N.L., Sybert, J.H. & Posey, C.H. 1959. Model tests of riprap scour protection. *Report 23, Rocky Mountain Hydraulic Laboratory.*

Det Norske Veritas 1977. Rules for the design, construction and inspection of offshore structures. Veritsveien 1, 1322 Hovek, Norway.

Drabkin, S. & Lacy, H. 1998. Prediction of settlement of structures due to pile driving. *Geotechnical Earthquake Engineering and Soil Dynamics III*, P. Dakoulas, M. Yegian & R. D. Holtz (eds). ASCE 2(75): 1496-1506.

Duncan, J.M., Evans L.T., Jr. & P.S.K Ooi 1994. Lateral load analysis of single piles and drilled shafts. *Journal of the Geotechnical Engineering Division*, ASCE 120(5): 1018-1033.

Dunnavant, T.W. & O'Neill, M.W. 1985. Performance, analysis and interpretation of a lateral load test of a 72-inch-diameter bored pile in overconsolidated clay. *Report UHCE* 85-4, 57 pp. Department of Civil Engineering, University of Houston, Texas.

Durgunoglu & Mitchell 1975. Static penetration resistance of soils: I-analysis. *Proceedings of the Specialty Conference on In Situ Measurement of Soil Parameters, Raleigh, North Carolina*, ASCE I.

Einstein, H.A. & Wiegel, R.L. 1970. A literature review on erosion and deposition of sediment near structures in the ocean. Report to the US Naval Civil Engineering Laboratory, Port Hueneme, California; Final *Report HEL* 21-6, 183 pp., Hydraulic Engineering Laboratory, College of Engineering. University of California, Berkeley.

Emrich, W.J. 1971. Performance study of soil sample for deep-penetration marine borings. *Sampling of Soil and Rock*, ASTM STP 483: 30-50.

Endley, S.N., Yeung, A.T. & Vennalaganti, K.M. 1997. Lateral loads on piles and piers. *Proceedings of the Texas Section ASCE, Spring Meeting, Houston, Texas, 4 April 1997*: 181-190.

Eurocode 2, (1992). Design of Concrete Structures

Eurocode 3, (1992). Design of Structures

Eurocode 7, (1994). Geotechnical Design. *Geotechnik*

European Committee for Standardization 1994. *Eurocode 7 Part I, Geotechnical Design - General Rules, 6th and Final Version.*
Evans, L.T. Jr. & J.M. Duncan 1982. Simplified analysis of laterally loaded piles. *Report UCB/GT/82-04*, University of California, Berkeley.
Fahey, M. & Carter, J.P. 1993. A finite element study of the pressuremeter test in sand using a nonlinear elastic plastic model. *Proceedings of the Canadian Geotechnical Journal*, (30): 348-362.
Feld, J. 1968. *Construction failures.* New York: Wiley.
Fleming, W.G.K. 1992. A new method of single pile settlement prediction and analysis. *Geotechnique*, 42(3): 411-425.
Focht Jr., J.A. & K.J. Koch 1973. Rational analysis of the lateral performance of offshore pile groups. *Proceedings of the V Offshore Technology Conference, Houston, Texas*, 2: 701-708.
Foundations Committee 1978. *Canadian Foundation Engineering Manual, Part 1, Properties of Soil and Rock.* Publications Office, Montreal, Quebec, Canada.
Francis, A.J. 1964. Analysis of pile groups with flexural resistance. *Journal of the Soil Mechanics and Foundations Division*, ASCE (3887): 1-32.
Franke, E. 1973. Principles for test-loadings of large bored piles by horizontal loads. *Proceedings of the VIII International Conference on Soil Mechanics and Foundation Engineering, Moscow, Russia*, 2.1: 97-104.
Franke, E. 1984. Seitendruck auf Pfähle in tonigen Böden Geotechnik. *Geotechnik 2/84*, (2): 73-84.
Franke, E. 1984. Einige anmerkungen zur anwendbarke des neuen sicherheitskonzepts im grundbau. *Geotechnik, 7 Jahrgang 1984*, (3): 144-149.
Franke, E. 1988. Group action between vertical piles under horizontal loads. *Proceedings of the I International Geotechnical Seminar on Deep Foundations on Bored and Auger Piles, Ghent, Belgium*: 83-93.
Franke, E. 1990 a. The Eurocode, safety approach as applied to foundations, Key-note lecture. *Proceedings of the IX European on Soil Mechanics and Foundation Engineering, Budapest, Hungary*: 173-182.
Franke, E. 1990 b. Neue Regelung der sicherheitsnachweise im zuge der Europäische bau-normung von der deterministische zur probabi- listische sicherheit auch im grundbau. *Bautechnik*: 67(7): 217-223. (Mitteilungen des instituts für grundbau und bodenmechanik, Eidge-nössische Technische Hochschule Zurich, nr. 136).
Franke, 1991. Eurocode safety approach as applied to single piles. *Proceedings of the IV International Conference on Piles and Deep Foundations, Stresa, Italy*: 13-18.
Fu, S.L., Petrauskas, C., Botelho, D.L.R., Carter, E.W., Basaldua, E.A. & Abbott, B.J. 1992. Evaluation of environmental criteria for Gulf of Mexico platform design. *Proceedings of the Offshore Technology Conference, Houston, Texas, 4-7 May 1992*, 1: 243-252.
Fukuoka, M. 1977. The effects of horizontal loads on piles due to landslides. *Proceedings of the Special Session 10-IX International Conference on Soil Mechanics and Foundation Engineering, Tokyo, Japan*: 1-16.
Gabr, M.A., Lunne, T. & Powell, J.J. 1994. p-y analysis of laterally loaded piles in clay using DMT. *Journal of the Geotechnical Engineering Division*, ASCE 120(5): 816-837.
Gazetas, G. & Mylonakis, G. 1998. Seismic soil-structure interaction: new evidence and emerging issues. *Geotechnical Earthquake Engineering and Soil Dynamics III*, Dakoulas, P., Yegian, M. & Holtz, R.D. (eds). ASCE 2(Geotechnical Special Publication 75): 1119-1174.
George, P. & Wood, D. 1977. *Offshore soil mechanics.* Cambridge: Cambridge University Engineering Department.
Georgiadis, M. 1983. Development of p-y curves for layered soils. *Proceedings of the Geotechnical Practice in Offshore Engineering*, ASCE: 536-545.
Gleser, S.M. 1953. Lateral load tests on vertical fixed-head and free-head piles. *Symposium on lateral load tests on piles*, ASTM Special Technical Publication 154: 75-101.

Gooding, T.J., Gore, J.R. & Gilbert, L.W. 1984. Field test of laterally loaded steel pile foundations for Louisiana Power and Light Company. (Unpublished report), 77 pp.

Griffis, L. 1990. The great American pyramid. *Civil Engineering*, ASCE 60(5): 56-58.

Gularte, R.C., Kelly, W.E. & Nacci, V.A. 1979. Scouring of cohesive material as a rate process. *Civil Engineering in the Oceans IV*, ASCE II: 848-863.

Ha, H.S. & O'Neill, M.W. 1981. Field study of pile group action: Appendix A' PILGR1 Users' Guide, *Report FHWA/RD-81/003, Federal Highway Administration*.

Hadjian, A.H., Fallgren, R.B. & Tufenkjian, M.R. 1992. Dynamic soil-pile-structure interaction, the state of the practice. *Piles Under Dynamic Loads*, Prakash, S. (ed.). ASCE (Geotechnical Special Publication 34): 1-26.

Haliburton, T.A. 1968. Numerical analysis of flexible retaining structures. *Journal of the Soil Mechanics and Foundations Division*, ASCE 94(SM6): 1233-1252.

Hansen, B. 1959. Limit design of pile foundations. *Bygningsstatiske Med-delelser Argang:xxx*, (2).

Hansen, J.B. 1961a. A general formula for bearing capacity. *Bulletin* 11: 38-46. Danish Geotechnical Institute, Copenhagen, Denmark.

Hansen, J.B. 1961b. The ultimate resistance of rigid piles against transversal forces. *Bulletin* 12: 5-9. Danish Geotechnical Institute, Copenhagen, Denmark.

Hardin, B.O. & Drnevich, P. 1972a. Shear modulus and damping in soils: measurement and parameter effects. *Journal of the Soil Mechanics and Foundations Division*, ASCE 98(SM6): 603-624.

Hardin, B.O. & Drnevich, P. 1972b. Shear modulus and damping in soils: design equations and curves. *Journal of the Soil Mechanics and Foundations Division*, ASCE 98(SM7): 667-691.

Hassiotis, S. & Chameau, J.L. 1984. Stabilization of slopes using piles. *Slope Stabilization, Report FHWA/IN/JHRP-84-8*, 181 pp. Purdue University, West Lafayette, Indiana.

Hetenyi, M. 1946. *Beams on elastic foundation*. Ann Arbor: The University of Michigan Press.

Hill, R. 1950. *The mathematical theory of plasticity*. Oxford: Clarendon Press.

Hognestad, E. 1951. A study of combined bending and axial load in reinforced concrete members. *Bulletin* 399, 128 pp. Engineering Experiment Station, University of Illinois, Urbana.

Holeyman, A., Bauduin, C., Bottiau, M., Debacker, P., De Cock, F., Dupont, E., Hilde, J.L., Legrand, C., Huybrechts, N., Mengé, P., Miller, J.P. & Simon, G. 1997. Design of axially loaded piles – Belgian practice. In De Cock & Legrand (eds), *Design of Axially Loaded Piles – European Practice*: 57-82.

Horne, M.R. 1971. *Plastic theory of structures*: 64. Cambridge: The M.I.T. Press.

Horvath, R.G. & Kenney, T.C. 1979 Shaft resistance of rock-socketed drilled piers. *Proceedings of the Symposium on Deep Foundations*, New York, New York. ASCE: 182-184.

Hrennikoff, A. 1950. Analysis of pile foundations with batter piles. *Transactions, ASCE* 115(2401).

Hvorslev, M.J. 1949. *Subsurface exploration and sampling of soils for civil engineering purposes*, Waterways Experiment Station, Corps. of Engineers, US Army, Vicksburg, Mississippi.

Ikeda, Y. & Matuzawa, H. 1998. Evaluation of coefficient of lateral subgrade reaction based on horizontal loading test for piles. *Proceedings First International Conference on site Characterization – ISC '98*, Atlanta, USA. 2: 825-830.

Imai, T. 1970. *Study on coefficient of lateral subgrade reaction of ground, Soils and Foundation*.

Ismael, N. 1990. Behavior of laterally loaded bored piles in cemented sand. *Journal of the Geotechnical Engineering Division*, ASCE: 1678-1699.

Jamiolkowski M., Ladd, C.C., Germaine, J.T., Lancellotta, R.L. 1985. New developments in field and laboratory testing of soils. Theme lecture, *XI International Conference on Soil Mechanics and Foundation Engineering*, San Francisco, California: 57-145.

Jamiolkowski M. & Belloti, R. & Lo Presti, D.C.F. & O'Neill, D.A. 1993. Anisotropy of small strain stiffness in ticino sand. *Soils and Foundations*.

Jamiolkowski, M., Ladd, C.C., Germaine, J.T. & Lancelotta, R.L. 1985. New developments in field

and laboratory testing of soils. *Proceedings of the XI International Conference on Soil Mechanics and Foundation Engineering, San Francisco, California*, 1: 57-152.
Jamiolkowski, M., Leroueil, S. & Lo Presti, D.C.F. 1991. Design parameters from theory to practice. *Proceedings of Geo-Coast 1991, Yokohama, Japan.*
Jamiolkowski, M. & Lo Presti, D.C.F. 1994. In situ testing and real soil behaviour. *Text of the Panel Discussion to be presented at the Plenary Session A-Soil Properties, Proceedings of the XIII International Conference on Soil Mechanics and Foundation Engineering, New Dehli, India.*
Japan Road Association 1976. Road bridge substructure design guide and explanatory notes. *Designing of Pile Foundations*, 144 pp.
Jardine, et al. 1992. Non linear stiffness parameters from undrained pressuremeter tests, *Canadian Geotechnical Journal, Vol. 29*: 436-447.
Johnson, G.W. 1982. Use of the self-boring pressuremeter in obtaining in situ shear moduli of clay. Thesis, University of Texas, Austin.
Karol, R.H. 1960. *Soils and soil engineering.* Englewood Cliffs: Prentice Hall.
Kaynia, A. M. & Kausel, E. 1982. Dynamic stiffness and seismic response of pile groups. *Research Report R82-03*, Department of Civil Engineering, Massachusetts Institute of Technology.
Kenley, R.M. & Sharp, D.E. 1993. Magnus foundation monitoring project instrumentation data processing and measured results. *Large Scale Pile Tests in Clay*, Thomas Telford, London, England: 28-50.
Kerisel, J.L. 1965. Vertical and horizontal bearing capacity of deep foundations in clay. *Bearing Capacity and Settlement of Foundations*: 45-51. Duke University, Durham, North Carolina.
Kjellman, et al. 1951. Testing the shear strength of clay in Sweden. *Geotechnique, Vol. II*, Nr. 3, June.
Kooijman, A.P. 1989. Comparison of an elastoplastic quasi three-dimensional model for laterally loaded piles with fields tests. *Proceedings of the III International Symposium, Numerical Models in Geomechanics (NUMOG III), Niagara Falls, Canada*, Elsevier Applied Science: New York: 675-682.
Kotthaus, M. & Jessberger, H.L. 1994. Centrifuge model tests on laterally loaded pile groups. *Proceedings of the XIII International Conference on Soil Mechanics and Foundation Engineering, New Delhi, India*: 639-644.
Kraft, L.M. Jr., Focht, J.A. Jr. & Amerasinghe, S.F. 1981. Friction capacity of piles driven into clay. *Journal of the Geotechnical Engineering* Division, ASCE 107(GT11): 1521-1541.
Kubo, K. 1964. Experimental study of the behavior of laterally loaded piles. *Report 12(2)*, Transportation Technology Research Institute, Japan.
Kubo, K. 1965. Experimental study of the behavior of laterally loaded piles. *Proceedings of the VI International Conference on Soil Mechanics and Foundation Engineering, Montreal, Canada*, II: 275-279.
Ladd, C.C., Foott, R., Ishihara, K., Schlosser & Poulos, H.G. 1977. Stress deformation and strength characteristics. *Proceedings of the IX International Conference on Soil Mechanics and Foundation Engineering, Toyko, Japan*, 2: 421-494.
Ladd, C.C. & Foott, R., 1974. New design procedure for stability of soft clays. *Journal of the Geotechnical Engineering Division*, ASCE (GT7): 763.
Laursen, E.M. 1970. Bridge design considering scour and risk. *Transportation Engineering Journal, Proceedings*, ASCE 96(TE2): 149-164.
Leonards, G.A. (ed.) 1971. *Foundation engineering.* New York: McGraw-Hill.
Long, J.H. 1984. The behavior of vertical piles in cohesive soil subjected to repetitive horizontal loading. Dissertation, University of Texas, Austin.
Long, M.M., Lambson, M.D., Clarke, J. & Hamilton, J. 1993. Cyclic lateral loading of an instrumented pile in overconsolidated clay at Tilbrook Grange. *Large Scale Pile Tests in Clay*, Thomas Telford, London, England: 381-404.

Long, J.H. & Reese, L.C. 1984. Testing and analysis of two offshore piles subjected to lateral load. *Laterally Loaded Deep Foundations: Analysis and Performance,* ASTM SPT 835, Langer, J.A., Mosely, E. & Thompson, C. (eds). Philadelphia: 214-228.

Long, J.H. & Vanneste, G. 1994. Effects of cyclic lateral loads on piles in sand. *Journal of Geotechnical Engineering,* ASCE 120(1): 225-244.

MacCamy R.C. & Fuchs, R.A. 1954. Wave forces on piles: a diffraction theory. *Tech. Memorandum 69,* US Army Corps of Engineers, Beach Erosion Board, Washington, D.C.

Majano, R.E., O'Neill, M.W & Person, G. 1998. Lateral split socket test in Franconia sandstone. *Proceedings of the Deep Foundations on Bored and Auger Piles (BAPIII):* 343-346. Edited by Van Impe, W.F. & Haegeman, W., Rotterdam: Balkema.

Mandolini, A. & Viggiani, C. 1993. Settlement predictions of piled foundations from loading tests on single piles. *Predictive Soil Mechanics,* Thomas Telford, London, England: 464-481.

Mansur, C.I. & Hunter, A.H. 1970. Pile tests-Arkansas River project. *Journal of the Soil Mechanics and Foundations Division,* ASCE: 1545-1582.

Matlock, H. 1970. Correlations for design of laterally loaded piles in soft clay. *Proceedings of the II Annual Offshore Technology Conference, Houston, Texas,* (OTC 1204): 577-594.

Matlock, H., Ingram, W.B., Kelly, A.E. & Bogard, D. 1980. Field tests of lateral load behavior of pile groups in soft clay. *Proceedings of the XII Annual Offshore Technology Conference, Houston, Texas,* (OTC 3871): 163-174.

Matlock, H. & Reese, L.C. 1962. Generalized solutions for laterally loaded piles. *Transactions,* ASCE 127: 1220-1248.

Matlock, H. & Ripperger, E.A. 1956. Procedures and instrumentation for tests on a laterally loaded pile. *Proceedings of the VIII Texas Conference on Soil Mechanics and Foundation Engineering,* Special Publication 29, Bureau of Engineering Research. University of Texas, Austin.

Matlock, H. & Ripperger, E.A. 1958. Measurement of soil pressure on a laterally loaded pile. *Proceedings,* ASTM, 58: 1245-1259.

Matlock, H., Ripperger, E.A. & Fitzgibbon, D.P. 1956. Static and cyclic lateral-loading of an instrumented pile. A report to Shell Oil Company, 167 pp. (unpublished).

Mattes, N.S. & Poulos, H.G. 1969. Settlement of single compressible pile. *Journal of the Soil Mechanics and Foundations Division,* ASCE (SM1): 189-207.

McCammon, G.A. & Asherman, J.C. 1953. Resistance of long hollow piles to lateral load. *Symposium on lateral load tests on piles,* ASTM Special Technical Publication (154): 3-11.

McClelland, B. & Focht, J.A. Jr. 1958. Soil modulus for laterally loaded piles. *Transactions,* ASCE 123: 1049-1086.

McVay, M., Zhang, L., Molnit, T. & Lai, P. 1998. Centrifuge testing of large laterally loaded pile groups in sands. *Journal of Geotechnical and Geoenvironmental Engineering,* ASCE 124(10): 1016-1026.

Meyer, B.J. 1979. Analysis of single piles under lateral loading. Thesis, University of Texas, Austin.

Meyerhof, G.G., 1970. Safety factors in soil mechanics. *Canadian Geotechnical Journal,* 7(4): 349-355.

Moore, W.L. & Masch, F.D. Jr. 1962. Experiments on the scour resistance of cohesive sediments. *Journal of Geophysical Research,* 67(4): 1437-1446.

Morison, J.R., O'Brien, M.P., Johnson, J.W. & Schaaf, S.A. 1950. The force exerted by surface waves on piles. *Petroleum Transaction, American Institute of Mining Engineers:* 189: 149-154.

Morrison, C.S. 1986. A lateral load test of a full-scale pile group in sand. Thesis, University of Texas, Austin.

Morrison, C.S. & Reese, L.C. 1988. A lateral-load test of a full-scale pile group in sand. *Geotechnical Engineering Report CR86-1,* 306 pp. US Army Engineer Waterways Experiment Station, Vicksburg, Mississippi.

Mosher, R.L. 1984. Load transfer criteria for numerical analysis of axially loaded piles in sand. US

Army Engineering Waterways Experimental Station, Automatic Data Processing Center, Vicksburg, Mississippi.
National Highway Institute (NHI) 1998. Load and resistance factor design (LRFD) for highway bridge substructures. *Reference Manual and Participant Workbook*, I.
Oakland, M.W. & Chameau, J.L. 1986. Drilled piers used for slope stabilization. *Report FHWA/IN/JHRP*-86-7, 305 pp. Purdue University, West Lafayette, Indiana.
O'Neill, M.W. 2000. Personal communication.
O'Neill, M.W. 1983. Group action in offshore piles. *Proceedings of the Specialty Conference on Geotechnical Engineering in Offshore Practice*, ASCE.
O'Neill, M.W. & Dunnavant, T.W. 1984. A study of the effects of scale, velocity, and cyclic degradability on laterally loaded single piles in overconsolidated clay. *Report UHCE* 84-7: 368 pp. Department of Civil Engineering, University of Houston, Texas.
O'Neill, M.W. & Gazioglu, S.M. 1984. An evaluation of p-y relationships in clays. A report to the American Petroleum Institute, *PRAC* 82-41-2. University of Houston, Texas.
O'Neill, M.W. & Murchison, J.M. 1983. An evaluation of p-y relationships in sands. A report to the American Petroleum Institute, *PRAC* 82-41-1. University of Houston, Texas.
Palmer, L.A. & Thompson, J.B. 1948. The earth pressure and deflection along the embedded lengths of piles subjected to lateral thrust. *Proceedings of the II International Conference on Soil Mechanics and Foundation Engineering, Rotterdam, The Netherlands*, 5: 156.
Parker Jr., F. & Cox, W.R. 1969. A method of the analysis of pile supported foundations considering nonlinear soil behavior. *Research Report* 117-1, Center for Highway Research. University of Texas, Austin.
Parker Jr., F. & Reese, L.C. 1971. Lateral pile-soil interaction curves for sand. *Proceedings of the International Symposium on the Engineering Properties of Sea-Floor Soils and their Geophysical Identification, Seattle, Washington*. University of Washington.
Peck, R.B. 1967. Bearing capacity and settlement: certainties and uncertainties. *Bearing Capacity and Settlement of Foundations, Department of Civil Engineers, Duke University, Durham, North Carolina*, Vesic, A.S. (ed.), 150 pp.
Peck, R.B. 1976. Rock foundations for structures. *Proceedings of the Specialty Conference on Rock Engineering for Foundations and Slopes, ASCE, New York, New York.*
Peck, R.B., Hanson, W.E. & Thornburn, T.H. 1974. *Foundation engineering*, 2nd edn. New York: Wiley.
Peterson, K.T. & Rollins, K.M. 1996. Static and dynamic lateral load testing of a full-scale pile group in clay. *Civil Engineering Department Research Report CEG.96-02*. Brigham Young University, Provo, Utah.
Portugal, J.C. & Sêco e Pinto, P.S. 1993. Analysis and design of piles under lateral loads. *Proceedings of the II International Geotechnical Seminar on Deep Foundations on Bored and Auger Piles (BAP II), Ghent, Belgium*: 309-313.
Posey, C.J. 1963. Scour at bridge piers; 2. Protection of threatened piers. *Civil Engineering*: 48-49.
Posey, C.J. 1971. Protection of offshore structures against underscour. *Journal of the Hydraulics Division, Proceedings*, ASCE 97(HY7): 1011-1016.
Poulos, H.G. 1968. Analysis of the settlement of pile groups. *Geotechnique*, 18: 449-471.
Poulos, H.G. 1971. Behavior of laterally loaded piles: II-pile groups. *Journal of the Soil Mechanics and Foundations Division*, ASCE 97(SM5): 733-751.
Poulos, H.G. 1979. Settlement of single piles in nonhomogeneous soil. *Journal of the Geotechnical Engineering Division*, ASCE 105(5): 627-641.
Poulos, H.G. 1988. Modified calculation of pile-group settlement interaction. *Journal of the Geotechnical Engineering Division*, ASCE 114(6): 697-706.
Poulos, H.G. & Chen, L.T. 1997. Pile response due to excavation-induced lateral soil movement. *Journal of Geotechnical and Geoenvironmental Engineering*, ASCE 123(2): 94-99.
Poulos, H.G. & Davis, E.H. 1968. The settlement behavior of single axially loaded incompressible piles and piers. *Geotechnique*, 18(3): 351-371.

Poulos, H.G. & Davis, E.H. 1980. *Pile foundation analysis and design*. New York: Wiley.
Poulos, H.G. & Hull, T.S. 1989. The role of analytical mechanics in foundation engineering. *Foundation Engineering, Current Principals and Practices*, ASCE 2: 1578-1606.
Poulos, H.G. & Mattes, N.S. 1969. The behavior of axially loaded end bearing piles. *Geotechnique*, 19(7): 285-300.
Prakash, S. 1962. Behavior of pile groups subjected to lateral load. Dissertation, University of Illinois, Urbana.
Price, G. & Wardle, I.F. 1981. Horizontal load tests on steel piles in London clay. *Proceedings of the X International Conference on Soil Mechanics and Foundation Engineering, Stockholm, Sweden*: 803-808.
Price, G. & Wardle, I.F. 1987. Lateral load tests on large diameter bored piles. *Contractor Report* 46, 45 pp. Transport and Road Research Laboratory, Department of Transport, Crowthorne, Berkshire, England.
Pugsley, A. 1966. *The safety of structures*. London: Arnold.
Quinn, A. 1961. *Design and construction of ports and marine structures*. New York: McGraw-Hill.
Radosavljevic, Z. 1957. Calcul et essais des pieux en groupe. *Proceedings of the IV International Conference on Soil Mechanics and Foundation Engineering, London, England*, 2: 56-60.
Ramshaw, C.L., Selby, A.R. & Bettess, P. 1998. Computed ground waves due to piling. *Geotechnical Earthquake Engineering and Soil Dynamics III*, Dakoulas, P., Yegian, M. & Holtz, R.D. (eds). ASCE 2 (Geotechnical Special Publication 75): 1484-1495.
Randolph, M.F. & Wroth, C.P. 1978. Analysis of deformation of vertically loaded piles. *Journal of the Geotechnical Engineering Division*, ASCE 104(12): 1465-1488.
Reese, L.C. 1958. Discussion of 'Soil modulus for laterally loaded piles.' By Bramlette McClelland & John A. Fotch, Jr. *Proceedings*, ASCE (1081), *Transactions*, ASCE 123: 1071-1074.
Reese, L.C. 1973. A design method for an anchor pile in a mooring system. *Proceedings of the Offshore Technology Conference, Houston, Texas*, 1(OTC 1745): 209-216.
Reese, L.C. 1984. Handbook on design of piles and drilled shafts under lateral load. *FHWA-IP*-84-11, 360 pp. US Department of Transportation, Federal Highway Administration.
Reese, L.C. 1989. Why foundations fail. *The Construction Specifier*, Alexandria: 66-71.
Reese, L.C. 1990. The action of soft clay along friction piles; bay mud revisited. *Proceedings of the Memorial Symposium in honor of Professor Harry Bolton Seed*, II: 134-153. University of California, Berkeley.
Reese, L.C. 1997. Analysis of laterally loaded piles in weak rock. *Journal of Geotechnical and Geoenvironmental Engineering*, ASCE 123(11): 1010-1017.
Reese, L.C. & Cox, W.R. 1968. Soil behavior from analysis of tests of uninstrumented piles under lateral loading. *Performance of Deep Foundations*, ASTM SPT 444: 161-176.
Reese, L.C., Cox, W.R. & Koop, F.D. 1968. Lateral-load tests of instrumented piles in stiff clay at Manor, Texas. A report to Shell Development Company, 303 pp. (unpublished).
Reese, L.C., Cox, W.R. & Koop, F.D. 1974. Analysis of laterally loaded piles in sand. *Proceedings of the VI Annual Offshore Technology Conference, Houston, Texas*, 2(OTC 2080): 473-485.
Reese, L.C., Cox, W.R. & Koop, F.D. 1975. Field testing and analysis of laterally loaded piles in stiff clay. *Proceedings of the VII Annual Offshore Technology Conference, Houston, Texas*, 2(OTC 2312): 672-690.
Reese, L.C. & Matlock, H. 1956. Nondimensional solutions for laterally loaded piles with soil modulus assumed proportional to depth. *Proceedings of the VIII Texas Conference on Soil Mechanics and Foundation Engineering*, 41 pp. University of Texas, Austin.
Reese, L.C. & Matlock, H. 1960. Numerical analysis of laterally loaded piles. *Proceedings of the II Structural Division Conference on Electronic Computation, Pittsburgh, Pennsylvania*, ASCE, 657-668.
Reese, L.C. & Matlock, H. 1966. Behavior of a two-dimensional pile group under inclined and eccentric loading. *Proceedings of the Offshore Exploration Conference, Long Beach, California*.

Reese, L.C. & Nyman, K.J. 1978. Field load test of instrumented drilled shafts at Islamorada, Florida. A report to Girdler Foundation and Exploration Corporation (unpublished), Clearwater, Florida.

Reese, L.C. & O'Neill, M.W. 1967. The analysis of three-dimensional pile foundations subjected to inclined and eccentric loads. *Proceedings of the Conference on Civil Engineering*, ASCE: 245-276.

Reese, L.C. & O'Neill, M.W. 1987. Drilled shafts: construction procedures and design methods. *Report FHWA-HI-88-042*, US Department of Transportation, Federal Highway Administration, Office of Implementation, McLean, Virginia.

Reese, L.C., O'Neill, M.W. & Smith, R.E. 1970. Generalized analysis of pile foundations. *Journal of Soil Mechanics and Foundations Division*, ASCE 96(SM1): 235-250.

Reese, L.C. & Wang, S.T. 1996. *Technical manual for documentation for program group*. Ensoft, Inc., Austin, Texas.

Reese, L.C., Wang, S.T. & Fouse, J.L. 1992. Use of drilled shafts in stabilizing a slope. *Proceedings of the Specialty Conference on Stability and Performance of Slopes and Embankments*, ASCE 2: 1318-1332. University of California, Berkeley.

Reese, L.C., Wang, S.T. & Long, J.H. 1989. Scour from cyclic lateral loading of piles. *Proceedings of the XXI Annual Offshore Technology Conference, Houston, Texas*, (OTC 6005): 395-401.

Reese, L.C., Wang, S.T. & Reuss, R. 1993. Tests of auger piles for design of pile-supported rafts. *Proceedings of the Deep Foundations on Bored and Auger Piles (BAPII)*: 343-346. Edited by Van Impe, W. F., Rotterdam: Balkema.

Reese, L.C. & Welch, R.C. 1975. Lateral loading of deep foundations in stiff clay. *Journal of the Geotechnical Engineering Division*, ASCE 101(GT7): 633-649.

Reese, L.C., Wright, S.G., Wang, S.T. & Walsh, M.A. 1989. Analysis of drilled shafts in a deep fill. *Proceedings of the Foundation Engineering, Current Principles and Practices, Evanston, Illinois*, ASCE 2: 1366-1380.

Reuss, R., Wang, S.T. & Reese, L.C. 1992. Tests of piles under lateral loading at the Pyramid Building, Memphis, Tennessee. *Geotechnical News*, 10(4): 44-49.

Roberts, J.N. 1992. Private communication to send partial draft of 'Shaft lateral load test, terminal separation,' describing tests performed by personnel of California Department of Transportation in San Francisco.

Robertson P.K., Campanella, R.G. & Wightman, A. 1983. SPT-CPT Correlations. *Journal of the Geotechnical Engineering Division*, ASCE 109(GT11): 1449-1460.

Robertson, P.K., Davies, M.P. & Campanella, R.G. 1989. SPT-CPT Design of laterally loaded driven piles using the flat dilatometer. *Geotechnical Testing Journal* (GTJODJ), March 1989. 12(1): 30-38.

Robertson, P.K. 1990. Soil classification using the cone penetration test. *Canadian Geotechnical Journal*, 27(1): 151-158.

Robertson, P.K. 1993. Design consideration for liquefaction. *US Japan Workshop*.

Robertson, R.N. 1961. The analysis of a bridge foundation with batter piles. Thesis, University of Texas, Austin.

Robinsky, E.I. & Morrison, C.E. 1964. Sand displacement and compaction around model friction piles. Canadian Geotechnical Journal, 1(2): 81-93.

Roesset, J.M. 1984. Dynamic stiffness of pile groups. Proceedings of the Analysis and Design of Pile Foundations, J. R. Meyer, (ed). ASCE National Convention, San Francisco, California.

Rollins, K.M., Peterson, K.T. & Weaver, T.J. 1998. Lateral load behavior of full-scale pile group in clay. Journal of the Geotechnical and Geoenvironmental Engineering, ASCE 124(6): 468-478.

Ruesta, P.F. & Townsend, F.C. 1997a. Evaluation of laterally loaded pile group at Roosevelt Bridge. Journal of the Geotechnical and Geoenvironmental Engineering, ASCE 123(12): 1153-1162.

Ruesta, P.F. & Townsend, F.C. 1997b. Prediction of lateral load response for a pile group. Transportation Research Board Record (1569): 36-47.

Rutledge, P.C. 1956. Design of Texas towers offshore radar stations. Proceedings of the VIII Texas Conference on Soil Mechanics and Foundation Engineering, 41 pp. University of Texas, Austin.

Sachs, P. 1978. Wind forces in engineering, 2nd edn: 24. Oxford: Pergamon Press.

Saul, W.E. 1968. Static and dynamic analysis of pile foundations. Journal of the Structural Division, ASCE (5936, ST5): 1077-1100.

Schmertmann, J.H. 1977. Report on development of a Keys limestone shear test for drilled shaft design. A report to Girdler Foundation and Exploration Corporation (unpublished), Clearwater, Florida.

Schmidt, H.G. 1981. Group action of laterally loaded bored piles. Proceedings of the X International Conference, Soil Mechanics and Foundation Engineering, Stockholm, Sweden: 833-837.

Schmidt, H.G. 1985. Horizontal load tests on files of large diameter bored piles. Proceedings of the XI International Conference, Soil Mechanics and Foundation Engineering, San Francisco, California: 1569-1573.

Scott, V.M. 1995. Interaction factors for piles in groups subjected to lateral loading. Thesis, University of Texas, Austin.

Seco e Pinto, P.S. & De Sousa Coutinho, A.G.F. 1991. Single pile and pile group tests under lateral load performed in Guadiana Bridge. Proceedings of the X European Conference on Soil Mechanics and Foundation Engineering, Florence, Italy (Balkema): 539-542.

Seed, H.B. & Reese, L.C. 1957. The action of soft clay along friction piles. Transactions, ASCE 122(2882): 731-754. (Reprinted: Seed, H.B. 1990. In Mitchell, J.K. (ed.), Vancouver: BiTech. Selected Papers, 1: 1-36).

Sherard, J.L., Dunnigan, L.P. & Decker, R.S. 1976. Identifying dispersive soils. Journal of the Geotechnical Engineering Division, ASCE 102(GT1): 69-86.

Shibata, T., Yashimi, A. & Kimura, M. 1989. Model tests and analysis of laterally loaded pile groups. Soils and Foundations, Japanese Society of Soil Mechanics and Foundation Engineering, 29(1): 31-44.

Shie, C.F. & Brown, D.A. 1991. Evaluation of the relative influence of major parameters for laterally loaded piles in three dimensional finite element models, 355 pp. Civil Engineering Department, Harbert Engineering Center. Auburn University, Alabama.

Skempton, A.W. 1951. The bearing capacity of clays. *Proceedings*, Building Research Congress, Division 1, London, England.

Sokolovskii, V.V. 1965. *Statics of granular media*. New York: Pergamon Press.

Speer, D. 1992. Shaft lateral load test terminal separation. California Department of Transportation, Sacremento (unpublished report).

Stevens, J.B. & Audibert, J.M.E. 1979. Re-examination of p-y curve formulation. Proceedings of the XI Annual Offshore Technology Conference, Houston, Texas, (OTC 3402): 397-403.

Stewart, D.P., Jewell, R.J. & Randolph, M.F. 1994. Design of piled bridge abutments on soft clay from loading from lateral soil movements. Geotechnique, London, England, 44(2): 277-296.

Stokoe II, K.H., Hwang, S.K., Lee, J.N.K. & Andrus, R.D. 1994. Effects of various parameters on the stiffness and damping of soils at small to medium strains. International Symposium on Prefailure Deformatioin Characteristics of Geomaterials, Sapporo, Japan. Shibuya, Mitachi & Miura (eds): 785-816.

Stokoe II, K.H. 1989. Personal communication.

Stokoe II, K.H., Lee, S.H.H. & Knox, D.P. 1985. Shear moduli measurements under true triaxial stresses. Proceedings of the Advances in the Art of Testing Soils Under Cyclic Loading Conditions, Geotechnical Engineering Division, ASCE Convention, Detroit, Michigan: 166-185.

Stokoe II, K.H., Mok, Y.S., Lee, N. & Lopez, R. 1989. In situ seismic methods: recent advances in testing, understanding and applications. XIV Conferenze di Geotecnia di Torino, Department of Structural Engineering, Politecnico di Torino, 1: (1-35).

Stokoe II, K.H., Nazarian, S., Rix, G.J., Sanchez-Salinero, I., Sheu, J.-C. & Mok, Y.-J. 1988. In situ seismic testing of hard-to-sample soils by surface wave method. Proceedings of the Specialty

Conference on Earthquake Engineering and Soil Dynamics II – Recent Advances in Ground Motion Evaluation, Park City, Utah, ASCE: 264-278.
Stokoe II, K.H., Wright, S.G., Bay, J.A. & Roesset, J.M. 1994. Geophysical characterization of sites. Rotterdam: Balkema.
Sullivan, W.R, Reese, L.C. & Fenske, C.W. 1980. Unified method for analysis of laterally loaded piles in clay. Numerical Methods in Offshore Piling, Institution of Civil Engineers, London, England: 135-146.
Sybert, J.H. 1963. Personal communication.
Szechy, C. 1961. Foundation failures. London: Concrete Publications, Ltd.
Terashi, M., Kitazume, M. & Kawabata, K. 1989. Centrifuge modeling of a laterally loaded pile. Proceedings of the XII International Conference on Soil Mechanics and Foundation Engineering.
Terzaghi, K. 1943. Theoretical soil mechanics. New York: Wiley.
Terzaghi, K. 1954. Anchored bulkheads. Transactions, ASCE 119.
Terzaghi, K. 1955. Evaluation of coefficients of subgrade modulus. Geotechnique, V: 297-326.
Terzaghi, K. 1956. Theoretical soil mechanics: 363-366. New York: Wiley.
Terzaghi, K., 1962. 'Does foundation technology really lag?' Engineering News-Record, 15 February 1962: 58-59.
Terzaghi, K. & Peck, R.B. 1948. Soil mechanics in engineering practice: 468. New York: Wiley.
Terzaghi, K., Peck, R.B. & Mesri, G. 1996. Soil mechanics in engineering practice, 3rd edn: 336. New York: Wiley.
Thompson, G.R. 1977. Application of finite element method to the development of p-y curves for saturated clays. Thesis, University of Texas, Austin.
Thurman, A.G. & D'Appolonia, E. 1965. Computed movement of friction and end bearing piles embedded in uniform and stratified soils. Proceedings of the VI International Conference on Soil Mechanics and Foundation Engineering, 2: 323-327.
Timoshenko, S. 1934. Theory of elasticity. New York: McGraw-Hill.
Timoshenko, S.P. 1941. Strength of materials, part II, advanced theory and problems, 2nd edn, 10th printing. New York: D. Van Nostrand.
Todeschini, C.E., Bianchini, A.C. & Kesler, C.E. 1964. Behavior of concrete columns reinforced with high strength steels. Proceedings of the Journal of the American Concrete Institute, 61(6): 701-716.
Tomlinson, M.J. 1995. Foundation design and practice, Singapore, 6th edn. City: Longman.
Turzynski, L.D. 1960. Groups of piles under mono-planar forces. Structural Engineer, 38(9): 286.
Van Impe, W.F. 1986. Evaluation of deformation and bearing capacity parameters of foundations, from static CPT-results. Proceedings of the IV NTI International Geotechnical Seminar, Singapore, Asia: 51-70.
Van Impe, W.F. 1991. Analysis of CFA pile behavior with DMT results at Geel test site. 4th DFI International Conference on Piling and Deep Foundations, Stresa, Italy 101-105.
Van Impe, W.F. 1988. Considerations on the auger pile design. Proceedings of the I International Geotechnical Seminar on Deep Foundations on Bored and Auger Piles, Ghent, Belgium 7-10 June 1988: 193-218.
Van Impe, W.F. 1991a. Deformations of deep foundations. Proceedings of the Session 3b-X European Conference on Soil Mechanics and Foundation Engineering, Florence, Italy, 26-30 May 1991: 1031-1062.
Van Impe, W.F. 1991b. Developments in pile design. General Report, Proceedings of the session 4– IV Deep Foundations Institute (DFI) International Conference on Piling and Deep Foundations, Stresa, Italy, 7-12 April 1991.
Van Impe, W.F. 1994. Influence of screw pile installation parameters on the overall pile behaviour. Workshop: Piled Foundations: Full Scale Investigations, Analysis and Design, Naples, Italy, 12-15 December 1994.

Van Impe, W.F. & De Clerq, Y. 1994. A piled raft interaction model. Proceedings of the DFI 94, V International Conference and Exhibition on Piling and Deep Foundations, Bruge, Belgium.

Van Impe, W.F. & Peiffer, H. 1997. Influence of screw pile installation on the stress state in the soil. International Seminar ERTC3, 17-18 April 1997, Brussel, Belgium: 3-19.

Van Weele, A.F. 1979. Some considerations with regard to bearing capacity of foundation piles. Geologie en Mijnbouw, 58: 405-416.

Vesic, A.S. 1970. Tests on instrumented piles, Ogeechee River site. Journal of Soil Mechanics and Foundations Division, ASCE 96(SM2): 561-584.

Vesic, A.S. 1961a. Bending of beams resting on isotropic elastic solids. Journal of Engineering Mechanics Division, ASCE 87(EN2): 35-53.

Vesic, A.S. 1961b. Beams on elastic subgrade and the Winkler's hypothesis. Proceedings of the V International Conference on Soil Mechanics and Foundation Engineering, Paris, France, 1: 545-550.

Vijayvergiya, V.N. 1977. Load-movement characteristics of piles. Proceedings of the Ports 77 Conference, Long Beach, California, ASCE.

Wang, S.T. 1982. Development of a laboratory test to identify the scour potential of soils at piles supporting offshore structures. Thesis, University of Texas, Austin.

Wang, S.T. 1986. Analysis of drilled shafts employed in earth-retaining structures. Dissertation, University of Texas, Austin.

Wang, S.T. &. Reese, L.C. 1986. Study of design method for vertical drilled shaft retaining walls. Research Report 415-2F. Center for Transportation Research, Bureau of Engineering Research. University of Texas, Austin.

Ward W.H., Samuels, S.G. & Butler, M.E. 1959. Further studies of the properties of London clay. Geotechnique, 14(2): 33-38.

Welch, R.C. & Reese, L.C. 1972. Laterally loaded behavior of drilled shafts. Research Report 3-5-65-89. Center for Highway Research. University of Texas, Austin.

Whitman, R.V. 1984. Evaluating calculated risk in geotechnical engineering. Journal of the Geotechnical Engineering Division, ASCE 110(2): 143-188.

Wilson, S.D. 1973. Deformation of earth and rockfill dams. Embankment Dam Engineering, Casagrande Volume, In Hirschfeld, R.C. & Poulos, S.J. (eds). New York: Wiley.

Woods, R.D. 1978. Measurement of dynamic soil properties. Proceedings of the Geotechnical Engineering Division Specialty Conference on Earthquake Engineering and Soil Dynamics, ASCE 1: 91-178.

Woods, R.D. & Stokoe II, K.H. 1985. Shallow seismic exploration in soil dynamics. Proceedings of the Richart Commemorative Lectures, Geotechnical Engineering Division, ASCE: 120-151.

Wright, S.J. 1977. Limit-state design of drilled shafts. Thesis, University of Texas, Austin.

Wroth, C.P. et al. (1979). Stress changes around a pile driven into cohesive soil. Proceedings of the Conference on Resent Developments in the Design and Construction of Piles, ICE, London, UK: 255-264.

Wroth, C.P. 1972. Some aspects of the elastic behavior of overconsolidated clay, Proceedings of the Roscoe Memorial Symposium, Foulis: 347-361.

Yamashita, K., Tomono, M. & Kakurai, M. 1987. A method for estimating immediate settlement of piles and pile groups. Soils and Foundations, Japanese Society for Soil Mechanics and Foundation Engineering, 27(11): 61-76.

Yegian, M. & Wright, S.G. 1973. Lateral soil resistance-displacement relationships for pile foundations in soft clays. Proceedings of the Offshore Technology Conference, Houston, Texas, II(OTC 1893): 663-676.

Author Index

Abbott, B.J. 189, 190
Alizadeh, M. 278
Allen, J. 88
Amerasinghe, S.F. 142, 146, 177
American Association of State Highway and Transportation Officials 188, 195, 197, 354, 358
American Concrete Institute 122, 296, 297
American National Standards Institute 187
American Petroleum Institute 16, 97, 145, 147, 148, 149, 150, 188, 189, 192, 194, 196, 214, 215, 223, 354, 355, 356
American Society of Civil Engineers 188
American Society for Testing and Materials 14, 240, 310, 311
Andrus, R.D. 52
Appel, G.C. 225
Aschenbrenner, R. 128
Asherman, J.C. 15
Asplund, S.O. 128
Atkins Engineering Services 192
Attkins, J.B. 192
Audibert, J.M.E. 97
Austin American-Statesman 198
Awoshika, K 100, 129, 131, 133, 134, 136, 137, 178, 179, 180, 181, 182, 183, 184

Baecher, G.B. 349
Baguelin, F. 16
Baldi, G. 352
Bang, S. 209
Banerjee, P.K. 152
Barltrop, N.D.P. 192
Basaldua, E.A. 189, 190
Bay, J.A. 52
Bauduin, C. 307, 310, 315
Belloti, R. 352
Bergfelder, J. 310
Bettess, P. 10
Bianchini, A.C. 114, 115
Bieniawski, Z.T. 103, 104
Bhushan, K. 97
Bogard, D. 154, 160, 161, 225, 314
Bolton, M.D. 176
Botelho, D.L.R. 189, 190
Bottiau, M. 307, 310
Bouafia, A.A. 303
Boughton, N.O. 252
Bowles, J.E. 175
Bowman, E.R. 58
Bransby, M.F. 225
Briaud, J.-L. 97
Broms, B.B. 12, 226, 369, 371, 375, 376, 377, 379, 381, 382, 383
Brown, D.A. 5, 12, 154, 164, 165, 166, 167, 225
Butler, M.E. 138

Campanella, R.G. 281, 352
Canadian Geotechnical Society 101, 153
Capozolli, L. 317
Carpentier, R. 354
Carter, E.W. 189, 190

Carter, J.P. 101, 176
CERC 192, 194
Chameau, J.L. 246
Charles, J.A. 252
Chen, L.T. 225
Chow, Y.K. 177
Clarke, J. 314
Coleman, R.B. 288
Comité Euro-International du Béton 114
Committee on Piles Subjected to Earthquake 267, 430
Costanzo, D. 354
Cox, W.R. 5, 6, 8, 52, 55, 64, 65, 66, 68, 72, 75, 84, 93, 129, 155, 273, 276, 319, 340, 344, 419, 420
Coyle, H.M. 142, 146, 148

D'Appolonia, E. 140
Davies, T.G. 152
Davis, E.H. 11, 12, 16, 140, 142, 152, 175
Davis, L.H. 288
Davisson, M.T. 278
Debacker, P. 307, 310
De Beer, E.E. 140, 187, 354
Decker, R.S. 66, 107
De Clerq, Y. 18
De Cock, F. 307, 310
Deere, D.V. 103
De Jonghe, A. 354
Desai, C.S. 225
De Sousa Coutinho, A.G.F. 313
de Sousa Pinto, N.L. 195
Det Norske Veritas 16, 214
Dixon, D.A. 155, 419, 420
Drabkin, S. 10

Drnevich, P. 176
Duncan, J.M. 13, 14, 92, 93
Dunnavant, T.W. 64, 65, 66, 97
Dunnigan, L.P. 66, 107
Dupont, E. 307, 310

Einstein, H.A. 196
Emrich, W.J. 217, 218, 219
Endley, S.N 14
Eurocode 2 114
Eurocode 3 116
Eurocode 7 1, 310, 441
Evans, Jr., L.T. 13, 14, 92, 193

Fahey, M. 176
Fallgren, R.B. 10
Feld, J. 348, 354
Fenske, C.W. 97
Fitzgibbon, D.P. 52
Fleming, W.G.K. 176
Focht, Jr., J.A. 16, 70, 142, 146, 152, 153, 154, 177, 225
Fong, P.T. 97
Foott, R. 52, 307
Fouse, J.L. 246, 249
Francis, A.J. 128
Franke, E. 155, 303, 305, 354, 356, 420
Fu, S.L. 189, 190
Fuchs, R.A. 192
Fukouka, M. 245

Gabr, M.A. 264
Garnier, J. 303
Gazetas, G. 10
Gazioglu, S.M. 97
George, P. 16
Georgiadis, M. 88, 90
Germaine, J.T 52, 137, 352
Ghionna, V.N. 352
Gilbert, L.W. 283
Gleser, S.M. 15
Gooding, T.J. 283
Gore, J.R. 283
Griffis, L. 238
Grime, D.B. 97
Grubbs, B.R. 276, 319
Gularte, R.C. 195

Ha, H.S. 225

Hadjian, A.H. 10
Haley, S. C. 97
Haliburton, T.A. 233
Hamilton, J. 314
Hancock, T.G. 288
Hansen, B. 127
Hansen, J.B. 57
Hanson, W.E. 72, 177, 232
Hardin, B.O. 176
Hassiotis, S. 246
Hetenyi, M. 11, 12, 14, 21, 22
Hilde, J.L. 307, 310
Hill, R. 226
Hognestad, E. 115
Holeyman, A.C. 307, 310
Horne, M.R. 116, 117, 118
Horvath, R.G. 102
Hrennikoff, A. 128, 129
Hull, T.S. 11
Hunter, A.H. 278
Huybrechts, N. 307, 310
Hvorslev, M.J. 218, 307
Hwang, S.K. 52

Ingram, W.B. 154, 160, 161, 314
Ishihara, K. 52
Ismael, N. 92, 290

Jamiolkowski, M. 52, 137, 280, 352
Japan Road Association 153
Jessberger, H.L. 303
Jewell, R.J. 224
Jezequel, J.F. 16
Johnson, G.W. 52
Johnson, J.W. 192

Kakurai, M. 176
Karol, R.H. 127
Kausel, E. 10
Kawabata, K. 303
Kaynia, A.M. 10
Kelly, A.E. 154, 160, 161, 314
Kelly, W.E. 195
Kenley, R.M. 317
Kenney, T.C. 103
Kerisel, J.L. 260
Kesler, C.E. 114, 115
Kimura, M. 158, 303, 421
Kitazume, M. 303

Knox, D.P. 352
Koch, K.J. 152, 153, 154, 225
Kooijman, A.P. 12
Koop, F.D. 5, 6, 8, 52, 55, 64, 65, 68, 72, 75, 84, 93, 273, 276, 340
Kotthaus, M. 303
Kraft, Jr., L.M. 142, 146, 177
Kubo, K. 100, 136, 137
Kulhawy, F.H. 101
Kumar, M. 12

Lacy, H. 10
Ladd, C.C. 52, 137, 307, 352
Lai, P. 303
Lambson, M.D. 314
Lancellotta, R.L. 52, 137, 352, 354
Laursen, E.M. 196
Lee, J.N.K. 52
Lee, L. J 97
Lee, N. 52, 352
Lee, S.H.H. 352
Legrand, C. 307, 310
Leonards, G.A. 219
Leroueil, S. 352
Long, J.H. 9, 65, 66, 76, 108, 251, 253. 254, 255, 256, 312
Long, M.M. 314
Lopez, R. 352
Lo Presti, D.C.F. 52
Lousberg, E. 354
Lunne, T. 264

MacCamy, R.C. 192
Majano, R.E. 108
Mandolini, A. 176
Mansur, C.I. 278
Masch, Jr., F.D. 195
Matlock, H. 8, 15, 16, 17, 36, 47, 52, 65, 67, 72, 73, 74, 75, 107, 129, 154, 160, 161, 162, 163, 164, 225, 269, 314, 343, 344, 410, 411, 412, 413, 414, 415, 416, 417, 418
Mattes, N.S. 140
McCammon, G.A. 15
McClelland, B. 16, 70
McVay, M. 172, 173, 303

Mengé, P. 307, 310
Mesri, G. 226, 229
Meyer, B.J. 97, 272, 288
Meyerhof, G.G. 351, 354
Miller, J.P. 307, 310
Mitchell, G.M. 192
Mok, Y.-J. 352
Mok, Y.S. 52
Molnit, T. 303
Moore, W.L. 195
Morison, J.R. 192, 193, 194
Morrison, C.E. 140, 141
Morrison, C.S. 165, 167, 168, 169
Mosher, R.L. 149
Murchison, J.M. 97
Murphy, B.S. 155, 419, 420
Mylonakis, G. 10

Nacci, V.A. 195
National Highway Institute 359, 360
Nazarian, S. 352
Nyman, K.J. 102, 293

Oakland, M.W. 246
O'Brien, M.P. 192
Ooi, P.S.K. 13, 14
O'Neill, D.A. 52
O'Neill, M.W. 64, 65, 66, 97, 108, 129, 131, 146, 147, 149, 150, 151, 152, 164, 177, 178, 225

Palmer, L.A. 14
Parker, Jr., F. 129, 132
Peck, R.B. 72, 103, 177, 225, 226, 229, 232, 348, 354, 363, 364
Peiffer, H. 309, 352
Person, G. 108
Peterson, K.T. 170, 171
Petrauskas, C. 189, 190
Portugal, J.C. 12, 286
Posey, C.H. 195
Posey, C.J. 196
Poulos, H.G. 11, 16, 52, 140, 142, 152, 153, 175, 176, 225
Powell, J.J. 264
Prakash, S. 156, 303, 382

Price, G. 264, 277
Pugsley, A. 351, 354

Quinn, A. 207

Radosavljevic, Z. 128
Ramshaw, C.L. 10
Randolph, M.F. 18, 176, 224
Reese, L.C. 5, 6, 8, 16, 17, 36, 52, 55, 64, 65, 68, 70, 72, 75, 82, 84, 93, 97, 98, 100, 102, 107, 129, 131, 132, 133, 134, 136, 137, 142, 146, 147, 149, 150, 151, 154, 158, 159, 164, 165, 208, 219, 225, 239, 241, 246, 249, 251, 253, 254, 255, 256, 263, 273, 276, 293, 294, 297, 312, 319, 340, 344, 410, 411, 412, 413, 414, 415, 416, 417, 418
Reuss, R. 9, 239, 241
Ripperger, E.A. 15, 52, 67, 343
Rix, G.J. 352
Roberts, J.N. 102
Robertson, P.K. 281, 352
Robertson, R.N. 129
Robinsky, E.I. 140, 141
Roesset, J.M. 10
Rollins, K.M. 168
Romualdi, J.P. 140
Ruesta, P.F. 171, 172, 173
Rutledge, P.C. 14

Sachs, P. 188
Samuels, S.G. 138
Sanchez-Salinero, I. 352
Saul, W.E. 128
Schaaf, S.A. 192
Schmertmann, J.H. 294
Schmidt, H.G. 156, 157, 310, 421
Scott, V.M. 154
Sêco e Pinto, P.S. 12, 286, 313
Seed, H.B. 142, 219
Selby, A.R. 10
Sharp, D.E. 317
Sherard, J.L 66, 107
Sheu, J.-C. 352

Shibata, T. 158, 303, 421
Shie, C.F. 5, 12, 225
Shields, D.H. 16
Simon, G. 307, 310
Skempton, A.W. 53, 60, 61, 62, 63, 70, 72, 146
Smith, R.E. 129
Smith, T.D. 97
Sokolovskii, V.V. 226
Speer, D. 295
Stevens, J.B. 97
Stewart, D.P. 224
Stokoe II, K.H. 52, 352
Sulaiman, I.H. 142, 148
Sullivan, W.R. 97
Sybert, J.H. 195, 196
Szechy, C. 348, 354

Taylor, R.J. 209
Terashi, M. 303
Terzaghi, K. 11, 12, 15, 54, 61, 68, 69, 127, 196, 225, 226, 229, 354, 376, 382
Thompson, G.R. 12, 56, 57
Thompson, J.B. 14
Thornburn, T.H. 72, 177, 232
Thurman, A.G. 140
Timoshenko, S.P. 14, 28, 61
Todeschini, C.E. 114, 115
Tomlinson, M.J. 175
Tomono, M. 176
Townsend, F.C. 171, 172, 173
Tufenkijian, M.R. 10
Turzynski, L.D. 128

Van Impe, W.F. 18, 51, 137, 138, 139, 140, 141, 174, 176, 280, 282, 309, 352
Vanneste, G. 108
Van Weele, A.F. 139
Vennalaganti, K.M. 14
Vesic, A.S. 149, 376
Viggiani, C. 176
Vijayvergiya, V.N. 146

Wallays, M. 354
Walsh, M.A. 225, 251, 253, 254, 255
Wang, S.T. 9, 65, 66, 107, 158, 159, 225, 239, 241, 246, 249, 251, 253, 254,

255, 256
Ward, W.H. 137, 138
Wardle, I.F. 264, 277
Weaver, T.J. 168
Welch, R.C. 66, 82, 107, 263, 343
Whitman, R.V. 349
Wightman, A. 281, 352
Wiegel, R.L. 196
Wilson, S.D. 252
Wood, D. 16
Woods, R.D. 10
Wright, S.G. 12, 225, 251, 253, 254, 255
Wright, S.J. 348, 351
Wroth, C.P. 18, 52, 137, 176
Yamashita, K. 176
Yashimi, A. 158, 303, 421
Yegian, M. 12
Yeung, A.T. 14

Zhang, L. 303

Subject Index

Active pile 2
Anchored bulkhead 231
Anchoring pile for a ship 207
Axially loaded single piles 137
   Analytical model 143
   Differential equation for analysis 142
   End bearing in cohesionless soils 149
   End bearing in cohesive soil 146
   Side resistance in cohesionless soil 147
   Side resistance on cohesive soil 145
   Stiffness curves for soil 140

Bending-moment curves 4
   Differentiation and integration 4, 341
   Examples from experiment 340
Bending moment, ultimate M_{ult} 111
   Reinforced-concrete section 119
   Steel H-section 116
   Steel pipe 118
Bending stiffness, $E_p I_p$ 111, 328
   Reinforced-concrete section 119
   Reinforced-concrete section, approximation 121
Boundary conditions at pile head
   Shear and moment 31

Shear and rotation 32
Shear and rotational restraint 32
Moment and deflection 53
Breasting dolphin 3, 203
Bridge foundations 3
Broms method 12

Calibration of test piles 325, 339
Case studies
   Cohesive soils with no free water 260
   Bagnolet 260
   Brent Cross 264
   Houston 263
   Japan 267
   Cohesive soil with free water 269
   Lake Austin 269
   Manor 273
   Sabine 272
   Cohesionless soils 276
   Arkansas River 278
   Garston 277
   Mustang Island 276
   Layered soils 283
   Alcácer do Sol 286
   Apapa 288
   Florida 288
   Talisheek 283
   Soil with cohesion and friction 290
   Kuwait 290
   Los Angeles 291
   Weak Rock 293
   Islamorada 293
   San Francisco 295
Centrifuge 303

Cone penetrometer 307
Consequences of the failure of a foundation 348
Constitutive modeling of in situ soil 5
Current Loading 194
Cyclic loading influence
   Clay 65
   Sand 66

Decay of modulus of soil, E_s 50
Design factors 363
Diameter effect 64
Difference-equation solution 29, 30
Differential equation for beam-column 21
Dilatometer 308
Dimensional analysis 36

Earth Pressures 223
Equivalent diameter for non-circular cross section 109

Factors of safety 354
   Global approach 354
   Load and resistance factors 358
   Partial safety factors 356
Failure of a foundation 248
Finite-element method 5, 12, 64
Forces from moving soil 224

Global approach to safety 354
Global loading 1
Ground settlement 338
   Lateral loading 339

Subject Index

Pile driving 338
Groups of piles under axial load 173
   Classical form of interaction factors 175
   Equivalent pier method 175
   Equivalent raft method 175
   Influence coefficients 173
   Interim recommendations 177
   Modified interaction factors 176
   Review by O'Neill 177
   Review by Van Impe 176
Groups of piles under lateral load, distribution of load to individual piles 125
   Method of prediction 129
   Review of theories
   Aschenbrenner 128
   Asplund 128
   Culmann 127
   Francis 128
   Hrennikoff 127
   Nokkentved 127
   Radosovljavic 128
   Reese & Matlock 129
   Reese & O'Neill 129
   Saul 128
   Turzynski 128
   Vamdepitte 127
   Wintergaard 127
Groups of piles under lateral load, efficiency of closely spaced piles 2, 151
   Comparisons of experiments with theory 160
   Experiments 158
   Method of prediction 152, 158
Groups of piles under lateral load, experiment with batter piles 178

High-rise structures 3

Ice Loading 197
Initial stiffness of p-y curves 53
Installation of piles 309, 332

Influence on soil properties 139
Instrumentation for piles 260, 322
Instrumentation for testing 313
Interaction with superstructure 42

Limit analysis 1
Limit-state conditions 1, 347
Length of pile, influence 30
Loading from waves 2
Loading from wind 2

Mat foundation supported by piles 237
Models for single piles under lateral load
   Elastic pile and elastic soil 11
   Rigid pile and plastic soil 12
   Characteristic load method 13
   Nonlinear pile and p-y model for soil 14
Mooring dolphin 3

Nondimensional coefficients 41
Nondimensional solution 42

Offshore platform 2, 213
Overhead signs 3, 199

Partial safety factors 356
Passive pile 2
p-y Curves
   Effect of installation on a batter 100, 136
   Examples from field experiment 4
   Experimental methods for acquisition 67
   Layered soils 87
   McClelland & Focht for clay 70
   Nondimensional methods for acquisition 68
   Sand above and below the water table 84
   Sloping ground 98
   Soft clay in presence of free water 72
   Soil with both cohesion and friction angle 91
   Stiff clay in presence of free water 75
   Stiff clay with no free water 82
   Stiffness of clay, ε_{50} 70
   Terzaghi recommendations 68
   Typical curve 4
   Typical set 44
   Weak rock 101
Peer review 353, 367
Penetrometer 304, 335
Piles in a settling fill 251
Pressuremeter 308
Program for testing under lateral load 304
Proof piles 304
Production piles 304

Quality control 353

Raked (batter) piles 1
Reaction modulus 3
Relative stiffness factor 40
Retaining wall supported by piles 226
Risk management 353

Safety be method of load and resistance factors 358
Safety coefficient 350
Scour of clay due to cyclic loading 7
Scour of soil (erosion) 195
Secant-pile wall 3
Serviceability load 1
Shearing force at bottom of pile 101
Ship impact 3, 198
Slope stabilization with piles 3, 245
Sign conventions 23, 24
Soil characterization 351
Soil resistance (reaction) 1, 4
Soil stiffness 1

Subject Index

Soil-structure interaction 1
Standard penetration test 308
Stress-deformation of soil 50
Stress-strain curve for concrete 114
Stress-strain curve for structural steel 114
Structural collapse 1
Subgrade modulus 59, 63
   Theoretical solution 60
Subsurface investigation 306, 328

Tangent-pile wall 3
Techniques for testing under lateral load 310, 334
Tests of piles under lateral loading 3, 310, 319

Types of lateral loading of piles
   Cyclic 7
   Dynamic 9
   Seismic 10
   Static 7
   Sustained 8

Ultimate soil resistance of p-y curves, pult 8, 54
   Cohesive soil 55
   Cohesionless soil 58

Validity of computations 364
   Pile group 366
   Single pile 364

Wave Forces 192
Wave Loading 189
Wind Loading 187
Winkler-type mechanisms 1